Our understanding of the 'hot' stars that form a halo around our galaxy is undergoing a renaissance: recent increases in the power of computers are now allowing a far more detailed and complete modelling of stellar evolution. A conference was held in Union College, New York, to gather experts in the field to re-examine the rôle of these hot stars and this volume draws together their articles to provide a timely review.

The articles show how advances in computer power have, in particular, allowed complex modelling of the core helium-burning and ultraviolet-bright stages. They go on to demonstrate how this modelling is leading to a better understanding of new observations of stars on the horizontal branch, both in the field and in globular clusters, as well as stars in later stages of stellar evolution.

Together these articles provide an up-to-date and comprehensive review for graduate students and researchers interested in the hot stars in the halo, especially the history of the halo and the evolution of old stellar populations of different metalicities.

Hot Stars in the Galactic Halo

A. G. Davis Philip

Hot Stars in the Galactic Halo

Proceedings of a Meeting,
held at Union College, Schenectady, New York
November 4-6, 1993
in honor of the 65th birthday of
A. G. Davis Philip

Edited by
SAUL J. ADELMAN
Department of Phyiscs, The Citadel

ARTHUR R. UPGREN
Van Vleck Observatory, Wesleyan University

CAROL J. ADELMAN
Institute for Space Observations, Charleston

Published by the Press Syndicate of the University of Cambridge
The Pitt Building, Trumpington Street, Cambridge CB2 1RP
40 West 20th Street, New York, NY 10011-4211, USA
10 Stamford Road, Oakleigh, Melbourne 3166, Australia

© Cambridge University Press 1994

First published 1994

Printed in Great Britain at the University Press, Cambridge

A catalogue record of this book is available from the British Library

Library of Congress cataloguing in publication data

ISBN 0 521 46087 5 hardback

Contents

Participants — xi
Preface — xiii
Foreword — xv
Acknowledgements — xvii

Introductory Papers

What is the Galaxy's Halo Population?
    Bruce W. Carney 3

Theoretical Properties of Horizontal-Branch Stars
    Allen. V. Sweigart 17

A Review of A-Type Horizontal-Branch Stars
    A. G. D. Philip 41

Surveys

A Progress Report on the Edinburgh-Cape Blue Object Survey
    D. Kilkenny, D. O'Donoghue, R. S. Strobie, A. L. Chen, C. Koen, and A. Savage 70

A 300 Square Degree Survey of Young Stars at High Galactic Latitudes
    J. Eamon Little, P. L. Dufton, F. P. Keenan, N. C. Hambly, E. S. Conlon, and L. Miller 79

The Isolation of A New Sample of B Stars in the Halo
    Kenneth J. Mitchell, Rex A. Saffer, and Steve B. Howell 82

A Northern Catalog of Candidate FHB/A Stars
    Timothy C. Beers, Ronald Wilhelm, and Stephen Doinidis 90

Recent Progress on a Continuing Survey of Galactic Globular Clusters for Blue Stragglers
    Ata Sarajedini 100

UV Observations with FAUST and the Galactic Model
    Noah Brosch 116

Hot Stars at the South Galactic Pole
    Phillip K. Lu 124

Clusters

Population II Horizontal Branches: A Photometric Study of Globular Clusters
    Kent A. Montgomery and Kenneth A. Janes 136

The Period-Shift Effect in Oosterhoff Type II Globular Clusters
    Márcio Catelan 149

Ultraviolet Observations of Globular Clusters
    Wayne B. Landsman 156

UV Photometry of Hot Stars in Omega Centauri
  Jonathan H. Whitney, R. W. O'Connell, R. T. Rood, B. Dorman, R. C. Bohlin, K. P. Cheng, P. M. N. Hintzen, W. B. Landsman, M. S. Roberts, A. M. Smith, E. P. Smith, and T. P. Stecher 163

Spectroscopic and UBV Observations of Blue Stars at the NGP
  David J. Bell, H. L. Detweiler, Kenneth M. Yoss, Stefano Casertano, Grant Bazan, Anurag Shankar, Rosa Murphy, and Sean Points 168

Population I Horizontal Branches: Probing the Halo-to-Disk Transition
  Randy L. Phelps, Kenneth A. Janes, and Kent A. Montgomery 175

Stars

Very Hot Subdwarf O Stars
  J. S. Drilling, T. C. Beers, and U. Heber 182

Quantitative Spectroscopy of the Very Hot Subluminous O-Stars: K648, PG1159-035, and KPD0005+5106
  Ulrich Heber, Stefan Dreizler, and Klaus Werner 187

Analyzing the Helium-Rich Hot sdO Stars in the Palomar Green Survey
  Peter Thejll 197

Late Type Companions of Hot sd O Stars
  Raul Jimenez, Peter Thejll, Rex Saffer, and Uffe G. Jørgensen 211

Hot Stars in Globular Clusters
  Sabine Moehler, Ulrich Heber, and Klaas S. de Boer 217

Faint Blue Stars from the Hamburg Schmidt Survey
  Stefan Dreizler, Uli Heber, S. Jordan, and D. Engels 228

Stellar Winds and the Evolution of sdB's to sdO's
  James MacDonald and Steven S. Arrieta 238

Halo Stars in the Vilnius Photometric System
  Vytas Straižys 242

Horizontal Branch Stars in the Geneva Photometric System
  B. Hauck 245

Zeeman Observations of FHB Stars and Hot Subdwarf Stars
  V. G. Elkin 249

What Does a FHB Star's Spectrum Look Like?
  C. J. Corbally and R. O. Gray 253

A Technique for Distinguishing FHB Stars from A-Type Stars
  Ronald Wilhelm, Timothy C. Beers, and Richard O. Gray 257

Elemental Abundances of Halo A and Interloper Stars
  Saul J. Adelman and A. G. Davis Philip 266

The Mass of Blue Horizontal Branch Stars in the Globular Cluster NGC6397
  Klaas S. de Boer, Jelena H. Schmidt, and Uli Heber 277

IUE Observations of Blue HB Stars in the Globular Clusters M3 and NGC6752
  C. Cacciari 282

Metallicities and Kinematics of the Local RR Lyraes: Lukewarm Stars in the Halo
  Andrew C. Layden 287

Baade-Wesselink Analyses of Field vs. Cluster RR Lyrae Variables
  Jesper Storm, Bruce W. Carney, Birgitta Nordström, Johannes Andersen, and David W. Latham 298

The Rotation of Population II A Stars
  William P. Bidelman 305

Horizontal-Branch Stars and Possibly Related Objects
  William P. Bidelman 306

A New Group of Post-AGB Objects - The Hot Carbon-Poor Stars
  E. S. Conlon 309

MK Classifications of Hot Stars in the Halo
  R. F. Garrison 314

Photometry of XX Virginis and V716 Ophiuchi and the Period Luminosity Relations of Type II Cepheids
  D. H. McNamara and M. D. Pyne 315

Rotation and Oxygen Line Strengths in Blue Horizontal Branch Stars
  Ruth C. Peterson, D. A. Crocker, and R. T. Rood 319

Miscellaneous

UBV CCD Photometry of the Halo of M31
  A. P. Fitzsimmons, F. P. Keenan, P. L. Dufton, J. E. Little, and M. J. Irwin 326

Can Stars Still Form in the Galactic Halo?
  Kenneth A. Janes 330

The Ultraviolet Imaging Telescope on the Astro-1 and Astro-2 Missions
  T. P. Stecher 340

Are Analogues of Hot Subdwarf Stars Responsible for the UVX Phenomenon in Galaxy Nuceli?
  B. Dorman, R. W. O'Connell, and R. T. Rood 341

A Survey for Field BHB Stars Outside the Solar Circle
  T. D. Kinman, N. B. Suntzeff, and R. B. Kraft 353

Post-AGB A and F Supergiants as Standard Candles
  Howard E. Bond 361

The Extended Horizontal-Branch: A Challenge for Stellar Evolution Theory
  Pierre Demarque 362

Astronomical Patterns in Fractals: The Work of A. G. Davis Philip on the Mandelbrot Set
  Michael Frame 363

Summary

Final Remarks
 T. D. Kinman 381

Author index 383
Subject index 385

Participants

Saul J. Adelman	The Citadel (U. S.)
Ralph Alpher	Union College (U. S.)
Timothy C. Beers	Michigan State University (U. S.)
David Bell	University of Illinois (U. S.)
William P. Bidelman	Case Western Reserve University (U. S.)
Howard Bond	STSCI (U. S.)
Noah Brosch	Tel Aviv University (Israel)
Carla Cacciari	Osservatorio Astronomico, Bologna (Italy)
Bruce Carney	University of North Carolina (U. S.)
Christopher J. Corbally	Vatican Observatory (U. S.)
Pierre Demarque	Yale University (U. S.)
H. L. Detweiler	Illinois Wesleyan University (U. S.)
Benjamin Dorman	University of Virginia (U. S.)
Stefan Driezler	Universitaet Erlangen-Nuernberg (Germany)
John S. Drilling	Louisiana State University (U. S.)
Vladimir Elkin	Special Astrophysical Observatory (Russia)
R. F. Garrison	University of Toronto (Canada)
Mike Frame	Union College (U. S.)
Bernard Hauck	Université de Lausanne (Switzerland)
Uli Heber	Universitaet Erlangen-Nuernberg (Germany)
Kenneth A. Janes	Boston University (U. S.)
David Kilkenny	SAAO (South Africa)
T. D. Kinman	NOAO/Kitt Peak National Observatory (U. S.)
Rebecca Koopmann	Yale University (U. S.)
Wayne Landsman	Hughes, STX (U. S.)
Andrew Layden	NOAO/CTIO (U. S.)
J. Eamon Little	Queen's University of Belfast (U. K.)
Phillip Lu	Western Connecticut State University (U. S.)
James MacDonald	University of Delaware (U. S.)
D. H. McNamara	Brigham Young University (U. S.)
Kenneth J. Mitchell	General Sciences Corporation (U. S.)
Sabine Moehler	Landessternwarte Heidelberg (Germany)
Kent A. Montgomery	Boston University (U. S.)
Benjamin Perry	Howard University (U. S.)
Ruth C. Peterson	Lick Observatory (U. S.)
Randy Phelps	Boston University (U. S.)
A. G. Davis Philip	Union College & Van Vleck Observatory (U.S.)
K. W. Philip	University of Alaska (U. S.)
Rex Saffer	STSCI (U. S.)
Ata Sarajedini	NOAO/Kitt Peak National Observatory (U. S.)
Theodore P. Stecher	NASA/Goddard Space Flight Center (U. S.)
Allen V. Sweigart	NASA/Goddard Space Flight Center (U. S.)
Peter Thejll	Niels Bohr Institute (Denmark)
Arthur R. Upgren	Van Vleck Obseratory (U. S.)
Andrew Vanture	Union College (U. S.)
Wayne Warren, Jr.	NASA/Goddard Space Flight Center (U. S.)
Jonathan H. Whitney	University of Virginia (U. S.)
Ronald Wilhelm	Michigan State University (U. S.)
Kenneth Yoss	University of Illinois (U. S.)

Preface

Greetings from the Warner & Swasey Observatory. We join with you in celebrating Dave Philip's 65th birthday. Dave is one of the many graduate students of whom we are justifiably so proud. He is also one of the early group, and thus is vividly remembered. The deep friendships that developed among that group, including also Art Upgren, Nick Sanduleak, Dick Herr, Bambang Hidajat, and Jack MacConnell, are testimony to the wonderful environment at the old Taylor Road Observatory, where our offices were. Funding levels alone tell us that those were halcyon days; the enduring friendships prove it!

Dave has exhibited splendidly the trait that so impresses me about our students, namely, an unflagging dedication to doing astronomy. Through good times and bad, their interest has never waned. And so, year after year, the cumulative contributions have been admirable.

Peter Pesch, William P. Bidelman
Cleveland, Ohio
November, 1993

I am pleased to welcome you to this meeting on behalf of Union College and its Physics Department. It is a well-deserved recognition of Professor A. G. Davis Philip which has brought together most, if not all, of those who contribute to his major field of astronomical interest.

Dave has strong local roots, inasmuch as he did his undergraduate work in physics at Union College, followed by a Masters degree at New Mexico State and a Ph. D. at Case Institute of Technology. His academic career includes stints at the University of New Mexico, and at the State University of New York at Albany, in the days when it had a Department of Astronomy. He has had a long connection with Dudley Observatory, beginning when it was in fact a working observatory and before it became a funding entity. I was on the Board of Dudley while he had was there. Now I am currently its Administrator.

His associations and honors are many. These include serving as a Shapley Lecturer for the American Astronomical Society and involvement with and serving as the President or Secretary of at least three IAU Commissions. He is currently organizing an IAU Symposium for next year at the General Assembly at The Hague. He has been instrumental in the success of the Astronomical Society of New York and the New York Astronomical Corporation. He is a Trustee of the Fund for Astrophysical Research, had been a visiting professor at many institutions, has his own press, L. Davis Press, and has organized the Institute for Space Observations, an umbrella organization for obtaining research grants. I understand that most of the members of this Institute are here.

At the present time he is a research professor at Union College and an adjunct professor at Wesleyan University. He must be one of the world's most active editors, not only of IAU Proceedings, but also publications of his own press, and the Contributions of Wesleyan's Van Vleck Observatory. For many years he edited Dudley Observatory reports. He is a fellow of the Royal Astronomical Society, and the American Association for the Advancement of Science. And to think that with all of this, he has stolen time to some fundamental things with Mandelbrot sets. Quite a record. Congratulations.

Ralph Alpher
Schenedtady, New York
November, 1993

Foreword

A. G. Davis Philip has been a pioneer in the study of faint blue stars in our Galaxy's halo. Holding a meeting to honor a colleague on his or her 65th birthday is a lovely and honored scientific and academic custom. It helps when a field like the subject of our meeting is active and there is a need for workers to come together and exchange results and ideas. We must thank Dave's parents for his being born at an appropriate time.

The rapid increase in the power of computers is leading to a more complete and detailed modeling of stellar evolution, especially through the core helium burning and ultraviolet-bright stages. This in turn helps astrophysicists better interpret the observations currently being obtained for stars on the horizontal branch both in the field and in globular clusters as well as for stars in subsequent stages of stellar evolution. This meeting re-examined the role of these hot stars in the halo, especially in the context of the halo's history and the evolution of old stellar populations of different metallicities.

The members of the Scientific Organizing Committee were Saul J. Adelman, The Citadel, Michael Frame, Union College, Donald S. Hayes, Pina Community College, Kenneth A. Janes, Boston University, Phillip K. Lu, Western Connecticut State University, Darrell J. MacConnell, Space Telescope Science Institute, Peter Pesch, Warner & Swasey Observatory, Case Western Reserve Univerity, Nickolai Samus, Sternberg Astronomical Insitute, Vytas Straižys, Institute of Theoretical Physics & Astronomy, Vilnius, Peter B. Steston, Dominion Astrophysical Observatory, Allen V. Sweigart, NASA Goddard Space Flight Center, Arthur R. Upgren, Van Vleck Observatory, Wesleyan University,

The chairs of the scientific sessions were Saul J. Adelman, Carla Cacciari, Kenneth A. Janes, Phillip K. Lu, David Kilkenny, Ruth C. Peterson, A. G. Davis Philip, Allen V. Sweigart, and Arthur R. Upgren.

Dave has organized many meetings. As we were on Dave's home turf, we have followed his pattern. We thank Cambridge University Press for publishing our Proceedings. Dave has had a considerable influence on such volumes.

Saul J. Adelman, Arthur R. Upgren
Co-Chairmen of the Scientific Organizing Committee

Acknowledgements

We would like to thank everyone who attended the conference for their mutual support with the production of this book. Mrs. Mary Bongiovanni helped with many important secretarial duties including getting the discusion comments typed. We are greatful for the help of Dr. Mei-Qin Chen and the use of her printer and Dr. William Denig for his help with latex. We appreciate the tolerance of The Citadel's Mathematics and Computer Science Department for SJA and CJA's frequent invasions of their domain.

We note the support of the International Science Foundation for grants to allow both Vladimir Elkin and Vytas Straižys to attend this meeting.

Introductory Papers

What is the Galaxy's Halo Population?

By BRUCE W. CARNEY

Deptartment of Physics & Astronomy, University of North Carolina, Chapel Hill, NC 27599, USA

Our current understanding of the Galaxy's halo population is reviewed, including its partial origins in accretion or merger events. Evidence is also presented for a metal-poor proto-disk population. While neither of these populations can be responsible for young, solar-metallicity early-type stars at large distances from the plane, consideration of both sources of metal-poor stars, the proto-disk and the accreted fraction, will be necessary to interpret the distributions of old, hot stars far from the Galactic plane.

1. Introduction

For a very long time, we have wondered about the structure and the history of our Milky Way Galaxy, and how similar it might be to other disk systems. The historical, or archaelogical, perspective has become wrapped up in the discussion of stellar "populations", the equivalent, perhaps, to the archaeologists' studies of dynasties in ancient kingdoms. Which one preceded and which succeeded? How far did their influences extend? And how did they affect the current state of affairs? Like the archaeologists, we have available for study only the surviving pieces of the ancient material, and we have been able to study only a tiny fraction of what is known. We have therefore been guided in our choices of what to study and how to interpret our findings by our own ideas of what the historical picture is like. And in the past few years, our ideas about what the halo is, where it came from, and how it is related to the solar neighborhood, have been in a state of flux.

It is not the purpose of this paper to present an extended review of the studies of the Galaxy's halo—the reader is advised to begin with Steve Majewski's (1993) excellent review article. Instead, we will begin with a quick review of how far from the plane disk-like populations may dominate the stellar densities, and how far one must look to be certain that the "traditional" halo population dominates. Then we will update the Majewski review to see what new insights on the halo population have been gained.

Before we proceed, however, be warned that the stellar populations approach will still be relied upon here, partly for taxonomic simplicity, like dynasties, but also because, similarly, it is the changes from one population to another that may reveal the historical dramas of the past. To this end, we must keep in mind the definition of a stellar population. For our purposes, a stellar population is an ensemble of stars and gas that share a similar history. The stars in a globular cluster would certainly qualify, then, as a population, but would all globular clusters also qualify as being members of a larger single population? No. The best perspective here is the elegant study by Bob Zinn (1985) of the spatial and chemical distributions of the Galaxy's globular clusters. He made it clear that there are in fact at least two separable populations of clusters: one that is metal-rich and concentrated in a disk and toward the Galactic center; and a second that is metal-poor and distributed spherically about the Galaxy. The two populations might have an historical link in that as the Galaxy was becoming enriched in heavy elements as it shrank, spun up, and flattened out. But we would still classify the "thick disk" globular cluster population as distinct from the "halo" globular clusters since they represent different epochs in the same history. And the two populations may not even have had an historical link at all. They could have evolved completely independently, in different but perhaps contemporary "kingdoms", as it were. The key here is to look at the metallicity distribution function: the numbers of clusters as a function of metallicity. In the simplest, closed box model of chemical evolution, gas with some initial metallicity begins producing stars, the stars enrich the gas in heavy elements, and the numbers of stars of a given metallicity rise as the metallicity increases. This continues until the gas begins to become depleted, whether through being locked away in long-lived stars or stellar remnants, or simply expelled by the energetics of star formation and death. Thus the halo globular clusters are, on average, probably metal-poor because most of the gas in the star-forming regions was expelled before significant enrichment took place. Had the gas not been lost, the mean metallicity of globular clusters would probably have been about the same as in the regions where gas loss is more difficult, like the disk or the bulge, where the mean metallicity is close to solar. That the disk and halo clusters have separate metallicity distribution functions, with peaks in [Fe/H] at values of about -1.6 and -0.5, rather than a continuum suggests (but does not prove) that they are separate populations with separate chemical and dynamical histories.

Metallicity is also not the only discriminant among populations. For example, had the Large Magellanic Cloud already been accreted by our Galaxy, we would now find a large number of field stars with mean metallicities of [Fe/H] ≈ -0.3, the mean metallicity of most of the LMC stars. The Galaxy itself has produced large numbers of stars with this same metallicity, but there would be in this case at least two post-merger populations, not one, because the histories of the stars would have been so different. The stellar dynamics would be bimodal throughout the metallicity regime [Fe/H] ≈ -0.3. Such a merger is one (Carney et al. 1989) explanation of the origin of the thick disk, although it is also very possible that the thick disk is simply the earlier stage of our current "thin disk" (see, for example, Ryan & Norris 1991). Disentangling the histories then involves the studies of metallicities, distributions, dynamics, and when possible, ages and detailed chemistries such as element-to-iron ratios (see, in particular, Edvardsson et al. 1993).

2. In Situ Studies

Studies of the halo population, which is a trace constituent in the solar neighborhood, have for decades relied upon either field stars selected with significant biases (i.e., proper motion samples to preferentially study high-velocity stars) or upon the globular clusters. As noted already, the clusters themselves comprise two, possibly distinct, histories, and in any case the number of clusters available for study is small. The selection biases on the vastly more numerous field stars (whose combined mass in the halo probably exceeds the combined mass in clusters by a factor of about 100) must either be well understood and modelled out, or samples must be selected without such biases. Norris (1986) was among the first to undertake such an unbiased study, utilizing samples of stars selected via spectroscopic or other means (e.g., variability to identify RR Lyrae variables) that did not involve a kinematic bias. His work and that of others led to a well-defined value for the asymmetric drift ($v_{drift} = -V$; $v_{rot} = V + \Theta_0$), where v_{rot} is the rotational velocity about the Galactic center, V is the velocity toward $l = 90$, $b = 0$ with respect to the Local Standard of Rest, and Θ_0 is the circular rotational velocity at the solar Galactocentric distance, which is about 220 km s^{-1}. Norris derived values for V, v_{drift}, v_{rot}, and the velocity ellipsoid (the velocity dispersions in the three orthogonal Galactic directions) as functions of metallicity, and thereby derived the values for the metal-poor solar neighborhood stars. With no data to controvert him, he assumed the most metal-poor stars in his sample defined the result for a single Galactic halo population. He found a slight prograde V velocity for the most metal-poor stars, of -192 ± 21 km s^{-1} for those with [Fe/H] < -2.28. He was unable to determine the distance from the plane at which this metal-poor population begins to dominate because his sample did not extend far enough.

A major step forward in this effort was the Ph.D. work of Steve Majewski. He used Kitt Peak 4-meter prime focus *UBVI* plates taken during two epochs, one from the mid 1970's and the second from the late 1980's. His calibrated photographic photometry enabled him to measure $\delta(U-B)_{0.6}$, the "normalized" ultraviolet excess defined by Sandage (1969) and calibrated in terms of [Fe/H] by Carney (1979). Much more impressively, however, Majewski also utilized quasars and galaxies in the field to determine proper motions for essentially all the stars down to a limiting magnitude of $B = 22.5$. His most distant stars are 25 kpc from the plane. The photometry leads also to distance estimates (under the assumption that all the stars are dwarfs, which at these faint magnitudes is a reasonable assumption), and hence both U and V velocities. The latter is the more important since it essentially is a measure of Galactic orbital angular momentum, and Majewski's (1992) plot of V vs. $\delta(U-B)_{0.6}$ was particularly noteworthy. As expected, at the distances closest to the plane (Z < 1.25 kpc), the stars all had small asymmetric drifts (small V velocities), and most were only slightly metal-poor. At the largest distances, Z > 10 kpc, the stars had large asymmetric drifts and generally low metallicities. There were two surprise treats. First, for Z > 5 kpc, the mean V velocity/asymmetric drift was so large, $<V> = -275 \pm 16$ km s^{-1}, that for the preferred value of Θ_0, the net rotation of the "high halo" was retrograde. The importance of this result is obvious: the sign of the high halo angular momentum vector differs from that of the disk, and hence it is unlikely that the high halo is a predecessor to the Galaxy's disk population(s). Second, Majewski found that the metal-poor population did not begin to dominate the number count statistics until Z values of 4 to 5 kpc are reached. Below that, disk stars dominate, although these are fairly "hot" (dynamically) disk stars whose W velocities carry them to such large distances from the plane.

What effect does this have on the interpretation of the numbers of hot stars far from the plane? Clearly models that predict the numbers of such stars must take into account the rather great extent of the disk population(s), and the rather low densities of the halo relative to them below, say, 5 kpc from the plane. But the disk stars at these great heights are not likely to be producing stars in the current epoch. This large-Z domain is that of the thick disk, and the thick disk is apparently very old. At one level, we may estimate its age from its constituent clusters: the thick disk globular clusters studied by Zinn (1985). If we use the recent proper motion work of Cudworth (1993) to select globular clusters with disk kinematics, the results of Carney *et al.* (1992) reveal that 47 Tuc and M107 both have ages like those of the metal-poor halo globular clusters, perhaps 12 to 14 Gyrs. Are these clusters representative of the ages of the entire thick disk? Carney *et al.* (1989) derived a histogram of dereddened $B - V$ color indices for their sample of proper motion stars whose metallicities and kinematics matched those of the disk globular clusters. Their sample had no bias against blue stars, but there was an obvious blue limit in the results, consistent with the color of the main sequence turn-off of the disk globular clusters. Apparently the thick disk is old and unlikely to harbor many hot stars other than the types found in the halo: blue stragglers, blue horizontal branch stars (although probably few in number given the relatively high metallicity of the thick disk), and rarer types. Other explanations must be found for metal-rich, massive/short-lived early type stars in the halo.

3. The Historical Perspective

3.1. *The Halo's Relation to the Disk*

The "normal" hot stars produced by the traditional halo population, such as the RR Lyraes and the blue horizontal branch stars, are thus expected to be relatively rare in the solar environs compared to disk stars, and become, again relatively, more common only at distances of several kpc above or below the plane. But have we made an unsupportable assumption? We have assumed there is only one "halo" (i.e., metal-poor and old) population that might produce such stars. Might there be more than one? After all, Majewski's results implied the high halo, defined here to be those objects with $Z > 4$ kpc, are on average in retrograde rotation and that therefore the majority of them were probably accreted by the Galaxy rather than formed during its earliest stages. Where are the metal-poor, old predecessors to the Galaxy's disk?

There is not yet consensus on the answer to this last question, but before addressing it, let us recall one of the first, most elegant ideas about the relation between the metal-poor halo and the metal-rich disk, that of Eggen *et al.* (1962; hereafter ELS). They found clear relations between the Galactic orbital angular momentum (the V velocity) and metallicity, implying a "spin-up" process, and another clear relation between the planar Galactic orbital eccentricity and the metallicity. Since the eccentricity was essentially an adiabatic invariant, the correlation between it and time/metallicity implied a short timescale for the "collapse" of the Galaxy from its spherical halo incarnation into its current rapidly rotating disk state. In recent years, the ELS data and model have been subjected to serious criticism, beginning with the cluster studies by Searle & Zinn (1978), and more recently with the field star studies of Ryan & Norris (1991), Carney *et al.* (1990b), and, of course, Majewski (1992). Further, the significant age spread among the metal-poor globular clusters (see, for example, Sarajedini & King 1989, VandenBerg *et al.* 1990 and Carney *et al.* 1992) is inconsistent with the rapid ELS timescale. The field star dispute is, perhaps, best summarized in Figures 1 and 2. These are taken from the expanded study of proper motion stars, begun by Carney & Latham (1987), and continued for over a decade in collaboration with John Laird and, recently, with Luis Aguilar. Briefly, the sample now contains 1452 F, G, and early K stars selected from the Lowell Proper Motion Catalog, although the proper motions used were taken from the Luyten NLTT catalogs, which are more precise and less biased. Photometry leads to distance and temperature estimates, while high-dispersion (R $\approx$ 30,000) but low signal-to-noise (typically 5 to 10) echelle spectra are used to obtain the radial velocities to high precision, as well as the metallicities by χ^2 fitting to grids of synthetic spectra (see Carney *et al.* 1987; Laird *et al.* 1988).

In Figure 1 we see the mean V velocities plotted against metallicity. The interpretation of this Figure is straightforward: it is consistent with the ELS model. If metallicity is a chronometer, so that more metal-rich stars are younger, then it appears that the Galaxy has indeed "spun-up". A problem is that the Galaxy did not, apparently, spin up or begin to do so until the metallicity level had reached [m/H] ≈ -1.4. Perhaps the metallicity enrichment timescale was so rapid that spin-up did not have time to occur. Another problem is the very low <V> values at the lowest metallicities: they are consistent with zero net angular momentum (for $\Theta_0 = 220$ km s^{-1}, at least). Of course, they do not appear to be consistent with Majewski's (1992) retrograde rotation, either. But study of Figure 1 is misleading: what is necessary, when more than one population is involved, are the data themselves, not the means.

The data are shown in Figure 2, and now it appears that the halo population, if we can identify it most reliably using only stars on retrograde orbits, extends to quite

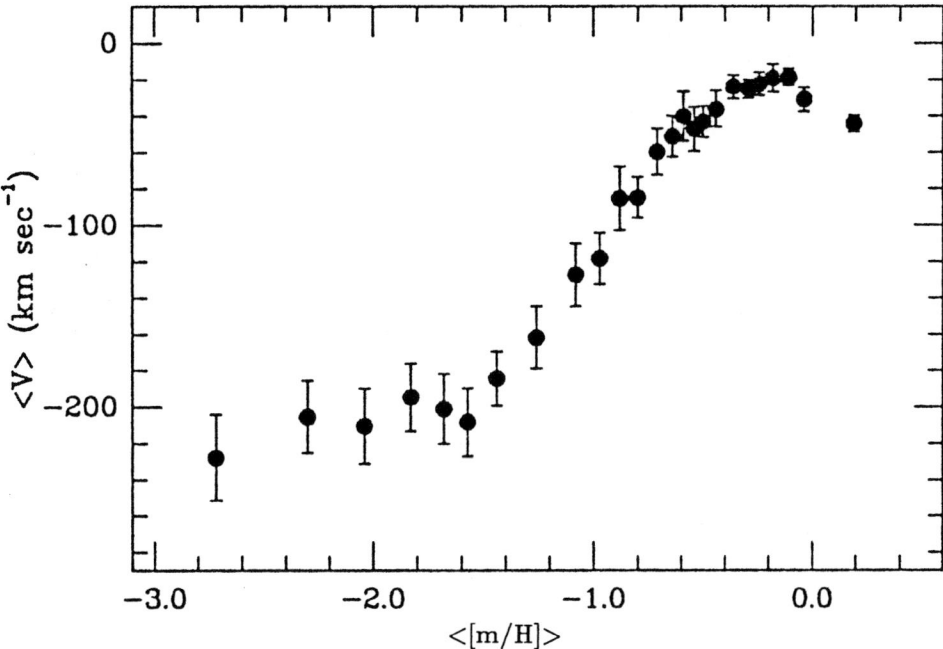

FIGURE 1. The relation between mean V velocity, $<V>$, vs. metallicity from a sample of proper motion stars. The 25 bins were chosen to each have equal numbers of stars.

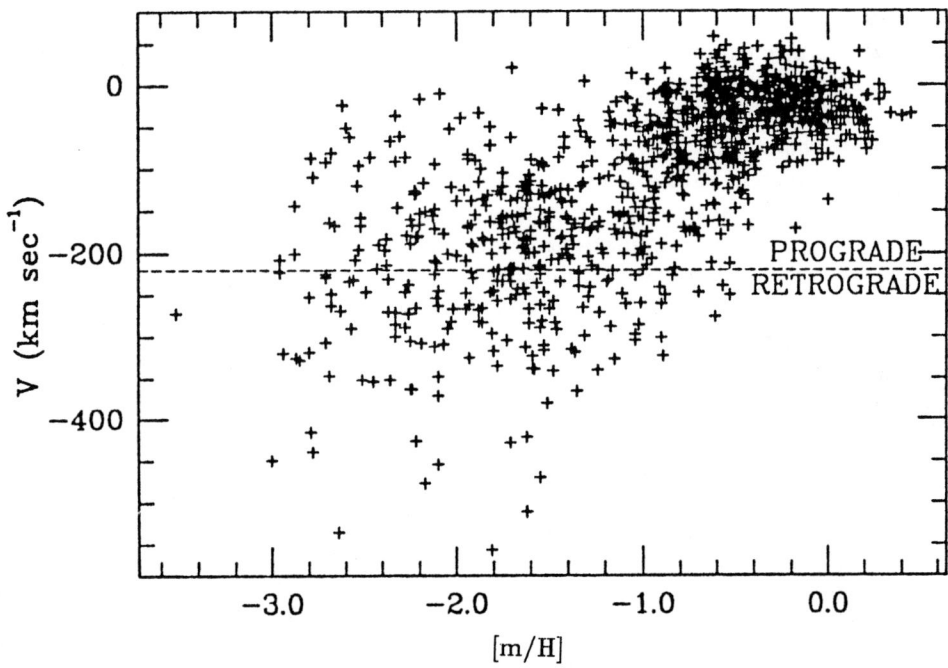

FIGURE 2. The individual data points from the sample of proper motion stars.

high metallicities. The trend seen in Figure 1 is not real: it is the blending in differing proportions of the metal-rich disk stars (V > −100 km s^{-1}) with the halo stars, which diminish in this sample relative to the disk stars as the metallicity increases. The halo population itself does not appear to have any obvious relation between V velocity (i.e., angular momentum) and metallicity, which is consistent with an accretion origin for the halo. In fact, this same trend may exist in the original ELS sample. Ryan & Norris (1993) have modelled the original selection effects in the two catalogs from which the ELS data were taken: one of high-velocity stars; the other of local disk stars. Assuming a dynamically hot halo with no relation between metallicity and kinematics, and a disk population with a monotonic relation between kinematics and chemistry (i.e., that the thick disk is the precursor of the thin disk), Ryan & Norris (1993) found that they could very satisfactorily reproduce the original ELS data, but obviously not their conclusions.

3.2. More Than One Metal-poor Population?

But is the halo population nothing more than (mostly) metal-poor stars accreted by our Galaxy? Are they the remains of loosely bound globular clusters or dwarf galaxies that formed stars and enriched their internal gas independently of what was happening within the Milky Way?

A completely independent halo population seems unlikely if for no other reason than it aggravates the "G dwarf problem", which is the failure of simple chemical evolution models to predict as few metal-poor stars in the solar neighborhood as are apparently observed. Removing even the few halo stars to independent origins outside the Milky Way certainly makes this problem only worse.

There have been some signs earlier, however, that have suggested that even the metal-poor stars may belong to two different populations with, presumably, two different histories. For example, Hartwick (1977) found that he could not readily model the distributions and kinematics of the local metal-poor ([Fe/H] ≤ −1) RR Lyraes with only one component. He needed two, one spheroidal as expected, and one flattened. Rodgers & Paltoglou (1984) found a suggestion that there is a "group" of comparable metallicity ([Fe/H] ≈ −1.5) globular clusters in retrograde rotation, presaging Majewski's (1992) result. (The fact that they had similar metallicities is interesting, but since the majority of the metal-poor clusters have a similar metallicity, this is not compelling.) The existence of such a sub-group means, of course, that even the metal-poor globular clusters might be divisible into separate sub-populations or even populations. Finally, Sommer-Larsen & Zhen (1990) studied a kinematically unbiased local sample of metal-poor stars ([Fe/H] ≤ −1.5), and, like Hartwick (1977), found they needed two populations, one spheroidal and one flattened, to provide a good fit to the data. The two populations are of comparable density in the solar neighborhood, but the spheroidal one has a greater total mass due to its greater spatial extent.

Three recent types studies have begun to reveal clearer signs of two metal-poor populations, one perhaps related to a disk population and the other to an accreted, spheroidal population.

First, there have been two detailed studies of the metal-poor globular clusters. Sidney van den Bergh (1993a, b, c) has compared the mean fundamental mode of the pulsation period (the "Oosterhoff class", where type I usually have $<P_{ab}>$ = 0.55 day, and type II have $<P_{ab}>$ = 0.65 day) against other properties of the clusters, particularly whether their radial velocities indicate prograde (i.e., disk or proto-disk) or retrograde (i.e., accreted) motion. There seems to be a fairly clean division, with the Oo I clusters preferring retrograde orbits, an extension of the Rodger & Paltoglou (1984) results. van den Bergh also found that the more distant clusters were more likely to be on plunging rather than circular orbits, and that this tendency is related to horizontal branch morphology. He interpreted these results as suggesting that both the Searle & Zinn (1978) accretion/merger model and a variant of the ELS continuum model for the formation of the halo may be operating. Perhaps the more quantitative results come from the study by Zinn (1993). He and his colleagues have made a thorough study of the morphology of globular cluster horizontal branch morphologies, using the quantity $C = [(B-R)/(B+V+R)]$, where B, V, and R are the numbers of horizontal branch stars blueward of, within, and redward of the instability strip. Everything else (such as [X/Fe], helium, rotation, etc.) being equal, C should depend on metallicity alone, with more metal-poor clusters having bluer horizontal branches (and more positive values of C). Clusters do not obey a simple such relation, so a "second parameter" must be operating. If this second parameter is age, Zinn could identify, at fixed [Fe/H], younger clusters by the C value, with redder horizontal branches (more negative C values) being indicative of younger ages. The relative ages he derived are supported by those derived by Sarajedini & King (1989), VandenBerg et al. (1990), and Carney et al. (1992). Zinn then divided up his sample into "old" and "young" clusters, with a shift in C of at least 0.4 required to move a cluster into the latter category. The probable age differences between the two groups were then somewhere between 1 and 3 Gyrs. He then looked for two key signatures of a disk population. Disks and proto-disks ought to have prograde rotation. The 46 metal-poor globular clusters in the "old" category showed a mean rotational velocity ($v_{rot} = V + \Theta_0$) of 70 ± 22 km s^{-1}. The 19 "young" metal-poor globular clusters showed $<v_{rot}> = -64 \pm 74$ km s^{-1}. Since the results came from a relatively small sample, and using only the measured radial velocities, large uncertainties in these results were expected, but they do at least suggest that there could be a retrograde component to the metal-poor stars, as Majewski (1992) had argued, and that, further, this accreted component might be younger than the possibly proto-disk sample. The second disk-like signature Zinn studied was the metallicity gradient. Disks are dissipative structures, and metallicity gradients are expected as a consequence of their formation. The "young" clusters showed no sign of a metallicity gradient, determined using the clusters' measured metallicities and their current Galactocentric distances, as might be expected from an accreted population (although other explanations for the lack of a gradient also exist—see Majewski 1993). The prograde rotation, "old" population, however, did show signs of a metallicity gradient. The clusters lying within 6 kpc of the Galactic center have a mean metallicity of -1.44 ± 0.06, while those lying between 6 and 15 kpc have $<[Fe/H]> = -1.80 \pm 0.07$, and those lying further than 15 kpc have $<[Fe/H]> = -1.93 \pm 0.10$. At least in the inner parts of the Galaxy, a metallicity gradient seems to be present for these older, prograde rotation clusters.

The second recent study is that of Norris (1993), who has been doing Monte Carlo simulations of large data samples, particularly of the Carney-Latham-Laird (CLL) sample. His procedure is to adopt relative number densities of the populations he will include in his model, and how their dynamics depend upon metallicity, in particular the asymmetric drift and the three components of the velocity ellipsoid. Having defined the relative mid-plane number densities of the constitutent populations and the relations each has between chemistry and kinematics, he can then use the Monte Carlo approach to generate synthetic surveys whose selection criteria match those of real observational programs, including limits on position, magnitude, color, and proper motion. While this procedure may seem somewhat arbitrary, there are already serious constraints upon what values he may adopt for the halo population, for example, and upon the relative densities of the thin disk, thick disk, and halo populations, largely through his previous work (Norris 1986) and that of Majewski (1992). His prior work, in collaboration with Sean Ryan

(Ryan & Norris 1991) was intended to determine if a separate, accreted thick disk (no relation between kinematics and metallicity) or an "extended disk" (i.e., the thick disk being a precursor to the thin disk, with a monotonic relation between kinematics and metallicity) better matched the CLL results (Carney et al. 1990a). The results were ambiguous, with neither model presenting a subjectively suitable match to the behavior of all 4 parameters, V, $\sigma^2(U)$, $\sigma^2(V)$, and $\sigma^2(W)$. In his new work (Norris 1993), he has adopted an infall of unenriched gas, the so-called "Best Accretion Model" of Lynden-Bell (1975), with the amount of such material measured by the ratio of the infalling mass to the original mass, as defined by Pagel (1989). This extra component means, mathematically, that there is one more free parameter that Norris can vary to obtain good fits to the observations. But it is physically plausible, and adds a larger amount of low-metallicity material to the disk component, where it is needed. Thus the metallicity distribution of the disk in this model is no longer symmetric about the mean, but has a strong tail to the lowest metallicities. The best fits result, in fact, when the tail is strong enough that in local samples, the density of the disk population may actually exceed or at least be comparable to that of the accreted halo population even at metallicities as low as [Fe/H] < −2.0. The very good match between Norris' models and the CLL data does not, of course, prove that such an "extended" disk (or, as Norris, dubs it, a "dual halo", but which I prefer to call a proto-disk) exists, but the results are very suggestive. In the words of the detective Nero Wolfe, "In a world of cause and effect, circumstance is suspect". Infall is, of course, not an outlandish idea. It must have occured at some level early on, and it is plausible that it did so at the level suggested by Norris' results. And of course recent infall may even be responsible for some of the massive, young stars seen far from the plane.

The final new contribution is as yet not even in preprint form, but rather still in preparation. This is the extension of the original CLL survey, from its original 900 stars to the current 1452 stars, from the original 6,000 echelle spectra to the current 23,000. Space does not permit a full discussion of the results from this new survey, but there are three points that must be raised and compared to the prior work by Majewski (1992), Zinn (1993), and Norris (1993).

First, following Majewski (1992), the new sample (hereafter CALL, for Carney, Aguilar, Latham, and Laird) is divided into "high" and "low" subsamples. With the proper motions, radial velocities, and distances, we have obtained U, V, and W velocity vectors, and these have been integrated within a model Galactic gravitational potential, following the same procedures outlined by Carney et al. (1990a). From these integrations, we have derived the mean apogalacticon, $<R_{apo}>$, and perigalacticon, $<R_{peri}>$ distances, as well as the mean maximum distance from the plane, $<|Z_{max}|>$. All of these represent, typically, the averages over 15 orbits. The "high" subsample includes those stars with $<|Z_{max}|> \geq 4.0$ kpc, while the "low" subsample has $<|Z_{max}|> \leq 2.0$ kpc. We must also add another criterion, one involving velocity, for our sample was selected from a proper motion catalog and hence is kinematically biased. As Ryan & Norris (1993) have shown quite convincingly, this type of bias works against disk-like kinematics, and even for the most metal-poor stars, which presumably have high velocities, it leads to a bias toward, for example, larger asymmetric drifts than an unbiased parent population might possess. Indeed, a sample with $<V> = -190$ km s^{-1} selected according to Lowell Catalog proper motion limits would result in a measured $<V>$ of about -220 km s^{-1}. One important consequence of this is that the remark made earlier about the nearly-zero net angular momentum of the most metal-poor stars (see Figure 1; $<V> \approx -220$ km s^{-1}) was inappropriate since it referred to such a biased sample. The mean V velocity of the most metal-poor stars, when this bias is removed, is thus probably mildly prograde.

Ryan & Norris (1993) estimate the correction to be $\Delta<V> = 30$ km s^{-1}, so that $<V>$ ≈ -190 km s^{-1}. Without invoking Monte Carlo modelling at this early stage in our analyses, however, we can avoid it somewhat by selecting only those stars with large U and W velocities, large enough to avoid large biases against their selection in the Lowell catalog. Thus for the "low" subsample, we require the stars to have $(U^2 + W^2)^{1/2} \geq$ 200 km s^{-1}. Such stars will easily be included in a proper motion catalog by virtue of their large U and W velocities. This not only means the V velocity component should not be affected by the proper motion bias, but that its overall value can have, physically, almost any value since the total kinetic energy content in the U and W motions is enough to keep the star at this Galactocentric distance (i.e., the net velocity is comparable to or greater than Θ_0). The results are very interesting: the "high" sample, like Majewski's, reveals retrograde rotation, $<v_{rot}> = -35 \pm 17$ km s^{-1}, whereas the "low" sample shows prograde rotation, $<v_{rot}> = +48 \pm 12$ km s^{-1}. If the "high" limit to $<|Z_{max}|>$ is increased to 5.0 kpc, $<v_{rot}>$ increases to -50 ± 21 km s^{-1}, in excellent agreement with Majewski's (1992) result of -55 ± 16 km s^{-1}. His result had been criticized on the grounds that his distance estimates may have been systematically in error, leading to the net retrograde rotation, but the CALL distances for both the "high" and "low" samples are determined using one algorithm, so the difference between the "high" and "low" subsamples is real. The "high" halo does indeed seem to be in net retrograde rotation and, therefore, at least the majority of it probably is an accreted component. The "low" halo, on the other hand, could be composed of a proto-disk component.

The second study is to search, like Zinn (1993), for a radial metallicity gradient, indicative of a dissipational process that may signal the formation of the disk. Using $<R_{apo}>$ as the radial variable, the "high halo" shows, as expected, no metallicity gradient, $d[m/H]/d<R_{apo}> = +0.003 \pm 0.014$ dex kpc^{-1}. The "low" halo, however, does manifest a metallicity gradient, $d[m/H]/d<R_{apo}> = -0.030 \pm 0.012$ dex kpc^{-1}, or about 0.5 dex from the inner part of the "halo" to the outer part, consistent with Zinn's (1993) findings for the "old" clusters.

Finally, Zinn (1993) began by dividing his sample into "young" and "old" metal-poor globular clusters. Can we find the same thing in our field stars? Ages of individual field stars are hard, but not impossible, to estimate. Schuster & Nissen (1989) have used model isochrones and $uvby\beta$ photometry to estimate ages for metal-poor stars, many of which have been studied by CALL. As they had already determined, there is no clear relation between the ages and the metallicities of the metal-poor field stars. Stars with [Fe/H] ≈ -1 were the same age as those with [Fe/H] < -2.

However, Figure 3 shows that the "high" halo stars appear to be younger than the "low" halo stars, just as Zinn (1993) found. While the statistics are not great due to the limited sample size, the "low" halo formally has $<[m/H]> = -1.71$ and $<t_9> = 15.7 \pm 0.3$ (N = 29, t_9 is the age in Gyrs), while the "high" halo has $<[m/H]> = -1.64$ and $<t_9> = 13.2 \pm 0.8$ (N = 5). (It even appears that there might be two age groups, with the younger one having a greater scale height that the older one.) The CALL survey, in other words, confirms that the "high" halo is in retrograde rotation and lacks a metallicity gradient, while the "low" halo is in prograde rotation and does show a metallicity gradient, and like Zinn's two samples that display these properties, the retrograding population is apparently younger than the prograding one. Clearly, a larger sample of metal-poor stars with measured stellar ages is needed, and the observations are already underway, in collaboration with Bill Schuster.

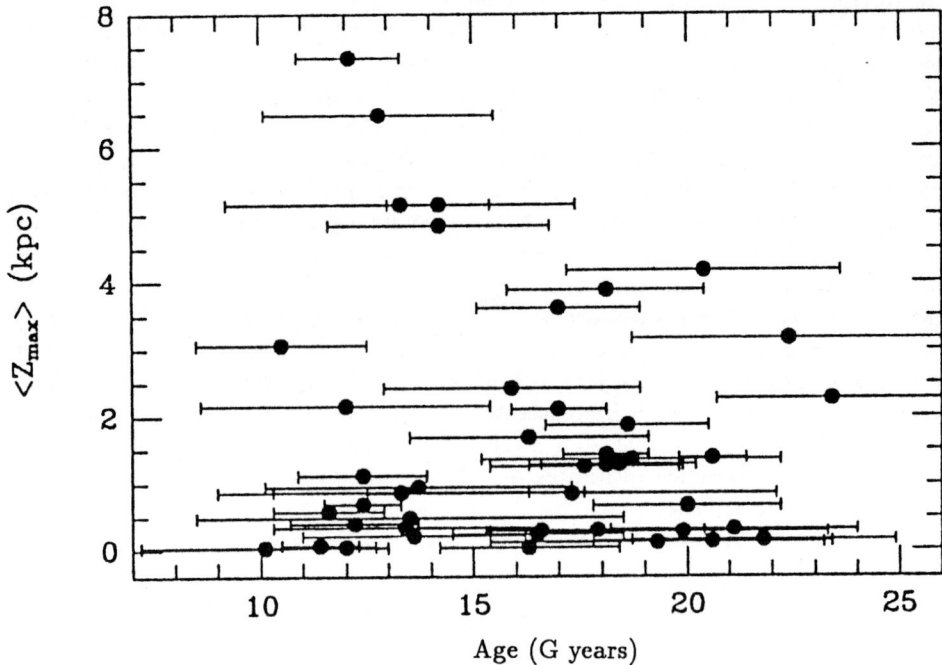

FIGURE 3. The ages obtained by Schuster & Nissen (1989) for individual field stars are plotted against their mean maximum distance from the Galactic plane.

3.3. *Is There a Very Metal-poor Disk Population?*

Heather Morrison's Ph.D. thesis work (see Morrison *et al.* 1990) showed clear evidence for disk stars with metallicities as low as [Fe/H] ≈ −1.6. The CALL sample now also shows evidence for a very metal-poor disk population, in spite of the very strong biases against stars with disk kinematics in the Lowell catalog. Our awareness of such stars originated as we began to model out the selection effects inherent in a hemisphere-limited, proper motion-selected sample, using a method devised by Luis Aguilar. Every star in the CALL sample with $\delta \geq -2°$ (below which the sample becomes incomplete) and with total proper motion $\mu \geq 0.23$ arcsec per year using the NLTT data (below which the Lowell catalog becomes incomplete) was assigned a weight for use in computing mean values of the velocities or any other quantity. To avoid some uncertainties in the distances derived from photometric parallaxes, we also excluded all stars whose dereddened $B - V$ color indices were within 0.04 mag of the estimated main sequence turn-off value, using globular clusters to calibrate the relation. This avoids turn-off stars in our sample, whose distances are somewhat uncertain and where subgiant contamination of the sample is at a maximum. Details will be in one of the papers now in preparation, but the point is that when we first divided the sample in 7 different metallicity bins, we obtained the results given in Figure 4. The circles represent the uncorrected results, and are thus similar to the results in Figure 1. The dots refer to the corrected means. The horizontal error bars refer to the width in metallicity of the 7 bins. The larger vertical error bars measure the velocity dispersions, whereas the smaller error bars signify the errors of the means.

As expected, the <V> values for the higher metallicity stars moved up closer to that zero velocity line. Indeed, all the <V> values increased, as expected. What was not expected was the very dramatic increase in the lowest metallicity bin, from <V> = −226

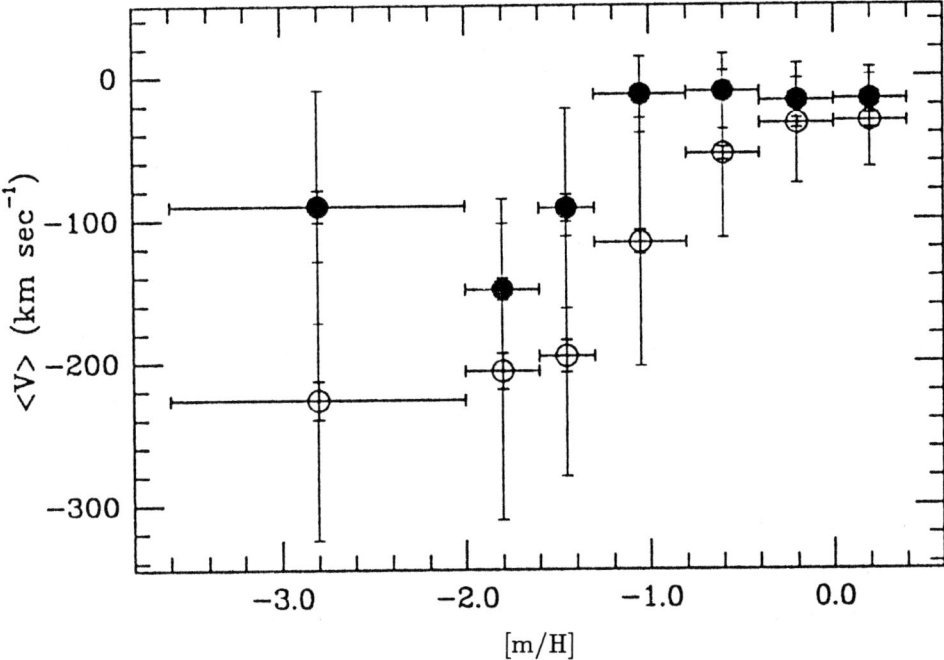

FIGURE 4. The resultant <V> velocities from a sample of stars restricted to $\mu \geq 0.23$ arcsec yr^{-1}, $\delta \geq -2°$, and colors at least 0.04 mag redder than cluster main sequence turn-offs of comparable metallicity. Circles are the raw data and dots are the "corrected" results. The lowest metallicity bin has been perturbed greatly by a few disk stars. Small vertical error bars are the errors of the means.

± 13 (uncorrected) to -90 ± 7 (corrected) km s^{-1} (errors are errors of the mean). Given the kinematically unbiased work of Norris (1986), and of Carney & Latham (1986), we had expected a value of <V> close to -200 km s^{-1}. The cause of the great change may be seen in Figure 5, where the logarithms of the weights are plotted against metallicity. The five stars with [Fe/H] < -2 and log W ≈ 5 are responsible, and they have high weights because they have disk kinematics. These may be true proto-disk stars. Removing them results in a mean (corrected) value of <V> = -198 ± 11 km s^{-1}, which is about what we had expected based on the Ryan & Norris (1993) Monte Carlo modelling. In fact, taking the weights as given, we may use them to estimate the true number of stars with [m/H] ≤ -2.0 which have disk-like kinematics vs. high-velocity halo-like kinematics in the solar neighborhood. We find the ratio is, very approximately, a factor of two, in reasonable agreement with the earlier results of Hartwick (1977) and Sommer-Larsen & Zhen (1990), who found the flattened and spheroidal components of the metal-poor stars to be of comparable density in the solar neighborhood. There does seem to be a very metal-poor proto-disk! The apparently large abundance of these stars in the solar neighborhood helps solve the "G dwarf problem" and also may explain why the halo star density far from the plane has been overestimated in the past (see Morrison 1993). Half or more of the local metal-poor stars are disk stars and hence do not travel far from the plane. The ELS concept of a continuum evolution from metal-poor to metal-rich disk stars is, apparently, applicable to the evolution of the disk and its relation to at least some, and perhaps even the majority, of the metal-poor stars in the solar neighborhood.

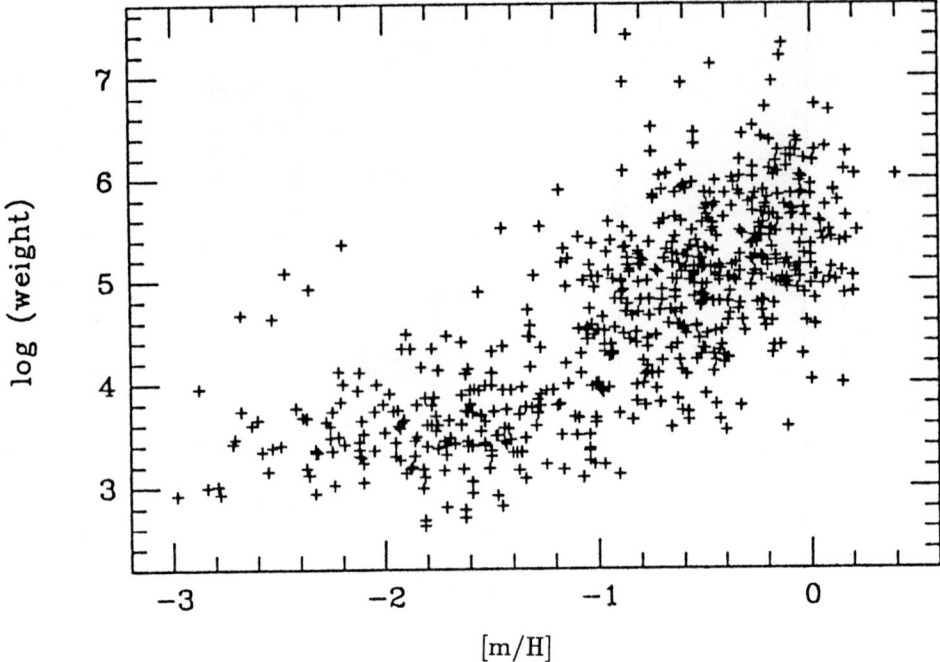

FIGURE 5. The computed "weights" for inclusion in the CALL survey of proper motion stars is plotted against each star's metallicity. Note the lowest metallicity, highest-weight stars.

4. Summary

The metal-poor, high-velocity population does not dominate in number density over the hotter disk populations until distances exceeding about 4 to 5 kpc are reached from the plane. Since the hotter disk populations are both very old and quite metal-rich, they may not produce a large number of blue horizontal branch or RR Lyrae stars. However, the halo population itself, as traditionally defined by very low metallicity, may actually have two components, the traditional high-velocity one, which now appears to be largely the result of accretion processes, and a lower-velocity, perhaps proto-disk one. Predictions of the expected numbers of blue horizontal branch and RR Lyrae variables vs. distance from the plane may require revision to account for the now-apparent complexity of the different stellar populations in the solar neighborhood, resulting from the rather complex history of our Galaxy.

REFERENCES

CARNEY, B. W., 1979, ApJ, 233, 211
CARNEY, B. W., AGUILAR, L., LATHAM, D. W., LAIRD, J. B., 1990a, AJ, 99, 201
CARNEY, B. W., LAIRD, J. B., LATHAM, D. W., KURUCZ, R. L., 1987, AJ, 94, 1066
CARNEY, B. W., LATHAM, D. W., 1986, AJ, 92, 60
CARNEY, B. W., LATHAM, D. W., 1987, AJ, 93, 116
CARNEY, B. W., LATHAM, D. W., LAIRD, J. B., 1989, AJ, 97, 423
CARNEY, B. W., LATHAM, D. W., LAIRD, J. B., 1990b, AJ, 99, 572
CARNEY, B. W., STORM, J., JONES, R. V., 1992a, ApJ, 386, 663

CUDWORTH, K. M., 1993, ASP Conf. Series, 49, 141
EGGEN, O. J., LYNDEN-BELL, D., SANDAGE, A., 1962, ApJ, 136, 762
EDVARDSSON, B., ANDERSEN, J., GUSTAFSSON, B., LAMBERT, D. L., NISSEN, P. E., TOMKIN, J., 1993, A&A, 275, 101
HARTWICK, F. D. A., 1977, ApJ, 214, 778
LAIRD, J. B., CARNEY, B. W., LATHAM, D. W., 1988, AJ, 95, 1843
LYNDEN-BELL, D., 1975, Vistas Astron., 19, 299
MAJEWSKI, S. R., 1992, ApJS, 78, 87
MAJEWSKI, S. R., 1993, ARAA, 31, 345
MORRISON, H. L., 1993, AJ, 106, 578
MORRISON, H. L., FLYNN, C., FREEMAN, K. C., 1990, AJ, 100, 1191
NORRIS, J. E., 1986, ApJS, 61, 667
NORRIS, J. E., 1993, preprint
PAGEL, B. E. J., 1989, in Beckman J. E., Pagel B. E. J., eds, Evolutionary Phenomena in Galaxies, Cambridge University Press, Cambridge, p. 201
RODGERS, A. W., PALTOGLOU, G., 1984, ApJ, 283, L5
RYAN, S., NORRIS, J. E., 1991, AJ, 101, 1835
RYAN, S., NORRIS, J. E., 1993, ASP Conf. Series, 49, 103
SANDAGE, A., 1969, ApJ, 158, 1115
SARAJEDINI, A., KING, C. R., 1989, AJ, 98, 1624
SCHUSTER, W. L., NISSEN, P. E., 1989, A&A, 222, 69
SEARLE, L., ZINN, R., 1978, ApJ, 225, 357
SOMMER-LARSEN, J., ZHEN, C., 1990, MNRAS, 242, 10
VAN DEN BERGH, S., 1993a, MNRAS, 262, 588
VAN DEN BERGH, S., 1993b, AJ, 105, 971
VAN DEN BERGH, S., 1993c, ApJ, 411, 178
VANDENBERG, D. A., BOLTE, M., STETSON, P. B., 1990, AJ, 100, 445
ZINN, R., 1985, ApJ, 293, 424
ZINN, R., 1993, ASP Conf. Series, 48, 38

Discussion

PHILIP: I was interested in your remark that the Horizontal Branch (HB) stars and disk stars, at modest distances from the galactic plane, were divided 50-50. In areas where I have done Strömgren photometry of all stars to a given limiting magnitude, I find the numbers of FHB stars and the numbers of apparently normal disk stars at distances of 2-4 kpc from the plane are split approximately 50-50.

CARNEY: That is certainly an interesting result, although if half the stars are really "normal disk" stars, then they are presumably very young. The nominal ages of the two "halo" groups, however, appear to be very old, inconsistent with normal young B stars. Thus, your 50-50 split at such large distances from the plane is really interesting.

PETERSON: The uniform enhancement of light elements (e.g. O, Mg) in all field stars with [Fe/H] < -1, as plotted, for example, in Wheeler, Sneden & Truran (1989, ARAA, 27, 279), seems rather surprising if, in fact, half of these stars have their origins in one type of halo and the other in another. It would seem very important as an additional constraint on the formation of the halo(s) to obtain α/Fe ratios for the six kinematically distinct stars in your sample.

CARNEY: That is a very good point. Of course, [α/Fe] depends primarily on the duration of nucleosynthesis. It does not seem far-fetched to me that the nucleosynthesis was similar in the proto-disk and the accreted halo. But no spectroscopic studies have been made of very metal-poor ([Fe/H]<-1.5) stars with disk kinematics. Thus we are in the process of doing echelle spectroscopy at McDonald Observatory to see what is going on. Maybe we will be surprised!

BEERS: In a previously published AAS abstract, (Sommer-Larsen, J., Beers, T. C., Alvarez, J., 1992, BAAS, 24, 1177) and a draft in preparation, Beers & Sommer-Larsen find (from an analysis of some 2000 non-kinematically-selected metal-poor stars) that at least 30 % of stars in the solar neighborhood with [Fe/H] < -1.5 can be kinematically associated with a disk-like population with large systematic rotation and low velocity dispersion.

CARNEY: That is the sample size we need to not only demonstrate the dual nature/origin of metal-poor stars. The sample also ought to be large enough to assess the relative dominance of the "proto-disk" and "accreted" stars at different metallicities. Have you looked into that?

BEERS: No not yet. To do this sort of analysis one might imagine concentrating on the intermediate metallicity range, say $-1.6 <$ [Fe/H] < -1.0, where the contribution of the halo stars is smaller. To do a definitive analysis of a multiple disk (proto + accreted) superposed on a (possibly) complex halo will require a same size sample, perhaps as large as 10,000 stars.

JANES: Is this the solution to the "G Dwarf Problem"?

CARNEY: I believe so. Rosie Wyse and Gerry Gilmore noted the "lack of local metal-poor" stars could be solved by including the thick disk. The proposed extension to very low metallicities, [Fe/H] < -2.0, in the disk/proto-disk population helps even further.

Theoretical Properties of Horizontal-Branch Stars

By ALLEN V. SWEIGART

Laboratory for Astronomy and Solar Physics, Code 681, NASA/Goddard Space Flight Center, Greenbelt, MD 20771, USA

The canonical theory of horizontal-branch (HB) evolution is reviewed with particular emphasis on the underlying assumptions. New calculations for the helium flash are presented to study the evolution of a globular cluster star from the tip of the red-giant branch (RGB) to the zero-age horizontal branch (ZAHB). The importance of the observed number ratio R_2 of asymptotic-giant-branch (AGB) to HB stars for testing the theoretical models is emphasized. We discuss the various approaches which have been taken to follow the evolution near the end of the HB phase. New models based on a parameterization of the overshooting efficiency are presented for both the core-helium-exhaustion and main HB phases.

1. Introduction

Theoretical HB models have many applications in the study of the galactic globular clusters. The determination of the globular-cluster helium abundance by the R-method, for example, depends on the theoretical HB lifetimes, while the determination of the globular-cluster ages from the observed magnitude difference between the HB and the main-sequence turnoff depends on the theoretical luminosity at which HB models evolve through the RR Lyrae instability strip. The morphology of the theoretical HB tracks must be known when studying the observed HB distributions and the well-known second parameter problem. This is also true for understanding the evolutionary status of the extended-HB and UV-bright stars found in many globular clusters. The theoretical luminosities and masses of the HB models within the instability strip are also needed for comparison with the observed properties of the RR Lyrae variables, as exemplified by the Sandage period-shift effect. Finally, by using the theoretical properties of the current "canonical" models for the HB, one can investigate the conditions under which such noncanonical effects as rotation might be important.

These applications raise a number of key questions which need to be addressed if one wishes to assess the reliability of the current theoretical models. What, for example, is meant by canonical HB theory, and what are the assumptions upon which it depends? What properties of the HB are predicted by theory, and what properties are actually determined by observational constraints? What, in fact, are the observational constraints for testing the theory?

Although detailed answers to all of these questions are beyond the scope of this paper, we would nevertheless like to address five relevant topics, beginning with the question: what is the ZAHB? This is not a trivial question. Although ZAHB models are frequently used in analyzing HB observations, their definition is somewhat ambiguous. The reason for this is that none of the available sets of HB sequences actually follow the evolution of a star from the tip of the RGB through the helium flash and onto the HB. Thus the structure of a ZAHB model depends on one's assumptions about the consequences of the helium flash.

For the second topic we will discuss the reliability of the current canonical values for the mass M_c of the helium core at the ZAHB phase and will summarize the arguments, both theoretical and observational, for a noncanonical change in these values of M_c. The third topic will present a brief description of canonical HB evolution with emphasis on the role of semiconvection. The fourth topic is concerned with the end point of the HB evolution when the central helium abundance Y_c is less than ≈ 0.1. Attempts to follow the evolution through this core-helium-exhaustion phase have been plagued with numerical difficulties which may be associated with a convective instability. Finally we will describe how the evolution of an HB star is affected if one relaxes the canonical assumption of instantaneous overshooting and instead uses a time-dependent formulation for the overshooting efficiency.

2. What is the zero age horizontal-branch?

The helium flash drastically alters the interior structure of a globular-cluster star from one characterized by a highly degenerate helium core and a single energy source, namely, a hydrogen-burning shell, to one characterized by a nondegenerate helium core and two energy sources, namely, a helium-burning core and a hydrogen-burning shell. The few available calculations for this phase indicate that the helium flash is a mildly violent event during which the helium burning briefly attains exceedingly high luminosities. The details of what happens during the helium flash are obviously important for the structure of the subsequent ZAHB models. In particular, we would like to answer the following questions. Does the helium flash lead to mixing between the core and the envelope? How much carbon is produced within the core? What is the timescale for the evolution from the tip of the RGB to the ZAHB? Does the hydrogen-shell profile relax during this pre-ZAHB evolution from its very steep profile at the tip of the RGB to its much broader profile during the HB phase?

To examine these questions, we have computed the evolution of a 0.83 $M_\odot$ star with a helium abundance Y of 0.23 and a heavy-element abundance Z of 0.001 through the helium flash and onto the HB. These model computations make the canonical assumptions that the helium flash is spherically symmetric and that the flash convection is adiabatic. Mass loss was included during the RGB evolution via a Reimers type formulation with a mass loss parameter η of 0.5. This yielded a final mass at the tip of the RGB of 0.645 $M_\odot$. The age at that time was 14.8 Gyr.

Figure 1 shows the track in the HR diagram during the evolution from the tip of the RGB to the ZAHB. The many gyrations in this track arise from the fact that the helium flash consists of a number of distinct flashes. As illustrated in Figure 2, there is first a main flash during which the helium burning approaches 10^{10} $L_\odot$ and then a series of secondary flashes of much lower amplitude. Prior to the main flash all of this star's energy output comes from the hydrogen shell. The sudden expansion of the outer layers of the core during the main flash effectively cools the hydrogen shell, thereby causing the shell's luminosity to suddenly drop at time $t = 0$ in Figure 2. It takes nearly 10^6 yr for the hydrogen shell to once again become a major energy producer. As a result, the hydrogen-shell profile in this star does not relax significantly prior to the ZAHB phase. It is interesting to note that the high rates of helium burning during the main flash do not lead to an increase in the surface luminosity. Rather, the surface luminosity at that time is rapidly decreasing. This apparent contradiction is due to the fact that nearly all of the energy from the helium burning goes into lifting the degenerate core out of its deep potential well.

Neutrino cooling of the helium core during the RGB phase is more effective near the center due to the higher densities at that point. As a result, the temperature maximum in the core and hence the site of the main helium flash lie off-center. This effect is illustrated in Figure 3, where we see that the main flash occurs off-center at the mass

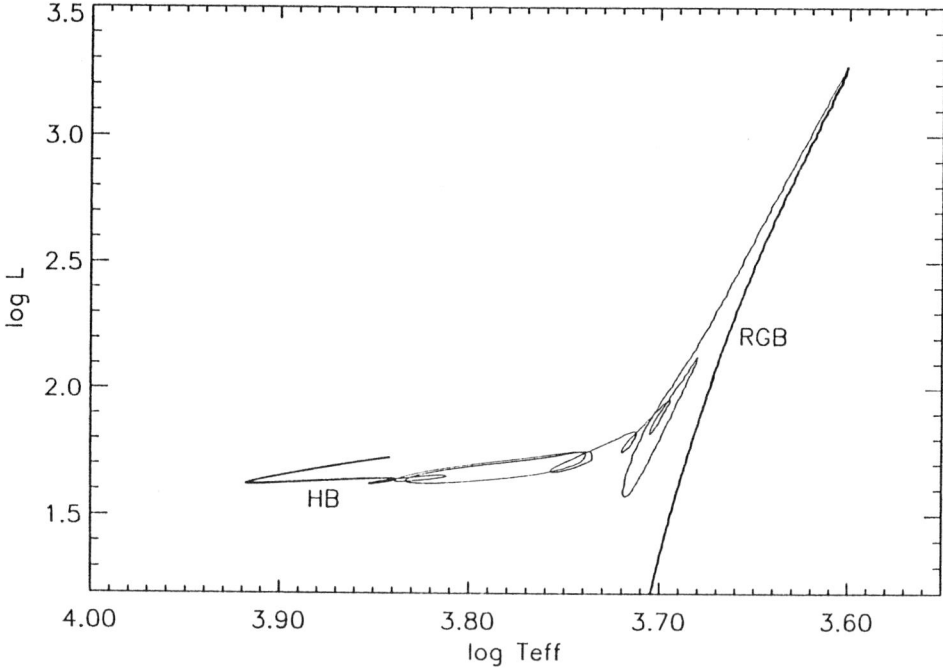

FIGURE 1. Evolutionary track of a 0.645 $M_\odot$ star during the helium flash. The evolution from the tip of the RGB to the ZAHB is plotted with a thin line in order to emphasize the short timescale of this phase. The evolution during the RGB and HB phases is also shown.

coordinate $M_r = 0.16\ M_\odot$. Each of the flashes in Figure 2 gives rise to a temporary convection zone which extends outward from the flash site. In the case of the main flash this convection zone narrowly misses the inner tail of the hydrogen shell, and hence the mixing of envelope material into the core does not take place. Note also that the convective envelope, shown by the shaded region outside the hydrogen shell, does not penetrate into the core. The main flash only removes the degeneracy of the layers exterior to the flash site. Consequently, an inner degenerate core containing $\approx 0.16\ M_\odot$ remains after the main flash subsides. Each subsequent secondary flash peals off another $\approx 0.03\ M_\odot$ from this inner degenerate core until eventually all of the degeneracy in the core is removed and the star settles onto the ZAHB at $t = 1.8 \times 10^6$ yr. Approximately 5 per cent of the core's helium fuel is consumed over the time interval in Figures 2 and 3.

From these results one can draw the following conclusions about the effects of a canonical helium flash: 1) mixing between the core and the envelope does not occur, 2) the central helium abundance at the ZAHB is ≈ 0.95, 3) a star requires $\approx 2 \times 10^6$ yr to evolve through the helium flash onto the ZAHB, and 4) during this evolution the hydrogen-shell profile does not significantly relax. Overall, these results confirm the earlier conclusions of Sweigart & Mengel (1981).

3. Canonical core mass M_c

Another important legacy that a ZAHB model retains from its prior evolution is the mass M_c of its helium core. Values of M_c are computed by evolving stars of various compositions up the RGB to the onset of the helium flash. The core mass at that time

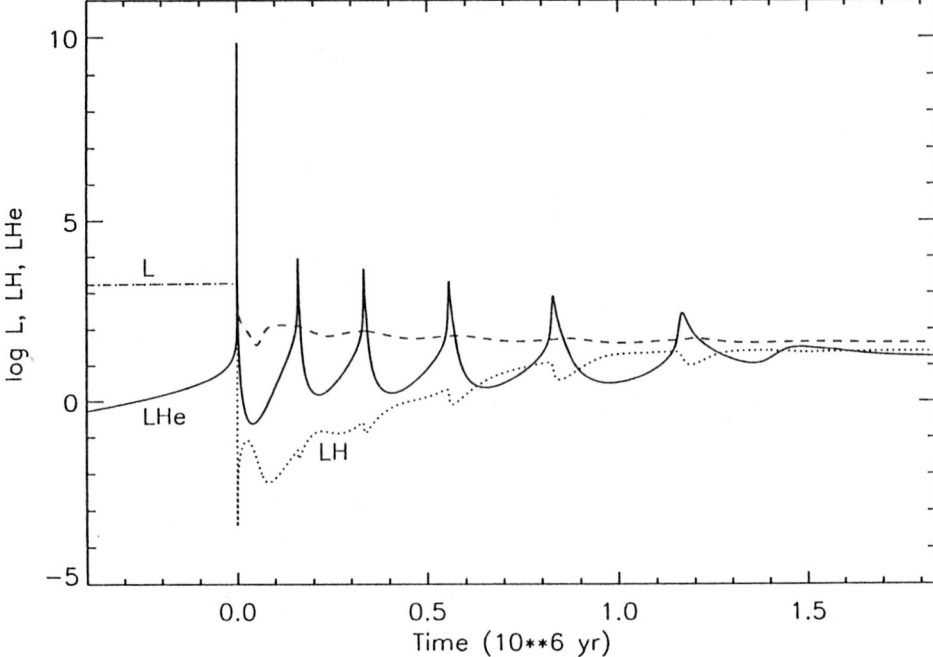

FIGURE 2. Time dependence of the helium-burning luminosity L_{He} (solid curve), the hydrogen-burning luminosity L_H (dotted curve), and the surface luminosity L (dashed curve) during the helium flash of a 0.645 $M_\odot$ star. The zero-point of the timescale corresponds to the peak of the main flash.

is then adopted as the core mass for the ZAHB model, a safe assumption since, as we have seen, very little hydrogen fuel is burned during the helium flash.

Many properties of the HB including the HB luminosity, lifetime and track morphology depend sensitively on the value of M_c, as can be readily appreciated from the tracks plotted in Figure 4. We see that the HB luminosity increases markedly with increasing M_c. From the HB sequences of Sweigart & Gross (1976) we find that an increase ΔM_c in the core mass leads to an increase $\Delta \log L_{HB}$ in the HB luminosity of

$$\Delta \log L_{HB} \approx 3 \Delta M_c \,. \quad (1)$$

HB models with larger core masses also have higher rates of helium burning and thus consume their helium fuel more rapidly. As a result, their HB lifetimes t_{HB} decrease by the amount

$$\Delta \log t_{HB} \approx -3 \Delta M_c \,. \quad (2)$$

The masses of the tracks in Figure 4 were adjusted to keep them near the instability strip. If the mass had instead been held constant, the tracks would have shifted rapidly to higher effective temperatures with increasing M_c.

The value of M_c is controlled by the thermal evolution of the helium core as a star ascends the RGB. The helium core during this phase gains energy from both compressional heating as it grows in mass and helium burning near the tip of the RGB. It losses energy by neutrino emission via the plasma process and thermal conduction by degenerate electrons. The net effect of the competition between these energy gains and losses is a gradual increase in the maximum core temperature until helium ignites at the tip of the

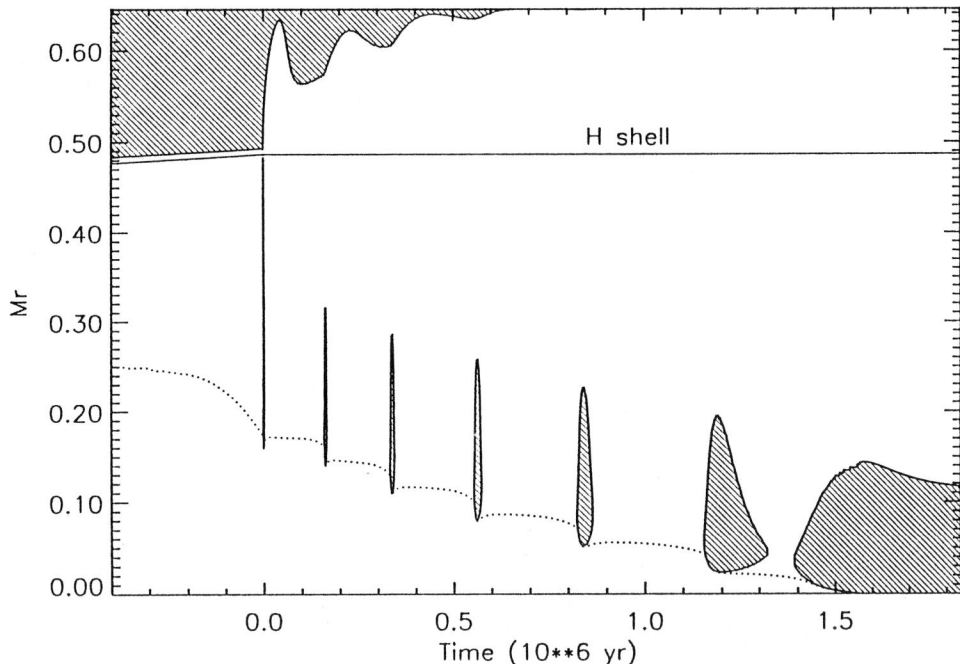

FIGURE 3. Location in mass M_r of the flash convection zones, the center of the hydrogen shell and the base of the convective envelope during the helium flash of a 0.645 $M_\odot$ star. The dotted curve gives the location of the temperature maximum within the core. Shaded areas are convective. The timescale is the same as in Figure 2.

RGB. Canonical RGB sequences predict the following dependence of M_c on composition and mass M (Buzzoni et al. 1983):

$$M_c = 0.476 - 0.221(Y - 0.3) - 0.009(\log Z + 3) - 0.023(M - 0.8). \quad (3)$$

Equation (3) implies that M_c should vary by only ≈ 0.02 $M_\odot$ among the globular clusters.

The above canonical results have been questioned on several grounds. Synthetic models for the HB morphology predict RR Lyrae pulsation periods that are systematically too small. A "noncanonical" increase in M_c of 0.02 - 0.03 $M_\odot$ would resolve this discrepancy (Lee, Demarque & Zinn 1990; Catelan 1992). A second argument is based on the Sandage period-shift effect, i.e., the observed decrease in the pulsation period at a fixed effective temperature with increasing metallicity (Sandage 1990). This effect could be explained if the core masses in the metal-poor Oosterhoff II clusters were substantially larger than those in the more metal-rich Oosterhoff I clusters (Caputo 1990). This would require a large noncanonical dependence of M_c on metallicity. Finally, it has been argued that the numerical algorithms used in RGB models for shifting the hydrogen shell and explicitly advancing the chemical composition are responsible for a systematic underestimate of M_c by ≈ 0.02 $M_\odot$ (Mazzitelli 1989).

There is also an observational argument for a "noncanonical" decrease in M_c. Ultraviolet Imaging Telescope (UIT) observations during the ASTRO-1 mission have found that the hot HB stars in several globular clusters lie ≈ 0.5 mag below the canonical ZAHB (Hill et al. 1992; Parise et al. 1994). A decrease in M_c of ≈ 0.05 $M_\odot$ would lower the ZAHB luminosity sufficiently to fit these UIT observations.

Despite the above arguments there are compelling reasons to believe that the canonical

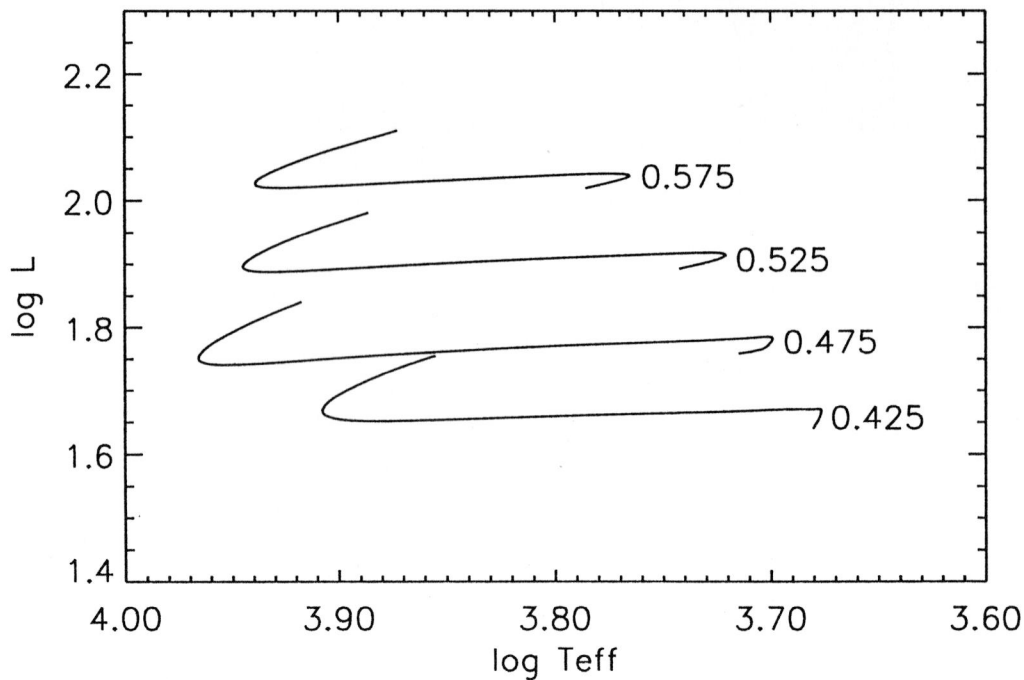

FIGURE 4. Dependence of the HB evolutionary tracks on the core mass M_c for $Y = 0.30$ and $Z = 0.001$ (Sweigart & Gross 1976). Each track is labeled by its value of M_c. The mass of each track was chosen so that the track would be near the instability strip.

values of M_c cannot be seriously in error. First of all, it is straightforward to demonstrate that the numerical algorithms mentioned above have a negligible effect on the value of M_c (Sweigart 1994). Moreover, a change in M_c of only 0.01 - 0.02 $M_\odot$ would require a large error in the input physics and therefore is not easily accomplished within the canonical framework. Perhaps the most compelling argument comes from observations of the RGB tip luminosity L_{tip} in the globular clusters (Frogel, Cohen & Persson 1983; Da Costa & Armandroff 1990). The RGB tip luminosity is particularly sensitive to the value of M_c and thus is an excellent discriminant for any deviation in M_c from its canonical values. Raffelt (1990) has analyzed the available observations of L_{tip} in terms of the luminosity difference between the tip of the RGB and the RR Lyrae variables. He finds good agreement with canonical predictions, especially in regard to the variation of this luminosity difference with metallicity.

It appears therefore that a significant change in the canonical values of M_c would require either a serious flaw in the canonical RGB input physics or some noncanonical effect such as rotation.

4. Canonical horizontal-branch evolution with semiconvection

In the preceding discussion we have outlined some of the physical processes which govern the pre-ZAHB evolution of a globular-cluster star. With this background let us now examine the structure of a typical ZAHB model in more detail and see how such a model evolves during the HB phase according to canonical theory. As a typical case we

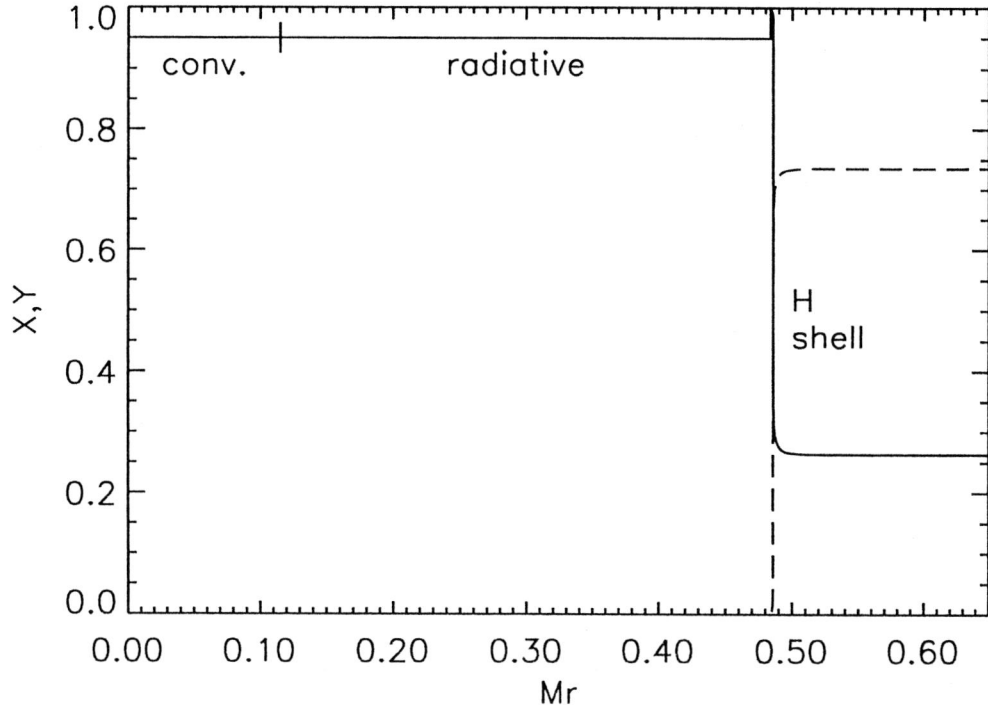

FIGURE 5. Composition profile within the interior of a 0.68 $M_\odot$ star at the ZAHB phase. The solid and dashed curves give the helium abundance Y and the hydrogen abundance X, respectively. The tick mark denotes the edge of the convective core. The mass coordinate M_r is in solar units.

have chosen a 0.68 $M_\odot$ star with the composition Y = 0.25 and Z = 0.001. None of the subsequent discussion should be affected by this choice of model parameters.

The interior structure of this star at the ZAHB phase is illustrated in Figure 5. The helium core has the expected helium abundance of 0.95 and a mass of 0.486 $M_\odot$. Most of the helium core is radiative except for the innermost 0.115 $M_\odot$ which is fully convective. Due to the high temperature dependence of the helium-burning reactions, nearly all of the helium luminosity is produced within the innermost $\approx 0.03\ M_\odot$. As a result, both the amount of helium fuel burned during the HB phase and, consequently, the HB lifetime are critically dependent on the size of the convective core. Fortunately one can easily locate the convective-core edge in a ZAHB model by applying the standard Schwarzschild criterion and looking for the point where the radiative and adiabatic temperature gradients are equal, as indicated in Figure 6.

At this point one might be tempted to think that it is straightforward to compute the evolution of an HB star. One merely has to locate the edge of the convective core in each model. However, there are a couple of surprises which make the HB evolution both more complicated and more interesting. The first surprise arises from the fact that the opacity of carbon-rich material exceeds the opacity of helium-rich material (Castellani, Giannone & Renzini 1971a). Thus, as the helium burning converts helium into carbon, the radiative gradient within the convective core begins to increase. As shown in Figure 7, the radiative gradient at the convective-core edge becomes more and more superadiabatic if one keeps the convective-core edge fixed in the mass coordinate M_r. It is

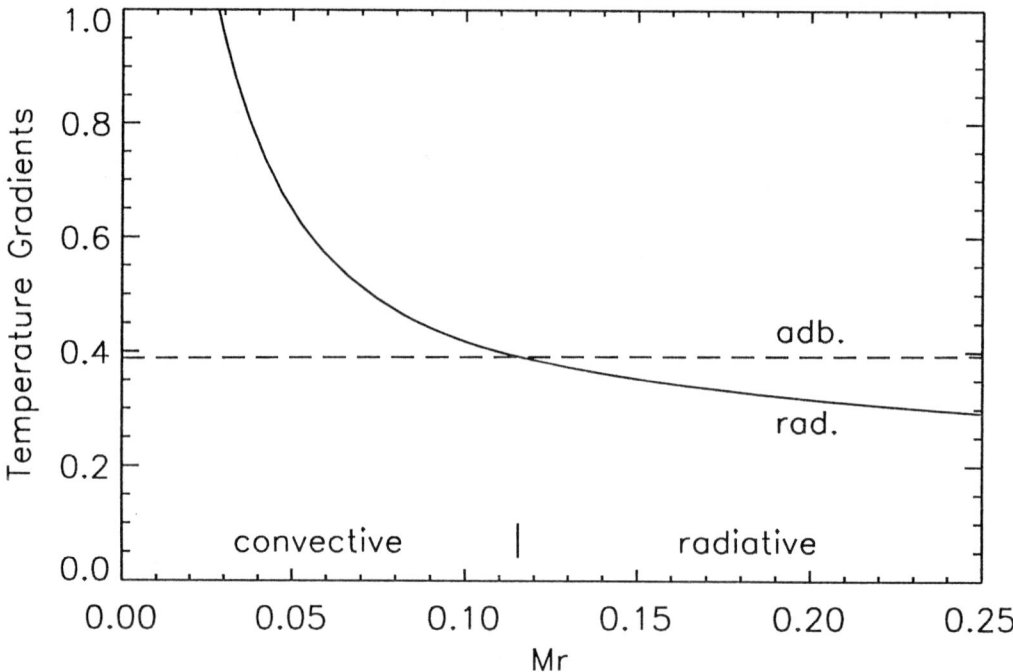

FIGURE 6. Variation of the radiative gradient (solid curve) and the adiabatic gradient (dashed curve) with mass M_r within the core of a 0.68 $M_\odot$ star at the ZAHB phase. The short vertical line at $M_r = 0.115$ $M_\odot$ marks the edge of the convective core.

quite plausible, however, that the strong superadiabaticity evident in Figure 7 will lead to convective overshooting and hence to an outward propagation of the convective-core edge. Unfortunately the efficiency of convective overshooting under such circumstances is entirely unknown. Canonical HB theory assumes that the convective overshooting is highly efficient and consequently that the convective core propagates outward until convective neutrality is restored at its edge (Castellani et al. 1971a). The sequence of events is shown more clearly in Figure 8. The solid curve in this Figure refers to a model in which the radiative gradient is superadiabatic at the convective-core edge. Under canonical theory the convective core will propagate outward along the dotted curve until the radiative gradient equals the adiabatic gradient. The point at which this occurs then defines the new convective-core edge. We see therefore that the evolution of a canonical model away from the ZAHB is characterized by a progressive growth in the size of the convective core and hence in the amount of helium fuel that the model can burn.

At this point one might again be tempted to think that one can compute the evolution of an HB star by merely following the outward propagation of the convective-core edge. However, there is still another surprise to come. Once the convective core exceeds a certain size, the radiative gradient no longer decreases as the convective-core edge propagates outward. This characteristic behavior of the HB models is illustrated schematically in Figure 9. We see that the radiative gradient reaches a minimum and then turns up with increasing M_r, once the convective core becomes sufficiently large. This behavior creates a real dilemma as an HB star tries to restore convective neutrality at the convective-core edge. The convective overshooting caused by the superadiabaticity of the radiative gradient will move the convective-core edge outward along the upward

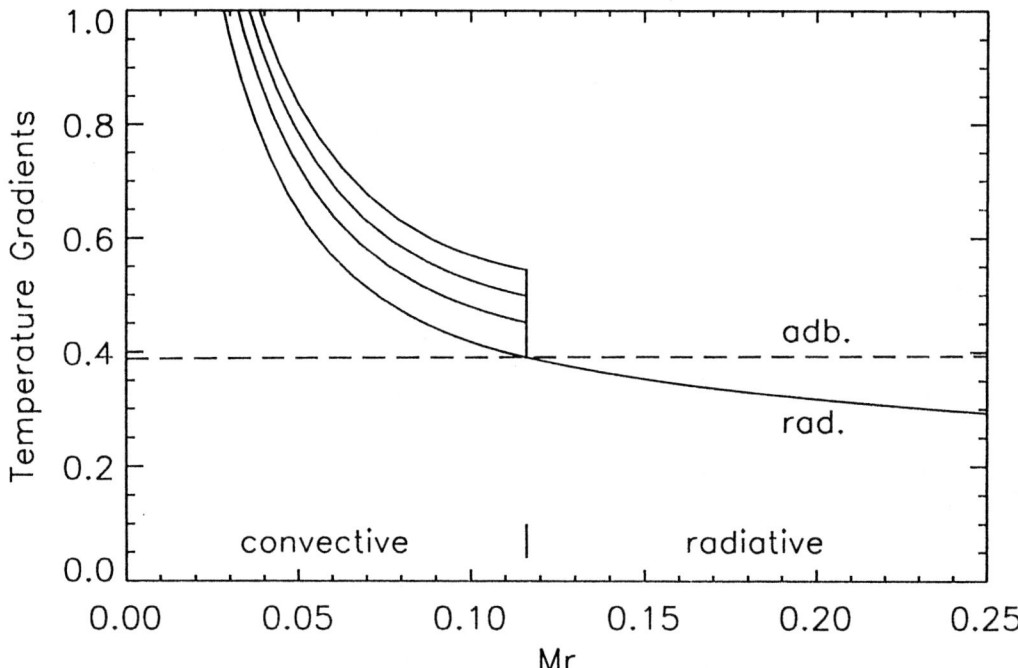

FIGURE 7. Increase in the radiative gradient (solid curve) within the convective core of a 0.68 $M_\odot$ star during the HB phase when the convective-core edge is held fixed at $M_r = 0.115\ M_\odot$. The dashed curve refers to the adiabatic gradient.

rising part of the radiative gradient curve in Figure 9 and hence will only increase the degree of superadiabaticity, just the opposite of what happens during the earlier evolution. Thus convective overshooting only makes the situation worse.

What does an HB star do? According to canonical theory an HB star resolves this dilemma by creating a transition zone between the fully convective core and the outer radiative region (Castellani et al. 1971b). This transition zone is referred to as a semiconvective zone because it has some of the properties of both a convective and a radiative region. It is like a convective region in that its temperature gradient is equal to the adiabatic gradient, and it is like a radiative region in that its composition varies from point to point. In fact, it is this variation in the helium abundance which maintains convective neutrality within the semiconvective zone. The run of the radiative gradient within an HB model with a semiconvective zone is presented in Figure 10.

Thus far we have discussed the physical effects which cause a semiconvective zone to form during the canonical evolution of an HB star. Let us now see how these effects change the interior structure of the core. Figure 11 gives the time dependence of the edge of the convective core and the outer edge of the semiconvective zone during the evolution of our 0.68 $M_\odot$ star. During the first 20×10^6 yr this star has only a convective core whose mass gradually increases from $0.115\ M_\odot$ to $0.171\ M_\odot$, i.e., by about 50 per cent. At $t = 20 \times 10^6$ yr the convective-core edge reaches the minimum in the radiative gradient curve (see Figure 9) where it remains fixed for the rest of the HB evolution. A semiconvective zone then develops and grows until it eventually contains $\approx 0.1\ M_\odot$.

An alternative way of illustrating the changes brought about by convective overshooting and semiconvection is to plot the composition distribution within the core at various

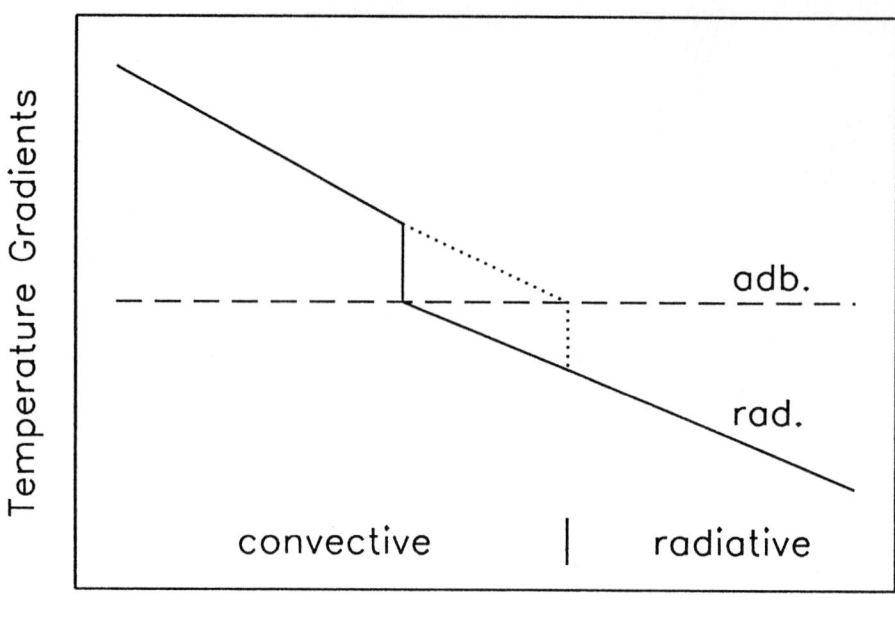

FIGURE 8. Convective overshooting induced by a superadiabaticity of the radiative gradient at the convective-core edge according to canonical theory.

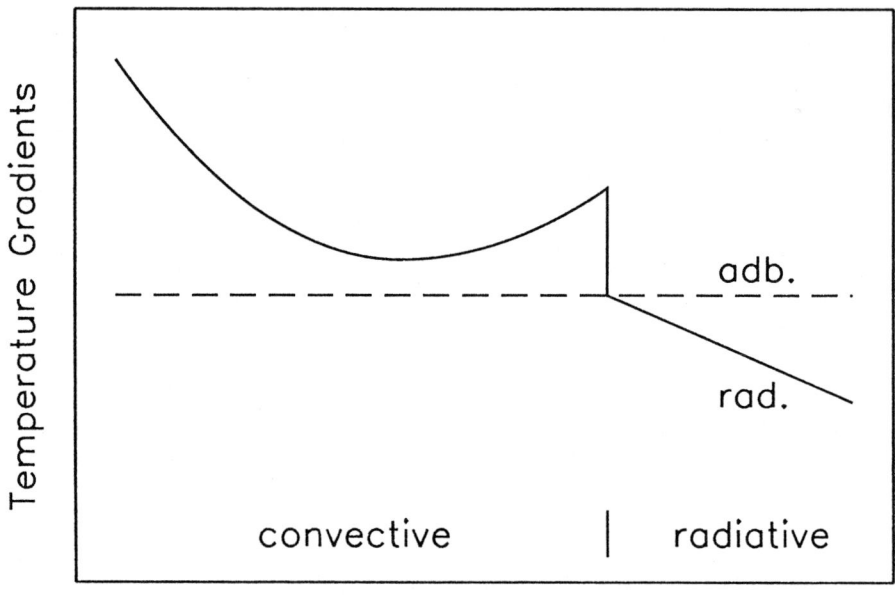

FIGURE 9. Variation of the radiative and adiabatic gradients with mass M_r within an enlarged convective core of an HB star.

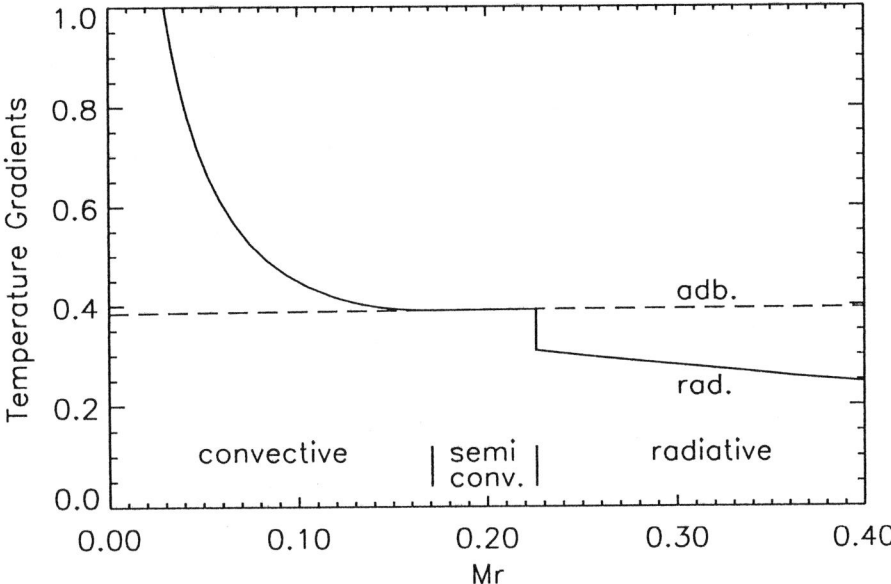

FIGURE 10. Variation of the radiative and adiabatic gradients with mass M_r within the core of a 0.68 $M_\odot$ star with a semiconvective zone. The semiconvective zone extends from $M_r = 0.171$ to 0.226 $M_\odot$.

times during the HB phase, as is done in Figure 12. Without convective overshooting an HB star would only be able to consume the helium fuel contained within its convective core at the ZAHB phase. Figure 12 demonstrates that convective overshooting and semiconvection increase the amount of helium fuel consumed during the HB phase by approximately a factor of 2.

How do we know that Figures 11 and 12 describe the evolution of real HB stars? One can address this question by using the observed number ratio $R_2 = N_{AGB}/N_{HB}$ of AGB to HB stars to determine the HB helium fuel consumption (Renzini & Fusi Pecci 1988). The lifetime of a star during the HB phase depends on the amount of helium fuel it is able to consume. Similarly, the lifetime during the following AGB phase depends on the amount of helium fuel that is left over at the end of the HB phase. Thus the number ratio R_2 or, equivalently, the ratio of the corresponding theoretical lifetimes provides a measure of the relative fuel consumption during these phases. If the theoretical models, for example, burn too little helium fuel during the HB phase, then the predicted HB lifetime would be too short and the AGB lifetime correspondingly too long. The predicted value of R_2 would then be larger than observed. Conversely too much helium fuel consumption during the HB phase would give too small a value of R_2. Thus by using the observed value of R_2 one can test a very fundamental aspect of the HB evolution (Renzini & Fusi Pecci 1988).

Let us suppose for the moment that convective overshooting and semiconvection do not occur and consequently that the edge of the convective core remains fixed in mass during the HB phase. We have computed this type of evolution for our 0.68 $M_\odot$ star, as shown in Figure 13. Since relatively little helium fuel is consumed during the HB phase in this case, the HB lifetime t_{HB} is only 54×10^6 yr. To determine the AGB lifetime t_{AGB}, we have evolved this 0.68 $M_\odot$ star up the AGB until the onset of the helium-shell flashes at $t = 95 \times 10^6$ yr. The results in Figure 13 give a theoretical AGB lifetime of 41

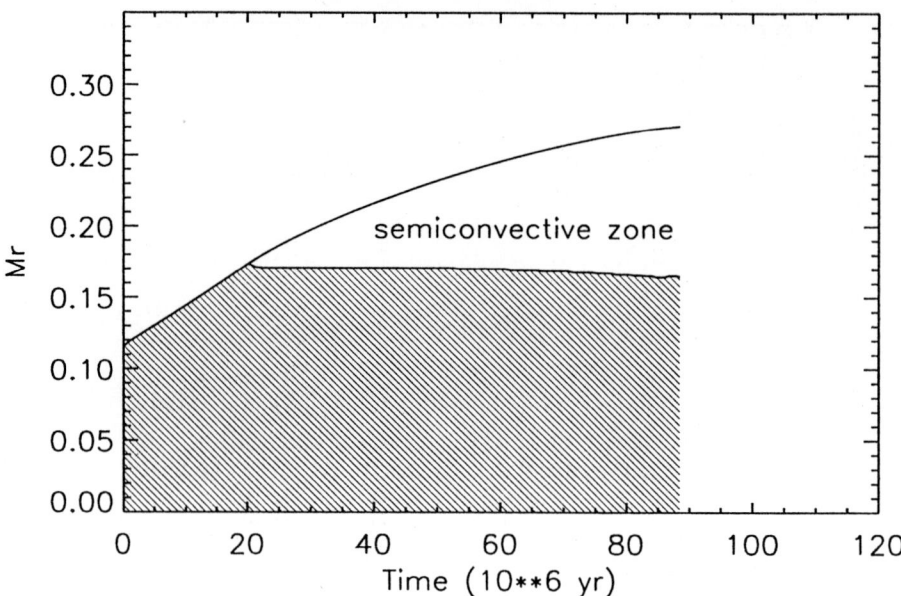

FIGURE 11. Time dependence of the location of the convective-core edge and the outer edge of the semiconvective zone during the evolution of a 0.68 $M_\odot$ HB star. Shaded areas are convective.

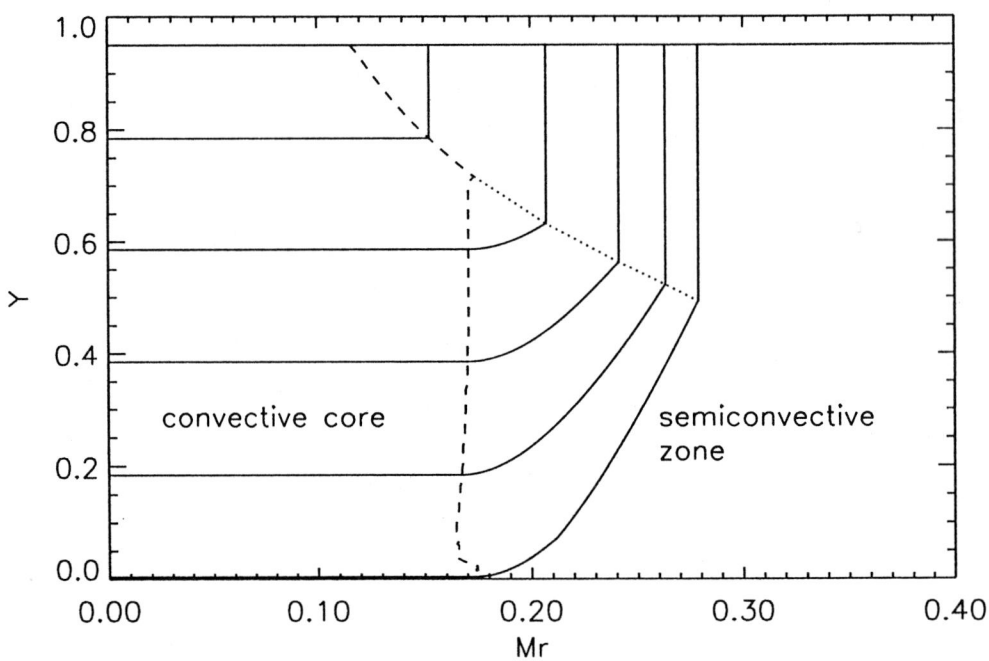

FIGURE 12. Core-helium distribution at various times during the evolution of a 0.68 $M_\odot$ HB star. The dashed and dotted curves denote the edge of the convective core and the outer edge of the semiconvective zone, respectively.

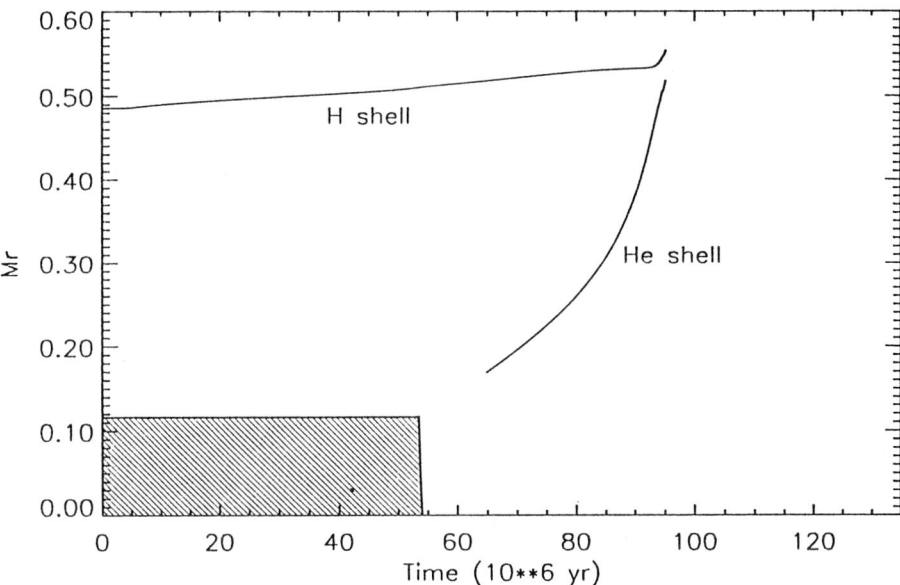

FIGURE 13. Evolution of a 0.68 $M_\odot$ star during the HB and AGB phases without convective overshooting and semiconvection. The two solid curves give the locations of the centers of the helium and hydrogen shells. Shaded areas are convective.

$\times 10^6$ yr and consequently a theoretical value for the ratio $R_2 = t_{AGB}/t_{HB}$ of 0.76. Since the observed value of R_2 in the globular clusters is 0.15 ± 0.01 (Renzini & Fusi Pecci 1988), it is clear that an HB star must live considerably longer than is shown in Figure 13 and therefore must consume considerably more helium fuel than is contained within its convective core at the ZAHB phase. We conclude that convective overshooting must play an important role in extending the size of the convective core in real HB stars.

An obvious question to ask is whether the canonical results shown in Figures 11 and 12 are consistent with the observed value of R_2. To answer this question, we must first consider how a star ends its HB evolution, which is the topic of the next section. Before turning to this topic, however, we would like to summarize the assumptions and deficiencies of canonical HB theory. The main assumptions are 1) that convective overshooting is highly efficient, 2) that an HB star is able to maintain convective neutrality at the convective-core edge and within the semiconvective zone, and 3) that the core-composition distribution varies smoothly with time. The main deficiencies are that canonical HB theory provides no timescale for the required convective overshooting and no description for how an HB star actually maintains the composition distribution within its semiconvective zone.

5. Core-helium-exhaustion phase

A number of numerical and theoretical difficulties arise during the final part of the HB evolution when the central helium abundance Y_c falls below ≈ 0.1. If one tries to use the canonical algorithms for semiconvection to follow the evolution into this core-helium-exhaustion phase, one finds that the convective core suddenly grows so large that it engulfs most of the semiconvective zone (Sweigart & Demarque 1973; Castellani et al. 1985). These so-called "breathing pulses" of the convective core can occur more than once before the central helium is finally exhausted. The models which undergo these breathing pulses are not convectively neutral at the convective-core edge and therefore are not consistent with the numerical algorithms used to compute them. Whether or not the breathing pulses reflect a real physical phenomenon or are merely numerical artifacts of the canonical assumption that convective boundaries can move instantaneously has been an open question for some time. Attempts to answer this question have generally been based on the predicted value of R_2, since the breathing pulses can significantly increase the amount of helium fuel consumed during the core-helium-exhaustion phase (Renzini & Fusi Pecci 1988).

To follow the evolution of an HB star through the core-helium-exhaustion phase, one must make some assumptions about how to treat the breathing pulses. Essentially four approaches have been taken in the published calculations. In the first approach one simply continues to apply the canonical algorithms used during the preceding HB evolution. Sequences computed in this fashion undergo breathing pulses which induce abrupt blueward fluctuations in the evolutionary tracks. A major drawback of this approach is that the numerical results depend quite sensitively on delicate details of the numerics. Small changes in the choice of time steps, convergence criteria, etc. can appreciably affect the timing and extent of the breathing pulses. Even with the same evolution code one can find substantial differences in the breathing pulses from one sequence to another. Moreover, as pointed out above, models with breathing pulses do not satisfy the requirements of the canonical algorithms, and therefore it is not clear what they really mean. In particular, there is no guarantee that such models will reproduce the observed value of R_2.

A second approach is to suppress the breathing pulses by imposing an ad hoc limit on the rate at which the convective core grows during the core-helium-exhaustion phase. The sudden outward movement of the convective core during a breathing pulse brings additional helium fuel into the center so rapidly that Y_c increases for a time. Under this approach one prevents such increases in Y_c by limiting the rate at which the convective core can grow. In some sense this approach does not really suppress the breathing pulses. Rather, it merely prolongs the time required for the convective core to grow. In fact, sequences computed with this approach often show a substantial increase in the size of the convective core just before the end of the HB phase. As a result, such sequences reproduce the main consequence of a breathing pulse, namely, the capture of additional helium fuel from the semiconvective zone, even though they do not explicitly undergo a breathing pulse. Besides being ad hoc, this approach shares some of the same drawbacks as the first approach. In particular, the models are not convectively neutral at the convective-core edge, and there is no requirement that the models consume the proper amount of helium fuel.

In the third approach one suppresses the breathing pulses by omitting the release of gravitational energy within the core. Dorman & Rood (1993) have recently shown that the canonical algorithms for semiconvection can be applied during the core-helium-exhaustion phase without encountering numerical difficulties if the breathing pulses are suppressed in this fashion. We have applied this approach to our 0.68 $M_\odot$ star to obtain the results given in Figure 14. Prior to the tick mark at $t = 88.5 \times 10^6$ yr the evolution was followed with a canonical algorithm for semiconvection. After the tick mark this algorithm was again applied but now without the release of gravitational energy. The calculations were terminated at $t = 100.2 \times 10^6$ yr when $Y_c = 0.004$. The evolution in Figure 14 proceeds smoothly without any hint of the breathing pulses, thus confirming the results of Dorman & Rood (1993). Note, however, that the semiconvective zone continues to grow throughout the time interval covered by Figure 14. Even at $Y_c =$

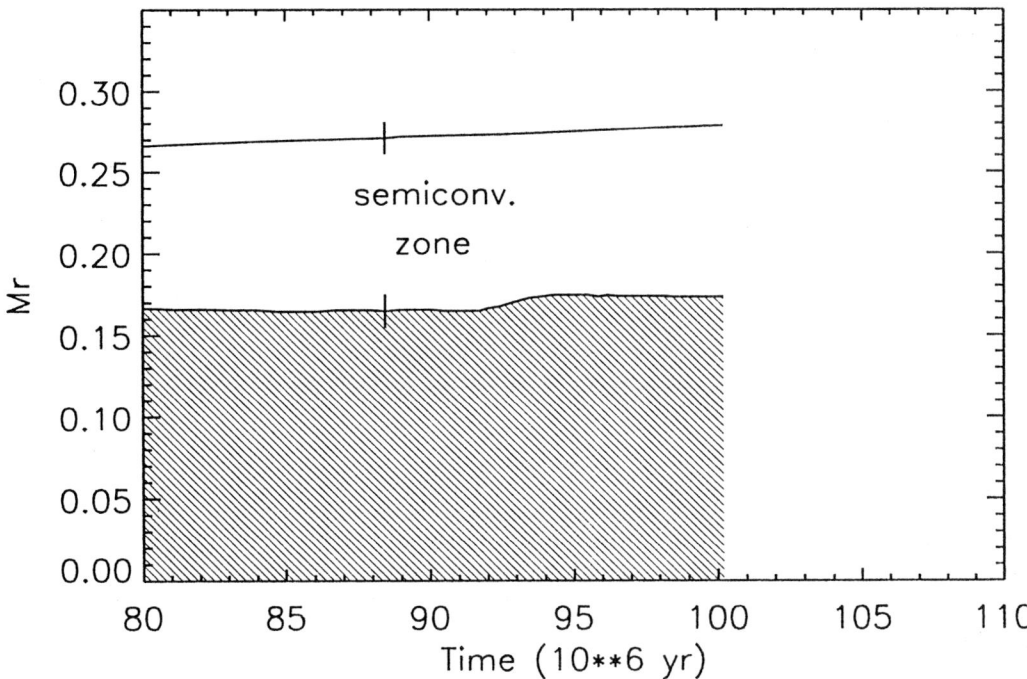

FIGURE 14. Time dependence of the location of the convective-core edge and the outer edge of the semiconvective zone during the core-helium-exhaustion phase of a 0.68 $M_\odot$ HB star. The evolution after the tick mark at $t = 88.5 \times 10^6$ yr was computed by omitting the release of gravitational energy. Shaded areas are convective.

0.004 there is still no indication that the semiconvective zone is about to recede, as it must as Y_c approaches 0 and the helium burning shifts outward from the center to a shell. These results predict that an HB star will consume a significant amount of helium fuel during the core-helium-exhaustion phase.

The main drawback of this approach concerns its assumption that the release of gravitational energy can be safely omitted from the models. As a star approaches the end of its HB phase, the core begins to contract, and, as a result, the release of gravitational energy makes an increasingly significant contribution to the outward energy flux. Since the extent of the semiconvective zone depends on the run of the radiative gradient which, in turn, depends on the outward flux, one needs to ask whether the numerical results could be affected in some way by this assumption, especially towards the end of the evolution in Figure 14. In addition, one cannot apply this approach to the very end of the HB phase, since the models will eventually diverge. At some point one must again include the release of gravitational energy (and hence turn off the semiconvection algorithm) which introduces some arbitrariness into the calculations.

The fourth approach is based on a parameterization for the efficiency of convective overshooting. The difficulties with the canonical algorithms during the core-helium-exhaustion phase arise at least in part because the models are free to instantaneously move the convective-core edge whenever the radiative gradient exceeds the adiabatic gradient. In real stars there must be some time dependence associated with this overshooting process. This suggests that one should try to follow the growth of the convective core during the core-helium-exhaustion phase by including an estimate for the overshooting efficiency in the model calculations. Unfortunately overshooting is a very complex phenomenon about which little is really known for stellar conditions (Renzini 1987). This fact represents the principal drawback to this approach. Nevertheless one can make some progress by parameterizing the overshooting efficiency - an approach that is often used in other areas, e.g., the determination of the RGB mass loss rate. Let us assume that the overshooting efficiency increases as the superadiabaticity of the radiative gradient increases at the convective-core edge. This is not an implausible assumption, since according to the canonical scenario it is this superadiabaticity which drives the overshooting. Since the overshooting material is carbon-rich and therefore heavier than the overlying material, it is also plausible to assume that the overshooting efficiency decreases as the difference in the mean molecular weight across the convective-core edge and thus the negative bouyancy force increase. These considerations suggest the following parameterization for the velocity v_c with which the convective-core edge propagates outward:

$$v_c = F_{ov}\, (\nabla_{rad} - \nabla_{ad})/(\mu_i - \mu_e), \quad (4)$$

where ∇_{rad} and ∇_{ad} are the radiative and adiabatic gradients, and μ_i and μ_e are the mean molecular weights interior and exterior to the convective-core edge. F_{ov} represents the free parameter governing the overshooting efficiency.

We have used equation (4) to compute the evolution of our 0.68 $M_\odot$ star through the core-helium-exhaustion phase. The results for a representative value of F_{ov} are shown in Figure 15. At $t = 88.5 \times 10^6$ yr the canonical algorithm for semiconvection was turned off, and equation (4) was then used to control the movement of the convective-core edge. We see that this star immediately underwent a substantial breathing pulse which engulfed the entire semiconvective zone. During this breathing pulse helium fuel was brought into the center more rapidly than it could be burned, thereby causing the temporary increase in Y_c shown in Figure 16. Although the evolution in Figures 14 and 15 looks strikingly different, there are, in fact, a number of similarities. In both cases the predicted HB lifetime is very close to 10^8 yr. Moreover, the final core-helium profiles, shown in Figure 17, agree quite well. Thus the different approaches in Figures 14 and 15 predict essentially the same fuel consumption during the HB phase.

It has sometimes been argued that breathing pulses do not occur because they lead to unrealistically low values of R_2. To examine this point, we have continued the evolution of the star in Figure 15 up the AGB to the onset of the helium-shell flashes. Figure 18 presents the results for the main HB, core-helium-exhaustion and AGB phases. The predicted HB and AGB lifetimes are 101.4×10^6 yr and 15.6×10^6 yr, respectively. This yields a predicted value for R_2 of 0.15 exactly as observed.

6. Horizontal-branch evolution with time-dependent overshooting

In the previous sections we have seen how canonical HB theory assumes that convective overshooting is highly efficient and how the canonical algorithms for semiconvection permit the convective-core edge to move as rapidly as needed to preserve convective neutrality. It is of some interest to know how the computed evolution of an HB star is affected by these assumptions. How, for example, would the HB lifetime, the HB fuel consumption, the value of R_2, etc., change if the convective overshooting was only moderately efficient or even inefficient? The fundamental obstacle to answering these questions, of course, is our lack of understanding of the overshooting process. However, one might gain some insight into these questions by evolving HB models in which the

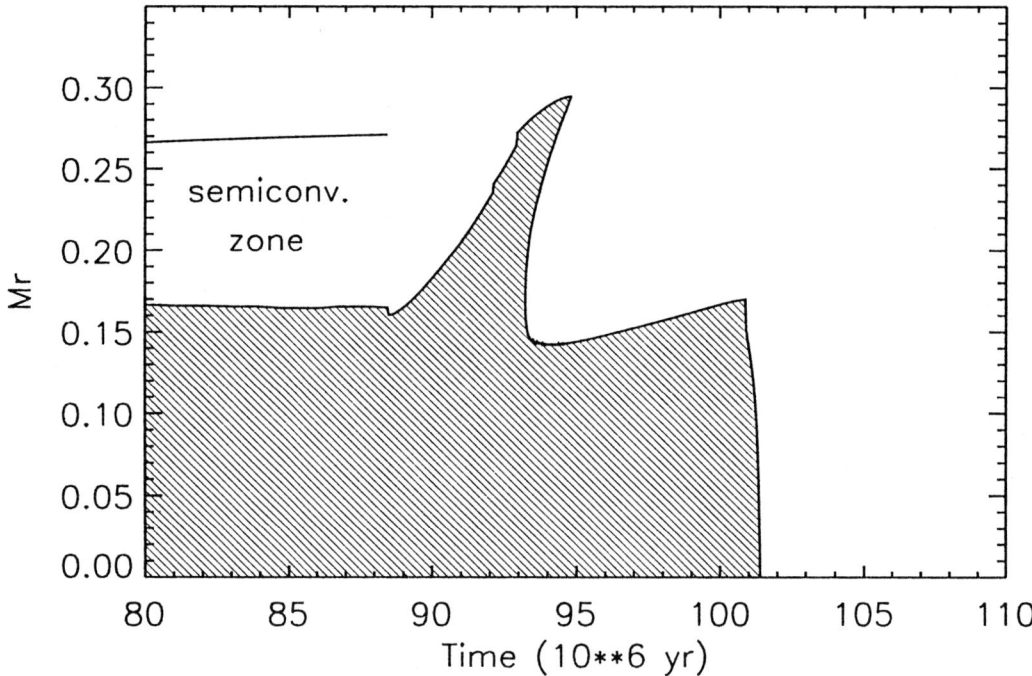

FIGURE 15. Time dependence of the location of the convective-core edge during the core-helium-exhaustion phase of a 0.68 $M_\odot$ HB star. The evolution after $t = 88.5 \times 10^6$ yr was computed by using eq. (4) with $F_{ov} = 10^{-5}$. Shaded areas are convective.

movement of the convective-core edge is controlled by equation (4). The purpose of this section is to present such calculations and to compare them with the canonical results.

Figure 19 shows the HB evolution of our 0.68 $M_\odot$ star, as computed from equation (4) with $F_{ov} = 10^{-5}$. We emphasize that these "overshooting" models contain no information about semiconvection. The only requirement is that the convective-core edge move in accordance with equation (4). The evolution in Figure 19 is characterized by a series of breathing pulses of gradually increasing amplitude which begin at about the time a semiconvective zone would have formed in the canonical calculations. While the evolution in Figure 19 looks superficially quite different from the canonical evolution in Figure 11, it is actually quite similar in many respects. For example, the predicted HB lifetime agrees with the canonical value of 10^8 yr. The upper envelope of the breathing pulses, which represents the outermost point from which the convective core has been able to capture helium fuel, coincides closely with the outer edge of the semiconvection zone in the canonical calculations. The lower envelope of the breathing pulses is located at the minimum in the radiative gradient (see Figure 9) and, not surprisingly, agrees with the canonical location of the convective-core edge. Moreover, the time dependence of the central helium abundance Y_c, shown in Figure 20, is nearly indistinguishable between the overshooting and canonical sequences.

The HB fuel consumption can be determined from the composition distribution within the core at the end of the HB phase. In Figure 21 we compare this composition distribution for the last model from the overshooting sequence in Figure 19 with that for the last model from the canonical sequence in Figure 14. We see that the breathing pulses produce a region of varying composition between $M_r = 0.150$ and 0.293 $M_\odot$ which looks

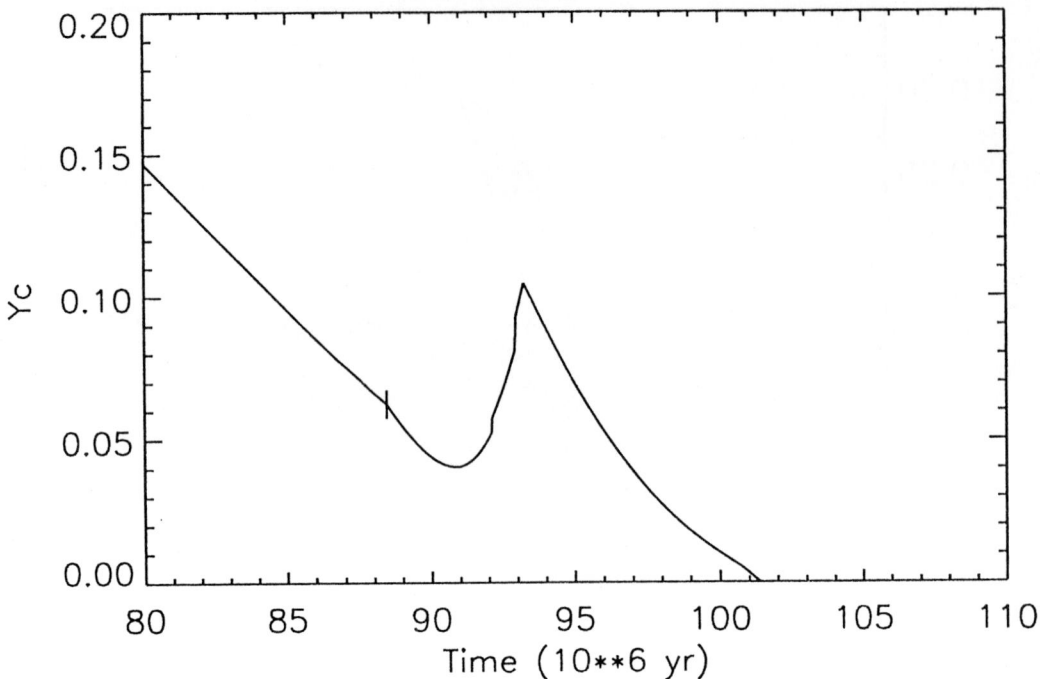

FIGURE 16. Time dependence of the central helium abundance Y_c during the evolution shown in Figure 15.

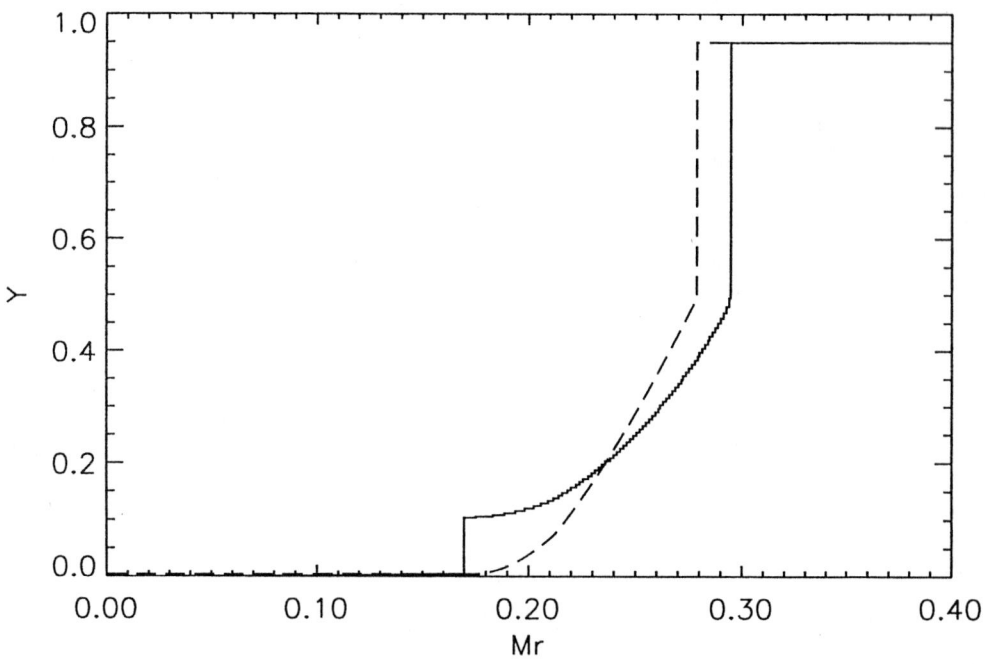

FIGURE 17. Core-helium distributions at the end of the evolution shown in Figure 14 (dashed curve) and Figure 15 (solid curve).

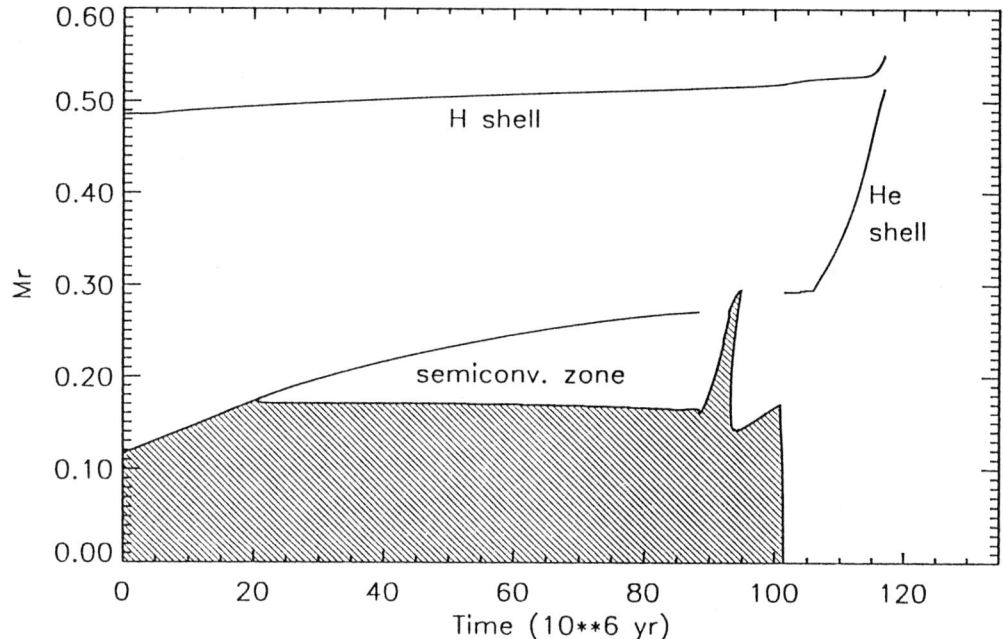

FIGURE 18. Evolution of a 0.68 $M_\odot$ star during the HB, core-helium-exhaustion and AGB phases. The HB phase was computed with a canonical algorithm for semiconvection while the core-helium-exhaustion phase is the same as shown in Figure 15. The two solid curves give the locations of the centers of the helium and hydrogen shells. Shaded areas are convective.

remarkably like the semiconvective zone in the canonical models. The similarity of the two curves in Figure 21 implies that both the overshooting and canonical models burn about the same amount of helium fuel during the HB phase.

An important check on these results is provided by the ratio R_2. To apply this check, we have evolved the overshooting sequence in Figure 19 up the AGB to the onset of the helium shell flashes. This yielded an AGB lifetime of 14.3×10^6 yr and consequently a value for R_2 of 0.14, in good agreement with the observed ratio.

Figure 22 shows that the evolutionary tracks for the overshooting and canonical sequences are quite similar in all but one respect. The evolution along the canonical track proceeds smoothly, since the canonical algorithms for semiconvection produce only gradual changes in the composition distribution. In contrast, the evolution along the overshooting track is noisy due to the many fluctuations caused by the individual breathing pulses. Such noise might actually be quite helpful in explaining the observed RR Lyrae period changes which are considerably greater than predicted by the canonical models (Sweigart & Renzini 1979).

The results in Figure 19 obviously depend on one's choice for the overshooting parameter F_{ov}. Figure 23 shows what happens if this overshooting parameter is increased by a factor of 5. The models then undergo many more breathing pulses, since it is much easier for the convective-core edge to move outward. In other respects the HB evolution is remarkably unchanged. The HB lifetime is still close to 10^8 yr, and the upper envelope of the breathing pulses still coincides with the outer edge of the semiconvective zone in the canonical models. Figure 23 does indicate, however, that the last several breathing pulses might be stronger than those during the earlier HB evolution. If the overshooting parameter is instead decreased by a factor of 10, one finds that the models undergo only a

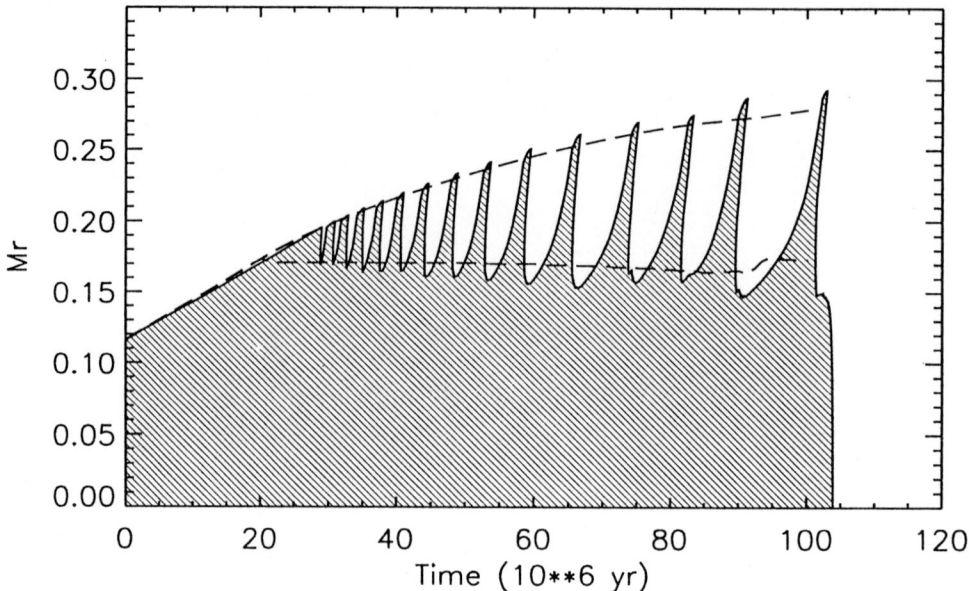

FIGURE 19. Time dependence of the location of the convective-core edge during the evolution of a 0.68 $M_\odot$ HB star from the ZAHB to central helium exhaustion. The evolution was computed by using eq. (4) with $F_{ov} = 10^{-5}$. The dashed curves give the location of the convective-core edge and the outer edge of the semiconvective zone as computed with the canonical algorithms for semiconvection (see Figures 11 and 14). Shaded areas are convective.

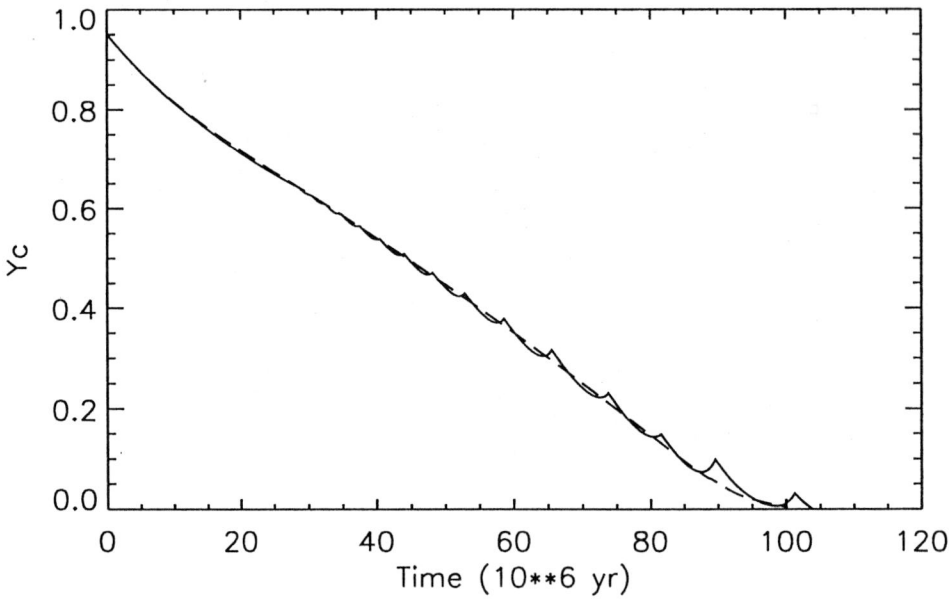

FIGURE 20. Time dependence of the central helium abundance Y_c during the evolution shown in Figure 19 (solid curve). The corresponding result for the canonical evolution in Figures 11 and 14 is shown by the dashed curve.

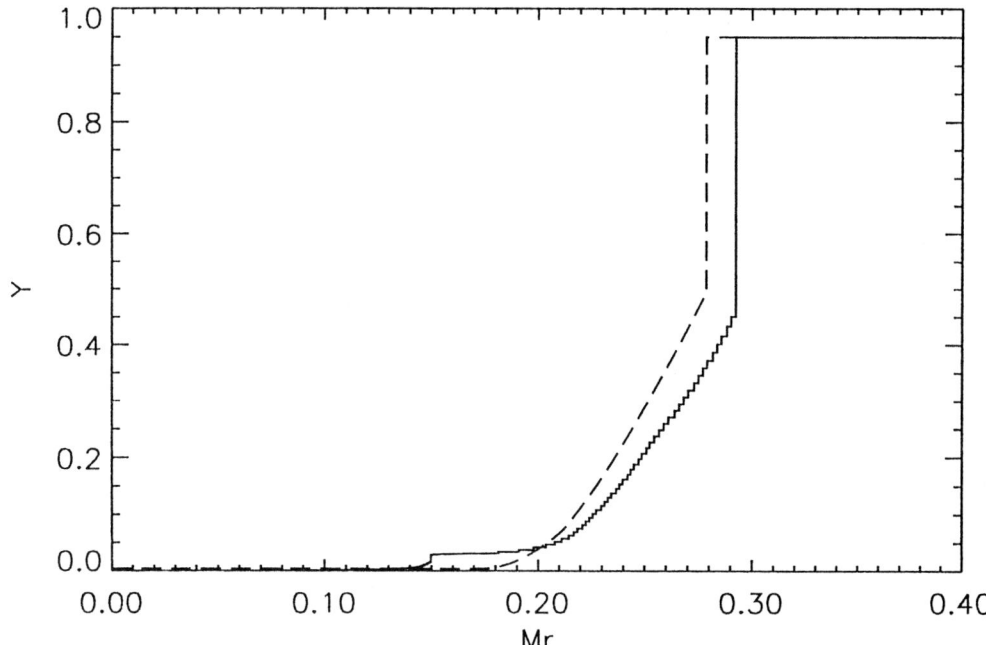

FIGURE 21. Core-helium distributions at the end of the evolution shown in Figure 14 (dashed curve) and Figure 19 (solid curve).

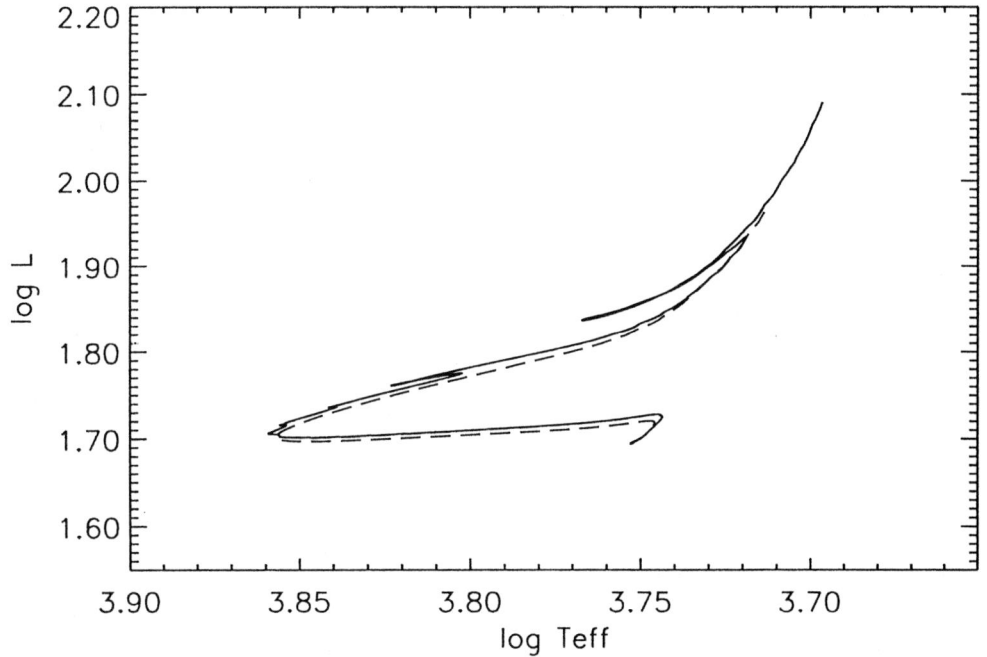

FIGURE 22. Comparison between the evolutionary tracks for a 0.68 M$_\odot$ HB star as given by canonical HB calculations (dashed curve) and the evolution shown in Figure 19 (solid curve).

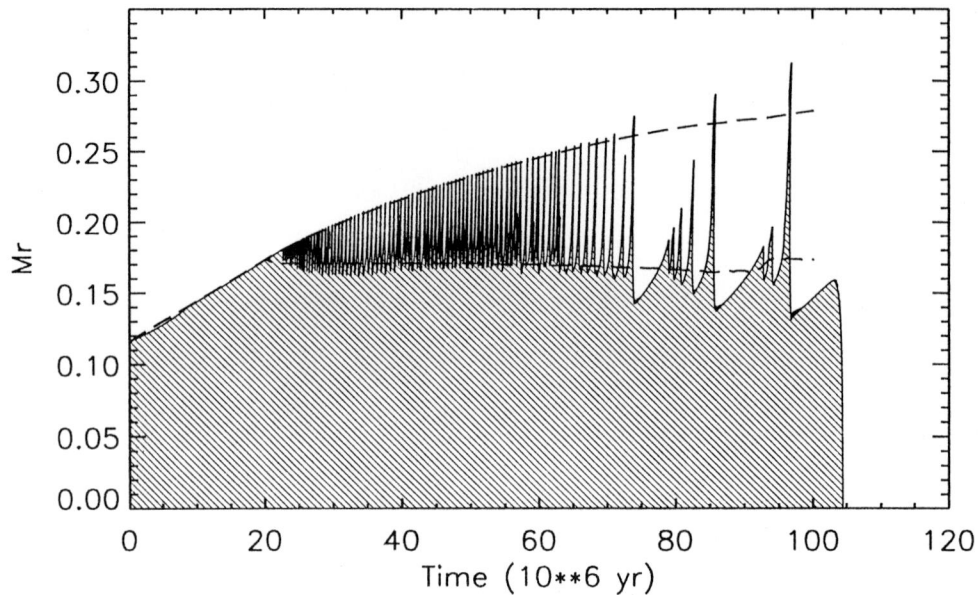

FIGURE 23. Same as Figure 19 but for $F_{ov} = 5 \times 10^{-5}$.

single breathing pulse. Even in this case, however, the HB lifetime and fuel consumption are almost identical to those for the sequence in Figure 19.

What conclusions can one draw from these results? It appears that HB stars know how much helium fuel they should consume during the HB phase. All that they require is that the convective overshooting be moderately efficient. It seems to make little difference whether the overshooting is provided by the breathing pulses as in Figure 19 or by the canonical algorithms for semiconvection as in Figure 11. The breathing pulses deposit carbon-rich material into the inner part of what would otherwise be called the semiconvective zone. This carbon-rich material then propagates outward, gradually being diluted by more helium-rich material, until it reaches the layer corresponding to the outer edge of the semiconvective zone in the canonical models. The net effect of the breathing pulses is to produce a region of varying composition that is close to convective neutrality, i.e., a semiconvective zone. Thus the breathing pulses or something like them might provide the mechanism by which HB stars change their composition distribution during the HB phase.

7. Conclusions

In the preceding sections we have reviewed some of the stellar evolution results for the RGB and HB phases of globular-cluster stars and, in particular, have discussed the assumptions which underlie canonical HB theory. These results can be summarized as follows:

1. The main consequence of the helium flash is to lift the degeneracy of the core by burning ≈ 5 per cent of the core's helium into carbon. Within the canonical framework mixing between the core and envelope does not occur.

2. The core mass at the ZAHB phase is well-determined by canonical RGB theory, i.e., standard input physics, no rotation, etc.

3. Canonical HB models with semiconvection burn substantially more helium fuel than models without convective overshooting, as is required to satisfy the constraint on the observed number ratio $R_2 = N_{AGB}/N_{HB}$ of AGB to HB stars.

4. HB models can be evolved through the core-helium-exhaustion phase either by suppressing the breathing pulses by omitting the release of gravitational energy or by permitting breathing pulses using time-dependent overshooting. Both approaches can yield the observed value of R_2.

5. HB evolution computed with breathing pulses closely approximates in the mean the evolution computed with canonical semiconvection. In particular, the HB lifetime and fuel consumption seem to be well-determined, if the convective overshooting is moderately efficient.

The author gratefully acknowledges the support of NASA RTOP 188-41-51 Task 03.

REFERENCES

BUZZONI, A., FUSI PECCI, F., BUONANNO, R., CORSI, C. E., 1983, A&A, 128, 94

CAPUTO, F. 1990, in Cacciari C., Clementini G., eds, Confrontation Between Stellar Pulsation and Evolution, ASP Conf. Ser., 11, 22

CASTELLANI, V., CHIEFFI, A., PULONE, L., TORNAMBE, A., 1985, ApJ, 296, 204

CASTELLANI, V., GIANNONE, P., RENZINI, A., 1971a, Ap&SS, 10, 340

CASTELLANI, V., GIANNONE, P., RENZINI, A., 1971b, Ap&SS, 10, 355

CATELAN, M., 1992, A&A, 261, 457

DA COSTA, G. S., ARMANDROFF, T. E., 1990, AJ, 100, 162

DORMAN, B., ROOD, R. T., 1993, ApJ, 409, 387

FROGEL, J. A., COHEN, J. G., PERSSON, S. E., 1983, ApJ, 275, 773

HILL, R. S., HILL, J. K., LANDSMAN, W. B., et al., 1992, ApJ, 395, L17

LEE, Y.-W., DEMARQUE, P., ZINN, R., 1990, ApJ, 350, 155

MAZZITELLI, I., 1989, ApJ, 340, 249

PARISE, R. A., MARAN, S. P., LANDSMAN, W. B., et al., 1994, ApJ, 423, 305

RAFFELT, G. G., 1990, ApJ, 365, 559

RENZINI, A., 1987, A&A, 188, 49

RENZINI, A., FUSI PECCI, F., 1988, ARAA, 26, 199

SANDAGE, A., 1990, ApJ, 350, 631

SWEIGART, A. V., 1994, ApJ, 426, in press

SWEIGART, A. V., DEMARQUE, P., 1973, in Fernie J. D., ed, Variable Stars in Globular Clusters and in Related Systems, Reidel, Dordrecht, p. 221

SWEIGART, A. V., GROSS, P. G., 1976, ApJS, 32, 367

SWEIGART, A. V., MENGEL, J. G., 1981, in Philip A. G. D., Hayes D. S., eds., IAU Colloq. 68, Astrophysical Parameters for Globular Clusters, Reidel, Dordrecht, p. 277

SWEIGART, A. V., RENZINI, A., 1979, A&A, 71, 66

Discussion

DEMARQUE: Is it possible to check on the physical reality of the core pulses by looking for changes in RR Lyrae periods?

SWEIGART: You raise an excellent point. One important difference between horizontal branch (HB) sequences computed with breathing pulses and those computed with canonical semiconvection is that the evolution is "noisy". The breathing pulses perturb the interior structure as they redistribute carbon from the central helium-burning regions, and this, in turn, causes fluctuations in the surface radius and hence the pulsation period. In fact, Renzini and I (1979, A&A, 71, 66) suggested some time ago that mixing events within the core could provide the noise needed to explain the observed RR Lyrae period changes. I have felt for a long time that the observed period changes could be an important diagnostic for understanding how mixing occurs within an HB star.

PHILIP: For a given globular cluster color-magnitude diagram the fact that there is a tight horizontal branch (and modern CCD CM diagrams are more precisely determined than the older diagrams) argues that the core mass must be remarkably constant. The total mass does vary for stars in the HB. What are the limits for variations in total mass for a given globular cluster?

SWEIGART: The observed color width of the HB in many globular clusters is considerably greater than the color width expected from a single evolutionary track, and consequently some quantity must vary from star to star along the HB in these clusters. Generally it is assumed that the total mass varies by a few 10^{-2} $M_\odot$ while the core mass is constant. However, it is also possible that the core mass might vary from star to star if internal rotation during the Red Giant Branch (RGB) phase was able to significantly delay the helium flash. A rotating star would have both a larger core mass and a brighter RGB tip luminosity and consequently would undergo more extensive mass loss along the RGB. The larger core mass and smaller total mass of a rotating star would also produce a spread in color along the HB. Unfortunately, the tightness of the HB in luminosity cannot be used to test for the constancy of the core mass. The figure I showed giving the variation in the HB luminosity with core mass applied to sequences whose masses were varied to keep their tracks within or near the instability strip. An increase in the core mass at either constant or decreasing total mass would spread out the HB in color without producing a substantial spread in luminosity.

BIDELMAN: What is the difference between a hot HB star and a cool one? Is it a difference in mass or a consequence of a star's evolution?

SWEIGART: Within a given globular cluster hot and cool stars are believed to differ in total mass. Between clusters metallicity is the main parameter determining whether a HB star is hot or cool.

CARNEY: Several of us have argued that Sandage's period shift is overestimated. Without getting into details of Baade-Wesselink methods, main sequence fitting, or how you define the temperature for the period shift analysis, there is a simple test. Use M_v vs. [Fe/H] to derive relative distances to (approximately) unevolved RR Lyrae stars, and from that derive relative M_k values. Using Sandage's steep slope will yield a slope for M_k vs. log P inconsistent with that observedand with all individual clusters studied to date.

SWEIGART: I agree.

DORMAN: Could presently unincorporated effects in the physical properties of stellar matter (specifically the equation of state) make any difference to the core masses?

SWEIGART: I believe you are referring to the possible effects of Coulomb interactions on the equation of state. Mazzitelli (1989, ApJ, 340, 249) has recently shown that the increase in the compressional heating caused by Coulomb interactions is offset by the increase in the specific heat at constant pressure and, hence, in the energy required to heat up the core material. He finds that Coulomb interactions cause no net change in the core mass.

A Review of A-Type Horizontal-Branch Stars

By A. G. DAVIS PHILIP[1,2]

[1]Union College, Physics Department, Schenectady, NY 12308, USA

[2]Insitute for Space Observations, 1125 Oxford Place, Schenectady, NY 12308, USA

An early spectroscopic observation of a bright field horizontal-branch (FHB) A-type star was that of the high velocity star, HD 161817, by Slettebak (1952). He noted that the spectrum showed Balmer lines slightly weaker and metal lines slightly weaker than normal. Nancy Roman (1955) published her list of High-Velocity stars and the following A-type stars in her list were later confirmed as FHB stars (HD 24000, 60778, 74721, 86986, 117880, and 161817). Interest in faint blue stars at high galactic latitudes had started with the survey by Humason & Zwicky (1947) (the HZ stars) and continued with surveys by Feige (1958) and Haro & Luyten (1962 and other papers). Schmidt surveys at the North and South Galactic Poles (Slettebak & Stock 1959, Philip & Sanduleak 1968, Slettebak & Brundage 1971) led to the finding of many new FHB stars. A search of the Michigan Survey Schmidt plates yielded a list of FHB stars (MacConnell et al. 1971). In more recent times large Schmidt surveys have been made by Pesch & Sanduleak (1989) and Preston, Schectman and Beers (1991a, b). Thousands of candidate FHB stars have been identified.

Oke, Greenstein & Gunn (1966) analyzed photoelectric spectrum scans of some highly evolved stars and proposed that HD 2857, 86986, 106223, 109995, 161817 and BD +17° 4708 were horizontal-branch A-type stars. Greenstein & Sargent (1974) discussed "The Nature of Faint Blue Stars in the Halo" and described the extended horizontal branch. From 1968 - 1973 Philip published lists of FHB stars, detected from four-color photometry of candidate A-type stars marked on Schmidt spectral plates of high galactic latitude fields. Drilling & Pesch (1973) identified FHB stars in two high galactic fields using the same methods. Kilkenny & Hill (in various combinations) have identified many FHB stars by four-color photometry.

High dispersion HB spectra have been analyzed by Kodaira (1964) [HD 161817], Wallerstein & Hunziker (1964) [HD 109995], Kodaira & Philip (1984a)[M 4-206, M 4-553, NGC 6397-48, HD 161817], Kodaira & Philip (1984b) [HD 2857, 14829, 130095, 161817 and PS 53 II], Adelman & Hill (1987) [HD 109995, 161817], Adelman & Philip (1990a) [HD 86986, 130095 and 202759], Adelman & Philip (1992) [HD 74721, 117880]. These analyses confirmed the low metallicity of the HB stars investigated. Danford & Lea (1981) obtained 17 Å mm^{-1} spectra of HD 2857, 60778, 86986, 109995 and 161817 and found [Fe/H] values of -0.05 to -1.8. Corbally & Gray (1994) have obtained spectra of the stars in the FHB list by Philip (1984). Radial velocities of FHB stars at the Galactic poles show a Z dispersion of over ± 100 km s^{-1}, a rigorous indication of Population II membership. Photoelectric scans of BHB and FHB stars show pairs of stars that seem to identical over the optical spectral range. Two Conferences on Faint Blue Stars have been published (Luyten 1964, Philip, Hayes & Liebert 1987) and in each a good summary of the status of the search and study of A-type horizontal-branch stars will be found. CCD techniques now allow photometry and spectroscopy of stars that could not be measured before. Much progress in this area should develop in the next few years.

1. Why study the characteristics of HB A-type stars?

The early history of the Galaxy can be studied by measuring the oldest stars. An important component of the Galaxy that falls in this class is that of globular clusters and the field analogs, namely field horizontal-branch stars (FHB). Candidate FHB stars are easily identified on Schmidt spectral plates. Followup Strömgren four color photometry can select the FHB stars among a group of A-type stars. For the brighter FHB stars there sometimes is confusion with some higher luminosity Population I A-type stars but for the more distant A-type stars the Population II classification is much more reliable. These stars are important for they allow us to measure photometrically, with greater precision, and spectroscopically, at higher dispersion, stars that are close analogs of the BHB stars. Modern, large spectroscopic surveys are now turning up thousands of candidate FHB stars. The further study of these newly identified stars will give us information concerning the distribution of Population II stars in the galactic halo, and their characteristics.

Theoretical work by Sweigart (1987, 1994) and others have outlined how a star evolves along the horizontal branch as a function of its mass, core mass, chemical composition and other parameters. By studying the position of stars on the horizontal branch in a CM-Diagram we can learn much about the variation of these parameters in globular clusters. Spectroscopic investigations reveal information concerning the chemical abundances of these old stars. Because the FHB stars are so numerous and easy to locate they are good probes of the galactic halo.

2. Early measures of field horizontal-branch stars

There are four, classic, prototype FHB stars, HD 2857, 86986, 109995, and 161817. These stars were proposed as FHB stars by Oke, Greenstein & Gunn (1966) who had made photoelectric spectrum scans of some of the brighter FHB candidates. The size of the Balmer discontinuity is a function of the temperature and log g. The slope of the continuum is a function of the temperature with only a small dependence on gravity. The Hγ profile can give a good estimate of temperature. Plots of the flux distribution can then be used to find the reddening and the gravity. They presented the following stars, listed in Table 1, as six possible HB stars. Later observations removed HD 106223 and BD +17° 4708 from the FHB class, but the remaining four stars became the prototypes of the FHB class.

HD 161817 has been a star of interest for many years. Albitzky (1933) obtained spectra of the star and noted that it had the very high velocity of -363 km s^{-1}. Svetlova (1946) stated that the star was probably an apparently bright intermediate white dwarf. Paul Merrill (1947) obtained a spectrum at 10 Å mm^{-1} with the 100-in telescope on Mt. Wilson. He said the spectrum resembled that of a sharp-lined main-sequence A4 star more closely than it did that of an intermediate white dwarf.

In 1952 Slettebak reported on spectroscopic observations of HD 161817. (In the 1940 and 1950's HD 161817 was sometimes referred to as Albitzky's star.) HD 161817 was one of the high velocity stars in Roman's (1955) Catalog of High Velocity Stars. Slettebak compared the spectrum of HD 161817 with that of η Oph (a main-sequence A2 star) and noted that the strengths of the hydrogen lines were slightly weaker in HD 161817, lines of the metals (Fe II, Cr II, Ni II, Mg II, and Si II) were slightly weaker while the metallic lines (Fe I, Ti II, Sr II, and Ca II) were about the same strength. People, at this time, suggested that the star was a main-sequence A-star, maybe located somewhat above the main-sequence or an intermediate white dwarf. Slettebak pointed out that the features of HD 161817 on his spectra were similar to those of the high-velocity subdwarf HD 140283. He suggested that HD 161817 was a star somewhere between a normal main-sequence star and a subdwarf.

Parenago (1958), in a paper entitled "A Revision of the Hertzsprung-Russell Diagram According to the Data on Near Stars", mentioned four λ Boo stars in his Table 13, namely λ Boo, 29 Cyg, γ Aqr, and 2 And. These stars are characterized as underluminous for A stars and had low radial velocities (ranging from -17 to 2 km s^{-1}). In Table 10, Giants

BD or Number	θ	Scans log g	Hγ θ	Type
2857	0.68	2.6	0.67	HB
86986	0.63	2.9	0.62	HB
106223	0.76	3.2	0.70	HB
109995	0.59	(3)	0.62	HB
161817	0.70	2.6	0.68	HB
17°4708	0.85	3.0	0.83	HB

TABLE 1. Scanner Results for FHB Stars

Star	Log T	Log g
HD 2857	3.87	2.6
HD 60778	3.92	3.0
HD 74721	3.97	3.0
HD 86986	3.88	2.8
HD 109995	3.90	2.8
HD 130095	3.91	3.1
HD 161817	3.88	3.0

TABLE 2. Parameters for Group A Stars

of the spherical system, Parenago listed 7 Sex and HD 161817 with spectral types of A1 V and sd A2, absolute magnitudes of +0.5 and +1.5 and radial velocities of +97 and -363 km s^{-1}.

Greenstein (1966) in "The Nature of the Faint Blue Stars". discussed photometric and spectroscopic observations of the HZ stars, the Tonantzintla stars and the Feige stars. (See Section 3 below on Surveys to find descriptions of these types of stars.) He found HZ 30 to be a B halo star, Tonantzintla 181 to be an A-type horizontal-branch star, Feige 65 to be a B-type and Feige 68 to be an A-type horizontal-branch star.

E. B. Newell (1969) studied 77 blue field stars which had been selected to have a high probability of being members of the galactic halo (Eggen 1970; Luyten 1957; Luyten & Anderson 1958, 1959; and Chavira 1958). The stars were observed with a special narrow-band system of filters which measured the Balmer jump, the slope of the Paschen continuum and the equivalent width of Hγ. Spectra were obtained for some of the program stars. In a (B-V) vs (U-B) plot of a representative sample of high-latitude blue stars Newell found that the distribution followed the two-color Population I relation closely but there were two gaps (at (U-B) = -0.4 and -0.9). He concluded that the gaps were a horizontal-branch phenomenon. Newell's analysis of the 77 program stars divided them into four groups. Group A was composed of true BHB stars, double energy source-stars in a post-red-giant phase, evolving to the asymptotic giant branch. Group BC was composed of true BHB stars, double energy-source stars in a post-red-giant phase, evolving to the blue, missing the asymptotic giant branch. Group D was composed of stars in an unknown evolutionary state. Group HL was composed of stars which were evolving to the sdO region. Among the high velocity field stars in his study were seven which were classified as members of Group A. Their parameters are shown in Table 2.

3. Surveys for faint blue stars

In the 1940 and 1950's there was a great deal of interest in the blue stars which were being found in the galactic halo. An important meeting was held in Strasbourg, France in 1964 (The First Conference on Faint Blue Stars (Luyten 1964)) and these proceedings summarized much of the survey work undertaken to find these stars. One of the first major searches was that of Humason & Zwicky (1947) (the HZ stars), who made the survey, originally, to find additional white dwarfs. Zwicky took plates of three high latitude fields in the UV, blue, and photovisual regions and found 48 blue stars. Humason took spectra of each of the stars. Included in the group were white dwarfs (DA to DO), stars with composite spectra, subluminous O, B or A-stars of the horizontal-branch type and some stars with spectra a good deal later than A0 (Greenstein in Luyten 1964).

Luyten made extensive searches for faint blue stars by blinking Schmidt plates taken in two different colors. The exposures were adjusted so that G-stars had images of equal intensity; then the blue and red stars were easily identified. Six thousand three hundred faint blue stars were found in the regions of the Hyades, Pleiades, Praesepe, M 67 and other galactic clusters, the North and South Galactic Poles and other selected regions. These reports were published mostly in the series, "A Search for Faint Blue Stars" by The Observatory, University of Minnesota. There is an extensive bibliography at the end of "The First Conference on Faint Blue Stars" which lists the various surveys made up to 1964.

Haro developed a three-color method of detecting blue stars. With the Tonantzintla Schmidt telescope and the Palomar 48-in Schmidt telescope he employed V, B and U filters and took a set of three exposures, moving the telescope slightly between exposures and ended up with three images side by side. For a main-sequence star of spectral type A5 the exposures were set so that the three images were of equal intensity. The following types of stars could be reliably identified on the plates; blue or ultraviolet stars, blue or ultraviolet galaxies, some U Geminorum type stars, some T Tauri stars and extremely red stars. In Luyten (1965) he reported "... have published a total of about 2200 stars. Haro & Luyten (1962) have examined plates taken with Haro's three-image method and have found 8700 faint blue stars. Feige (1958) has searched some Palomar Schmidt survey plates and has published a list of 112 blue stars found on them."

An early paper comparing the positions of Population II and Population I stars in the Chalonge-Divan λ, D, ϕ_b system was published by Berger et al. (1958). In their Figure 3 they showed HD 86986 and 161817 falling above the main-sequence in the λ, D plane. When Strömgren created the four-color photometric system (Strömgren 1952, 1955, 1963) the new system made use of some of the parameters of the Chalonge-Divan system. The four-color system became one of the best ways of discriminating among the different types of early-type stars of Population I and Population II.

Klemola (1962) discussed the mean absolute magnitudes of the blue stars at high galactic latitudes. He found two stars which had been classified by Slettebak et al. (1961) as horizontal-branch stars (BD+32° 2188 and +39° 4926). Slettebak et al. (1961) had noted that the spectra of these two stars had sharp Balmer lines and a sharp density break at the Balmer limit when observed at low dispersion. However, they pointed out that it was not possible to separate these two stars reliably from similar-appearing subdwarf A stars and high luminosity B and A stars. BD +32°2188 has UBV colors making it a B-type star so it is outside the subject area of this paper. BD +39°4926 has four-color indices (Philip 1968) of 0.169, 1.599, 0.049 [(b-y), c_1, m_1] which places it far outside the range of horizontal-branch A-type stars. It is a star of higher luminosity. In Klemola's Table I HD 2857 is listed as A2 spectral type. No mention was made of its being a FHB star. Mean absolute magnitudes were computed from the components of the proper motion. He concluded that many of the high-latitude B stars between tenth and twelfth magnitude are similar to the horizontal-branch stars in globular clusters.

S. W. McCuskey (1965) discussed the distribution of common stars in the galactic plane and presented a series of plots showing the space distribution of stars of different spectral types. He had set up a series of LF regions in the galactic plane in which stars were classified as to spectral type from Schmidt spectral plates. Their magnitudes were determined from photographic UBV photometry. I became interested in this work and set up a program of HLF regions to investigate the stellar distribution perpendicular to the galactic plane. There is a summary of this work in Philip (1972). A stellar density analysis was completed in 1 HLF 2 (at l = 170°, b = -30°) complementing that of Upgren (1962, 1963). In each area a dip was found in the luminosity function at M_v = 1.5 - 2.0. This structure in the luminosity function was caused by the presence of the early-type FHB stars in the galactic halo which meant there were additional stars at $M_v \approx 1.5$. Schmidt spectral plates were taken in many high-galactic latitude regions (using the Schmidt telescopes of Warner and Swasey Observatory, Tonantzintla Observatory and Cerro Tololo Inter-American Observatory) and finding charts of stars of spectral type A7 and earlier were published. The majority of the A stars on these lists were measured in the four-color system and this work identified many candidate FHB stars. Finding charts for 71 of the brighter FHB stars, selected from these lists and from the literature, were published in the VVO Contributions No. 2 (Philip 1984).

Kinman (1965) and Kinman et al. (1965, 1966) made an extensive survey of RR Lyrae stars with the Lick 20-in Carnegie Astrograph by taking plates at intervals of a few hours in fields at the North Galactic Pole and towards the galactic center but at a galactic latitude of +30°. An exposure time of 30 minutes was selected because it is short compared to the periods of most RR Lyrae stars and such exposures would reach B magnitudes near 19.0. One of his surveys covered three fields near the North Galactic Pole. In his Figure 8 (Kinman et al. 1966, reproduced here as Figure 1) he plots the space density of RR Lyrae stars of type a with period > 0.44 days as a function of height above the galactic plane. Each cross in the plot represents the space density determined from four RR Lyrae stars. The dotted curve is the $\log \rho$ versus z relation for a spherical halo. This RR Lyrae distribution has been copied to Figure 4 of Philip (1972, reproduced here as Figure 2) which shows the distribution of A-type Population I stars at the NGP, as well as FHB stars at the North and South Galactic Poles. At a z distance of 3 - 4 kpc from the galactic plane the FHB stars have a stellar density approximately ten times that of the RR Lyrae stars.

Other Schmidt surveys located horizontal-branch star candidates. MacConnell et al. (1971) searched the Michigan Schmidt Survey plates to find stars of special interest and they published a list of stars that had the characteristics of horizontal-branch stars. Bond & Philip (1973) presented an additional list of FHB stars. Philip (see references in Philip 1978) published lists of 60 candidate FHB stars found by a combination of locating candidate stars from Schmidt spectral plates (obtained at the Tonantzintla Observatory and Cerro Tololo Inter-American Observatory) and followup four-color photometry. Finding charts and positions were published for 71 of the brighter FHB stars in Philip (1984).

Preston, Schectman & Beers (1991a, b) made a Schmidt objective-prism spectral survey using a narrow interference filter centered on the Ca II H and K lines. The HK survey identifies extremely metal-poor stars but there are a number of false candidates, thus the authors have been observing stars from their catalog using the three-color UBV system. They show two-color diagrams for their HB stars and compare this distribution with that of BHB stars in globular clusters. Some of the HB stars fall close to the main-sequence

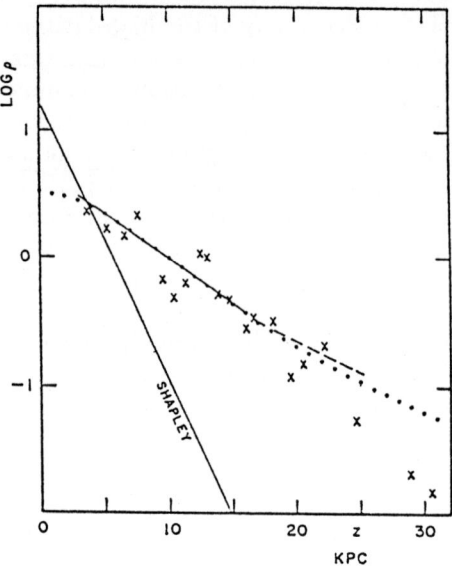

FIGURE 1. The logarithm of the space density (ρ) in stars kpc^{-3} of the RR Lyrae stars of type a as a function of height (z) above the galactic plane. (Kinman et al. 1966)

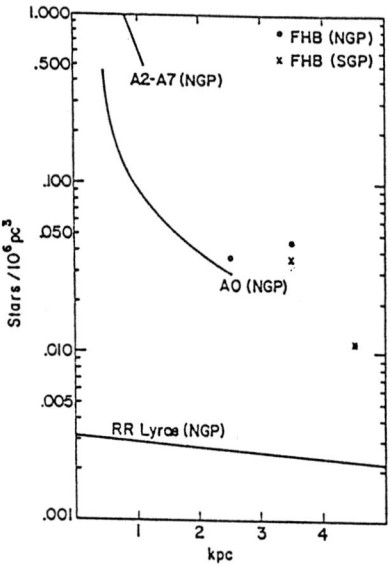

FIGURE 2. The stellar density (stars per 10^6 parsec^3) as a function of the distance from the galactic plane in parsecs. (Philip 1972)

in the two-color diagram, thus UBV photometry will not completely separate HB stars from main-sequence stars. Beers et al. (1988) list a large number of FHB candidates.

Pier (1982, 1983) has measured 234 A and B stars in the southern galactic halo, selected from the Curtis Schmidt survey of Preston, Schectman & Beers. The majority of the stars lie in the region occupied by field and globular cluster horizontal-branch stars in a (U-B), (B-V) diagram. Preston et al.'s Figures 2 and 3 are shown here as Figures 3 and

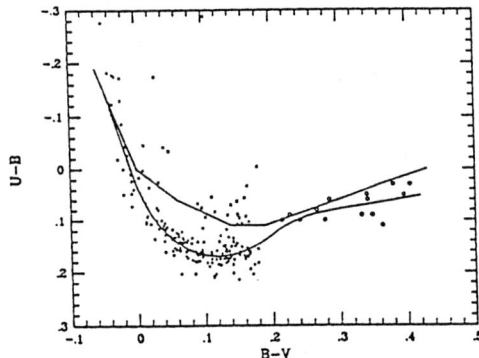

FIGURE 3. Two-color diagram for HB stars. The upper solid curve represents Johnson's (1966) main-sequence relation. The lower solid curve is the relation derived from high-latitude BHB candidates. Filled circles represent the high-latitude BHB candidates, open circles represent RR Lyrae stars of group 1. (Preston et al. 1991a)

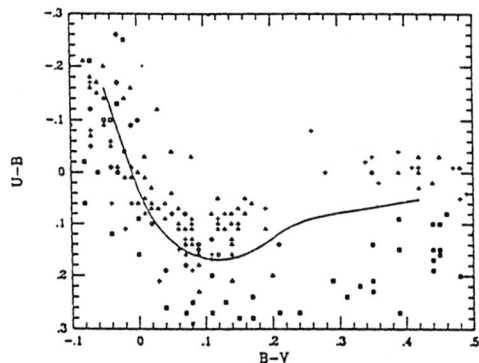

FIGURE 4. Two-color diagram for HB stars. Observations are of BHB stars by Cannon & Stobie (1973) [+], Newell et al. (1969) [diamonds], Sandage (1969) [triangles], and Menzies (1974) [boxes]. (Preston et al. 1991a).

4. However, three-color photometry does not separate out the Population II component as well as four-color photometry does. (See Figures 16 and 17 in Section 5 for an example of the four-color diagrams.)

4. Observations of the FHB stars

In the check of references for this paper the SIMBAD database was searched at Strasbourg via the remote login service. The FHB stars on the list of FHB candidates were checked for their bibliographic references and a spreadsheet was made listing all the authors and the stars they worked on. A plot was made showing the number of stellar references per author for the group of stars contained in the spreadsheet and this plot is shown in Figure 5. The ISO group (Philip, Hayes & Adelman) has contributed a little over 1/5 of the total measures. Approximately 200 astronomers have been involved in measuring early-type halo stars over the last 24 years.

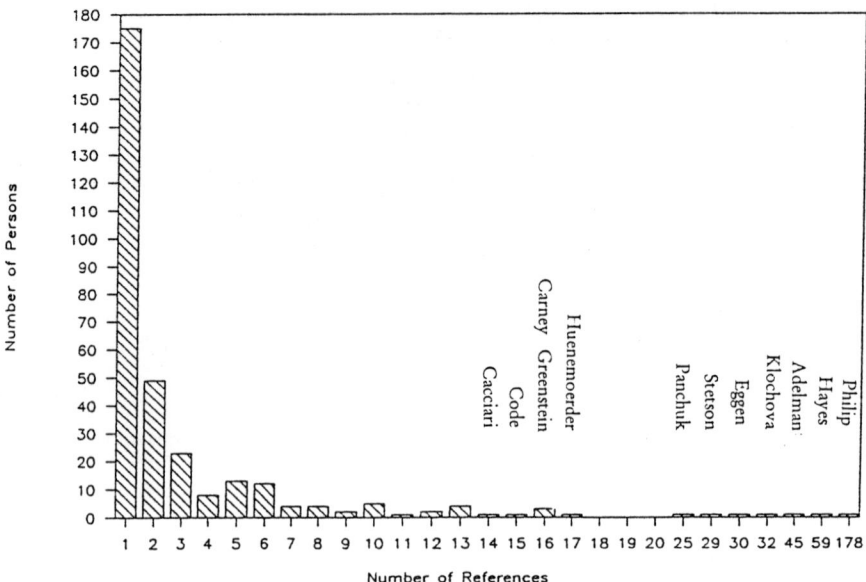

FIGURE 5. The number of references per person found from searching the SIMBAD database on FHB stars.

4.1. Photometric Observations

John Graham did some of the first four-color photometry of the halo Population II stars. He, (Graham 1970), measured the Feige (1958) blue stars in the four-color system and classified Feige nos. 2, 8, 32, 42, 60, 68, 96, and 97 as FHB stars. Feige 3 and 16 were possible FHB stars. His criterion for classifying a star as a FHB star was that in the color range $0 < $ (b-y) $ < 0.2$ the c_1 index $ > 1.200$.

Greenstein & Sargent (1974) discussed the nature of the faint blue stars in the halo. Their sample included 189 blue stars selected from various star surveys. From their slit spectra they classified the stars into seven different groups. Two of these groups were globular cluster type A and B stars. In their table A2 they listed 32 B and A-type field horizontal-branch stars. Seven of the HZ stars appear in the table as do 13 Feige (1958) stars, and an assortment of NGP, Tonantzintla, PHL, GD, HD and BD stars. They found a dispersion in radial velocity of ± 126 km s^{-1}, which is in agreement with their classification of these stars as Population II stars. They proposed that sdB and sdO stars might be stars on an extended horizontal branch

Kilkenny & Hill (1975) presented plots of four-color indices which showed where stars of different spectral classes and population types fall. These plots are reproduced here as Figures 6 and 7 and indicate why the four-color system is such a good photometric system for identifying horizontal-branch A-type stars. In the c_1 vs (b-y) diagram the main sequence is represented by the heavy line. The horizontal-branch stars are indicated by the curved dotted lines in the upper middle of the diagram. For stars near A0 the horizontal-branch intersects the main sequence, but as the spectral type goes from A0 to F0 the δc_1 index becomes larger and thus is a better discriminant for the FHB stars. In the (b-y) vs. m_1 diagram, again the heavy line indicates the position of main-sequence

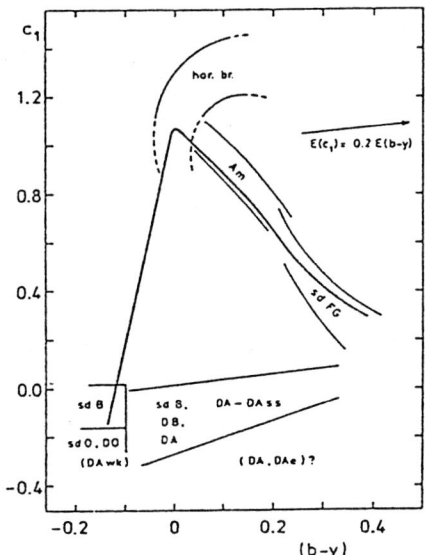

FIGURE 6. c_1 versus (b-y). The ZAMS is indicated by the heavy line; other types of stars are indicated. (Kilkenny & Hill 1975)

stars. The horizontal-branch stars in this diagram occupy a region orthogonal to the main sequence near (b-y) = 0.

In the period up to 1978, four-color photometry of FHB A-type stars was summarized in Philip (1978). A list of photometric observations of FHB and BHB stars is presented in Table I of that paper. More recently, Stetson, in an important series of papers, has investigated early-type high-velocity stars in the solar neighborhood. In Stetson (1991) a continuing report is made of four-color and $H\beta$ photometry of stars in Stetson's high-velocity star catalog. In his Table 6 he presents a list of 31 possible FHB stars. Nine of these stars were already known as FHB stars; the remaining 22 stars will be found below in Table 3.

Photometry in other photometric systems has confirmed the classification of stars selected as Population II stars on the basis of four-color photometry. Philip & Hauck (1984) compared-four-color and Geneva System measures of MK standard stars and found that the two systems agreed well. Hauck (1994) reports on a set of FHB stars, identified from four-color photometry which are confirmed in Geneva photometry. Straižys & Kaslauskas (1993) have published a catalog of Vilnius photoelectric photometry and many early-type FHB stars are identified.

$H\beta$ photometry gives us a good index for luminosity determinations for early-type stars. In Figure 8 the $H\beta$ measures of Stetson (from Table 3 above) are combined with $H\beta$ measures taken at the North Galactic Pole (Philip & Tifft 1971). The heavy line indicates the position of the Zero-Age Main-Sequence in this diagram. Most of the points, representing the FHB stars, fall well below this line, an indication of their higher luminosity. Recent CCD $H\beta$ measures of BHB stars in the Globular Cluster, M 92, show the same effect (Philip 1994). Ultraviolet observations of HB stars are discussed in Section 7.

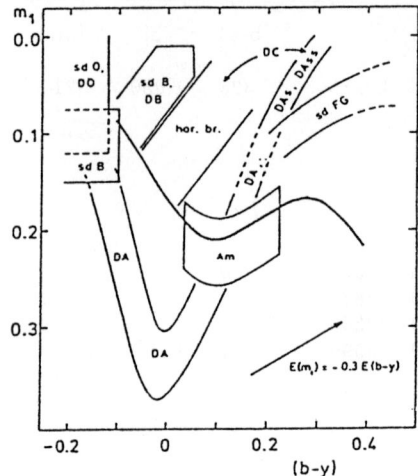

FIGURE 7. m_1 versus (b-y). The ZAMS is indicated by the heavy line; other types of stars are marked. (Kilkenny & Hill 1975)

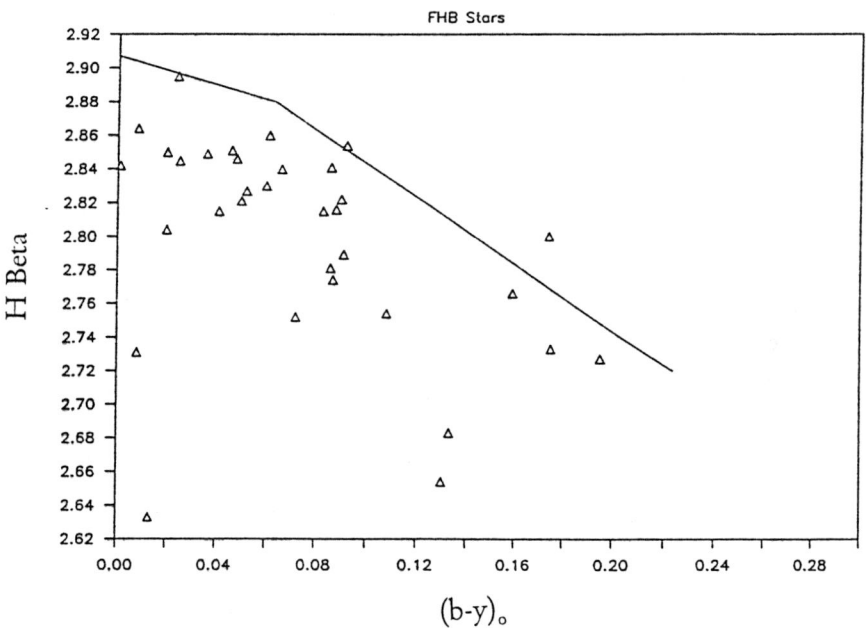

FIGURE 8. Hβ vs (b-y) for FHB Stars.(Stetson 1991, Philip & Tifft 1971)

4.2. Spectroscopic Observations

Burbidge & Burbidge (1956) reported on the chemical compositions of five stars (HD 84123, 106223, 161817, 29 Cyg, and λ Boo). They did a curve of growth analysis relative to 95 Leo and they found that HD 84123 had normal composition, 29 Cyg had slight underabundance ratios, HD 161817 had moderate underabundance and λ Boo and HD 106223 had the largest underabundances. But HD 161817 had a small rotational velocity while λ Boo, 29 Cyg and HD 106223 had rotational velocities near 100 km s^{-1}. They

Name	(b-y)	[c]	[m]	β
HD 203854	0.328	1.030	0.214	2.810
HD 87112	0.001	1.161	0.115	2.842
HD 167105	0.036	1.253	0.131	2.849
HD 8376	0.092	1.255	0.132	2.854
HD 87074	0.091	1.255	0.133	2.798
HD 252940	0.159	1.183	0.138	2.766
SAO 82681	0.046	1.289	0.142	2.851
HD 93329	0.060	1.303	0.141	2.830
HD 128801	-0.005	1.058	0.108	2.816
HD 203563	0.020	1.142	0.128	2.804
HD 180903	0.174	1.220	0.147	2.800
HD 4850	0.048	1.272	0.147	2.846
HD 13780	0.088	1.267	0.145	2.816
HD 16456	0.175	1.025	0.145	2.733
HD 31943	0.083	1.209	0.167	2.815
HD 106304	0.025	1.157	0.122	2.845
HD 130201	0.061	1.232	0.127	2.860
HD 176387	0.195	1.036	0.119	2.727
HD 78913	0.066	1.274	0.138	2.840
HD 213468	0.020	1.232	0.133	2.850

TABLE 3. New FHB Stars from Stetson

Stars in Stetson's table which were already known as FHB stars were HD numbers 60778, 74721, 86986, 109995, 117880, 130095, 139961, 161817, 202759, and 214539.

quote Greenstein's (1956) remark that stars with high velocities seem to have small rotational velocities. Greenstein summarized his spectra of FHB stars as having 1): small stellar rotation, 2): lines slightly weak, for example Mg II, 3): the interstellar K line is often seen, with a velocity corresponding to the nearby gas.

Kodaira (1964) used the 100-in telescope on Mt. Wilson to obtain six spectra of HD 161817 from $\lambda\lambda = 3650 - 6700$ Å. He gave an extensive table of equivalent widths of stellar lines from $\lambda\lambda = 3358$ to 6562 Å and then did an analysis to find an effective temperature of 7630 K, $\log g = 3.0$ (if He/H = 1/6) or $\log g = 3.2$ (if He/H = 1/60) and a mean underabundance of metals relative to the Sun of -1.11. He concluded that HD 161817 is located on or slightly above the horizontal branch.

Wallerstein & Hunziker (1964) obtained six spectra of HD 109995 on the 100- and 120-inch telescopes covering the region from $\lambda\lambda = 3800 - 6600$ Å. They performed a comparative curve-of-growth analysis relative to Sirius and Vega and found HD 109995 to have a metal deficiency of -1.0 relative to Vega, and HD 161817 to have a metal deficiency of -0.4. They concluded that HD 109995's metal deficiency was consistent with membership in the halo population and that it could be a horizontal-branch star.

Philip (1969, 1970) used the 100-in on Mt. Wilson and the 84-in at KPNO to obtain spectra of the brighter FHB stars. The z velocity dispersion for seven NGP FHB stars was ± 113 km s^{-1}, confirming their membership in Population II. The velocity dispersion for six FBH in the 1 HLF 2 area was ± 109 km s^{-1}. These velocity dispersions are in agreement with the Greenstein & Sargent (1974) velocity dispersion of ± 126 km s^{-1} for 18 FHB stars.

Rodgers (1972) obtained spectra of high-velocity A stars at 10 and 6.7 Å mm^{-1} with the 74-in coudé spectrograph. For seven FHB stars (HD 60778, 74721, 86986, 109995, 117880,

HD Number	T_{eff}	log g	Abundance
2857	7450	2.6	-1.8
60778	8750	3.3	-0.5
86986	8150	3.4	-1.0
109995	8890	3.4	-0.5
161817	7750	3.1	-1.0

TABLE 4. Parameters for 5 FHB Stars

Star	T_{eff}	log g	log A	Type
HD 2857	7700	2.9	-1	FHB
HD 14829	9300	3.35	-1	FHB
HD 130095	9200:	3.4:	-1	FHB
HD 161817	7700	2.9	-1	FHB
PS 53 II	9500	3.3	-1	FHB
HD 107369	8000	2.1	0	Hi Lum.
HD 214539	9800	1.6	-1	Hi Lum.
HD 130156	7700::	4.2::	0	Main Seq.
HD 184779	7500:	4.0:	0	Main Seq.
PS 37 II	8100::	4.2::	0	Main Seq.
HD 202759	7400:	2.8	-1	RR Lyr.
M 4 206	9250	3.27	-0.4	BHB
M 4 553	9000	3.00	-0.4	BHB
6397 48	9200	3.05	-1.4	BHB

TABLE 5. Atmospheric Parameters for Group A Stars

130095 and 161817) he found that the Ca abundance was down from solar abundance (except for a low-weight determination of [Ca] = 0.0 for HD 74721)

Kodaira (1973) revised his chemical abundances for the FHB stars HD 86986, 109995 and 161817, using new gf values for Mg, Si, Fe, and Ni. The mean abundance of elements from Na to Ba (except for S) were -1.4 for HD 86986, -1.7 for HD 109995 and -1.2 for HD 161817.

Danford & Lea (1981) determined abundances for five of the brighter FHB stars (HD 2857, 60778, 86986, 109995, and 161817 from the coudé feed of the 2.1-m telescope at Kitt Peak. Table 4 summarizes their results:

Kodaira & Philip (1984a, b) obtained high-dispersion spectra with the Cassegrain-Image Tube Spectrograph on the CTIO 4-m telescope of BHB stars in the globular clusters M 4 and NGC 6397 and some FHB candidates. Table 5 summarizes their findings.

For the globular cluster BHB stars Kodaira and Philip employed two models, a hot and a cool model, to determine the atmospheric parameters. The cool models gave lower abundances and they represented the Hγ profile and the photometric indices (b-y) and c_1 but they could not match m_1. The hot models were chosen since they matched all the observed parameters.

Philip and Samus obtained spectra of FHB stars using the 6-meter telescope in the Caucasus. After these observing runs were over Panchuk observed a few additional stars and sent the plates to Schenectady. John Lee (1985) measured spectra of HD 161817, HD 86986, and HD 109995 from tracings obtained by Philip at Kitt Peak National

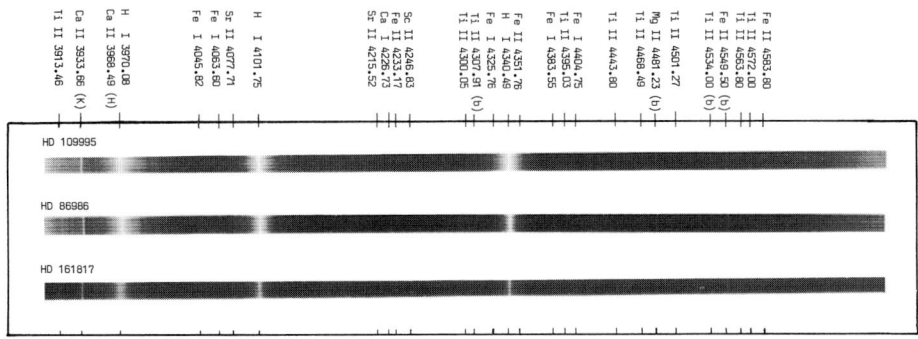

FIGURE 9. Six-meter spectra (at 9 Å mm^{-1}) of three FHB stars.

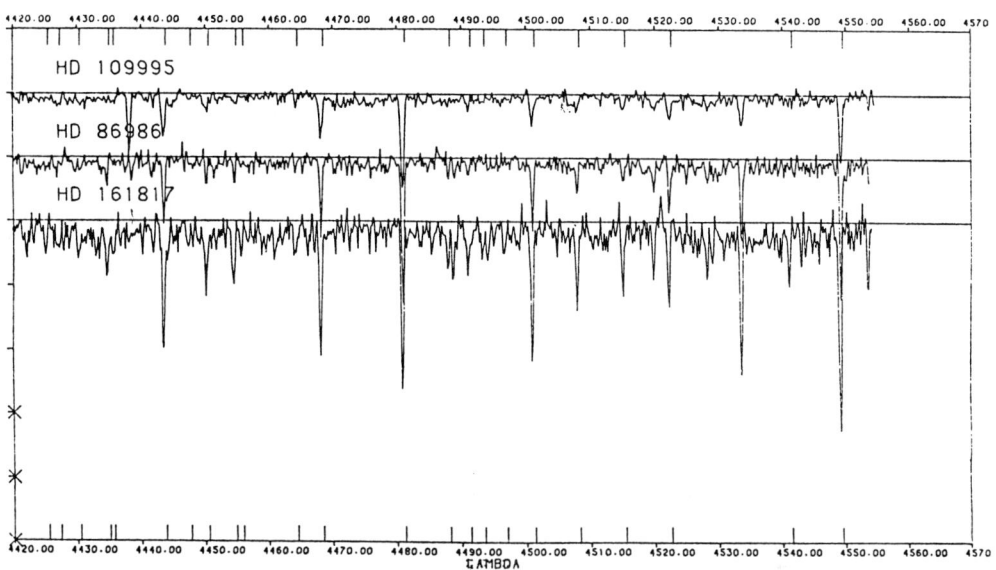

FIGURE 10. Tracings of the three stars in Figure 9, from λ 4420 to λ 4570 Å.

Observatory. A copy of portions of the spectral tracings of the three stars is shown in Figure 9. Photographic reproductions of the spectra are shown in Figure 10.

Klochova & Panchuk (1985) obtained spectra (at 9 Å mm^{-1}) of four FHB stars (HD 744721, 86986, 109995, and 161817) using the 6-meter telescope. They found that the metal abundances of all four stars were the same. The abundances relative to normal A stars in open clusters was log ϵ = -1.5 ± 0.22.

Peterson (1985) has measured the rotational velocities of BHB stars in several globular clusters. Some of these measures are summarized in Table 6.

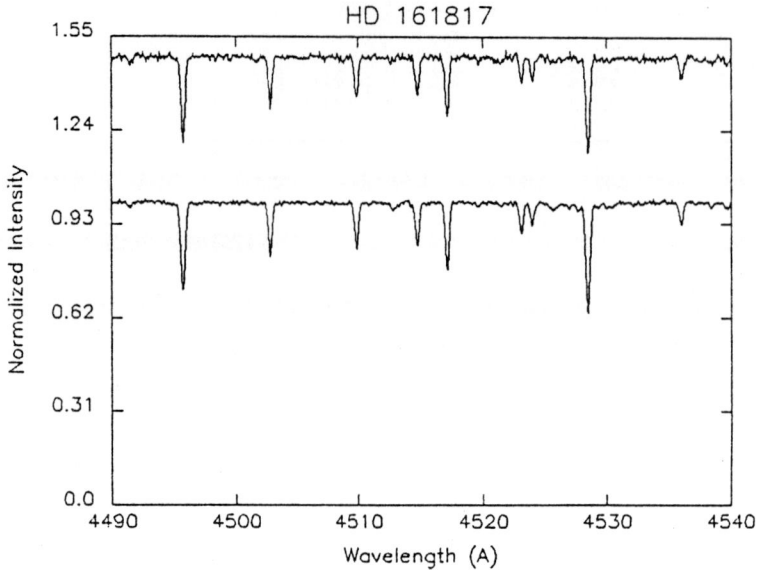

FIGURE 11. A comparison of DAO coadded (upper) and KPNO CCD spectra of HD 161817. The DAO spectrum is the coaddition of 6.8 Å mm^{-1} IIa-O spectrograms. (Adelman & Philip 1990b)

	M 4	M 3	M 5	NGC 288	M 13
HB color ratio B/B+R	0.44	0.55	0.76	1.00	1.00
Number of Stars	9	4	7	7	6
Av. v sini (km s^{-1})	9.6	12.0	10.4	14.4	21.0

TABLE 6. Rotational Velocities of BHB Stars

Adelman & Hill (1987) coadded about 10 6.8 Å mm^{-1} DAO spectrograms of HD 109995 and HD 161817 to increase the signal-to-noise level. Their data forms the basis of the spectral atlas of Adelman, Fisher & Hill (1987). This data has recently been reanalyzed by Adelman & Philip (1994).

Adelman & Philip (1990a, b, 1992) and Philip & Adelman (1991) have used the coudé feed on the 2.1-m telescope at Kitt Peak to obtain high-resolution spectra of several FHB candidate stars in several wavelength regions. Figure 11 shows two spectra (Upper DAO, Lower KPNO) of HD 161817 between λ 4490 and λ 4540. Spectra of the FHB stars HD 86986, 130095, 202759, 161817, 109995, 74721, and 117880 were obtained and analyzed. Table 7, below contains the logarithm of the abundances of Fe/H for each of the stars measured. This work has been extended by Adelman & Philip (1994).

Luck (1991) discussed the determinations of abundances for Population II stars in the field and in globular clusters. If one looks at the compendium of abundance determinations in Cayrel de Strobel et al. (1985) a great range of [Fe/H] values for stars which have been measured many times is found. For example, for HD 161817 the [Fe/H] values

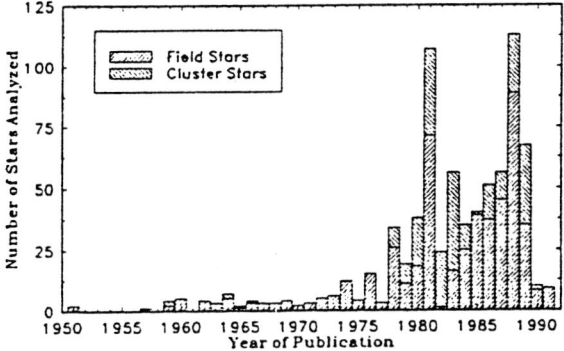

FIGURE 12. The number of metal-poor stars analyzed per year. (Luck 1991)

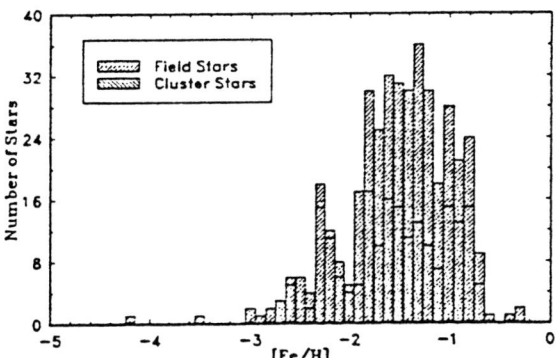

FIGURE 13. The cumulative [Fe/H] distribution from the field and cluster metal-poor stars. (Luck 1991)

range from -0.54 to -1.6, a variation greater than an order of magnitude. Luck found 141 references containing 750 analyses of 404 individual metal-poor stars. He showed two plots describing the number of metal-poor stars analyzed per year and the [Fe/H] distribution for field and cluster metal-poor stars. These figures are displayed here as Figures 12 and 13. He noted that a number of stars in the database have been analyzed multiple times. He found for these stars that the standard deviation, instead of decreasing, remains nearly constant as the number of abundance determinations increases. For HD 122563 (with 20 analyses) there is no trend in the effective temperature but in gravity there is a trend with time. The derived gravities decreased from 1972 to 1983 and then the gravities increased by nearly an order of magnitude and stayed constant until 1990 and then returned to the 1983 value. Luck concluded, "It is my opinion that the larger problems with the metal-poor analyses lies in the gravities."

4.3. Scanner observations

The first scanner observations published of FHB stars have already been mentioned (Oke, Greenstein & Gunn 1966). In the same year, Jones (1966) published a scan of HD 161817. Christensen (1978), in an investigation of energy distributions of metal-poor stars, published scans of eleven stars in the horizontal-branch star section of his Table V. (These stars were HD 85504, 74721, 109995, 117880, 60778, 86986, 161817, 106223,

Star	Fe I	Fe II	Mg I
HD 74721	-6.10	-5.68	-5.60
HD 86986	-6.45	-6.26	-5.80
HD 109995	-6.43	-9.18	-5.85
HD 130095		-6.73	
HD 161817	-6.04	-5.98	-5.54
HD 202759	-6.75	-6.97	

TABLE 7. Iron Abundances for Selected FHB Stars

KPNO		Palomar	
HD 2856	1HLF2 S 14	M 5 II 82	1HLF2 18 21
HD 12293	SS 287 I	M 5 III 17	1HLF2 17 17
HD 14829		M 13 SA 16	1HLF2 17 24
HD 24000		M 13 SA 18	1HLF2 18 98
HD 60778		M 13 SA 477	1HLF2 17 136
HD 64488		M 13 SA 531	1HLF2 13 110
HD 74721		M 92 VI 3	1HLF2 S 70
HD 86986		M 92 X 5	1HLF2 S 14
HD 109995		M 92 XII 6	
HD 117880		M 92 XII 26	
HD 130095		M 15 IV 4k	
HD 161817			
HD 202759			

TABLE 8. FHB and BHB stars scanned at KPNO and Palomar.

205539, 46703, and BD -15° 4515.) The 5-m telescope, with the Oke multichannel scanner was used by Hayes and Philip to obtain scans of BHB stars in globular clusters and some of the brighter FHB stars. The FHB and BHB stars observed are listed in Table 8.

In Figure 14 scans of seven of the brighter FHB stars are shown (from Figure 1 of Philip & Hayes 1983). The scans are arranged in order of increasing (b-y) and are normalized at $\lambda = 5000$ Å. In Figure 15 the scans of four BHB stars are shown with the scans of three FHB stars (Hayes & Philip 1983). The scans are arranged in three groups and within a group the magnitude differences between common wavelength points is 0.02 mag or less. This supports the idea that the FHB stars are photometric analogs of the BHB stars in globular clusters.

5. Globular cluster BHB stars

Graham & Doremus (1968) published the first four-color study of stars in a globular cluster (NGC 6397). Philip (1973) published four-color indices for BHB stars in M 4. The indices for BHB stars fell in areas of the four-color diagrams occupied by the FHB stars. There are two plots in Philip (1978) showing $(c_1)_o$ vs $(b-y)_o$ and $(m_1)_o$ vs $(b-y)_o$ diagrams. These figures are reproduced here as Figures 16 and 17. The distribution of stars in each set of diagrams is similar, with the scatter for the BHB stars being a bit larger, due to the fainter magnitudes of the BHB stars. CCD four-color observations have been made in several globular clusters by Philip (1990). With the greater internal accuracy of the CCD photometric measures it has been possible to divide the BHB stars

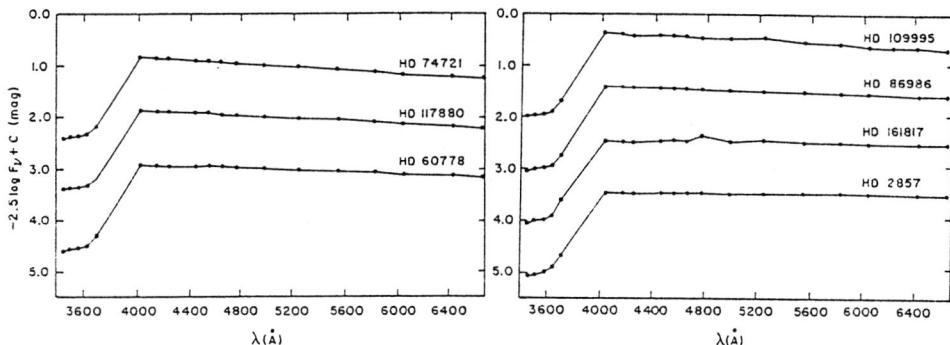

FIGURE 14. Scans of seven of the brighter FHB stars arranged in order of (b-y) color. (Philip & Hayes 1983)

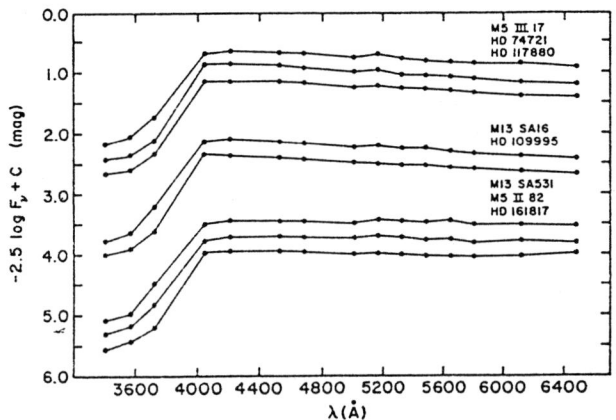

FIGURE 15. Comparison of scans of BHB with FHB stars. In each group of stars, the difference in the magnitude of one scan point from the similar point on the other scans in the group is 0.02 mag or less. (Hayes & Philip 1983)

into two groups. One group follows the ZAHB very closely and the other group falls in a loose aggregation about 0.2 mag above the first group. The interpretation that has been made is that the lower group represents stars that are evolving to the blue on the horizontal branch and that the upper group represents stars that are more evolved and are heading back towards the asymptotic giant branch.

6. Population I stars at high galactic latitudes

Philip (1968, 1974), Rodgers (1971), Rodgers, Harding & Sadler (1981), Tobin & Kilkenny (1981), and Stetson (1981a) have all pointed out that there exist apparently normal early-type stars at high galactic latitudes. Rodgers (1971) and Rodgers, Harding & Sadler (1981) found from measures of SGP stars that approximately half were metal-rich, high-gravity stars, one quarter were metal-poor, low-gravity stars (Population II horizontal-branch) and one quarter were metal-poor, high-gravity stars (which could be blue stragglers). Tobin & Kilkenny (1981) found that faint OB stars at high galactic

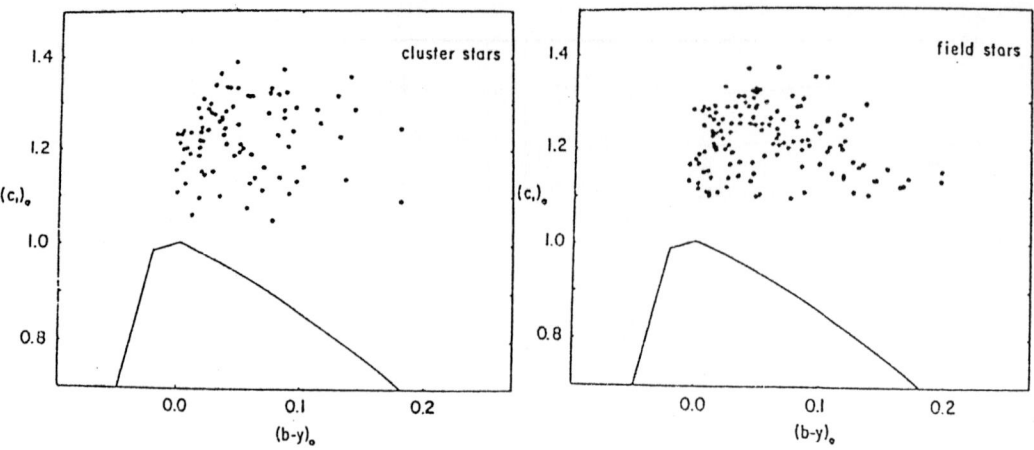

FIGURE 16. c_1 versus (b-y) for field and globular cluster horizontal-branch stars. (Philip 1978)

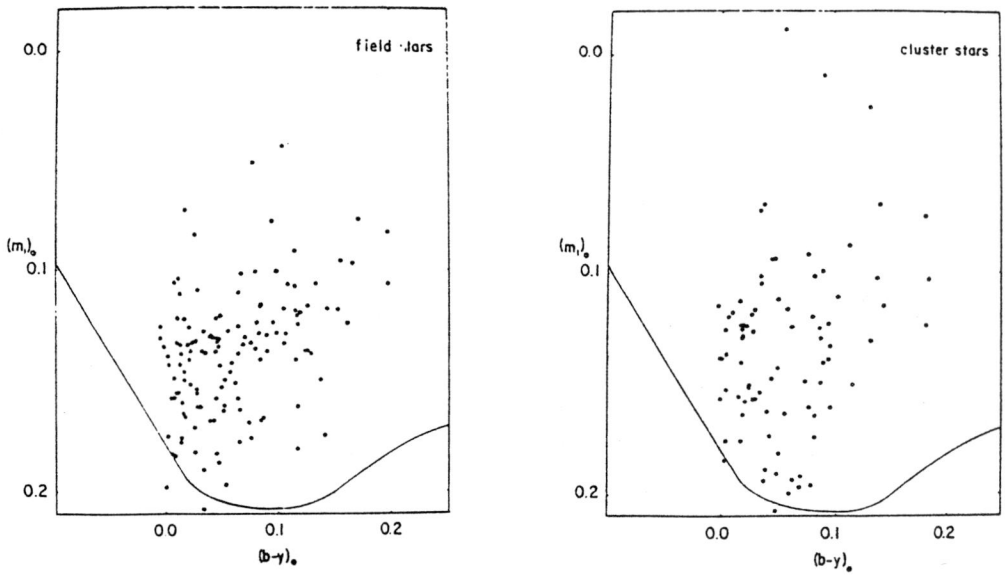

FIGURE 17. m_1 versus (b-y) for field and globular cluster horizontal-branch stars. (Philip 1978)

latitudes were quite similar to OB stars closer to the galactic plane. Stetson searched the Smithsonian Observatory Star Catalog for stars with reduced proper motions and spectral types that would make their transverse velocities be about 100 km s^{-1} or greater. A catalog of 371 stars was presented of which 168 are spectral types B and A. In Stetson (1981b) four-color and Hβ photometry was presented for 78 stars in the high-velocity catalog. He found that five of the A0 stars were probably FHB stars, two more stars were probable FHB, one third of the main-sequence A stars and nearly all of the F stars appear

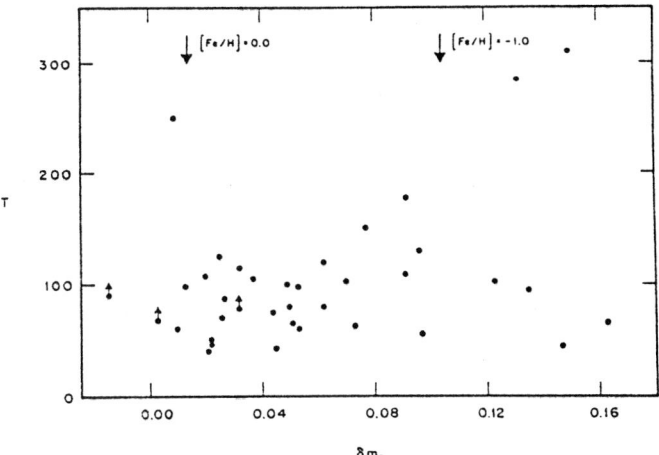

FIGURE 18. Transverse velocity, T, versus δm_1 for F stars. [Fe/H] = 0 and -1.0 are indicated by arrows. (Stetson 1981b)

to be true high-velocity objects and some seem to have solar metal abundances. Figure 18 (Figure 5 of Stetson 1981b) presents δm_o vs T (the transverse velocity) and shows that there are high velocity stars of solar abundance in the sample. In a further analysis (Stetson 1983) obtained spectra of 124 of the stars in his high-velocity star catalog. The radial velocities presented support the existence of an excess of high-velocity main-sequence A stars. In one section of his paper Stetson states, "Finally, from Fig. 10 of Sandage (1981) we estimate that local stars with a w velocity of $\sim$ 60 km s^{-1} only reach distances of $\sim$ 1 kpc, while stars with $|w| \sim$ 120 km s^{-1} reach to 3 kpc from the galactic plane. To travel as far as 5 - 8 kpc into the galactic halo a star would need a velocity of 150-200 km s^{-1}. Thus, if high-velocity A stars have a Gaussian velocity distribution with $\sigma \sim$ 60 km s^{-1}, Philip's (1968, 1974) and Pier's (1982) assertion that few such stars would be found beyond 3 kpc is supported".

Pier (1983) also found apparently normal stars from 1 to 3 kpc below the galactic plane, but the numbers were less than in Rodgers' work. He concludes that these stars do not reach much further into the halo than 2 - 3 kpc and are members of an old disk population with the scale height (1400 pc) found by Gilmore & Reid (1983).

Rodgers & Roberts (1993) reported on spectrophotometric and radial velocity measures of stars of spectral type < F0 in two fields at a galactic latitude of -45°. They found 80 stars with [Fe/H] > -0.5. The metal-rich population has a vertical scale height of $\approx$ 600 pc and a space density at the Sun equal to 1/225 of the young thin disk A-star population.

7. Ultraviolet observations

de Boer & Wesselius (1980) observed blue stars at high galactic latitudes with ANS. Of seven possible horizontal-branch stars two had uv energy distributions with an excess at 1550 Å and a large Balmer jump. These two stars were HD 130095 and BD +0° 145. HD 97859 (classified as a FHB by Stalio (1974)) matches a normal B6 star. HZ 24 (classified as a FHB by Greenstein & Sargent (1974)) matches a B8.5 star with m_{3300}

too bright. HZ 45 (classified as a FHB by Greenstein & Sargent (1974)). F 15 and F 18 (classified by Newell (1973) as a FHB) match a B9.5 star but F 15 has high m_{1800} and F 18 has a high m_{2200} (depending on the reddening assumed). F 41 (classified as FHB by Greenstein & Sargent (1974), but with a note that the Mg lines were too strong) matches the distribution of an A0 star but it is 0.55 mag below the nominal level.

Huenemoerder, de Boer & Code (1984) published the uv spectra of 17 FHB stars at 5 Å resolution between 1150 Å and 3200 Å. In the A-star range these stars were HD 130095, 214539, 117880, 213468, 139961, 74721, 57336, 109995, 86986, 60778, 161817, 2857, and 130056. They noted that HD 214539 is not a horizontal-branch star but a high-luminosity, low gravity star. HD 57336 has strong C I and Al II near 1663 Å and its C(19-V) color and Mg II index match the values for a Population-I star. The paper presents the 17 uv spectra in their Figure 1, arranged in order of decreasing intrinsic blueness, which is copied here as Figure 19. They published an interesting figure (their Figure 6, copied here as Figure 20) which compares the uv flux of FHB stars with that of Population I stars. An average HB spectrum was computed by averaging the spectra of HD 60778, 74721, 86986, 109995, 130095, and 139961 and a second average spectrum was computed by averaging the spectra of HD 41695, 65810, 79439, 107966, and 108765. Between 1300 Å and 1700 Å, the FHB spectrum falls well above the average Population I spectrum. The Population I spectrum shows absorption lines due to Si II, Al II, C I, Sc II, Si III and O I while the FHB spectrum shows no strong absorption lines. Cacciari (1985) published a catalogue of IUE ultraviolet fluxes for 36 metal-poor field halo stars. Cacciari et al. (1987) used the energy distributions in Christensen (1978) and Cacciari (1985) to derive effective temperatures and surface gravities. One of the best ways of confirming the identification of Population II A-type stars is to obtain uv spectra.

Ultraviolet observations have been made of HB stars in two globular clusters. Aurière, Adams & Seaton (1983) obtained spectra of M 15 743 and 751 with IUE covering the spectral range of 1000 to 3000 Å. They determined effective temperatures of 17,500 K for the two stars so these two are B-type BHB stars. de Boer & Code (1981) observed two blue HB stars in M 13 and found that model atmospheres of 15,000 K $< T_{eff} <$ 20,000 K and log g = 4.5 best characterize the spectra. Again, these two stars were of spectral type B.

The ISO group has continued the work of obtaining uv spectra of FHB stars. Philip et al. (1987) published uv spectra of four FHB stars (HD 2857, HD 14829, HD 117880, and SS 287 I). Philip et al. (1990) published uv spectra for six stars, four of which matched the uv criteria for FHB stars (HD 60825, HD 64488, BD +01° 514, and BD +01° 548). Two additional stars, which had four-color indices matching those of the FHB class, had uv spectra showing absorption lines and thus are not members of Population II. HD 60825 was anonymously blue for its C(19-V) color. Later observations of HD 60825 imply that it is a Population I star.

8. λ Bootis stars

λ Boo stars share one of the characteristics of FHB stars, namely, low metallicity. Morgan et al. (1943) showed spectra of λ Boo and defined a new class of star characterized by a weak Ca II line, hydrogen lines similar to early A-type dwarfs and extreme weakness of the strongest metallic line, Mg II (λ 4481). However, over the years the λ Boo class became a collection of a diverse group of stars. Gray (1988) discussed the spectral classification of λ Boo stars and "as a first step towards an improved working definition of the λ Bootis class, shell and protoshell stars, supergiants, helium-weak stars, and stars that show other obvious "chemical" peculiarities such as the Ap stars should be excluded

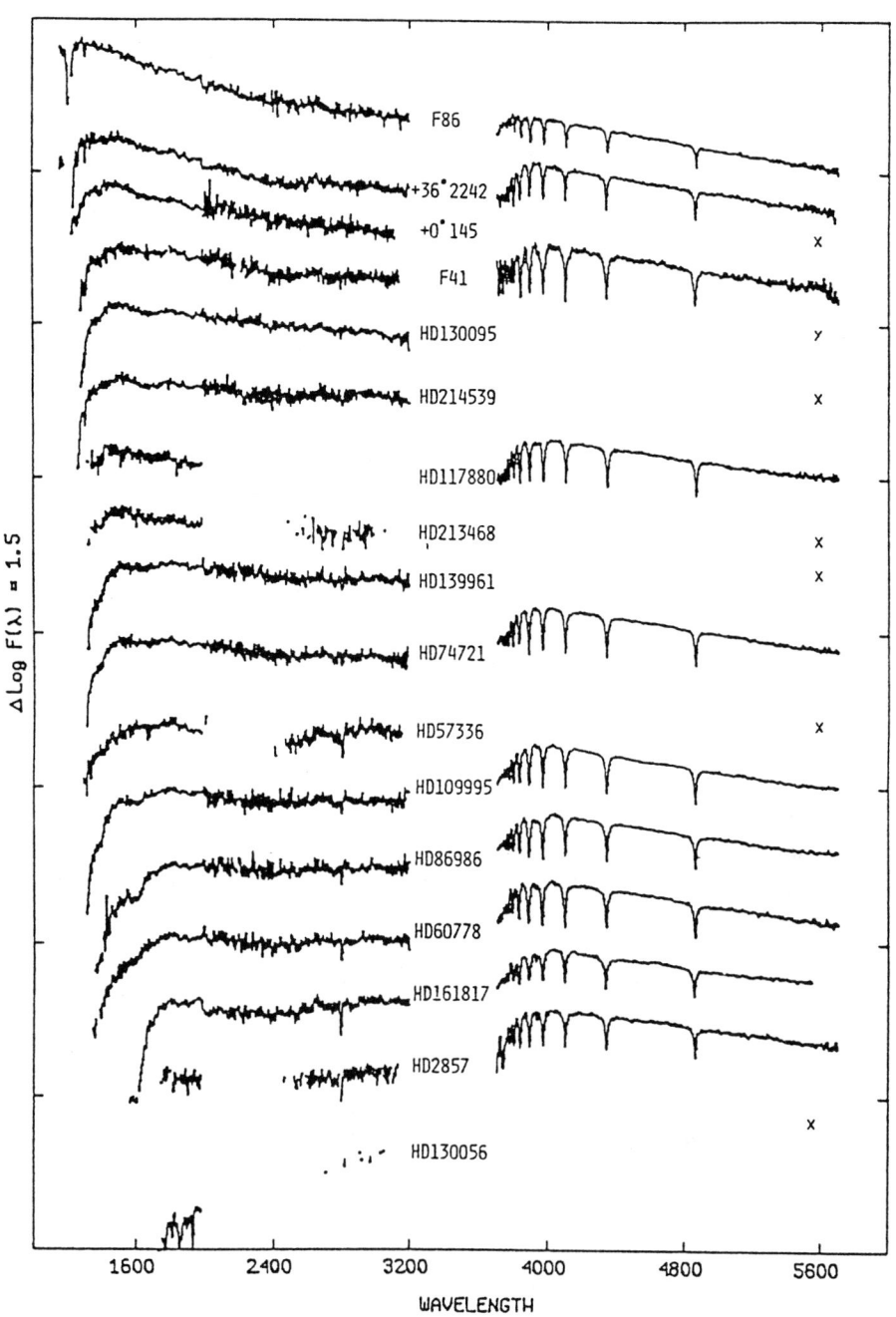

FIGURE 19. Ultraviolet (IUE) and visual (PBO) spectra for FHB stars, in order of decreasing intrinsic blueness. (Huenemoerder et al. 1984)

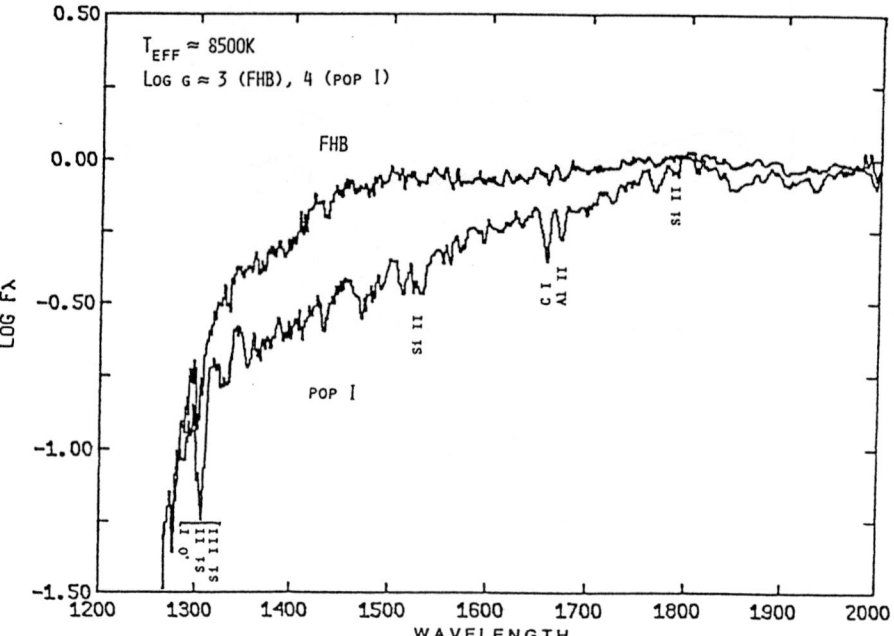

FIGURE 20. Average spectrum for six FHB stars (upper) and five Population I stars (lower) with mean temperatures and log surface gravities of 8600 K, 3.1 and 8500 K, 3.9, respectively. The excess flux for the FHB stars below 1800 Å is largely due to the lower metallicity of the FHB stars. (Huenemoerder et al. 1984)

from the class even though they show weak λ 4481 lines. Population II field horizontal-branch (FHB) and "intermediate Population II" A type stars should also be excluded from the λ Bootis class". He presented a table of 16 λ Boo stars. All of these stars have weak Mg II λ 4481 lines, show broad hydrogen-line wings, the K line type, and the metallic-line type are earlier than the hydrogen-line type. A good proportion of the stars show peculiar hydrogen-line profiles. He found two classes of λ Boo stars based on the hydrogen lines. The NHL class has normal hydrogen-line profiles (early-A dwarf) and the PHL class has peculiar hydrogen-line profiles with weak cores and broad but often shallow wings. In the four-color system λ Boo stars have δc_1 indices that place them in the main-sequence band. FHB stars, on the other hand, have large δc_1 indices (about 0.2 mag) that place them well outside the main-sequence band. At A0 the horizontal-branch crosses the main-sequence so at A0 this discrimination in δc_1 is lost. Gray did not find any examples of evolved λ Boo stars in his survey of the literature.

Faraggiana et al. (1990) considered uv spectra as good indicators of the λ Boo class. They described the spectrum of λ Boo in the uv as having the following characteristics: The spectral class derived from the 1200 - 2000 Å range is A2 in contrast to the A0 class from the optical spectrum, most of the metal lines are weakened, C I at 1657 Å is easily detected on IUE spectra, and there is a broad absorption feature (of unknown origin) centered at λ 1600. In their paper they say, "The five stars which are definitely λ Boo type are HD 31295, HD 105058, HD 110411, and HD 199786. These stars have been selected as λ Boo stars from various sources, ... They are distinguished from weak-lined Population II (horizontal-branch) stars by the fact that they have normal space velocities

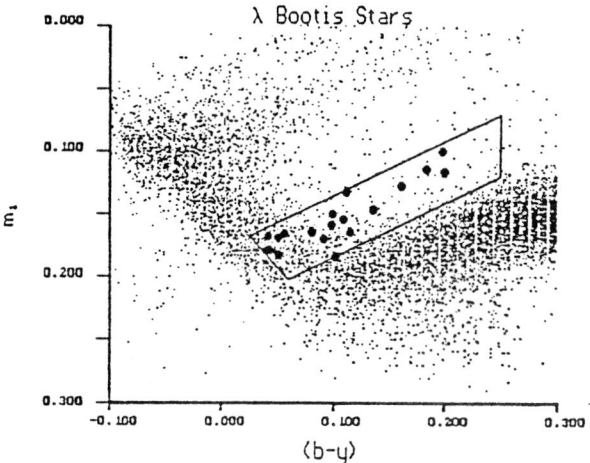

FIGURE 21. m_1 versus (b-y) showing the location of λ Boo stars. (Gray 1991)

and moderately large rotational velocities." In their conclusions they list the following criteria for the definition of a homogeneous group of λ Boo stars.
- "- The ratio C I/A II: if its value is greater than 1.5
- - The bump at λ1600: if it is detectable
- - The discrepancy between the MK type and the UV type: if the Sp(UV)-Sp(MK) is greater than two spectral subclasses
- - The value of Δm_1 must be less than -0.020 but the star must not be in the (b-y) vs c_1 diagram in the domain of the horizontal-branch A stars."

The λ Boo stars that I measured at the NGP had near normal c_1 indices but low m_1 indices. Thus, in photometry there does not seem to be any problem in distinguishing λ Boo stars from FHB stars. A number of people had pointed out that λ Boo stars are moderate rotators, unlike the FHB stars which are sharper-lined stars with low rotational velocities.

Recently Gray (1991) has described a method for selecting λ Boo candidates by means of the combination of Strömgren photometry and objective-prism spectra of field stars. High-quality objective-prism spectra (for example those in the Houk survey) allow A0 - A3 stars to be selected reliably. In the Houk catalogs the λ Boo type stars are classified with luminosity types of II, II/III or III. When the λ Boo stars are plotted in the (b-y), m_1 diagram they are found to populate a rectangle, above the main-sequence relation. This is shown in Gray's Figure 1 (reproduced here as Figure 21) where the λ Boo stars are represented by the large dots. FHB stars occupy the same area in this diagram. Other criteria, such as rotational and radial velocities, uv spectra, and detailed atmospheric abundances, will help distinguish one type from the other.

9. The identification of FHB stars

Some luminous stars of Population I fall in similar regions in the photometric diagrams to the regions occupied by Population II early-type stars. I have found that when faint FHB candidates are identified by four-color photometric measures they almost always turn out to be assigned correctly to Population II. But for bright A-type FHB candidates quite often the candidate turns out to be some other class of star. However, performing other measures will help make the classification of Population II more correct. Ultraviolet spectra provide a very good means of identifying FHB stars due to the distinct differences in the uv flux distribution and the presence or absence of lines. For a group of stars, the determination of the velocity dispersion, while it does not definitely identify every individual star in the group as a member of Population II, can definitely place the group in the Population II class. High-dispersion spectra provide a good way to distinguish Population II from Population I stars if an abundance determination is done. Low dispersion spectra can be used to determine the equivalent width of the Ca II line (λ 3933.66). In a (B-V) vs Ca II equivalent width plot the FHB stars fall well below the Population I relation. FHB stars are slow rotators so the spectral lines should not be rotationally broadened if the star is to be a Population II FHB star. The maximum rotational velocities so far found by Peterson (1985, 1994) in any globular cluster are about 30 km s^{-1}, so FHB candidates with rotational velocities well above this limit are suspect.

Hayes & Philip (1979), Philip (1985), and Philip & Adelman (1991) discuss aspects of the identification of FHB stars.

10. Future Work

A major unsolved problem in the galactic halo is the precise description of the evolutionary history of the apparently normal, early-type stars found far from the galactic plane. Some of these stars may be blue stragglers. Other investigators have suggested that an encounter between the Milky Way and a third Magellanic-type object produced these stars (Rodgers *et al.* 1981).

Although the FHB stars seem to be close photometric analogs of the BHB stars in globular clusters there are some spectroscopic differences that have been found between the two groups of stars. As has been pointed out, the [Fe/H] values for FHB stars show a great scatter and it would be an interesting project for someone to obtain high-dispersion spectra of all the confirmed FHB stars and then perform a modern stellar abundance analysis on each of the stars. Much of the scatter in the published [Fe/H] values can be due to the differences in the observed spectra, taken with different spectrographs on different telescopes and then followed by analysis programs that use different assumptions in the reduction of the data.

In an interesting paper given at this meeting, Corbally & Gray (1994) presented a report of spectra obtained of the FHB stars listed in VVO Contribution No. 2. Two of the stars had the expected spectral characteristics of horizontal-branch stars but the remaining stars showed broader hydrogen lines than expected. Among these stars were some of the classical FHB stars. Corbally & Gray suggested that this might be a sign of a greater He abundance. More work should be done on stars in the FHB star group. Spectra should be obtained of the newly classified FHB stars and spectra at higher dispersion should be taken of all the brighter stars. It would also be interesting to obtain spectra of BHB stars in globular clusters to match the Corbally & Gray studies so the two groups of stars can be compared.

REFERENCES

ALBITZKY, A. V., 1933, Poulkova Obs. Circ. No. 7
ADELMAN, S. J., FISHER, W. A., HILL, G., 1987, Publ. Dom. Astrophs. Obs. Victoria, 16, 203
ADELMAN, S. J., HILL, G., 1987, MNRAS, 226, 581
ADELMAN, S. J., PHILIP, A. G. D., 1990, MNRAS, 247, 132

ADELMAN, S. J., PHILIP, A. G. D., 1990, PASP, 102, 842
ADELMAN, S. J., PHILIP, A. G. D., 1992, MNRAS, 254, 539
ADELMAN, S. J., PHILIP, A. G. D., 1994, in Adelman S. J., Upgren A. R., Adelman C. J., eds, Hot Stars in the Halo, Cambridge University Press, p. 266
AURIÈRE, M., ADAMS, S., SEATON, M. J., 1983, MNRAS, 205, 571
BEERS, T. C., PRESTON, G. W., SCHECTMAN, S. A., 1988, ApJS, 67, 461
BERGER, J., CHALONGE, D., DIVAN, L., FRINGANT, A. -M., WESTERLUND, B., 1958, J. Obs., 41, 100
BOND, H. E., PHILIP, A. G. D., 1973, PASP, 85, 332
BURBIDGE, E. M., BURBIDGE, G. R., 1956, ApJ, 124, 116
CACCIARI, C., 1985, A&AS, 61, 407
CACCIARI, C., MALAGNINI, M. L., MOROSSI, C., ROSSI, L., 1987, A&A, 183, 314
CAYREL DE STROBEL, G., BENTOLILA, C. HAUCK, B., DUQUENNCOUDEOY, A., 1985, A&AS, 59, 145
CHAVIRA, E., 1958, Bol. Obs. Tonantzintla y Tacubaya, 17, 15
CHRISTENSEN, C. G., 1978, AJ, 83, 244
CORBALLY, G., GRAY, R. O., 1994, in Adelman S. J., Upgren A. R., Adelman C. J., eds, Hot Stars in the Halo, Cambridge University Press, Cambridge, p. 253
DANFORD, S. C., LEA, S. M., 1981, AJ, 86, 1909
DE BOER, K. S., CODE, A. D., 1981, ApJ, 243, L33
DE BOER, K. S., WESSELIUS, P. R., 1980, AJ, 85, 1354
DRILLING, J. S., PESCH, P., 1973, AJ, 78, 47
EGGEN, O. J., 1970, Vistas Astro., 12, 367
FEIGE, J., 1958, ApJ, 128, 267
FARAGGIANA, R., GERBALDI, M., BÖHM, C., 1990, A&A, 235, 311
GILMORE, G., REID, N., 1983, MNRAS, 202, 1025
GRAHAM, J. A., 1970, PASP, 82, 1305
GRAHAM, J. A., DOREMUS, C., 1968, AJ, 73, 226
GRAY, R. O., 1988, AJ, 95, 220
GRAY, R. O., 1991, in Philip A. G. D., Upgren A. R, Janes K. A., eds, Precision Photometry, L. Davis Press, Schenectady, p. 309
GREENSTEIN, J. L., 1956 in Neyman J., ed, Third Berkeley Symposium on Mathematical Statistics and Probability, University of California, Berkeley, p. 11
GREENSTEIN, J. L., 1966, ApJ, 144, 496
GREENSTEIN, J. L., SARGENT, A. I., 1974, ApJS, 28, 157
HARO, G., LUYTEN, W. J., 1962, Bol. Obs. Tonantzintla y Tacubaya, 22, 35
HAUCK, B., 1994, in Adelman S. J., Upgren A. R., Adelman C. J., eds, Hot Stars in the Halo, Cambridge University Press, Cambridge, p. 245
HAYES, D. S., PHILIP, A. G. D., 1979, PASP, 91, 71
HAYES, D. S., PHILIP, A. G. D. 1983, ApJS, 53, 759
HUENEMOERDER, D. P., DE BOER, K. S., CODE, A. D., 1984, AJ, 89, 851
HUMASON, M. L., ZWICKY, G., 1947, ApJ, 105, 85
IRIARTE, B., CHAVIRA, E., 1957, Bol. Obs. Tonantzintla y Tacubaya, 16, 3
JONES, D. H. P., 1966, in Lodèn K., Lodèn L. O., Sinnerstad U., eds, IAU Symposium No. 24, Spectral Classification and Multicolour Photometry, Academic Press, NY, p. 141
KILKENNY, D., HILL, P. W., 1975, MNRAS, 173, 625
KINMAN, T. D., 1965, ApJS, 11, 199
KINMAN, T. D., WIRTANEN, C. A., JANES, K. A., 1965, ApJS, 11, 223

KINMAN, T. D., WIRTANEN, C. A., JANES, K. A., 1966, ApJS, 13, 320
KLEMOLA, A. R., 1962, AJ, 67, 740
KLOCHOVA, V. G. PANCHUK, V. E., 1985, AZh, 62, 552
KODAIRA, K., 1964, Z. Astrophysics, 59, 139
KODAIRA, K., 1973, A&A, 22, 273
KODAIRA, K., PHILIP, A. G. D., 1984a, ApJ 278, 201
KODAIRA, K., PHILIP, A. G. D., 1984b, ApJ 278, 208
LEE, J. T., 1985, Master's Thesis, Wesleyan University
LUCK, R. E., 1991, in Michaud G., Tutukov, A., eds, IAU Symposium No. 145, Evolution of Stars: The Photospheric Abundance Connection, Kluwer, Dordrecht, p. 247
LUYTEN, W. J., (multiple), various reports in Observatory of Minnesota
LUYTEN, W. J., 1957, A Search for Faint Blue Stars, University of Minnesota Observatory, Minneapolis, No. 9
LUYTEN, W. J., 1964, First Conference on Faint Blue Stars, The Obs. of University of Minnesota
LUYTEN, W. J., 1965 in Blaauw A., Schmidt M., eds, Galactic Structure, University of Chicago Press, Chicago, p. 393
LUYTEN, W. J., ANDERSON, J. H., 1958, A Search for Faint Blue Stars, University of Minnesota Observatory, Minneapolis, No. 12
LUYTEN, W. J., ANDERSON, J. H., 1959, A Search for Faint Blue Stars, University of Minnesota Observatory, Minneapolis, No. 18
MACCONNELL, D. J., FRYE, R. L., BIDELMAN, W. P., BOND, H. E., 1971, PASP, 83, 98
MCCUSKEY, S. W., 1965, in Blaauw A., Schmidt M., eds, Galactic Structure, University of Chicago Press, Chicago, p. 1
MERRILL, P. W., 1947, PASP, 59, 256
MORGAN, W. W., KEENAN, P. C., KELLMAN, E., 1943, An Atlas of Stellar Spectra, University of Chicago, Chicago
NEWELL, E. B., 1969, Ph.D. Thesis, Yale University
NEWELL, E. B., 1973, ApJS, 228, 26
OKE, J. B., GREENSTEIN, J. L., GUNN, J., 1966, in Stein R. F., Cameron A. G. W., eds., Stellar Evolution, Plenum Press, New York, p. 399
PARENAGO, P. P., 1958, Soviet Astron. −AJ, 2, 151
PESCH, P., SANDULEAK, N., 1989, ApJS, 71, 549
PETERSON, R. C., 1985, in Philip A. G. D., ed, Horizontal-Branch and UV-Bright Stars, L. Davis Press, Schenectady, p. 85
PETERSON, R. C., 1994, in Adelman S. J., Upgren A. R., Adelman C. J., eds, Hot Stars in the Halo, Cambridge University Press, Cambridge, p. 319
PHILIP, A. G. D., 1968, AJ, 73, 1000
PHILIP, A. G. D., 1969, ApJ, 158, L113
PHILIP, A. G. D., 1970, AJ, 75, 246
PHILIP, A. G. D. 1972, in Haug U., ed, The Role of Schmidt Telescopes in Astronomy, Hamburger Sternwarte, p. 117
PHILIP, A. G. D., 1973, ApJ, 182, 517
PHILIP, A. G. D., 1974, ApJ, 190, 573
PHILIP, A. G. D., 1978, in Philip A. G. D., Hayes D. S., eds, The HR Diagram, Reidel, Dordrecht, p. 209
PHILIP, A. G. D., 1984, Contr. of Van Vleck Obs., 2, 1
PHILIP, A. G. D., 1985, in Philip A. G. D., ed, Horizontal-Branch and UV Bright Stars, L. Davis Press, Schenectady, p. 41
PHILIP, A. G. D., 1990, in Philip A. G. D., Hayes D. S., Adelman S. J., eds, CCDs in Astronomy.

II, L. Davis Press, Schenectady, p. 107
PHILIP, A. G. D., 1994, unpublished
PHILIP, A. G. D., ADELMAN, S. J., 1991, PASP, 103, 63
PHILIP, A. G. D., HAUCK, B., 1984, in Garrison R. F., ed, The MK Process and Stellar Classification, University of Toronto, Toronto, p. 243
PHILIP, A. G. D., HAYES, D. S., 1983, ApJS, 53, 751
PHILIP, A. G. D., HAYES, D. S., ADELMAN, S. J., 1990, PASP, 102, 649
PHILIP, A. G. D., HAYES, D. S., HUENEMOERDER, D. P., 1987, PASP, 99, 54
PHILIP, A. G. D., HAYES, D. S., LIEBERT, J. W., 1987, IAU Colloquium No. 95, The Second Conference on Faint Blue Stars, L. Davis Press, Schenectady
PHILIP, A. G. D., SANDULEAK, N., 1968, Bol. Obs. Tonantzintla y Tacubaya, 30, 253
PHILIP. A. G. D., TIFFT, L. E., 1968, AJ, 76, 567
PIER, J. R., 1982, AJ, 87, 1515
PIER, J. R., 1983, ApJS, 53, 791
PRESTON, G. W., SCHECTMAN, S. A., BEERS, T. C., 1991a, ApJS, 76, 1001
PRESTON, G. W., SCHECTMAN, S. A., BEERS, T. C., 1991b, ApJ, 375, 121
RODGERS, A. W., 1971, ApJ, 165, 581
RODGERS, A. W., 1972, MNRAS, 157, 171
RODGERS, A. W., HARDING, P., SADLER, E., 1981, ApJ, 244, 912
ROMAN, N., 1955, ApJS, 2, 195
SLETTEBAK, A., 1952, ApJ, 115, 576
SLETTEBAK, A. BAHNER, K., STOCK, J., 1961, ApJ, 134, 195
SLETTEBAK, A., BRUNDAGE, R. K., 1971, AJ, 76, 338
SLETTEBAK, A., STOCK, J., 1959 Hamburg-Bergedorf Astr. Abh. 5, No. 5
STALIO, R., 1974, A&A, 31, 89
STETSON, P. B., 1981a, AJ, 86, 1337
STETSON, P. B., 1981b, AJ, 86, 1882
STETSON, P. B., 1983, AJ, 88, 1349
STETSON, P. B., 1991, AJ, 102, 589
STRAIŽYS, V., KAZLAUSKAS, A., 1993, Baltic Astron., 2, 1
STRÖMGREN, B., 1952, AJ, 57, 200
STRÖMGREN, B., 1955, in Beer A., ed, Vistas Astron., 2, p. 1336
STRÖMGREN, B.. 1963 in Strand K. Aa., ed, Basic Astronomical Data, Vol. III of Stars and Stellar Systems, University of Chicago Press, Chicago, p. 123
SVETLOVA, L. P., 1946, Astron. J. - USSR, 23, 147
SWEIGART, A. V., 1987, ApJS, 65, 95
SWEIGART, A. V., 1994, in Adelman S. J., Upgren A. R., Adelman C. J., eds., Hot Stars in the Halo, Cambridge University Press, Cambridge, p. 17
TOBIN, W., KILKENNY, D., 1981, MNRAS, 194, 937
UPGREN, A. R., 1962, AJ, 67, 37
UPGREN, A. R., 1963, AJ, 68, 194
WALLERSTEIN, G., HUNZIKER W., 1964, ApJ, 140, 214

Discussion

GARRISON: The stars you have identified as FHB stars have always made me wonder about their origin and whether there are other criteria that come to bear which limit the sample. For example, there may be solar abundance HB stars which you are missing. If I gave you spectra of two stars, one of $2M_\odot$ and just post main sequence, the other one of less than $1\ M_\odot$ and post AGB, but both with solar abundance, would you be able to tell them apart, without radial velocity information or location information? This is important - see Corbally & Gray in this volume. Can a low mass, post AGB star successfully and completely mimic a high mass ($2M_\odot$) star before it becomes a red giant?

PHILIP: I have been identifying stars which match the photometric qualities of BHB stars in globular clusters. They have high c_1 indices, low m_1 indices, and low metal abundances. Radial velocity studies help confirm that they are indeed members of Population II. So this process will not identify the solar abundance HB stars that you propose. In my 4-color work on faint A stars at the galactic poles a group of apparently normal, Population I A type stars has been identified. Rodgers and his collaborators have been studying these and other, normal metallicity A type stars in a series of papers.

YOSS: Is not Garrison asking about the A type stars with solar abundances?

PHILIP: I believe Bob was asking about a more evolved, Population I star, that would be in the horizontal branch. As I have said, there are apparently normal main sequence A type stars at high galactic latitudes, far off the galactic plane.

Surveys

A Progress Report on the Edinburgh–Cape Blue Object Survey

By D. KILKENNY[1], D. O'DONOGHUE[2], R. S. STOBIE[1], A. L. CHEN[2], C. KOEN[1], AND A. SAVAGE[3]

[1]South African Astronomical Observatory, Observatory 7935, SOUTH AFRICA

[2]Department of Astronomy, University of Cape Town, Rondebosch 7700, SOUTH AFRICA

[3]Anglo-Australian Observatory, PO Box 296, Epping, NSW 2121, AUSTRALIA

A brief review is given of the Edinburgh-Cape blue object survey and the progress to date. The completed survey should cover around 10,000 square degrees at high galactic latitudes in the southern hemisphere to a limiting magnitude of B $\sim$ 18 mag. Low dispersion (100 Å mm^{-1}) spectroscopy and UBV photometry are being obtained for all objects brighter than B = 16.5 mag. Preliminary results for some classes of object and for some of the more interesting individual discoveries are described.

1. Introduction

The publication of the Palomar-Green (PG) catalogue of ultraviolet-excess objects (Green, Schmidt & Liebert 1986) has been of fundamental importance to the study of evolved hot stars. Many new and interesting objects have been studied as a result of the PG survey, and new classes of object have been discovered – such as the pulsating hot white dwarf (or "PG1159") stars. One result of the success of the PG survey has been to increase the imbalance between the numbers of evolved hot objects known in the northern and southern hemispheres; the Edinburgh-Cape survey was started as an attempt to redress the balance and, hopefully, to provide many new southern objects for further investigation.

2. The EC survey – a brief description and status report

Fairly detailed descriptions of the procedures being followed in the EC survey have been given by Stobie et al. (1987, 1992). Briefly, U and B plates are being obtained with the 1.2m UK Schmidt telescope on the Anglo-Australian site at Siding Spring in Australia. The U plates are 60 min exposures on IIaO with a UG1 filter; the B plates, which are 15 min exposures on IIaO with a GG385 filter, go deeper than the U plates (to B $\sim$ 20 mag) and are used for star/galaxy separation. The plates are digitised by the COSMOS machine (MacGillivray & Stobie 1984) at the Royal Observatory, Edinburgh which produces parameters related to position, size and shape of the images. As noted above, the B plates are used for star/galaxy separation, which is largely successful brighter than B $\sim$ 18 mag on the EC survey plates. The natural system B magnitudes and (U-B) colours are then used to select the blue objects (see Stobie et al. 1992 for a more detailed description). Typically, we find about 50–60 very blue objects per field ($\sim$28.5 square degrees) and all objects are visually inspected to eliminate spurious images such as plate flaws, images affected by satellite and asteroid trails and the like.

As currently envisaged, the EC survey will cover the $\sim$10,000 square degrees of sky south of (plate centres) $\delta = -15°$ at galactic latitudes in excess of 30°. To date, we have

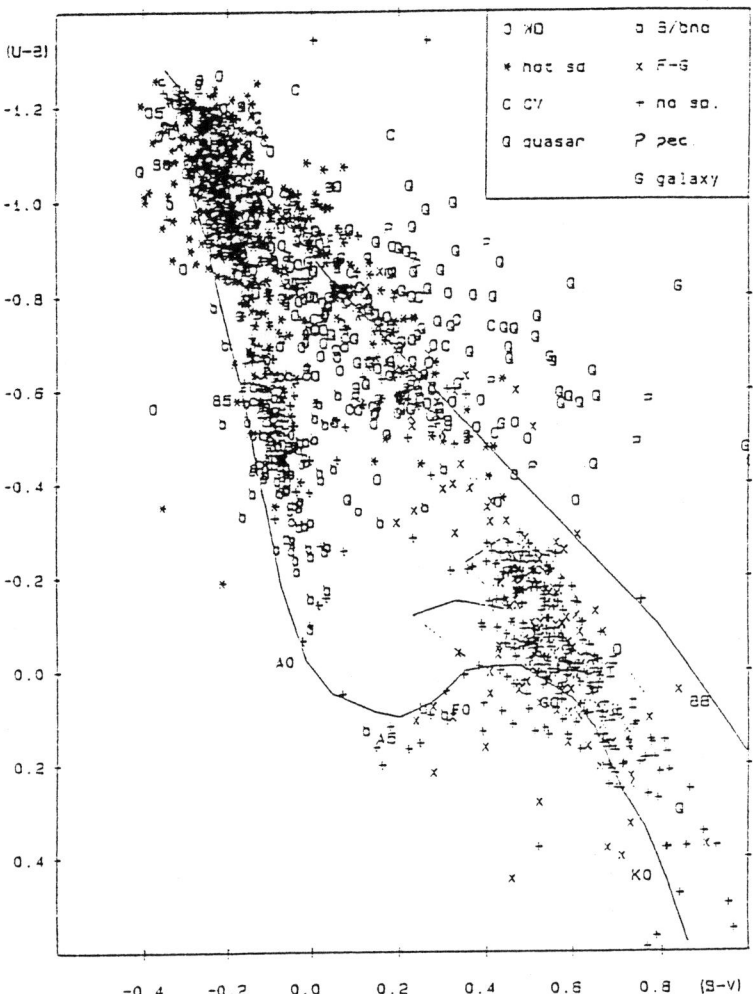

FIGURE 1. Two-colour diagram (U–B/B–V) for the EC stars. The intrinsic colour line is shown and the spectral types O5, B0, B5, A0, etc. indicate the approximate intrinsic colours of those types. 'BB' indicates the black-body line; many of the objects above this line are quasars or galaxies with prominent spectral emission features.

plate pairs for over 230 fields (about 70 % of the total in the survey) almost 200 of which have been processed by COSMOS for blue object selection.

The second stage of the survey is carried out using telescopes of the South African Astronomical Observatory (SAAO) and involves obtaining at least one UBV measurement (see Figure 1) and a low-dispersion (100 Å mm^{-1}) spectrogram for objects brighter than 16.5 mag. The photometry gives us:

- a preliminary "classification" of each object. It is clear from the UBV data that some of our sample are metal-weak F–G stars (sdFG) and we do not, in general, pursue these stars further;

- confirmation in each field that we are reaching our target limiting magnitude (B=16.5 mag);

- a possibility of detecting variability – this is not applicable to most objects, since initially we are only making single observations, but we have detected some interesting variables in a few cases when extra UBV measures have been made;

- the possibility of calibrating retrospectively the photographic material. Darragh O'Donoghue has already put considerable effort into using the UBV photoelectric photometry we have obtained to investigate the photographic U,B photometry which forms the basis of the EC survey (O'Donoghue et al. 1993). His conclusions are that the 1σ errors in B are ∼0.2 mag and in (U-B) are ∼0.16 mag (cf. ∼0.38 in the PG survey) and that the photographic selection of blue objects appears to work well. (The reason that the 1σ errors in (U-B) are smaller than the errors in B (or U) is that the photographic errors are correlated). The relatively large numbers of sdFG stars which we detect (around 17 % of all objects selected) are picked up because of the non-Gaussian distribution of errors in the photographic photometry; we are seeing the tip of a long "tail" in the error distribution of the photometry of ∼30,000 sdFG stars per plate.

The spectroscopy is used for:

- classification (DA, DB, CV, sdB, sdO, Quasar, ... , pec, ...) – a good range of typical spectrograms can be found in Stobie et al. 1992);

- radial velocities – in principle, we should be able to get velocities accurate to about 10–20 km s^{-1} from our data, although we have not attempted this is any systematic way, as yet. These data should be useful for statistical studies and, for example, testing the suspected binary stars in the survey;

- crude analysis (temperature, surface gravity). Although our spectra are really not of adequate signal/noise or resolution, we can use them for preliminary analysis, as has been done by Kilkenny et al. (1991) for a sample of apparently normal B stars at high galactic latitudes detected in the EC survey.

By the end of 1992, we had spectroscopy and photoelectric photometry for over 1600 objects brighter than about B = 16.5 mag and either photometry or spectroscopy (but not yet both) for a further 300 objects; this is about equivalent to the northern PG survey (Green, Schmidt & Liebert 1986). Overall, the EC survey should detect about 20,000 stars brighter than B ∼ 18 mag, and we should find about 5000 stars in our follow-up of stars brighter than B = 16.5 mag. We have, thus, completed about a half of "phase I" (obtaining and processing the photographic material) about a third of "phase II" (follow-up UBV and spectroscopy) and have only really just begun to work on "phase III" (detailed investigations of interesting objects, either as individuals or classes). Some phase III results will be described below.

Since the whole survey will take several years to complete, it is our plan to publish the results of phases I and II in zones with around 50–70 fields in each zone. The zones will be: galactic latitudes $\geq +30°$, -30° to -40°, -40° to -50°, -50° to -60°, -60° to -70°, and -70° to -90°. A further zone covering that part of the South Galactic Cap with plate centres in the range $0° \geq \delta \geq -10°$ was planned, but will depend on the available time. The first zone (b $\geq +30°$ and south of $\delta = -15°$ – that part of the North Galactic Cap not covered by the PG survey) is essentially completed and is being prepared for publication.

From objects observed spectroscopically and for which we have reduced data, we find the following percentages (excluding the sdFG stars, which comprise about 17 % of the initial numbers):

Type of object	Number	(%)
Hot subdwarfs	659	50
White dwarfs	210	16
Cataclysmic variables	28	2
Quasars & QSOs	88	7
Galaxies	9	1
Apparently normal B stars	138	10
Horizontal-branch stars	120	9
Peculiar/other	66	5

We are finding more horizontal-branch stars than the PG survey, probably because our cut-off in (U-B) is somewhat redder. We also find a somewhat smaller percentage of white dwarfs, possibly because some of these stars (perhaps DC types, for example) are, as yet, unclassified and have been lumped in the "peculiar/other" category. The few galaxies we find are all strong emission-line objects where the bulk of the light comes from the nucleus, making them appear star-like on the coarse-grained IIaO Schmidt plates, which is presumably why they are not eliminated in the star/galaxy separation process. It is highly likely that some of the apparently normal B stars will turn out to be hot subdwarfs or horizontal-branch stars when detailed atmospheric analyses can be performed.

3. Results – classes of object

So far, we are not really in a position to make definitive statistical studies (one difficulty is that we have not yet made quantitative measures of the completeness of the survey) but we have carried out some preliminary investigations.

An initial study of the bright Quasars and QSOs in our sample has been carried out by Savage et al. (1993). A comparison with the Hewitt & Burbidge (1987) catalogue suggests that the EC survey could be incomplete for bright QSOs by up to 30 % (but the numbers are small and preliminary studies of white dwarf samples suggest a statistically complete sample of these objects to a limit of B = 16.4 mag.) On the other hand, out of the total sample of 72 EC bright QSOs detected at the time the Savage et al. study was done, only 10 were previously known – indicating the general incompleteness of bright southern QSO samples. Taking the completed EC fields in the North Galactic Cap (61 fields, 48 QSOs, ∼1500 square degrees), Savage et al. (1993) find a number density of about twice that found in the PG survey and more comparable with the results of Goldschmidt et al. (1992) from a much smaller area and sample of objects.

To date, there are 28 candidate cataclysmic variables (CVs) in the EC survey material. Their spectra show some combination of H, He I and He II in emission or broad Balmer absorption. An-Le Chen and others have carried out "high-speed" photometry and, in some cases, time-resolved spectroscopy and have so far confirmed 22 objects as CVs. Seven of these were previously known and the fifteen new objects include six nova-like variables and three dwarf novae. By comparison with the Ritter (1990) catalogue, we estimate the EC survey to be over 80 % complete in detecting CVs.

Over 130 EC stars appear to be normal B stars and therefore at some considerable distance from the galactic plane. Of course, it is possible that detailed analyses will show some of these stars to be actually hot subdwarfs or horizontal-branch stars, for example. A preliminary analysis is possible with the EC survey photometry and spectroscopy, and Kilkenny et al. (1990) have given a list of 20 candidates for further study; four appear to be B stars at large distances from the plane ($z \geq 5$ kpc), twelve have $1 \leq z \leq 4$ kpc and four are hot subdwarfs ($\log g \geq 4.5$). This sample of objects, selected from 33 completed EC fields, leads the authors to estimate that the total survey should discover a few hundred B stars with $1 \leq z \leq 4$ kpc and a few tens with $z \geq 5$ kpc. Thus, the Leonard & Duncan (1988, 1990) mechanism for cluster (binary-binary collision) ejection is plausible for the former stars and the Dyson & Hartquist (1983) model of star formation in high-velocity gas clouds is possible for the latter (since both mechanisms can only produce relatively small numbers of high-latitude B stars). A further sample of 30 apparently normal B stars is currently being prepared for publication.

4. Results – some individual objects

Only one of our candidate cataclysmic variables, EC19314-5915, has been found to be an eclipsing system. Buckley et al. (1992) have shown the system to have a period of 4.75 hr and that the spectrum shows emission lines characteristic of a dwarf nova or nova-like variable, but that there are also absorption lines present which are typical of a late G star. It appears that there is a third star in the system (other than the pair which form the CV binary) with spectral type ~G8, which yields a distance for the system of ~600 pc.

The spectrum of the object EC11575-1845 shows the C III/N III blend and Balmer lines in emission and was originally regarded as a cataclysmic variable. Photometry, however, shows the object to have a smoothly varying light curve (no "flickering") with a period of 7.86 hours (see Figure 3), very reminiscent of LSS 2018 (Drilling 1985). We now believe that the system consists of an sdO primary of very high temperature (~85,000 K to 100,000 K) and a late K- or early M-type secondary which is exhibiting an enormous reflection/heating effect. A preliminary analysis has been given by Chen et al. (1993) showing the object to be indeed very similar to LSS 2018, which is recognised as the central star of the planetary nebula DS1; EC11575-1845 is, however, not known to have any associated nebulosity.

A number of our white dwarf discoveries with (B-V) colours typical of the region of instability ($0.15 < (B-V) < 0.25$) have been investigated for variability. So far, only one, EC23487-2424, has been found to be a ZZ Ceti star. Stobie et al. (1993) have presented data from several nights which show periods in the range 800 – 1000 sec (see Figure 4). Two frequencies (near 1.0 and 1.2 mHz) are definitely present and at least two others are probably present. With a maximum amplitude of 0.24 mag, EC23487-2424 is a large amplitude ZZ Ceti star.

5. For the future

It is likely that for several years, a substantial fraction, probably the majority of our effort will go into completing phase II of the survey – the preliminary photometry and spectroscopy. But it is already the case that an increasing effort is being applied to phase III – the detailed follow-up work – as I have tried to indicate above. We are also

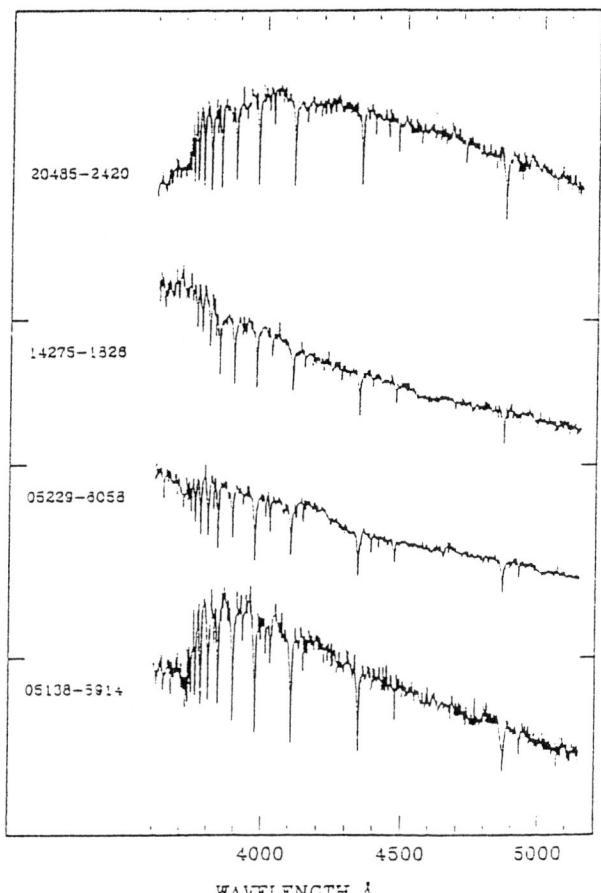

FIGURE 2. Spectrograms at 100 Å mm^{-1} for four apparently normal B stars from the EC survey with $z \geq 5$ kpc (Kilkenny et al. 1991). The spectrograms are flux-calibrated and ordinate marks indicate zero flux levels. Note that the criterion used by Moehler et al. (1990) for normality – comparable strengths of the He I lines at 4388 and 4471 Å – suggests that 14275–1826 and 05138–5914 could be subluminous.

exploring the possibility of extending phase II of the survey to fainter stars. We have had one run with FLAIR, a fibre array on the UK Schmidt telescope (see Watson et al. 1991, for example) – the data are still being reduced. Also, Mike Bessell at the Australian National University (ANU) has used the ANU 2.3m telescope to look at the fainter objects in a number of selected fields and reports finding a substantial increase (perhaps not unexpectedly) in the percentage of Quasars/QSOs in the EC survey fields. Results will hopefully soon be available from these areas.

A project such as the one described here would not be possible without considerable support. We gratefully acknowledge the staff at the UK Schmidt telescope on the Anglo-Australian site for the high quality photographic plates, and the staff of the COSMOS facility at the Royal Observatory, Edinburgh, for assistance with plate measurement. Successive Directors of SAAO have supported the survey with generous allocations of time on three telescopes.

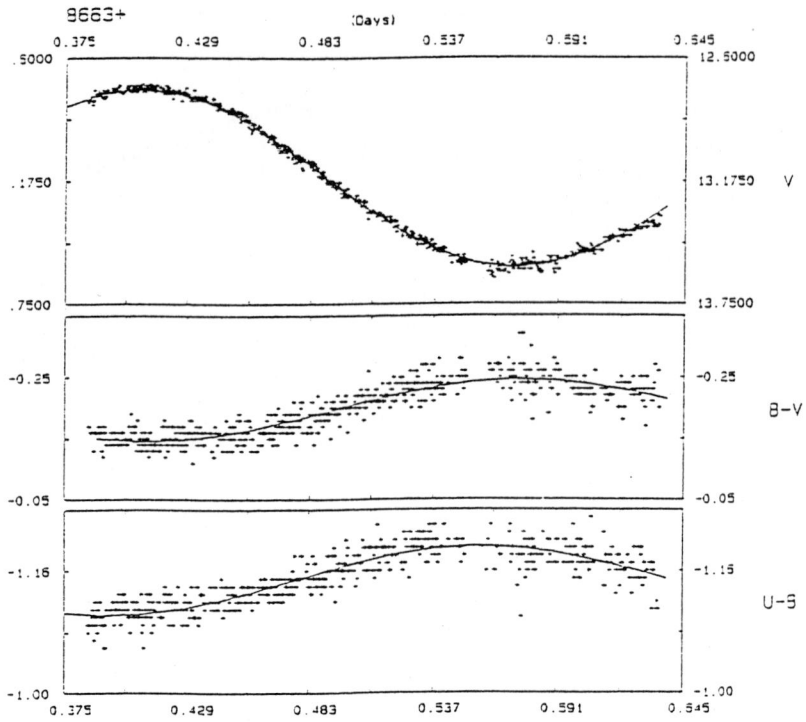

FIGURE 3. UBV photometry of EC11575-1845 obtained on 1992 Feb 10 using the University of Cape Town photometer on the SAAO 0.75m telescope

REFERENCES

BUCKLEY, D. A. H., O'DONOGHUE, D., KILKENNY, D., STOBIE, R. S., REMILLARD, R. A., 1992, MNRAS, 258, 285

CHEN, A., O'DONOGHUE, D., STOBIE, R. S., KILKENNY, D., ROBERTS, G., VAN WYK, F., 1993, in Kilkenny D., Lastovica E., Menzies J. W., eds, in Precision Photometry, SAAO, p. 200

DRILLING, J. S., 1985, ApJ, 294, L107

DYSON, J. F., HARTQUIST, T. W., 1983, MNRAS, 203, 1233

GOLDSCHMIDT, P., MILLER, L., LA FRANCA, F., CRISTIANI, S., 1992, MNRAS, 256, 65p

GREEN, R. F., SCHMIDT, M., LIEBERT, J., 1986, ApJS, 61, 305

HEWITT, A., BURBIDGE, G., 1987, ApJS, 63, 1

KILKENNY, D., O'DONOGHUE, D., STOBIE, R. S., 1991, MNRAS, 248, 664

LEONARD, P. J. T., DUNCAN, M. J., 1988, AJ, 96, 222

LEONARD, P. J. T., DUNCAN, M. J., 1990, AJ, 99, 608

MACGILLIVRAY, H. T., STOBIE, R. S., 1984, Vistas Astron., 27, 433

MOEHLER, S., RICHTLER, T., DE BOER K. S., DETTMAR, R. J., HEBER, U., 1990, A&AS, 86, 53

O'DONOGHUE, D., STOBIE, R. S., CHEN, A., KILKENNY, D., KOEN, C., 1993, in Kilkenny D., Lastovica E., Menzies J. W., eds, Precision Photometry, SAAO, p. 72

RITTER, H., 1990, A&AS, 85, 1179

SAVAGE, A., CANNON, R. D., STOBIE, R. S., KILKENNY, D., O'DONOGHUE, D., CHEN, A.,

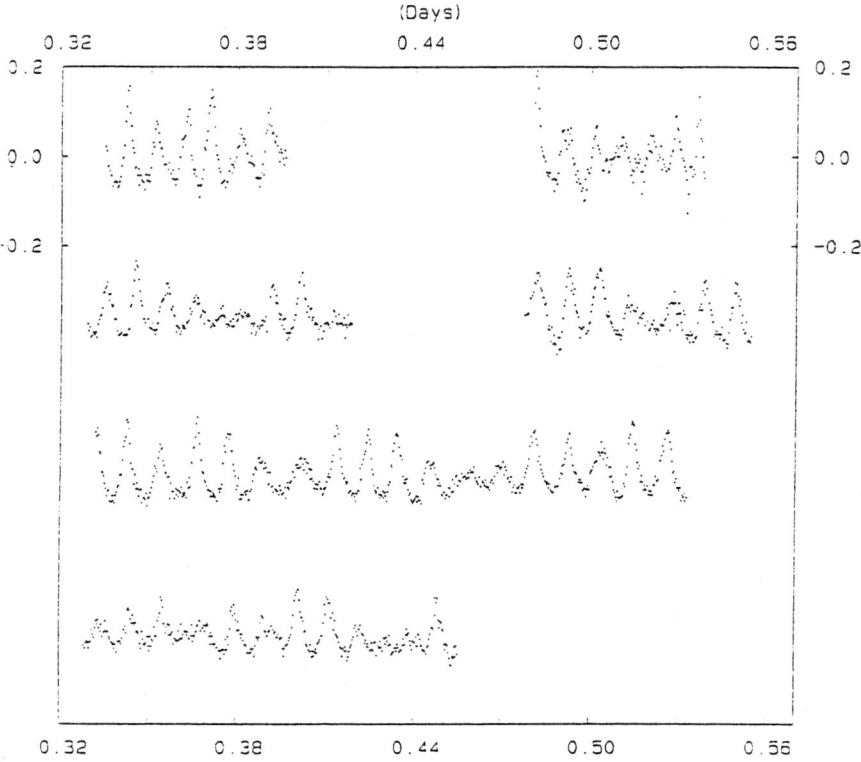

FIGURE 4. "White light" observations of EC23487−2424 on six nights (Stobie et al. 1993). The resolution of the plotted points is 30 seconds (binned from 10 second observations). The abscissa is in units of fractional intensity.

1993, Proc. Astron. Soc. Australia, 10, 265

STOBIE, R. S., MORGAN, D. H., BHATIA, R. K., KILKENNY, D., O'DONOGHUE, D., 1987, in Philip A. G. D., Hayes D. S., Liebert J. W., eds, IAU Colloq. 95, The Second Conference on Faint Blue Stars, p. 493

STOBIE, R. S., CHEN, A., O'DONOGHUE, D., KILKENNY, D., 1992, in Warner B., ed, Variable Stars and Galaxies, ASP Conf. Ser., 30, 87

STOBIE, R. S., CHEN, A., O'DONOGHUE, D., KILKENNY, D., 1993, MNRAS, 263, L13

WATSON, F. G., OATES, A. P., SHANKS, T., HALE-SUTTON, D., 1991, MNRAS, 253, 222

Discussion

SAFFER: In the sample of Bergeron, Saffer & Liebert (approximately 130 DA white dwarfs) (1992, ApJ, 394, 228), 10 % of the candidates proved upon analysis to be subdwarf B stars, misclassified in the Palomar Green Survey in the low S/N followup spectroscopy. This would bring the relative numbers of DAs in the Palomar Green and Edinburgh-Cape surveys into somewhat closer agreement.

KILKENNY: OK, that helps. Also, some stars in our "unclassified" category could be DC stars, for example, and the numbers are not really directly comparable because the Edinburgh Cape survey goes somewhat redder than the Palomar Green (so, more BHB stars are found in the Edinburgh Cape survey).

DORMAN: What is the solution for the mass and temperature of the sd in the system you called a K dwarf - subdwarf binary pair?

KILKENNY: EC 11575-1845 has a preliminary solution by An-LeChen which gives 0.4 $M_\odot$ and 90,000 K for the hot subdwarf primary, although I should stress that these are preliminary; the temperature, for example, is probably in the range 80,000 K $\leq$ T $\leq$ 100,000 K.

DRILLING: Could your paucity of white dwarfs with respect to the PG survey be due to our not being at the center of the disk?

KILKENNY: I do not know. The white dwarfs we find must certainly be very close by.

KINMAN: Are the B and U plates taken together? Do you find many variables?

KILKENNY: As far as I know, the U and B plates for each area are taken sequentially, but that still implies a difference of about an hour between mid-exposures. We are finding some variables (W UMa Stars) but it is not possible to say at present what percentage of our sample might be variable.

PHILIP: Has the availability of Kodak plates affected the survey plans, or do they have a good stock of plates for the Schmidt?

KILKENNY: We recently heard from the Schmidt unit that IIaO plates would be no longer produced. As I understand the position, these are enough plates (or a final order will be made). Otherwise, we might have to start using film-based emulsions, for example.

A 300 Square Degree Survey of Young Stars at High Galactic Latitudes

By J. EAMON LITTLE,[1] P. L. DUFTON,[1]
F. P. KEENAN,[1] N. C. HAMBLY,[1]
E. S. CONLON[1] AND L. MILLER[2]

[1] The Queen's University of Belfast, Belfast BT7 1NN, N. Ireland, U. K.

[2] Royal Observatory, Edinburgh, Scotland EH9 3HJ, U. K.

Final results from model atmosphere analyses of all young blue stars in a 300 square degree region of the galactic halo are presented. One star in our sample has an evolutionary age shorter than the flight-time required to reach its current halo position according to contemporary disc ejection models, implying the existence of some 200 such objects in the galaxy as a whole. A lower limit of 10,000 has also been estimated for runaway B-stars in the galactic halo.

1. Introduction

Since the first photographic survey of the galactic halo (Humason & Zwicky 1947) it has been known that hot stars, normally associated with star forming regions in the spiral arms, exist at high galactic latitudes. Blaauw (1961) considered this problem and proposed that these objects were runaway stars, ejected from binary systems following a supernova explosion. More recently, models that envisage ejection occurring from the cores of open clusters have been developed (Leonard & Duncan 1988; Clarke & Pringle 1991). There remains, however, some faint blue stars (Greenstein & Sargent 1974; Keenan et al. 1982; Tobin 1984; Kilkenny et al. 1991; Conlon et al. 1992) at high latitudes, whose apparent magnitudes imply they are at great distances above the plane (e.g., PG0832+676 at $z = 18$ kpc, Brown et al. 1989), which cannot be explained by simple disc ejection models.

Recently, Hambly et al. (1993) presented results for blue objects in the magnitude range $12 \leq V \leq 16.5$ in the 300 square degree region $b = 45° - 65°$; $l = 285° - 355°$. Here we supplement these data with observations of brighter targets in the same region to estimate the space density of B-type stars in the halo.

2. Observations and Analysis

High dispersion (5 Å mm^{-1}) optical spectra of 27 UKST bright survey stars ($V < 12$) identified as potential OB-type objects from UBV colours (Mitchell et al. 1990) were obtained during observing sessions between 15 and 22 April 1992 at the William Herschel (WHT) and Jacobus Kapetyn (JKT) Telescopes on La Palma.

Initial inspection of the spectra revealed 21 of our stars to be of spectral type A or later and these objects will not be discussed further. The remaining six spectra were normalised by fitting low order polynomials before determining metal and non-diffuse helium line equivalent widths by using non-linear least squares fitting (see Little et al. 1993 for details). A model atmosphere analysis was then performed using the line blanketed models of Kurucz (1991) to determine stellar atmospheric parameters and chemical compositions from which the evolutionary tracks of Maeder & Maynet (1988) were used to deduce solar mass ratios ($M/M_\odot$), absolute luminosities (log $L/L_\odot$) and hence distances (Z) above the galactic plane (see Table 1). Also listed in Table 1 are

Star	$M/M_\odot$	log $L/L_\odot$	Z (pc)	T_{ev}(max) (Myr)	T_f(min) (Myr)
HD 121968	11.5	4.09	3140	15	24
BD +02° 2711	6.0	3.00	1343	33	37
HD 120086	8.5	3.57	640	16	12
HD 118246	8.1	3.82	1270	200	31
UKST 1315+002	4.0	2.75	2980	200	34
HD 129956	3.2	2.34	128	200	10

TABLE 1. The UKST bright star survey results

minimum flight-times (T_f) for stars to attain their current Z-heights (assuming they have been ejected vertically out of the disc) and their maximum projected stellar evolutionary ages (T_f) (Maeder & Maynet 1988). This approach indicates that two objects: HD 121968 and BD +02° 2711 cannot be explained from disc ejection, although further investigation (see Conlon 1992 for details) reveals a possible disc ejection of BD +02° 2711 according to contemporary models. The five fainter objects investigated in paper I all have evolutionary times consistent with disc ejection.

3. Discussion

A 300 square degree region of the galactic halo in the magnitude range $6 \leq V \leq 16.6$ at galactic latitude b ~ 50° has been surveyed to detect young blue stars. Eleven main sequence objects have been identified, of which one (HD 121968) cannot be accounted for in terms of disc ejection models, implying the existence of some 200 such stars. The remaining stars in our sample with $z < 5$ kpc have been used to put a lower estimate of 10,000 to the amount of disc ejected runaway stars in the galaxy.

Finally, plotted in Figure 1 is the mass distribution for halo runaway stars having $z \leq 5$ kpc. Both supernova ejection (Blaauw 1961; Leonard & Dewey 1993) and binary-binary ejection (Leonard & Duncan 1988, 1989; Clark & Pringle 1992) models predict lower mass stars will receive larger ejection velocities, a claim supported by the negative mass gradient.

REFERENCES

BLAAUW, A., 1961, Bull. Astron. Inst. Netherlands, 15, 265
BROWN, P. J. F., DUFTON, P. L., KEENAN, F. P., BOKENSBERG, A., KING, D. L., PETTINI, M., 1989, ApJ, 339, 397
CLARKE, C. J., PRINGLE, J. E., 1992, MNRAS, 255, 423
CONLON, E. S., DUFTON, P. L., KEENAN, F. P., MCCAUSLAND, R. J. H., HOLMGREN, D. E., 1992, ApJ, 400, 273
DYSON, J. E., HARTQUIST, T. W., 1983, MNRAS, 203, 1233
GREENSTEIN, J. L., SARGENT, A. I., 1974, ApJS, 28, 157
HUMASON, M. L., ZWICKY, F., 1947, ApJ, 105, 85
KEENAN, F. P., DUFTON, P. L., MCKEITH, C. D., 1982, MNRAS, 200, 673
KILKENNY, D., O'DONOGHUE, D., STOBIE, R. S., 1991, MNRAS, 248, 664
KURUCZ, R.L., 1991, in Philip A. G. D., Upgren A. R., Janes K. A., eds, Astrophysics of the Galaxy, L. Davis Press, Schenectady

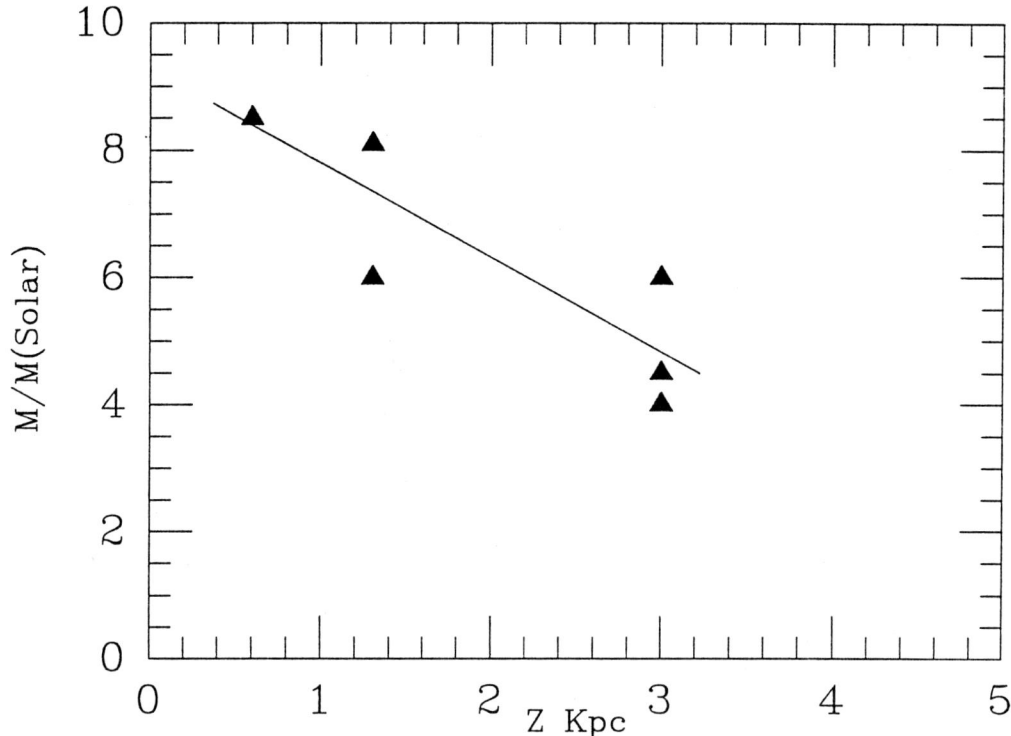

FIGURE 1. Plot of mass versus z-height for runaway stars in the UKST survey

LEONARD, P. J. T., DUNCAN, M. J., 1988, AJ, 96, 222
MAEDER, A., MAYNET, G., 1988, A&AS, 76, 411
MITCHELL, P. S., MILLER, L., BOYLE, B. J., 1990, MNRAS, 224, 1
TOBIN, W., 1984, A&AS, 56, 221

Discussion

JANES: In a survey such as yours, where you have covered a substantial area, if there are stars that have very recently formed, wouldn't you expect them to be clumped? It would seem incredible if stars were forming uniformly throughout the halo.

LITTLE: It is as yet unresolved whether these stars are being formed in the halo or acquire their large distances from the plane by some mechanism unknown to us. If they do form in the halo it is unlikely that they do so in isolation and work is currently underway to discover if clumping does occur or if lower mass stars lie in the vicinity of these objects.

The Isolation of a New Sample of B Stars in the Halo

By KENNETH J. MITCHELL[1], REX A. SAFFER[2], AND STEVE B. HOWELL[3]

[1]General Sciences Corporation, 6100 Chevy Chase Drive, Laurel, MD 20707, USA

[2]Space Telescope Science Institute†, 3700 San Martin Drive, Baltimore, MD 21218, USA

[3]Planetary Science Institute, 2421 E. 6th Street, Tucson, AZ 85719, USA

A new, complete sample of 22 faint B stars is presented. The B stars were isolated during low-resolution spectrophotometric observations of blue- and ultraviolet-excess objects selected by the US survey. Spectral classification of the B stars was aided by Strömgren colors derived from the spectrophotometry, and by atmospheric model fitting. The integral number counts derived from this sample show a shallow increase with magnitude. Further analysis of the sample must await a more accurate sub-classification of these stars.

1. Introduction

There are three major types of B stars being found in surveys for faint blue objects: horizontal branch B stars, apparently normal B stars, and post-asymptotic giant branch (post-AGB) B stars (e.g., Greenstein & Sargent 1974; Conlon et al. 1991). The evolved horizontal branch and post-AGB B stars could be expected to be found throughout the Galactic halo. However, the existence of some apparently normal B stars at relatively large distances from the Galactic plane poses potential problems for the canonical picture of star formation, evolution, and kinematics in the Galaxy (Tobin 1987; Keenan 1992). To obtain clues as to the origin and evolutionary history of the three different types of halo B stars, there is a need for studies of complete samples.

In this paper we present some preliminary results concerning the isolation of a new, complete sample of 22 faint B stars. The sample currently serves as a reliable candidate list for higher-quality, follow-up spectroscopy to accurately sub-classify the individual stars into the three B-star types mentioned above. Although relatively small, the present sample of B stars is of potential interest due to: (1) its faint magnitude completeness limit at $B \sim 18$; and (2) its "red" color completeness limits at $(U-V) = 0$ and $(B-V) = 0$. Section 2 of this paper reviews the methods employed to select and isolate this new B star sample, while Section 3 presents some initial analysis of the sample.

2. Selection and Spectroscopy of Faint Blue Star Candidates

2.1. The US Survey

The B stars in the present sample were initially selected for their blue and ultraviolet excess (B-UVX) as part of the US survey for faint blue, starlike objects (Usher et al. 1988, and references therein; Usher & Mitchell 1990). The US survey was conducted using 3-color (UBV) Palomar 1.2m Schmidt photographic plates, and the selection is nominally complete for objects with colors $(U-V) < 0$ or with ultraviolet excess relative to the blackbody line in the $(U-B)$, $(B-V)$ two color diagram. The claimed faint magnitude

† Operated by the Association of Universities for Research in Astronomy, Inc., for the National Aeronautics and Space Administration

completeness limits for the US survey fields range from B = 16.9 to B = 18.7. A literature search has helped to recover some bright, color-excess objects missed by the US survey due to problems with the recognition of bright, color-excess candidates on the 3-color plates (Mitchell & Usher 1994). The nominal bright magnitude completeness limit is therefore taken to be at B = 10.0.

The US survey covers ∼ 206 square degrees of sky within the central regions of seven high galactic-latitude ($|b| > 30°$) fields. Within this area, and within the magnitude completeness limits, the US catalogs can be expected to provide nominally complete samples of stellar objects with colors bluer and/or more ultraviolet than halo F/G subdwarfs and A stars. Even late B stars should therefore be catalogued with some completeness in the US survey.

2.2. The Spectrophotometric Survey

A spectrophotometric survey of the US catalogs reaching B-magnitude limits between B = 16.5 and B = 18.3 has been conducted using the KPNO 2.1m telescope and the Intensified Image Dissector Scanner and GoldCam instruments. Wide entrance apertures were used to obtain low resolution, moderate signal-to-noise spectra, with spectral coverage extending blueward of the Balmer jump. Most of the spectra have a spectral resolution of ≈ 14 Å and spectral coverage $\lambda\lambda 3500$–7000 Å. The quality of the resultant flux- and wavelength-calibrated spectra has allowed the stars to be classified into the major sub-groups (cf. Greenstein & Sargent 1974; Green et al. 1986) through visual inspection and comparison with observations of bright flux standards which spanned a wide range of hot star types. Preliminary samples of both QSOs and hot stars from the US survey have been previously reported in Mitchell et al. (1984) and Mitchell et al. (1987).

2.3. Isolation of the New B Star Sample

Although late B and early A star candidates were readily recognized in the survey spectrophotometry, the low spectral resolution and moderate signal-to-noise preclude an accurate visual MK classification. Strömgren colors and model atmosphere fitting (discussed below) have therefore been used to help separate the later B stars in the present sample from the cooler early A stars and the hotter early B and subdwarf B (sdB) stars. The current sample of B stars is bounded above at $T_{eff} = 22{,}500$ K, with further analysis needed to separate the hottest B stars from the sdB stars. The full sample of 23 US B stars is presented in Table 1; the 22 B stars with B > 10.0 comprise the complete sample. The B magnitudes in Table 1 have been corrected for extinction and vignetting where appropriate.

2.4. Numerically Convolved Stromgren Colors

The wide spectral coverage has allowed calibrated Strömgren color indices to be derived from the spectrophotometry through numerically convolved filters (Howell 1986). Although somewhat noisy relative to conventional photometry, these Strömgren colors for the US hot stars do show the expected trends with spectral type in Strömgren color-color diagrams (e.g., Kilkenny & Hill 1975). The Strömgren $[c_1]$ vs. (b–y) color-color diagram for the current B star sample, as well as for a number of A and F stars that have been observed, is shown in Figure 1. A few of the B stars show offsets in the direction of reddening (i.e., to more positive (b–y) values) which are larger than would be deduced from the Burstein & Heiles (1982) maps.

Figure 1 has been used to help with the final A/B star separation. Those A/B stars with $[c_1] > 0.9$ and (b–y) > −0.01 have been classified as A stars, a criterion which is

TABLE 1. US Halo B Stars with $T_{eff} < 22{,}500$ K

US #	B	$[c_1]$	Color T_{eff}	Model T_{eff}	σ_T	Model $\log g$	σ_g	Notes
96	15.1	0.195	19,600	17,700	500	4.4	0.1	
361	15.2	1.015	10,000	14,700	400	3.9	0.4	
437.9	14.6	0.422:	15,000	21,500	4700	4.5	0.7	(1)
710	17.4	0.252	18,000	19,100	800	5.0	0.2	
808	15.4	0.679	12,200	15,100	400	4.7	0.2	
1051	12.7	0.416	15,200	15,800	300	4.0	0.1	
1269	17.0	0.731	12,000	16,100	1200	4.4	0.4	
1280	16.4	1.217	9,800	14,900	300	5.2	0.3	
1749	17.3	0.916	11,000	14,300	100	3.8	0.1	
1810	11.9	0.685	12,200	14,900	200	4.3	0.1	
1909	17.0	1.073	9,800	14,100	100	3.8	0.2	
2550	14.0	0.271	17,700	17,900	800	4.3	0.2	
2941	16.3	0.753	11,700	15,500	1000	4.4	0.4	
3420.9	9.8	0.361	16,200	18,900	1200	4.0	0.2	(1)
3785	12.6	0.252	18,400	16,600	500	4.0	0.1	
3798	15.2	0.800	11,600	14,800	400	4.2	0.3	
3826	11.2	0.840	11,000	14,700	300	4.2	0.2	
3977	16.5	0.303	17,100	15,100	500	3.9	0.4	
3978	16.1	1.057	9,900	15,400	600	5.2	0.3	
3986	14.0	0.847	11,100	15,900	800	4.7	0.2	
SBS10	10.7	0.787	11,800	16,600	300	3.4	0.1	(2)
Feige 71	10.6	1.085	10,000	–	–	–	–	(2,3)
PB6998	11.5	–	10,000	–	–	–	–	(2,4)

Notes to Table 1
(1) These stars were selected in searches of the US survey plates subsequent to the original US survey (Mitchell & Usher 1994).
(2) These stars are in the US survey area, but were not selected by the US survey (most likely due to their bright magnitudes). They were found to be classified as B stars in the literature. Only SBS10 has been observed in the current spectrophotometric survey.
(3) Strömgren colors are from Graham (1970).
(4) Temperature estimate based on spectral type of "late B" given in Berger & Fringant (1980).

consistent with published results (Crawford 1978). This photometric separation can be expected to produce some classification error at the A/B star boundary of the present sample due to noise. However, we expect that approximately as many early A stars will be misclassified as late B stars as vice versa. Only better data can improve these spectral classifications.

Color temperatures have been derived for the US B stars using the observed reddening-free Strömgren $[c_1]$ color index and the theoretical temperarture–$[c_1]$ relations of Lester et al. (1986). The $[c_1]$ values and temperatures are listed in Table 1.

2.5. Model Atmosphere Fitting

The Balmer and any detected He I lines in the individual B star spectra were fit with a grid of LTE model atmospheres and synthetic spectra. The model grid (Bergeron et al. 1992; Saffer et al. 1994) is calculated for 20,000 K $< T_{eff} <$ 40,000 K and 4.0 $< \log$

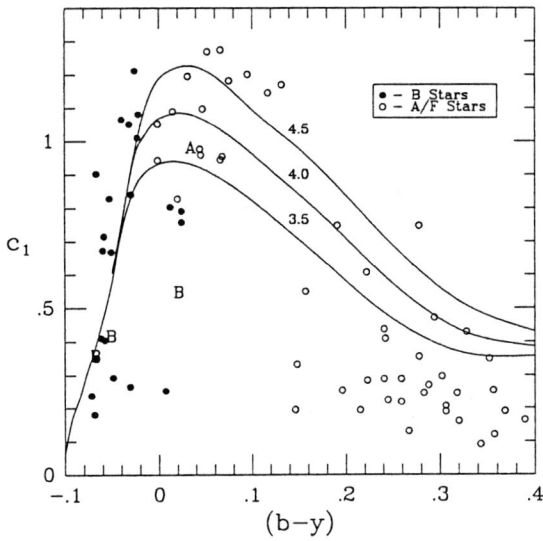

FIGURE 1. Strömgren $[c_1]$, (b–y) diagram for a sample of B, A, and F stars from the US survey. The Strömgren colors were numerically convolved from the spectrophotometric observations. The curves are theoretical relations from Lester et al. (1986), and are labeled with their log g values. The colors of four standard stars, used as flux and Strömgren-color calibrators, are marked by letters. The B and A stars follow the expected relations fairly well. The F star colors are known to suffer from systematic calibration problems.

g < 6.0, and has been convolved to the appropriate spectral resolutions. A non-linear, least-squares fitting technique produces synthetic profiles from the model grid which fit the observed data. An example of the model-fitting can be seen in Figure 2.

The temperature and gravity values which result from the model fitting are given in Table 1 and are plotted in the T_{eff} – log g diagram in Figure 3. The moderate S/N of the data in the line profiles result in moderately large fit errors. As can be seen through a comparison of the color and model temperatures in Table 1, the extrapolation of the model grid to temperatures below 20,000 K has produced an offset in the fitted T_{eff} and log g values for the US B stars. For this reason, the absolute placement of the B stars in Figure 3 is uncertain. In this study, the model fitting results have been used to help eliminate sdB and B stars with T > 22,500 K from the current sample; separation of the hottest B stars from the sdB stars is proceeding in ongoing work.

The relative placement of the stars in Figure 3 should be more reliable than their absolute placement. For example, a few of the lower gravity stars would be good candidates for post-AGB B stars. Two stars appear to have anomalously large surface gravities, which might be indicative of either (a) an attempt to fit lower temperature A stars with the hotter B star model atmosphere grid, or (b) the noise in the line profiles of the data.

3. Preliminary Results and Discussion

Table 1 presents derived parameters for the complete sample of 22 cooler B stars with 10.0 < B < 17.5, and with temperatures between the cool B star limit at spectral type A0 and ≈ 22,500 K. Although relatively small, the US B star sample is unique for its inclusion of the late B stars and its faint magnitude limit. The current data and analysis do not allow all of these stars to be accurately sub-classified into the three major types of

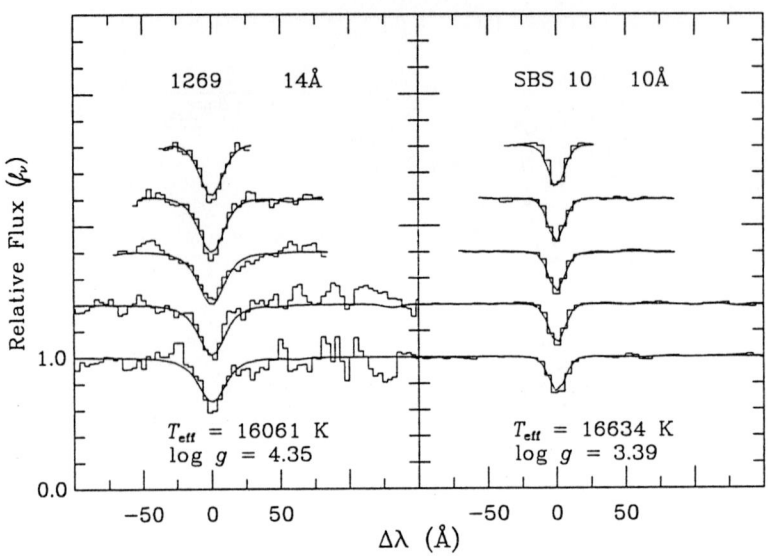

FIGURE 2. Model atmosphere fits to the Balmer lines of 2 US B stars. Smooth lines are the fitted synthetic line profiles. The Hβ (bottom) through H8 (top) profiles are fitted simultaneously using non-linear, least-squares techniques. Some moderately noisy data produce larger errors, but the fits generally appear satisfactory. The model grid is being extrapolated in temperature to fit these cooler B stars.

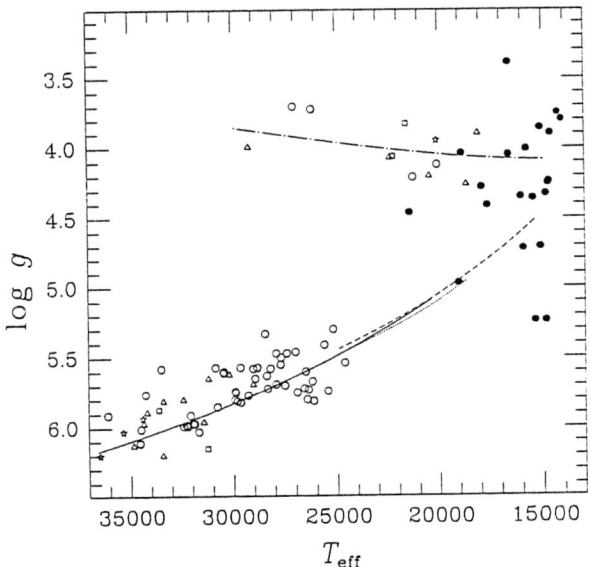

FIGURE 3. Placement of the US B stars in the T_{eff} – log g diagram based on the atmospheric model fitting results. The US B stars are indicated by solid circles. Open symbols represent PG hot stars analyzed by Saffer et al. (1994). Lines represent the main sequence, horizontal branch, and extended horizontal branch. Extrapolation of the model grid to cooler temperatures has produced offsets in the locations of the US B stars in T_{eff} and log g, although their relative placement should be reliable. Two stars appear to have anomalously large surface gravities which could be an indication that they are really A stars.

B stars being found in the halo, i.e., horizontal branch, apparently normal, and post-AGB stars. The lack of B stars within the present sample with B magnitudes fainter than 17.5 is not due to a lack of survey coverage to fainter limits. The spectrophotometric sample is complete over ≈ 119 square degrees at B = 17.8, and ≈ 88 square degrees at B = 18.0. This suggests a possible turnover in the counts of faint B stars at B ~ 17.5. The faintest B stars in this sample could be expected to be comprised of mainly horizontal branch B stars, which have the faintest absolute magnitudes for spectral types earlier than ≈ B9.

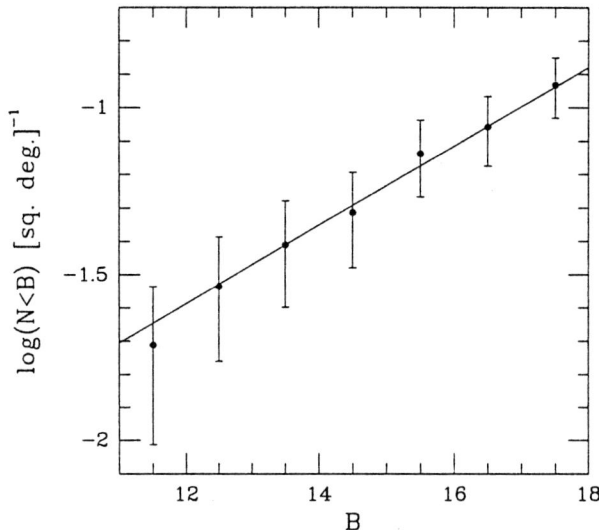

FIGURE 4. The integral number counts for the US B stars. The error bars show the Poisson counting errors that obtain at various limiting magnitudes, and are not independent. A relatively flat slope of +0.12 is indicated.

Knowledge of the completeness characteristics of the US and spectrophotometric surveys allows the integral number counts shown in Figure 4 to be derived. Although the error bars are large, the counts indicate a shallow increase with magnitude, exhibiting a slope ($d \log N(< B)/dB$) of only +0.12. The lack of an accurate value for the integral count at B = 10.0 will tend to make the present estimate of the slope steeper than it really is. This is the shallowest number count slope observed of any of the major hot star types isolated from the US survey, including DA white dwarfs, sdB stars and sdO stars (Mitchell et al. 1987). This would normally indicate a highly flattened spatial distribution centered on the Galactic disk. However, the unknown mix within the US B star sample of the three different types of B stars, each with its own characteristic absolute magnitude and, perhaps, spatial distribution, makes any interpretation of the integral counts highly uncertain at this point.

The surface density of B stars from the US sample can be compared to that recently derived from another B star sample. Kilkenny et al. (1991) have reported a preliminary sample of ≈ 20 B stars with 10.0 < B < 16.5 isolated within 924 square degrees of the Edinburgh-Cape faint blue object survey. Stobie et al. (1987) indicate that the selection cutoff for this survey is slightly redder than (U–B) = −0.40. This cutoff can be translated into an approximate Strömgren color completeness limit of $[c_1] < +0.60$ for their B star sample. The temperature upper limit currently imposed on the US B star sample corresponds to an approximate color limit of $[c_1] > +0.07$. The Edinburgh-

Cape sample contains 12 B stars within the following combined color and magnitude limits: $+0.07 < [c_1] < +0.60$ and $10.0 < B < 16.5$. The resulting surface density of $(12/924) \approx 0.013$ per square degree compares to a surface density of $(6/206) \approx 0.029$ per square degree for the US B stars within these same combined magnitude and color limits. The accuracy of these surface densities is low due to the small numbers of stars involved. Some of the apparent surface density discrepancy may be due to the fact that the Edinburgh-Cape sample is largely free of horizontal branch B stars, thus suggesting that $\approx 50\%$ of the US B stars could be horizontal branch B stars.

REFERENCES

BERGER, J., FRINGANT, A. -M., 1980, A&AS, 39, 39
BERGERON, P., SAFFER, R. A., LIEBERT, J. W., 1992, ApJ, 394, 228
BURSTEIN, D., HEILES, C., 1982, AJ, 87, 1165
CONLON, E. S., DUFTON, P. L., KEENAN, F. P., MCCAUSLAND, R. J. H., 1991, MNRAS, 248, 820
CRAWFORD, D. L., 1978, AJ, 83, 48
GRAHAM, J. A., 1970, PASP, 82, 1305
GREEN, R. F., SCHMIDT, M., LIEBERT, J., 1986, ApJS, 61, 305
GREENSTEIN, J. L., SARGENT, A. I., 1974, ApJS, 28, 157
HOWELL, S. B., 1986, AJ, 91, 171
KEENAN, F. P., 1992, QJRAS, 33, 325
KILKENNY, D., HILL, P. W., 1975, MNRAS, 173, 625
KILKENNY, D., O'DONOGHUE, D., STOBIE, R. S., 1991, MNRAS, 248, 669
LESTER, J. B., GRAY, R. O., KURUCZ, R. L., 1986, ApJS, 61, 509
MITCHELL, K. J., HOWELL, S. B., USHER, P. D., 1987 in Philip A. G. D., Hayes D. S., Liebert J. W., eds, The Second Conference on Faint Blue Stars, IAU Colloq. No. 95, L. Davis Press, Schenectady, NY, p. 513
MITCHELL, K. J., USHER, P. D., 1994, in preparation
MITCHELL, K. J., WARNOCK, A., III, USHER, P. D., 1984, ApJ, 287, L3
SAFFER, R. A., BERGERON, P., KOESTER, D., LIEBERT, J., 1994, ApJ, submitted
STOBIE, R. S., MORGAN, D. H., BHATIA, R. K., et al.. 1987, in Philip A. G. D., Hayes D. S., Liebert J. W., eds, The Second Conference on Faint Blue Stars, IAU Colloq. No. 95, L. Davis Press, Schenectady, NY, p. 493
TOBIN, W., 1987, in Philip A. G. D., Hayes D. S., Liebert J. W., eds, The Second Conference on Faint Blue Stars, IAU Colloq. No. 95, L. Davis Press, Schenectady, NY, p. 149
USHER, P. D., MITCHELL, K. J., 1990, ApJS, 74, 885
USHER, P. D., MITCHELL, K. J., WARNOCK, A., III, 1988, ApJS, 66, 1

Discussion

KILKENNY: Just a comment. The EC spectral types so far have been largely classified at the telescope so that stars we have called 'B' may well be later classified 'HB' (and possibly vice versa) when we make a more careful classification. So the statistics are somewhat uncertain as yet.

MITCHELL: Our statistics are also currently uncertain. We have not sub-classified the US Survey B stars yet. It will be interesting to compare the surface densities of the various B-star samples after the classifications become more firm.

LU: How large an area and where has the B-star survey been conducted? Anywhere near SGP? In the 40 square degree area at the SGP, no B-star brighter than 16.0 has been found.

MITCHELL: We have surveyed 206 square degrees, mostly in the Northern Galactic Polar Cap. Our statistics would indicate that you should have found about 3 B stars with $10.0 < B < 16.0$ over 40 square dgrees.

A Northern Catalog of Candidate FHB/A Stars

By TIMOTHY BEERS[1], RONALD WILHELM[1], AND STEPHEN DOINIDIS[2]

[1] Department of Physics and Astronomy, Michigan State University, E. Lansing, MI 48824, USA

[2] Department of Electrical Engineering, New Mexico State University, Las Cruces, NM 88003, USA

We discuss the assembly of a catalog of candidate field horizontal-branch and A-type stars in the northern Galactic hemisphere. After reviewing the techniques employed, we discuss what has been learned about the nature of these stars from previous analyses of catalogs of this type, and discuss what questions might profitably be explored with the complete database.

Every hour of every day I'm learning more
The more I learn the less I know about before
The less I know the more I want to look around
Digging deep for clues on higher ground

— "Higher Ground" (UB40, 1993)

1. Introduction – The Search Goes On

The snippet of verse above nicely captures the theme of this meeting. We astronomers in attendance at this meeting, and the many colleagues who could not make it, are all part of a search to understand the origin and evolution of faint blue stars in the disk and halo of our Galaxy. Others in this volume have summarized the history of this endeavor. We would like to discuss some possible directions for the future. Although a complete understanding of the origin of some of the hottest members of this blue star population (the sdB and sdO stars) is still beyond our grasp, we believe we are rapidly approaching consensus on the nature of the dominant constituent of this population, the field horizontal-branch (hereafter, FHB) stars. It is time to consider how we can exploit our understanding of these stars to yield clues about the formation and evolution of our Galaxy.

Stars in the horizontal-branch stage of their evolution present particularly attractive probes of the Galactic halo, in part due to their high intrinsic luminosities, but also due to their great numbers and the relative ease with which a large sample of these stars can be selected. We, along with our colleagues George Preston and Stephen Shectman of the Carnegie Observatories, have undertaken a survey which is particularly well-suited to the discovery of FHB (and other hot) stars. The first published catalog in this series (Beers, Preston & Shectman 1988, hereafter referred to as FHB I) identified some 4400 candidate FHB/A stars, primarily in the southern Galactic hemisphere. Followup spectroscopy and photometry of these stars is still far from complete, but many interesting results pertaining to the formation and evolution of the halo and thick-disk populations have already been obtained (Doinidis & Beers 1989; Preston, Shectman & Beers 1991b; Beers *et al.* 1992).

With the present catalog, we extend our sample to include an additional 3500 FHB/A candidates, primarily located in the northern Galactic hemisphere. Below we describe the methods used to identify candidate FHB/A stars, to obtain relative brightness estimates, and to measure positions. We conclude with synopsis of the analyses which have been completed to date, and a discussion of work in progress or to be undertaken in the near future.

2. Candidate Selection

Details of our objective-prism/interference-filter technique are given in Beers, Preston & Shectman (1985), FHB I, and Beers, Preston & Shectman (1992). The objective-prism plates in the present catalog were obtained with the Burrell Schmidt Telescope at Kitt Peak National Observatory, using the 4 degree prism and IIa-O emulsions, and a narrow-band interference filter which isolates roughly a 150 Å band centered on the H and K lines of Ca II. Note that for moderately hot stars the strength of the Hϵ line, which is nearly coincident with the Ca II H line, dominates, thereby enabling rather clear delineation of this particular class of star. Ninety minute exposures, widened to 0.2 mm, permit the identification of candidates to roughly a B magnitude of 15 to 15.5. Figure 1 shows the positions of the complete set of survey plates obtained to date. The dark squares in Figure 1 show the plates which are either included in the present catalog or have already had candidate FHB/A stars listed in FHB I; the open squares indicate the positions of plates which have yet to be scanned for candidates. The FHB/A candidate stars from these plates will be presented in a future catalog.

The stars included in the present catalog show weak or absent Ca II K but strong Hϵ, based upon visual inspection of the survey plates. When the Ca II K feature is discernible, the stars are typed class A. When no Ca II K feature is apparent, they are classified as type AB. Figure 2 is a cartoon illustrating the difference between classes A and AB. This distinction is more difficult to make for the fainter objects, where weak absorption at K can be more easily lost in the plate noise. A selection criteria which is based on the relative appearance of these two lines does lead to a discrimination against the more metal-rich stars, but will capture the numerous metal-weak halo stars which dominate more than a few kpc from the Galactic plane.

3. Brightness Estimates

As in FHB I, we assign discrete estimates of the brightness of each star based on the relative densities of spectra on each individual plate. Seven brightness classes are used: very bright (1), bright (2), medium bright (3), medium (4), medium faint (5), faint (6), and very faint (7). Stars which are judged to be ideally exposed are grouped into brightness class (4); stars brighter or fainter than the ideal case are divided by the classifier (in this case, Beers) into the remaining classes. As a result, the true apparent magnitudes of the catalog stars within any brightness class will vary from plate to plate.

Only a limited amount of followup photometry is presently available for the stars in our catalog (Doinidis & Beers 1990; Doinidis & Beers 1991). Figure 3 (left) is a box plot of the photoelectric V magnitudes measured for some 120 stars, grouped by their brightness classifications. A one-sigma spread in the magnitude estimates on the order of 0.5 magnitudes is seen.

We had hoped to take advantage of the large catalog of photographic photometry which is now available in the form of the Space Telescope Guide Star Catalog Volume 1.1 (hereafter, GSC). In time, as the calibration program now underway is completed, we can expect to obtain reasonable estimates of the apparent magnitudes for the stars

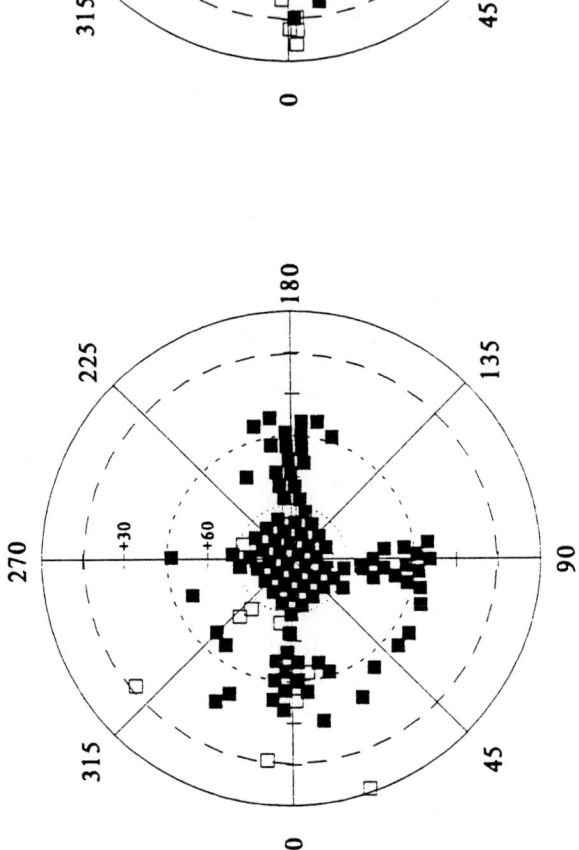

Figure 1. Positions of HK interference-filter/objective-prism plates on polar plots of the north and south Galactic hemispheres. The squares are drawn roughly to scale. Filled squares indicate plates which have been visually scanned for FHB/A candidates.

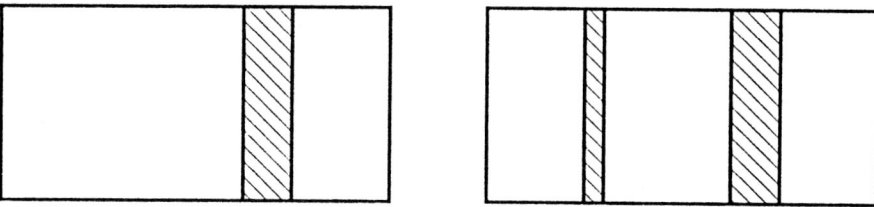

FIGURE 2. Examples of the appearance of stars classified as AB (left) or A (right)

in our catalog. For the present, however, Figure 3 (right) shows that the scatter in the GSC magnitudes for the stars identified as FHB/A candidates is too great to represent an improvement over our eyeball estimates.

4. Positions

We have obtained coordinates for each of our candidates in the present catalog using the two-axis measuring engine at the Case Western Reserve University. Our internal estimates of the accuracy of these coordinates indicates an rms scatter of roughly two seconds of arc. We have used these positions to identify our stars in the GSC. We first generated a map of the region surrounding each candidate, then verified its identification from photographic enlargements of the objective-prism plates. The final position is taken to be that given in the GSC, which should be accurate to on the order of one second of arc. Some stars identified on the plates were too faint to be included in the GSC, so our original positions were used.

5. Previous Analyses

Investigations based on the stars in FHB I have already led to a number of interesting results. For example, an analysis of the two-point correlation function of the field horizontal-branch (FHB) candidates in this first catalog by Doinidis & Beers (1989) indicated an excess of close pairs, which may be due to the presence of recently-disrupted groups in the Galactic halo. A number of other researchers have reported the possible existence of similar loose groups of FHB (Sommer-Larsen & Christensen 1987; Arnold & Gilmore 1992) and non-FHB stars (Majewski 1992; Poveda, Allen & Schuster 1992; Côté et al. 1993) in the halo, which lends some credence to the notion that additional such groups will be identified in the future.

Broadband UBV photometry of some 1300 of the candidates in FHB I has been obtained by Preston, Shectman & Beers (1991a). Analysis of a subset of these stars by Preston, Shectman & Beers (1991b) indicates the existence of a small, but statistically significant, color gradient as a function of distance from the Galactic center. With a smaller data batch the minor color shift, amounting to 0.025 magnitudes in $(B-V)_o$, would have never been detected. These authors interpret this gradient to be the result

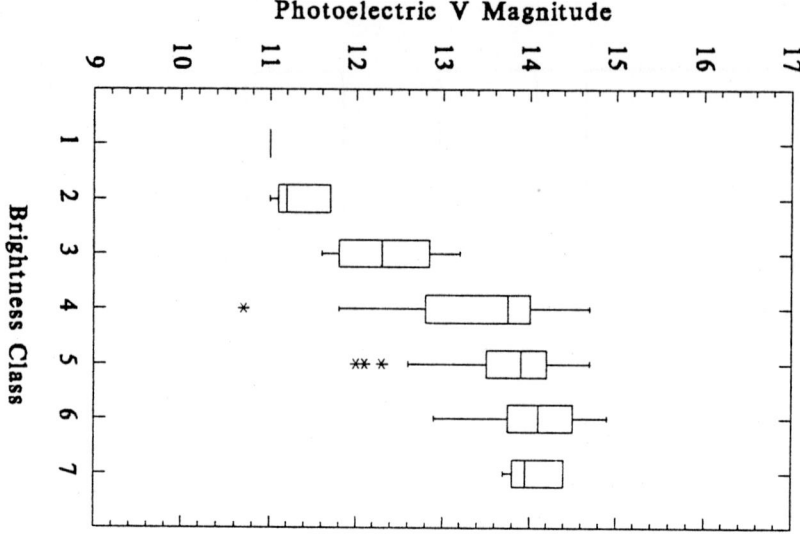

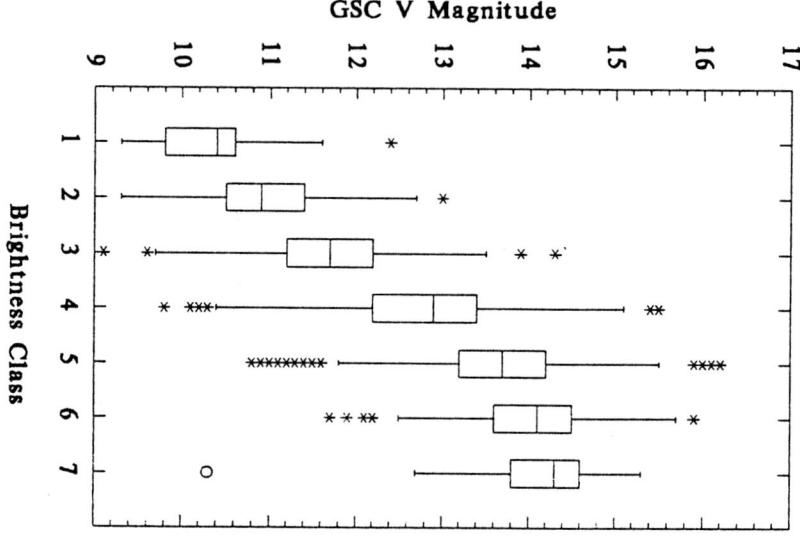

Figure 3. Boxplots of apparent V magnitudes for FHB/A candidates obtained (left) with a photoelectric photometer and (right) with the Guide Star Catalog. Brightness classes are as explained in the text. The horizontal line within each box corresponds to the location of the median. The box corresponds the the inter-quartile range. The whiskers extend to the farthest data point not considered an outlier. Stars represent outliers.

of an increase in the mean age of the halo by some 2–3 Gyr over the range $2 < R < 12$ kpc. In lockstep with the variation of mean color, the *dispersion* of FHB colors exhibits a marked increase with Galactocentric distance, which suggests an outward increase in age dispersion as well.

Followup spectroscopy of some 550 FHB/A candidates by Beers *et al.* (1992) revealed a mixture of halo and thick-disk kinematic populations. The higher gravity A-type stars may represent members of the halo and/or thick-disk blue straggler populations (but see comments below). Medium-resolution (1 Å) optical spectroscopy at low signal-to-noise ($S/N \approx 10 - 20$) cannot, in and of itself, be used to uniquely separate stars on the horizontal branch from hot stars on or near the main sequence. One requires additional information which may be used to constrain the surface gravity of the star in question. Beers *et al.* (1992) describe how UBV photometry may be utilized to distinguish between the bona-fide FHB stars and the higher-gravity A-type stars. A similar method, which replaces the Johnson U with a Strömgren u band for greater sensitivity to surface gravity, is described by Kinman (1994).

6. Future Analyses

With the addition of the northern FHB/A candidates in the present catalog, and the 3000-4000 candidates we expect to add shortly from a second southern catalog, the total number of candidates available from this survey (to date) will be in excess of 10,000. What are the issues we might hope to address with this sample? Several suggest themselves immediately.

6.1. *What is the nature of the halo, and the halo/thick-disk transition ?*

Carney (1994) reviews recent results which suggest the halo may be considerably more complex than once (not so long ago!) imagined. For example, Zinn (1993) and Norris (1993) consider the possibility that "the" halo may in fact be a superposition of a "young" halo (which might have arisen from relatively recent mergers with dwarf-like galaxies) and an "old" halo (which might be identified with a more ancient collapsed ELS-like component). Beers & Sommer-Larsen (1994) show from an analysis of some 2000 non-kinematically-selected stars that the metal-weak tail of the thick-disk population may extend to arbitrarily low abundances – i.e., at least 30 % of the stars in the solar neighborhood with [Fe/H] < -1.5 can be kinematically associated with the thick-disk population. Majewski (1992) comments on the possibility of a retrograde halo population. Unfortunately, all of these samples are either too small in number or concentrated in too small a volume of the Galaxy to draw definitive conclusions. For example, 75 % of the stars in the Beers & Sommer-Larsen database are located less than 2 kpc from the sun.

The FHB/A stars in the present survey cover the apparent magnitude range $11 \le V \le 15$. For the expected range of absolute magnitudes on the horizontal-branch, this translates to distances from the sun $2 < r < 8$ kpc. At the low-latitude limits of the survey, there are stars in this sample which reach all the way to the Galactic bulge, and out to some 15 kpc from the Galactic center. At intermediate- to high-latitudes, survey stars populate $2 < |Z| < 8$ kpc; distances from the Galactic plane which straddle the transition zone from regions dominated by thick-disk stars to those dominated by halo stars. The kinematics of this second group of stars should reveal the presence (or lack) of a retrograde halo population of field stars, and allow us to test for possible changes in the kinematic properties of the thick disk with distance from the plane.

6.2. What is the distribution of matter in the inner halo of the Galaxy ?

As discussed by Thomas (1989), the motions of tracer particles in the Galactic potential provides information on the distribution of mass. For example, the run of line-of-sight velocity dispersion as a function of distance from the Galactic center may be used to determine the equivalent circular velocity, which in turn constrains the mass distribution within the Galaxy. A number of different mass profiles are examined by Thomas, but the differences in predicted rotation speeds between radically different models never exceeds 20 km s^{-1} in the inner halo. Due to the limited number of globular clusters, measurement errors on the equivalent circular velocity will never be driven much below 10 km s^{-1}. Fortunately, the much more numerous FHB stars in our catalogs may be used to pull observational errors on this quantity down to a few km s^{-1}.

6.3. What is the total mass of the Galaxy ?

The best (model-independent) constraints on the total mass of the Galaxy are provided from the limiting escape velocities of bound stars (Leonard & Tremaine 1990). Although high-proper-motion surveys are a natural place to begin such an investigation, difficulties in modelling the selection biases limit the usefulness of extreme velocity stars discovered in this manner. A large sample of FHB stars provides an ideal data batch with which to explore the high-velocity tail of the halo. Rather than simply identifying a few extreme ($v_{rad} > 400$ km s^{-1}) stars, one could look for evidence, in radial-velocity space, for the limit in total energy beyond which significant numbers of stars are no longer found. The few stars exceeding this limit may truly be unbound or simply perturbed by recent interactions. Leonard & Tremaine estimate that a sample of at least 200 stars with radial velocities exceeding 250 km s^{-1} are required to adequately constrain models for the high-velocity tail of the Galaxy.

6.4. Halo Blue Stragglers or Young (ish) Metal-Poor Stars ?

One surprise from our survey may have important implications for the formation of the Galactic halo, and perhaps even the thick disk. Preston, Beers & Shectman (1994, hereafter PBS) identify a sample of 176 blue ($0.15 \leq (B-V)_o \leq 0.35$), metal-poor ([Fe/H] ≤ -1.0) main-sequence stars (hereafter BMPs) among the candidate FHB/A stars from the first southern catalog. The space density of this species within 2 kpc of the sun is determined to be on the order of 450 kpc^{-3}. These authors obtain a measurement of the specific frequency, defined as the number of BMPs per field horizontal-branch star, on the order of 10, at least an order of magnitude larger than the specific frequency found among the Galactic globular clusters which have been searched for blue stragglers all the way to their centers. The kinematics of the BMPs are also unique. PBS obtain a Galactic rotation of $V_{rot} = 130$ km s^{-1}, and a velocity ellipsoid of $(\sigma_r \approx 90, \sigma_\phi \approx 90, \sigma_\theta \approx 90)$ km s^{-1}. These values appear distinct from either the halo or thick-disk populations.

PBS argue that the above results indicate that the great majority of BMPs cannot be the field analogs of the blue stragglers found in halo globular clusters. Rather, they suggest that the BMPs are stars near the main-sequence turnoff which were formed in one or more Milky Way satellites, perhaps similar to Fornax or Carina, that were captured by the Galaxy during the past 3 to 10 Gyrs. These results would reinforce earlier arguments by Rodgers, Harding & Sadler (1981) and by Lance (1988) on the basis of more limited data for less metal-deficient early-type stars. If this hypothesis proves correct, the BMPs may provide a key part of the puzzle of Galactic formation and evolution. The disrupted (or even only partially disrupted) dwarf galaxies might provide the energy and angular momentum required to thicken a pre-existing thin disk, without necessarily destroying it in the process (see Quinn, Hernquist & Fullagar 1993). At the same time, the tidal debris stripped from such a galaxy might populate the metal-weak tail of the thick disk, which Freeman (1993) estimates to have a mass on the order of 10^8 $M_\odot$, similar to the inferred mass for the parent population of the BMPs derived by PBS.

7. Work in Progress

We are almost finished with our (visual) examination of the extant set of HK survey plates. Before taking any additional plates, it is our plan to make sure that long-term spectroscopic and photometric followup observations of the candidate FHB/A is well under way. Due to the enormous numbers of stars to be examined, we would welcome interested parties to inquire about participation in this effort.

Broadband photometry and spectroscopy of a subset of the northern candidates is presently being obtained, albeit at a slower pace than we desire. Ultimately, we should have a northern sample with which to confirm the color gradient for halo FHB stars noted by Preston, Shectman & Beers (1991b) based on the southern catalog. Refinement of techniques for estimation of surface gravity and metal abundances for FHB and A stars (Wilhelm, Beers & Gray 1994) will allow a more definitive separation of these populations.

Spectroscopy of stars in the southern photometric sample is now underway; we expect to soon have a sample of on the order of 1000 FHB/A stars with which detailed models of the gravitational potential in the inner Galactic halo may be constrained. It will be of particular interest to investigate the change of the velocity ellipsoid for FHB stars from the inner to outer halo by a comparison with data now coming available for FHB stars with Galactocentric distances from 20 to 60 kpc (e.g., Norris & Hawkins 1991; Arnold & Gilmore 1992).

The authors would like to express their thanks to former Michigan State University undergraduate students Julie Ann Kage and Eric Sullivan, and former graduate student Jon Truax, who participated in acquiring plates for the northern survey, and to Peter Pesch and Charles Knox, who provided access to the Case Western Reserve University measuring engine. T. C. B. acknowledges support provided by the National Science Foundation in the form of grants AST 90-1376 and AST 92-22326, and supplements to these grants from the REU program. Finally, the authors would like to credit the inspiration, assistance, and encouragement of Dr. George Preston and Dr. Stephen Shectman of the Carnegie Observatories, without whom this (northern) survey would have never been undertaken.

REFERENCES

ARNOLD, R., GILMORE, G., 1992, Halo blue horizontal branch stars: spectroscopy in two fields. MNRAS, 257, 225–239

BEERS, T. C., SOMMER-LARSEN, J., 1994, Kinematics of metal-poor stars in the galaxy. ApJS, submitted

BEERS, T. C., PRESTON, G. W., SHECTMAN, S. A., 1985, A search for stars of very low metal abundance. I. AJ, 90, 2089–2102

BEERS, T. C., PRESTON, G. W., SHECTMAN, S. A., 1988, A catalog of candidate field horizontal-branch stars. I. ApJS, 67, 461–501 (FHB I)

BEERS, T. C., PRESTON, G. W., SHECTMAN, S. A., 1992, A search for stars of very low metal abundance. II. AJ, 103, 1987–2034

BEERS, T. C., PRESTON, G. W., SHECTMAN, S. A., DOINIDIS, S. P., GRIFFIN, K. E., 1992, Spectroscopy of hot stars in the galactic halo. AJ, 103, 267–296

CARNEY, B., 1994, What is the galaxy's halo population? in Adelman S. J., Upgren A. R., Adelman C. J., eds, Hot Stars in the Halo, Cambridge University Press, Cambridge, p. 3

CÔTÉ, P., WELCH, D. L., FISCHER, P., IRWIN, M. J., 1993, preprint

DOINIDIS, S. P., BEERS, T. C., 1989, Evidence for clustering of field horizontal-branch stars in the galactic halo. ApJ, 340, L57–L60

DOINIDIS, S. P., BEERS, T. C., 1990, Photoelectric UBV photometry of northern stars from the HK objective-prism survey. PASP, 102, 1392–1399

DOINIDIS, S. P., BEERS, T. C., 1991, Photoelectric UBV photometry of northern stars from the HK objective-prism survey. II. PASP, 103, 973–986

FREEMAN, K. C., 1993, The dynamical history of the galactic disk. in Majewski S., ed, Galaxy Evolution: The Milky Way Perspective, ASP Conf. Ser., vol. 49, pp. 125–138

KINMAN, T. D., SUNTZEFF, KRAFT, R. P., 1994, in Adelman S. J., Upgren A. R., Adelman C. J., eds, Hot Stars in the Halo, Cambridge University Press, Cambridge, p. 353

LANCE, C. M., 1988, Young, high-velocity A stars. II. Misidentified, ejected, or unique? ApJ, 334, 927–946

LEONARD, P. J. T., TREMAINE, S., 1990, The local galactic escape speed. ApJ, 353, 486–493

MAJEWSKI, S. R., 1992, A complete, multicolor survey of absolute proper motions to $B \approx 22.5$: structure and kinematics. ApJS, 78, 87–152

NORRIS, J. E., 1993, Population studies. XII. The duality of the galactic halo. preprint

NORRIS, J. E., HAWKINS, M. R. S., 1991, Population studies. X. Constraints on the mass and extent of the galaxy's dark corona. ApJ, 380, 104–115

POVEDA, A., ALLEN, C., SCHUSTER, W., 1992, Moving clusters among galactic halo stars. in Barbuy B., Renzini A., eds, The Stellar Populations of Galaxies, IAU Symp. 149, Kluwer, Dordrecht, p. 471

PRESTON, G. W., SHECTMAN, S. A., BEERS, T. C., 1991a, Photoelectric UBV photometry of stars selected in the HK objective-prism survey. ApJS, 76, 1001–1031

PRESTON, G. W., SHECTMAN, S. A., BEERS, T. C., 1991b, Detection of a galactic color gradient for blue horizontal-branch stars of the halo field and implications for the halo age and density distributions. ApJ, 375, 121–147

PRESTON, G. W., BEERS, T. C., SHECTMAN, S. A., 1994, The space density and kinematics of metal-poor blue main sequence stars near the solar circle. ApJ, submitted

QUINN, P. J., HERNQUIST, L., FULLAGAR, D. P. 1993, Heating of galactic disks by mergers. ApJ, 403, 74–93

RODGERS, A. W., HARDING, P., SADLER, E., 1981, The nature of the metal-rich stellar population in the galactic halo. ApJ, 244, 912–918

SOMMER-LARSEN, J., CHRISTENSEN, P. R., 1987, Discovery of a star cluster in the galactic halo. MNRAS, 225, 499–503

THOMAS, P., 1989, Spatial distributions and dynamics of the galactic globular cluster system. MNRAS, 238, 1319–1343

WILHELM, R., BEERS, T. C., GRAY, R. O., 1994, A technique for distinguishing FHB stars from A-type stars. in Adelman S. J., Upgren A. R., Adelman C. J., eds, Hot Stars in the Halo, Cambridge University Press, Cambridge, p. 257

ZINN, R., 1993 The galactic halo cluster systems: evidence for accretion. in Smith G. H, Brodie J. P., eds, The Globular Cluster - Galaxy Connection, ASP Conf. Ser., vol. 48, pp. 38–47

Discussion

BIDELMAN: The Beers, Preston & Shectman HK Survey for metal-poor stars has picked up a good many known or suspected RR Lyrae stars, but these have not usually been so-indicated.

Since variable stars often have rather inaccurate positions, a few of the identifications which I have done may be in error.

BEERS: We are certainly indebted to you for taking the time to search your own extensive database for the known or suspected variable stars we have recovered in the HK survey. Also, we would not be able to easily measure accurate positions for our (northern) stars without access to the CWRU measuring engine.

CARNEY: Can you say something provocative about the blue metal-poor main sequence stars?

BEERS: My response is in the text version of the proceedings.

Recent Progress on a Continuing Survey of Galactic Globular Clusters for Blue Straggler Stars

By ATA SARAJEDINI

Kitt Peak National Observatory, National Optical Astronomy Observatories,
P. O. Box 26732, Tucson, AZ 85726, USA

The results of a continuing photometric survey of selected Galactic globular clusters are presented. The primary aim of this survey is to increase the sample of globular clusters known to contain blue straggler stars. When combined with those from the literature, there are now $\approx$ 700 blue stragglers identified in 29 globular clusters. The spatial and photometric properties of these stars are analyzed and discussed.

1. Introduction

In his study of the Galactic globular cluster M3, Sandage (1953) discovered a group of stars lying brightward and blueward of the main sequence turnoff. As such, he was the first investigator to recognize these anomalous stars as a new class of object. Several years later, Burbidge & Sandage (1958) isolated a significant number of these "blue straggler" stars (BSSs) in the open cluster NGC 7789. In addition to coining the term, they also noted the existence of BSSs in other open clusters such as M67 and h & χ Persei. Approximately 15 years later, Arp & Hartwick (1971) reported the detection of BSSs in the metal-rich globular cluster M71. By 1988, when the work described herein was begun, only *four* Galactic globular clusters were known to contain BSSs: M3, M71, ω Cen (Da Costa *et al.* 1986), and NGC 5466 (Nemec & Harris 1987). On the other hand, many more open clusters were observed to harbor BSSs. Why was this the case? Are BSSs more prevalent in open clusters relative to globular clusters or was there some sort of selection effect at work? To answer this question, it seemed appropriate to launch a survey of Galactic globular clusters in search of BSS candidates. Most of this work was carried out as part of the author's doctoral dissertation at Yale University.

The strategy has been to obtain CCD images in the B and V passbands in order to construct color-magnitude diagrams (CMDs) for as many globular clusters as possible. In many cases, CMDs of off-cluster regions have also been constructed to assess the degree of field contamination. After BSS candidates are identified in each cluster, we attempt to study their properties.

With the recent explosion of interest in this field, several excellent reviews have been published including those by Sarajedini & Da Costa (1991), Fusi Pecci *et al.* (1992), and Stryker (1993). The results presented here are an extension of the work described previously in Sarajedini & Da Costa (1991) and Sarajedini (1992a, 1992b, 1993a, and 1993b).

2. Observations and Reductions

The observations of the program clusters in this continuing study have been obtained using the 0.9-m telescopes at Cerro Tololo Inter-American Observatory (CTIO) and Kitt Peak National Observatory (KPNO). As noted above, the data consist of CCD images situated as close as possible to the cluster center and photometered using the

DAOPHOT/DAOPHOT II (Stetson 1987) crowded-field photometry package. The resultant magnitudes are edited and matched to form instrumental colors. These are then transformed to the standard system using observations of standard stars drawn from the lists of Graham (1982) and Landolt (1983, 1993). Because the images are reduced independently, each star is detected and measured a number of times. To qualify as a genuine BSS candidate, a star must be recovered on two or more frame pairs and have B and V errors typically less than 0.05 mag.

Off-cluster photometry has also been obtained for a significant fraction of the program clusters. All of these field CMDs have essentially identical structure, which consists of a column of stars with $0.5 \lesssim B - V \lesssim 0.9$ extending to redder colors at fainter magnitudes. However, the region of interest, the color range of the BSSs, contains few if any stars, indicating that the BSSs seen in the cluster CMD are most likely members of the cluster. This is important because noncluster stars appearing in the blue stragglers region of the cluster CMD can give the misleading impression of being BSSs.

3. Color-Magnitude Diagrams

The heart of this investigation is the construction of high precision CMDs for the program clusters. Several of these have already been presented in the papers cited at the end of §1. In this section, our aim is to give the reader a taste of how these diagrams are used.

3.1. NGC 5053

Figure 1a shows our CMD for NGC 5053 based on 13 B and 13 V images taken with the KPNO 0.9-m telescope. Only stars with frame-to-frame standard errors less than 0.05 mag in V and 0.071 mag in $B - V$ are plotted.

This cluster has already been observed in $(g, g-r)$ by Nemec & Cohen (1989, hereafter NC89) using the Palomar 5-m telescope. They identified 24 BSS candidates and showed that the spatial distribution of the bright BSSs is more centrally concentrated than the cluster subgiant stars in the same magnitude range. This behavior indicates that the masses of these bright BSSs are, on average, higher than stars at the turnoff ($\approx 0.8 M_\odot$). Moreover, NC89 compared multi-mass King models to the radial distribution of these BSSs to derive a mean mass of $1.3 \pm 0.3 M_\odot$. Of the 24 BSSs isolated by NC89, we recover 23 of them; one of their BSSs is located off of our CCD frames. In addition, we find BSS #24 to have a color of $B - V = 0.59 \pm 0.02$ placing it well outside of the blue straggler region. Note that, BSS #24 has been detected on 9 of the 13 frame pairs with a frame-to-frame standard deviation of only 0.06 in $B - V$. This observed dispersion is consistent with the photometric error and implies that BSS #24 is probably not a variable. When combined with the three new candidates discovered, the total number of BSSs in our photometry of NGC 5053 is 25.

The other three panels of Figure 1 show our photometry divided into three radial bins in the same manner as Figure 3 of NC89. Note that the morphologies of the BSS sequences in the four panels of Figure 1 are remarkably similar to those in the corresponding panels of Figure 3 in NC89. Thus, the fact that we have confirmed the existence and CMD morphology of the NC89 BSSs strenghtens their conclusions regarding the masses of these stars.

3.2. NGC 6584

The only previously published photometry for this globular cluster is a study of its RR Lyrae variables by Millis & Liller (1980). Recently, Sarajedini & Forrester (1994) have

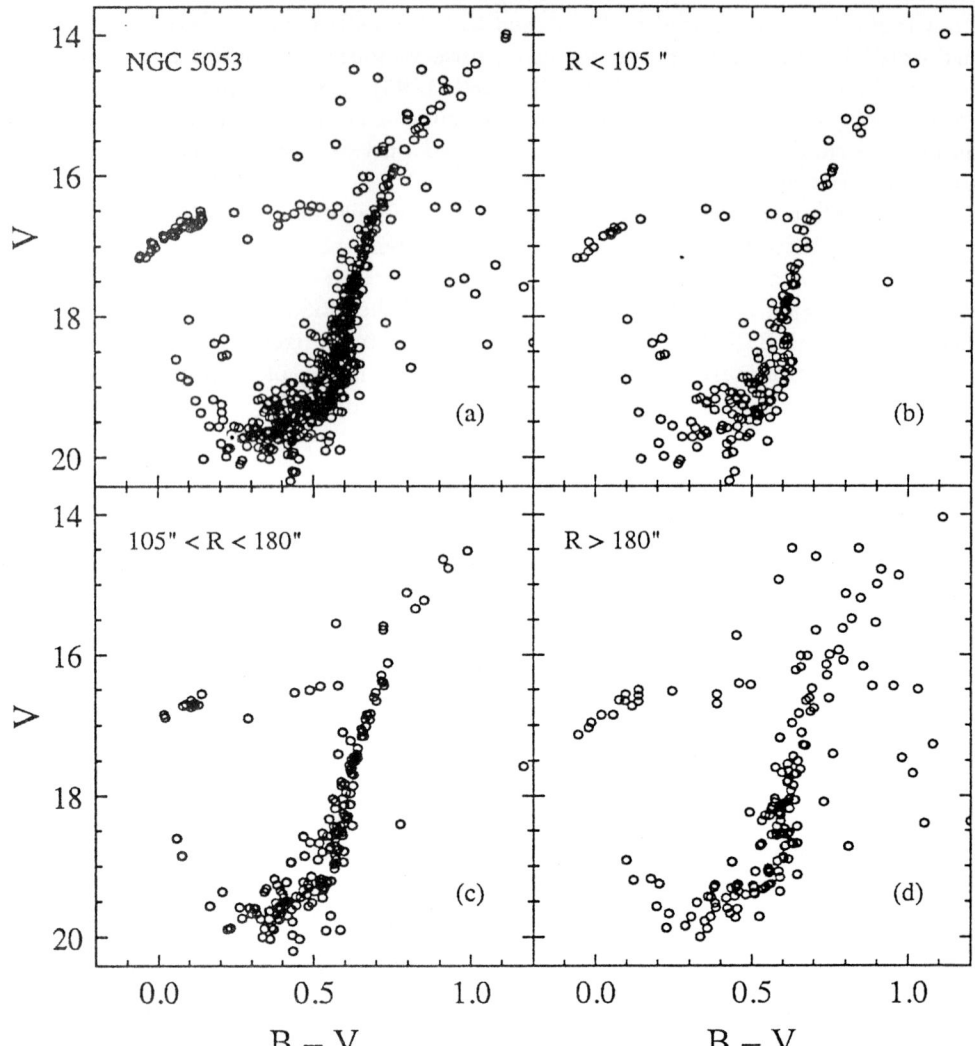

FIGURE 1. (a) The color-magnitude diagram of NGC 5053. Only stars with frame-to-frame dispersions less than 0.05 mag in V and 0.071 mag in $B - V$ are plotted. (b)–(d) Same as (a) except that the photometry is divided into radial bins. Compare to Figure 3 of Nemec & Cohen (1989).

obtained CCD photometry for ≈ 3000 stars in the direction of NGC 6584. Figure 2 presents the cluster CMD divided into three radial bins, plus the CMD of an off-cluster region. These data are based on 10 B and 10 V images of the cluster taken with the CTIO 0.9-m telescope. Only stars with 3 or more detections and frame-to-frame standard errors less than 0.05 mag in V and 0.071 mag in $B - V$ are plotted.

The cluster CMD reveals the presence of 33 BSS candidates. Furthermore, the off-cluster CMD contains no stars with $B - V \lesssim 0.5$ mag implying that the majority of the BSSs are probably cluster members. It also seems that the BSSs are preferentially located relatively close to the cluster center. As in the case of NGC 5053, this possibility

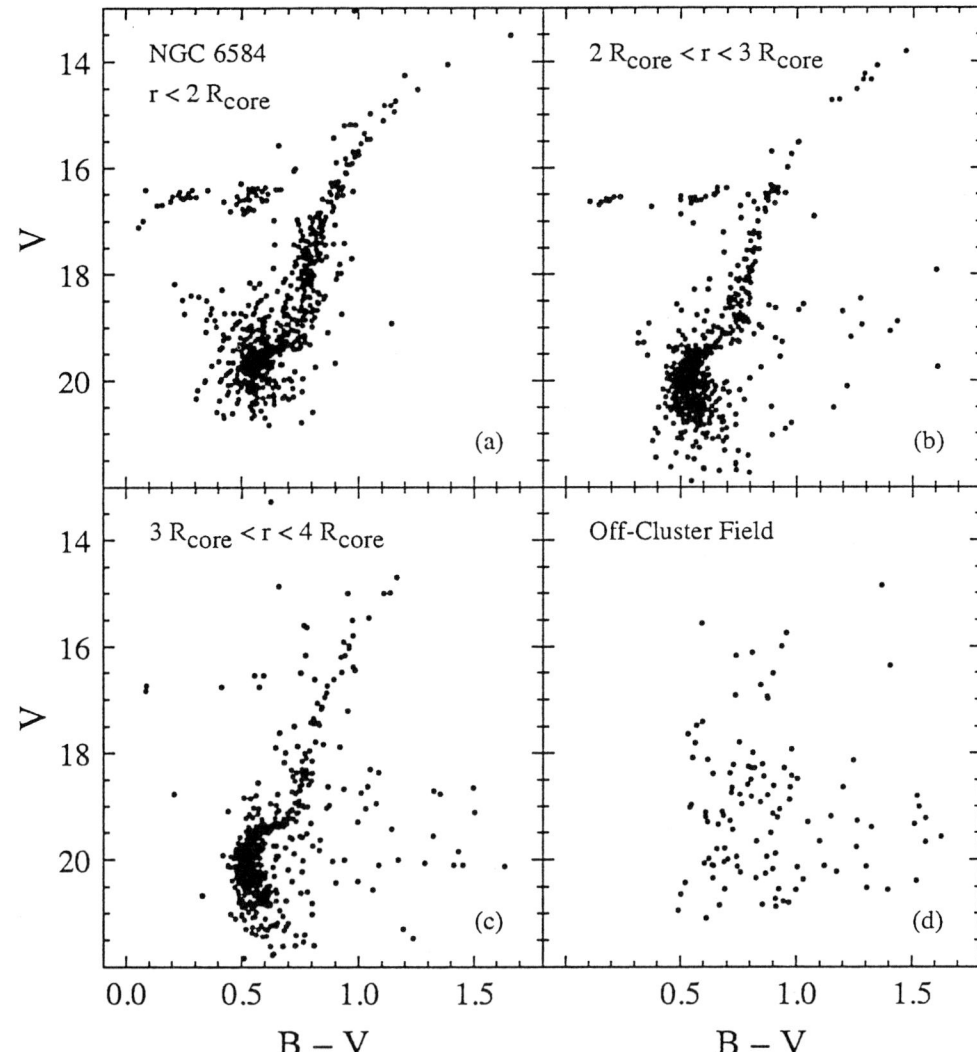

FIGURE 2. (a)–(c) Radial color-magnitude diagrams for the globular cluster NGC 6584 from the work of Sarajedini & Forrester (1994). Note the significant population of blue straggler stars. (d) The color-magnitude diagram for a region located outside of the cluster tidal radius. This diagram has no stars with $B - V \lesssim 0.5$ mag implying that the majority of BSSs are probably cluster members.

can be tested by comparing the BSS radial distribution to that of the subgiants, which serve as a set of "control" stars. Figure 3 shows such a comparison with the dashed line representing the BSSs and the solid line indicating the subgiant branch stars. It is clear from this diagram that, again as for NGC 5053, the BSSs are more centrally concentrated (i.e., more massive) than the subgiant stars. In fact, a similar phenomenon has been confirmed in just about every other Galactic globular cluster known to contain BSSs (Sarajedini 1993b), except perhaps M3 (see Bolte *et al.* 1993 and Ferraro *et al.* 1993). We return to this point in §8.

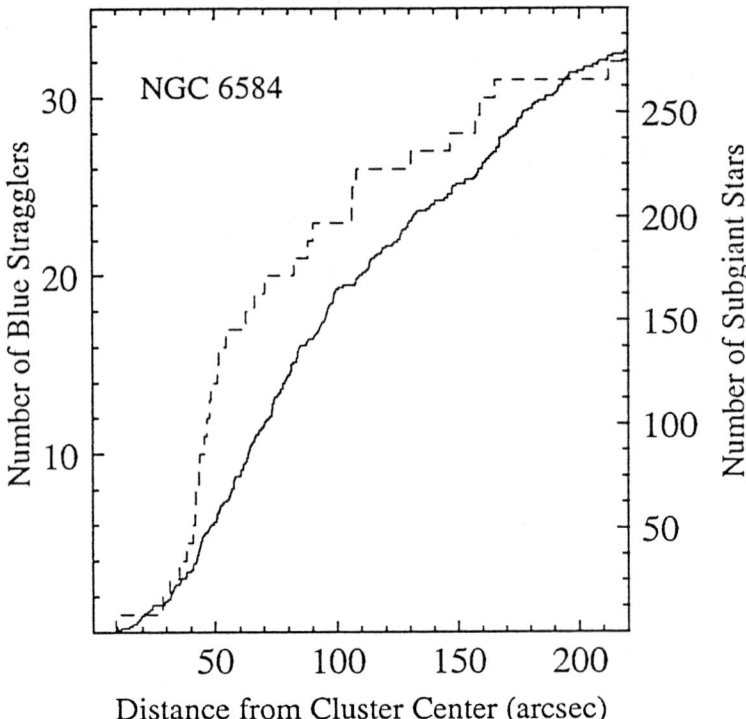

FIGURE 3. Cumulative radial distributions for the blue stragglers (dashed line) and the subgiant branch stars (solid line). At each radial distance, the proportion of blue stragglers is greater than subgiants indicating that the former are more massive than the latter.

4. Formation of Blue Stragglers

Leonard (1989) provides an excellent summary of the various blue straggler formation scenarios. In addition, two of the more prominent theories have been described in detail by Sarajedini (1993b). In the present work, we will summarize these two formation mechanisms as well as a third process which has been suggested previously.

The most prominent theory is that BSSs are the result of a binary star mass transfer/merger scenario (McCrea 1964; Eggen & Iben 1989). In short, this increases the mass of one component and makes it appear brighter and bluer. Furthermore, given enough time, the binary system can coalesce and eventually merge to form a single star. This theory makes several observationally testable predictions. First, the BSSs should have masses *greater* than stars at the cluster turnoff, but *less* than twice this mass. These limits are precisely what the observational data indicate for the masses of BSSs in globular clusters. Second, depending on their orientation, binary systems should exhibit photometric and radial velocity variations. Indeed, Mateo *et al.* (1990) have detected photometric variability in some of the BSSs of NGC 5466. We note that, if the binary system has already merged, it will not show these variations. Thus, chances are that all traces of its binary heritage will be lost.

Another formation mechanism considers BSSs as products of stellar collisions. The work of Benz & Hills (1987, see also Hills & Day 1976) has shown that encounters between typical globular cluster main sequence stars can produce a single massive zero-age main sequence (ZAMS) star which has been completely mixed. As one might expect, these types of collisions occur preferentially in high density clusters. Since the two original stars must have produced some helium in their cores due to nuclear burning, the resultant star will have an enhanced abundance of helium in its envelope and thus appear brighter than expected based solely on its increased mass. From the work of Bailyn (1992), this increase in luminosity can range from ≈ 0.15 to ≈ 0.3 in M_V.

Thirdly, BSSs can result from the mixing of fresh hydrogen into the stellar core thus replenishing the available nuclear fuel (Wheeler 1979; Saio & Wheeler 1980). In principle, this is a simple concept. However, in practice, the observed color and magnitude range among the globular cluster BSSs implies that each star has undergone a *different* amount of mixing perhaps even at a different rate. It would seem difficult to accomplish this within the framework of conventional stellar structure theory.

5. Cluster Compilation

Table 1 lists the Galactic globular clusters which are known to contain BSSs. We also include their metallicities from Zinn & West (1984) and Galactocentric distances (R_{GC}) from the compilation of Zinn (1985). There are two exceptions to this. The metallicity of Ruprecht 106 is taken from Da Costa *et al.* (1992, and recently corroborated by Sarajedini 1994a), and the R_{GC} is that of Chaboyer *et al.* (1992). The metallicity of NGC 5053 is that derived by Armandroff *et al.* (1992). Finally, the cluster central densities (Log ρ_o) measured in $M_\odot/pc^3$ all come from the work of Webbink (1985).

Note that the globular clusters in Table 1 span the entire range of metal abundance from the most metal-rich (M71) to the most metal-poor (M15). They have R_{GC} values from ≈ 2 Kpc to as much as ≈ 18 Kpc, and central densities which span the entire range of globular cluster values. The point is that globular clusters that contain BSSs have no other distinguishing features. It stands to reason therefore that *all* Galactic globular clusters contain BSSs. They have not been identified in all such systems because we have not looked hard enough. As mentioned previously, the BSSs are preferentially located near the cluster center. Some photometric studies do not even include the regions near the cluster center. Consequently, they miss most, if not all, of the BSSs.

6. The Blue Straggler HR Diagram

To better understand the relationship between the blue straggler populations of different globular clusters, it is important to construct an HR Diagram for these BSSs. To this end, we must adopt distance moduli and reddenings for all of the clusters listed in Table 1. However, six of these clusters cannot be included for the following reasons. The photometric data for 47 Tuc, M15, and M4 were obtained with HST; as such, they are only precise at the ≈ 10 % level. Both NGC 6366 and NGC 4372 have been omitted because of their excessively high reddening values and the presence of significant differential reddening. The BSSs of M71 were observed in U and V. Finally, ω Cen has not been included because the possible existence of a metallicity range among the BSSs could complicate their interpretation. This leaves us with a total of 580 BSSs in 22 Galactic globular clusters.

The distance moduli are calculated in a two step process. First, the M_V of the RR Lyrae variables is computed using the $[Fe/H]$ value listed in Table 1 and the relation between $M_V(RR)$ and $[Fe/H]$ derived by Lee (1990), i.e., $M_V(RR) = 0.17[Fe/H] + 0.79$. Then, adopting a value for the apparent magnitude of the horizontal branch ($V(HB)$ in Table 2) and assuming $M_V(RR) \approx M_V(HB)$, we calculate the apparent distance modulus of each cluster.

Cluster	[Fe/H]	R_{GC} (Kpc)	Log ρ_o ($M_\odot/pc^3$)	N(BSS)	BSS Reference
47 Tuc	−0.71	6.4	5.02	21	Paresce et al. (1991)
NGC 288	−1.40	10.1	2.02	38	Bolte (1992)
NGC 3201	−1.61	7.8	3.12	31	Brewer et al. (1993)
NGC 4372	−2.08	6.2	2.33	25	Sarajedini (1992b)
Ruprecht 106	−1.69	17.9	1.22	32	Buonanno et al. (1990)
M68	−2.09	9.2	2.88	7	Sarajedini (1992b)
NGC 5024	−2.04	18.2	3.18	4	Heasley & Christian (1991)
NGC 5053	−2.41	17.0	0.58	25	This Work
ω Cen	−1.59	5.6	3.35	5	Da Costa et al. (1986)
M3	−1.66	10.6	3.86	135	Ferraro et al. (1993)
NGC 5466	−2.22	15.6	1.25	48	Nemec & Harris (1987)
IC 4499	−1.50	13.4	1.63	24	Sarajedini (1993a)
Palomar 5	−1.47	14.9	−0.50	8	Smith et al. (1986)
NGC 5897	−1.68	6.8	1.82	34	Sarajedini (1992a)
NGC 6101	−1.81	11.4	2.00	27	Sarajedini & Da Costa (1991)
M4	−1.33	5.4	3.91	≈5	Bailyn & Mader (1993)
M107	−0.99	3.0	3.27	26	Ferraro et al. (1991)
NGC 6366	−0.99	4.2	2.40	27	Harris (1993)
NGC 6397	−1.91	5.2	4.20	18	Lauzeral et al. (1992)
NGC 6535	−1.75	3.4	3.63	9	Sarajedini (1994b)
NGC 6584	−1.54	7.4	3.16	33	Sarajedini & Forrester (1994)
NGC 6723	−1.09	2.4	3.09	9	Sarajedini (1992b)
M55	−1.82	3.3	2.64	5	Sarajedini (1992b)
M71	−0.58	5.9	3.24	≈15	Richer & Fahlman (1988)
NGC 6934	−1.54	10.1	3.79	3	Sarajedini (1992b)
M72	−1.54	11.4	2.74	17	Sarajedini (1992b)
M15	−2.15	9.5	5.26	9	Ferraro & Paresce (1993)
Palomar 12	−1.14	13.5	1.38	20	Stetson et al. (1989)
NGC 7492	−1.82	16.5	1.07	27	Côté et al. (1991)

TABLE 1. Galactic Globular Clusters Containing BSSs

As for the reddening, we adopt the values listed in the compilation of Zinn (1985) and tabulated in Table 2. There are three exceptions to this: Ruprecht 106 (Da Costa et al. 1992), IC 4499 (Sarajedini 1993a), and NGC 6535 (Sarajedini 1994b).

Figure 4 shows the HR Diagram for 512 BSSs in our sample with $B - V$ photometry. Note that there are 68 BSSs in M3 which only have $V - I$ photometry (Ferraro et al. 1993). Open circles represent clusters with $[Fe/H] > -1.6$ while plus signs represents those with $[Fe/H] \leq -1.6$. For comparison purposes, we also show in this figure a metal-poor (Z = 0.0001, Y = 0.23) ZAMS (solid line), a metal-rich (Z = 0.004, Y = 0.23) ZAMS (dashed line) both from the Revised Yale Isochrones (Green et al. 1987), the M92 (Stetson & Harris 1988, solid line) and 47 Tuc (Hesser et al. 1987, dashed line) fiducial sequences, and the M92 blue HB (Buonanno et al. 1985, solid line) along with the 47 Tuc red HB (Hesser et al. 1987, dashed line). These have been placed in Figure 4 assuming $[Fe/H]$ values from Zinn & West (1984), reddenings from Zinn (1985), and $V(HB)$s from Armandroff (1989) combined with the Lee (1990) distance scale.

Several features of Figure 4 deserve comment. First, the BSSs are distributed over a significant range in color. One of the reasons for this is the metallicity range of the clusters; it is evident that the plus signs ($[Fe/H] \leq -1.6$) tend to be bluer than the

Cluster	$E(B-V)$	$V(HB)$	$V(HB)$ Reference
NGC 288	0.04	15.31	Bolte (1992)
NGC 3201	0.21	14.76	Cacciari (1984)
Ruprecht 106	0.20	17.85	Buonanno et al. (1990)
M68	0.03	15.61	Harris (1975)
NGC 5024	0.00	16.94	Armandroff (1989)
NGC 5053	0.01	16.52	This Work
M3	0.01	15.65	Fusi Pecci et al. (1992)
NGC 5466	0.00	16.62	Nemec & Harris (1987)
IC 4499	0.15	17.68	Sarajedini (1993a)
Palomar 5	0.03	17.35	Sandage & Hartwick (1977)
NGC 5897	0.06	16.35	Sarajedini (1992a)
NGC 6101	0.04	16.60	Sarajedini & Da Costa (1991)
M107	0.31	15.65	Ferraro et al. (1991)
NGC 6397	0.18	12.90	Alcaino et al. (1987)
NGC 6535	0.44	15.73	Sarajedini (1994b)
NGC 6584	0.11	16.53	Sarajedini & Forrester (1994)
NGC 6723	0.00	15.47	Sarajedini (1992b)
M55	0.06	14.35	Lee (1977)
NGC 6934	0.15	16.90	Sarajedini (1992b)
M72	0.04	16.84	Sarajedini (1992b)
Palomar 12	0.02	17.13	Stetson et al. (1989)
NGC 7492	0.00	17.63	Buonanno et al. (1987)

TABLE 2. Adopted Cluster Parameters

open circles ($[Fe/H] > -1.6$). As we shall see later, stellar evolution combined with a range of masses and formation times also acts to increase the color range of the BSSs. Second, there appear to be some BSSs which lie blueward of the metal-poor ZAMS. This phenomenon is probably due to uncertainties in the adopted reddening values (Sarajedini 1993b; see also Da Costa et al. 1986).

7. Blue Straggler Luminosity Function

The luminosity function (LF) of globular cluster BSSs is an important diagnostic tool which can yield clues to their origin and evolution (Nemec & Harris 1987; Sarajedini & Da Costa 1991; Sarajedini 1992a). The LF of all 580 BSSs compiled herein is morphologically similar to that of the 409 BSSs in Sarajedini (1993b). The reader is referred to Figure 4 and § 7 of that paper for a detailed discussion of the LF.

Based on the discussion in § 4 regarding the formation of BSSs through stellar collisions, one might expect the LF of BSSs to vary with cluster central density (Log ρ_o). We have divided our sample so that there are 283 BSSs in clusters with Log $\rho_o < 2.4$ and 297 in those with Log $\rho_o > 2.4$. The LFs of these datasets are plotted in Figures 5a and 5b. These have been scaled to unit area with error bars based on Poisson statistics. Because the fainter portions of these LFs may be adversely affected by selection effects (Sarajedini & Da Costa 1991; Fusi Pecci et al. 1992), we apply a K-S test to the regions with $M_V < 3.5$ and find that the LFs in Figures 5a and 5b are different at the >99 % confidence level. To illustrate this difference, Figure 5c shows the result of subtracting these two LFs.

If BSSs are formed via the binary mass transfer/merger or the collisional formation

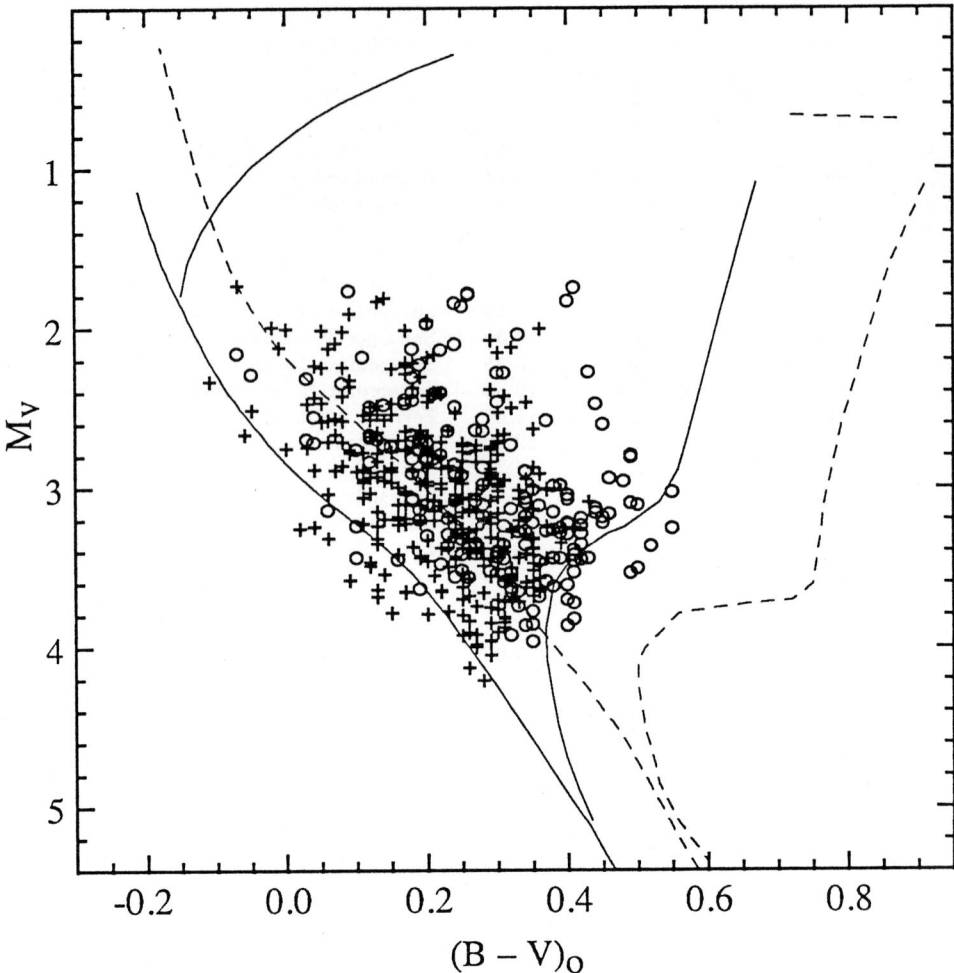

FIGURE 4. HR Diagram of 512 globular cluster blue straggler stars. Open circles represent clusters with $[Fe/H] > -1.6$ while plus signs represent those with $[Fe/H] \leq -1.6$. The blue stragglers in metal-poor clusters appear to be bluer on-average than those in metal-rich clusters.

scenarios, the largest mass they can have is twice that of the turnoff (i.e., $\approx 1.6 M_\odot$). Furthermore, based on the Revised Yale Isochrones (Green *et al.* 1987), the luminosity on the ZAMS for a star with $[Fe/H] = -1.6$, $Y = 0.23$, and a mass of $1.6 M_\odot$ is $M_V \approx 2.6$. As a result, the BSSs with $M_V < 2.6$ are probably evolved stars (see Sarajedini & Da Costa 1991; Sarajedini 1993b). However, it appears that high density globular clusters have a greater fraction of BSSs with $M_V < 2.6$ relative to clusters of low density (see Figure 5c). This is probably because some of the bright BSSs in dense clusters are the result of stellar collisions, which, as we discussed above, are more common in dense stellar environments and produce brighter BSSs than binary mass transfer.

To close out this section, we note that the LF of BSSs in high density clusters (5b) shows evidence of bi-modality with a "dip" at $M_V = 2.9$. One possible explanation for this is that high density clusters produce fewer BSSs with $M_V = 2.9$ relative to clusters of low density. However, this appears unlikely because there seems to be nothing special

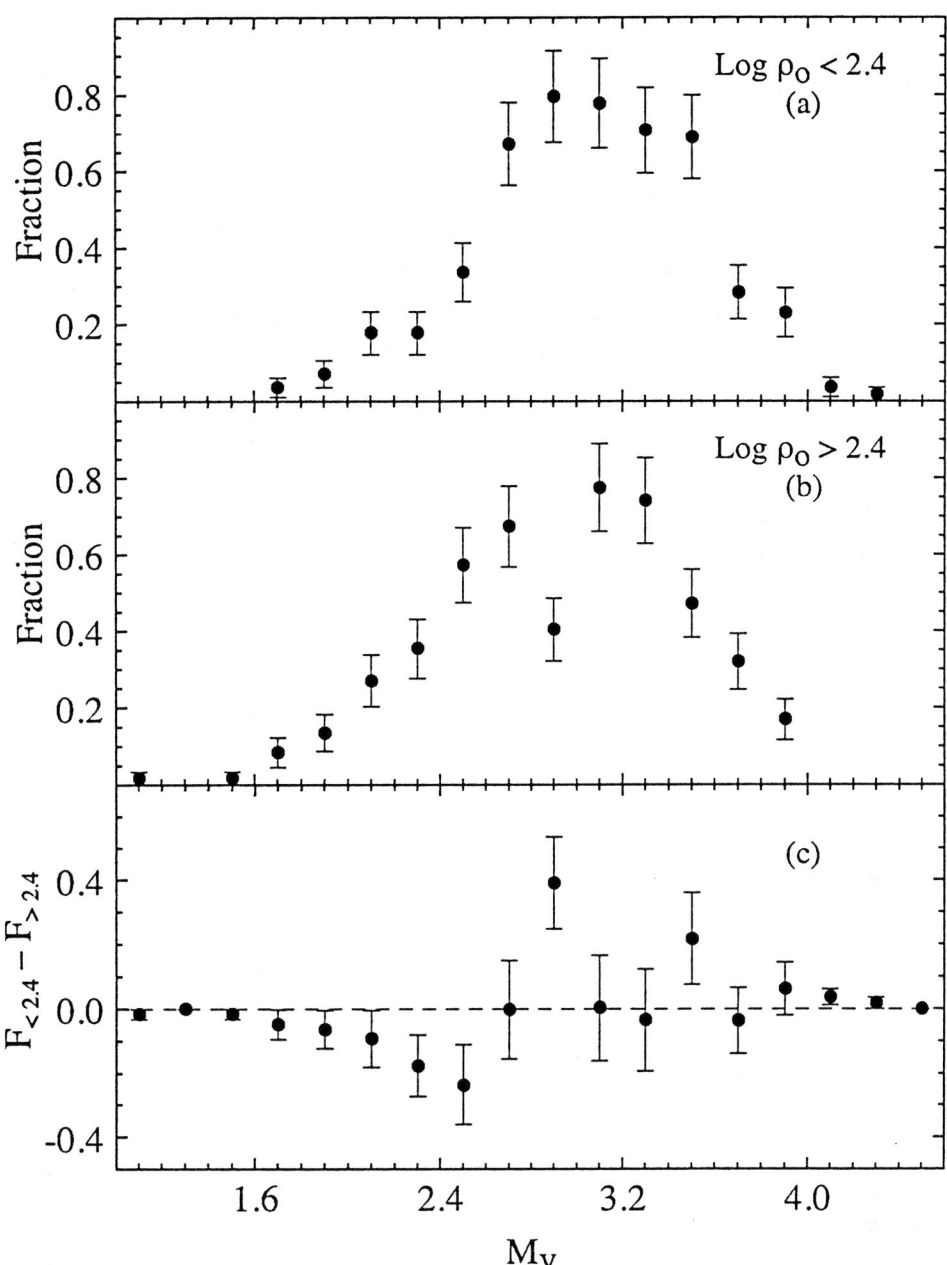

FIGURE 5. Luminosity functions for the 580 blue stragglers in the present sample divided into two groups: those in clusters with Log ρ_o < 2.4 (5a) and those in clusters with Log ρ_o > 2.4 (5b). Error bars reflect 1σ Poisson noise. The lower panel (5c) represents the numerical difference of these luminosity functions.

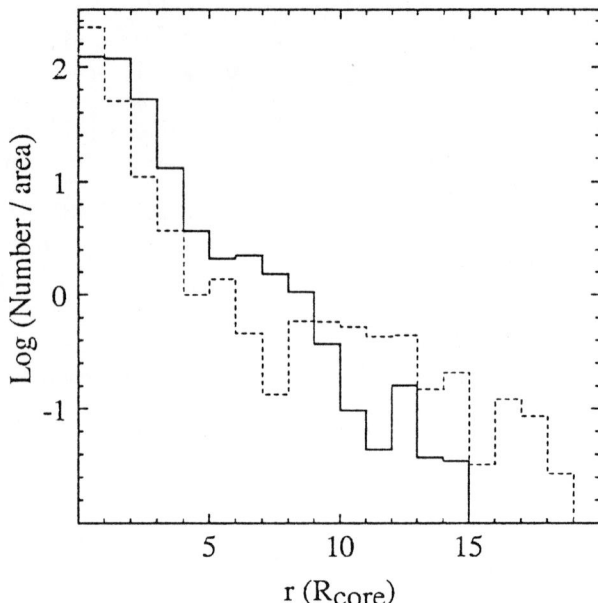

FIGURE 6. The number of blue stragglers (dotted line) and subgiant branch stars (solid line) per unit area as a function of radial distance. These distributions include 554 blue stragglers from 19 of the globular clusters in Table 2 and 972 subgiants taken from NGC 5897, IC 4499, and NGC 6584.

about this luminosity. Another explanation might be that BSSs in high density clusters evolve faster through this region than those in low density clusters. Again, the reasons for this are not completely clear.

8. Radial Distribution of Blue Stragglers

In § 3, we discussed the propensity of BSSs to be located near the cluster center in NGC 5053 and NGC 6584. We now consider all known globular cluster BSSs as a whole and examine their radial distribution. For each cluster in Table 2 except NGC 5024, Pal 5, and M55, we have calculated the distance of each BSS from the center of its respective cluster using positional information provided by the authors. The dotted line in Figure 6 shows the number of BSSs per unit area as a function of radial distance measured in terms of each cluster's core radius as given by Webbink (1985). For comparison, the solid line represents the subgiant stars in the globular clusters NGC 5897 (Sarajedini 1992a), IC 4499 (Sarajedini 1993a), and NGC 6584 (Sarajedini & Forrester 1994). Not surprisingly, the distribution of BSSs is more centrally concentrated than that of the subgiants. Note also that in NGC 6397, one of the blue stragglers is located less than 1" from the dynamical center of the cluster; five of the blue stragglers are positioned within 5" of the center (Aurière et al. 1990).

Since we have cluster positions for the majority of BSSs, it is interesting to construct LFs for BSSs located in different regions of the clusters. These are plotted in Figure 7. We have arbitrarily divided the 554 BSSs into two bins - those located inside 2 core radii (solid line) and those located outside 2 core radii (dotted line). It is clear from Figure 7 that the BSSs close to the center are peaked at a higher luminosity than those further

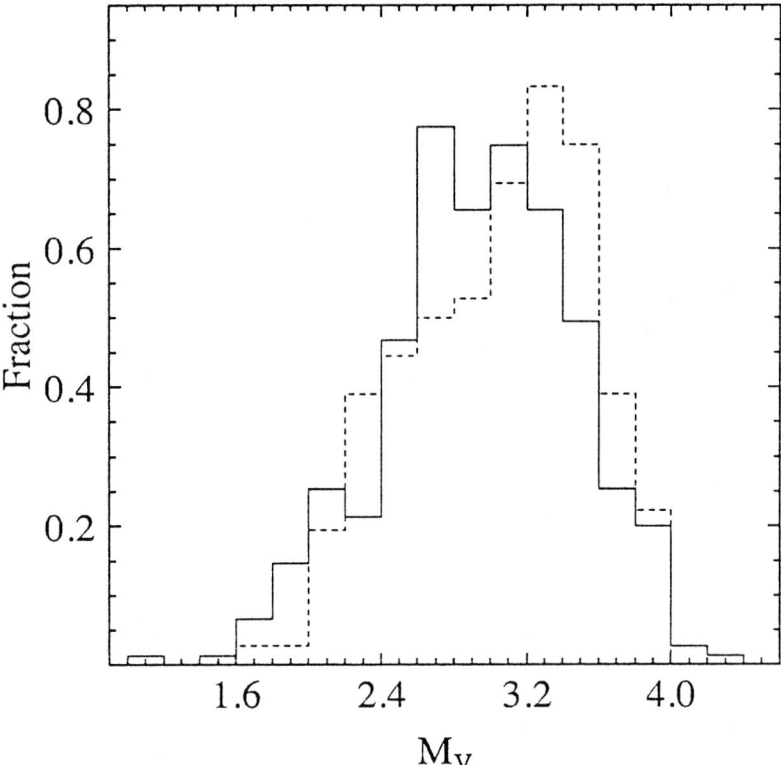

FIGURE 7. The luminosity functions of blue stragglers located inside (solid line) and outside (dotted line) 2 core radii. These have been scaled to unit area. Note that the blue stragglers close to the cluster center are peaked at a higher luminosity than those further away.

away. In addition, the extremely bright BSSs ($M_V \lesssim 2.0$) are preferentially located inside 2 core radii. This trend of decreasing luminosity with increasing radial position has been noticed before in individual globular clusters such as NGC 5053 (Nemec & Cohen 1989) and NGC 6101 (Sarajedini & Da Costa 1991).

9. Conclusions

We have presented the results of a continuing survey of Galactic globular clusters in search of blue straggler stars. There are now a total of $\approx$ 690 blue stragglers in 29 globular clusters. It is no longer true that only a select few globular clusters contain blue stragglers. On the contrary, blue straggler stars are now a legitimate component of probably all globular clusters. Their mean masses are definitely greater than stars at the turnoff (i.e., $\gtrsim 0.8 M_\odot$). Furthermore, $\approx$ 20 % of the blue stragglers have luminosities *greater* than a twice-the-turnoff-mass ZAMS star. This indicates that we are probably seeing the effects of blue straggler evolution away from the ZAMS or, perhaps more appropriately, the zero-age blue straggler sequence.

It is a great pleasure to thank the observing support staffs at CTIO and KPNO for making the observations reported herein possible.

REFERENCES

ALCAINO, G., BUONANNO, R., CALOI, V., CASTELLANI, V., CORSI, C. E., IANNICOLA, G., LILLER, W., 1987, The CM Diagram of the Nearby Globular Cluster NGC 6397. AJ, 94, 917–947

ARMANDROFF, T. E., 1989, The Properties of the Disk System of Globular Clusters. AJ, 97, 375–389

ARMANDROFF, T. E., DA COSTA, G. S., ZINN, R. J., 1992, Metallicities for the Outer-Halo Globular Clusters Pal 3, 4, and 14. AJ, 104, 164–177

ARP, H. C., HARTWICK, F. D. A., 1971, A Photometric Study of the Metal-Rich Globular Cluster M71. ApJ, 167, 499–509

AURIÈRE, M., ORTOLANI, S., LAUZERAL, C., 1990, Blue Stragglers at the Centre of the Post-Core-Collapse Globular Cluster NGC 6397. Nature, 344, 638–640

BAILYN, C. D., 1992, Are There Two Kinds of Blue Stragglers in Globular Clusters? ApJ, 392, 519–521

BAILYN, C. D., MADER, V., 1993, HST Observations of Blue Stragglers in M4. in Saffer R., ed, Blue Stragglers, ASP Conference Series, in press.

BENZ, W., HILLS, J. G., 1987, Three-Dimensional Hydrodynamical Simulations of Stellar Collisions. I. Equal-Mass Main-Sequence Stars. ApJ, 323, 614–628

BOLTE, M., 1992, CCD Photometry in the Globular Cluster NGC 288. I. Blue Stragglers and Main-Sequence Binaries. ApJS, 82, 145–165

BOLTE, M., HESSER, J. E., STETSON, P. B., 1993, CFHT Observations of Globular Cluster Cores I: Blue Straggler Stars in M3 (NGC 5272, GC1339+286). ApJ, 408, L89–L92

BREWER, J. P., FAHLMAN, G. G., RICHER, H. B., SEARLE, L., THOMPSON, I., 1993, CCD Photometry in the Globular Cluster NGC 3201. AJ, 106, 2158–2171

BUONANNO, R., CORSI, C. E., FUSI PECCI, F., 1985, The Giant, Asymptotic, and Horizontal Branches of Globular Clusters. II. Photographic Photometry of the Metal-Poor Clusters M15, M92, and NGC 5466. A&A, 145, 97–117

BUONANNO, R., CORSI, C. E., FERRARO, I., FUSI PECCI, F., 1987, CCD Photometry in Globular Clusters. II. NGC 7492. A&AS, 67, 327–340

BUONANNO, R., BUSCEMA, G., FUSI PECCI, F., RICHER, H., & FAHLMAN, G. G., 1990, Ruprecht 106: A Young Metal-Poor Galactic Globular Cluster. AJ, 100, 1811–1840

BURBIDGE, E. M., SANDAGE, A., 1958, The Color-Magnitude Diagram for the Galactic Cluster NGC 7789. ApJ, 128, 174–184

CACCIARI, C., 1984, BV Photometry of RR Lyrae Variables in the Globular Cluster NGC 3201. AJ, 89, 231–262

CHABOYER, B., SARAJEDINI, A., DEMARQUE, P., 1992, Ages of Globular Clusters and Helium Diffusion. ApJ, 394, 515–522

CÔTÉ, P., RICHER, H. B., FAHLMAN, G. G., 1991, Star Counts, Blue Stragglers, and Mass Segregation in NGC 7492. AJ, 102, 1358–1370

DA COSTA, G. S., ARMANDROFF, T. E., NORRIS, J. E., 1992, The Metal Abundance and Age of the Globular Cluster Ruprecht 106. AJ, 104, 154–163

DA COSTA, G. S., NORRIS, J. E., VILLUMSEN, J. V., 1986, The Blue Stragglers of Omega Centauri. ApJ, 308, 743–754

EGGEN, O. J., IBEN, I., 1989, Starbursts, Blue Stragglers, and Binary Stars in Local Superclusters and Groups. II. The Old Disk and Halo Populations. AJ, 97, 431–457

FERRARO, F. R., PARESCE, F., 1993, HST/FOC Observations of the Inner Regions of the Galactic Globular Cluster M15. AJ, in press

FERRARO, F. R., CLEMENTINI, G., FUSI PECCI, F., BUONANNO, R., 1991, CCD Photometry of Galactic Globular Clusters. III. NGC 6171. MNRAS, 252, 357–377

FERRARO, F. R., FUSI PECCI, F., CACCIARI, C., CORSI, C. E., BUONANNO, R., FAHLMAN, G. G., RICHER, H. B., 1993, Blue Stragglers in the Galactic Globular Cluster M3: Evidence for Two Populations. AJ, 106, 2324–2334

FUSI PECCI, F., FERRARO, F., R. CORSI, C. E., CACCIARI, C., BUONANNO, R., 1992, On the Blue Stragglers and Horizontal Branch Morphology in Galactic Globular Clusters: Some Speculations and a New Working Scenario. AJ, 104, 1831–1849

GRAHAM, J. A., 1982, UBVRI Standard Stars in the E-Regions. PASP, 94, 244–265

GREEN, E. M., DEMARQUE, P., KING, C. R., 1987, The Revised Yale Isochrones and Luminosity Functions. Yale University Observatory, New Haven

HARRIS, H. C., 1993, The Globular Cluster NGC 6366: Its Blue Stragglers and Variable Stars. AJ, 106, 604–612

HARRIS, W. E., 1975, New Color-Magnitude Data For Twelve Globular Clusters. ApJS, 29, 397–429

HEASLEY, J. N., CHRISTIAN, C. A., 1991, Photometry of the Outer Halo Globular Clusters NGC 5024 and NGC 5053. AJ, 101, 967–979

HESSER, J. E., HARRIS, W. E., VANDENBERG, D. A., ALLWRIGHT, J. W. B., SHOTT, P., STETSON, P. B., 1987, A CCD Color-Magnitude Diagram Study of 47 Tucanae. PASP, 99, 739–808

HILLS, J. G., DAY, C. A., 1976, Stellar Collisions in Globular Clusters. Astrophys. Lett., 17, 87–93

LANDOLT, A. U., 1983, UBVRI Photometric Standard Stars Around the Celestial Equator. AJ, 88, 439–460

LANDOLT, A. U., 1992, UBVRI Photometric Standard Stars in the Magnitude Range $11.5 < V < 16.0$ Around the Celestial Equator. AJ, 104, 340–371

LAUZERAL, C., ORTOLANI, S., AURIÈRE, M., MELNICK, J., 1992, Radial Distribution of Blue Stragglers and Surface Brightness Profile in the Post-Core-Collapse Globular Cluster NGC 6397. A&A, 262, 63–72

LEE, S.-W., 1977, The C-M Diagram of the Globular Cluster M55. A&AS, 29, 1–8

LEE, Y.-W., 1990, On the Sandage Period Shift Effect Among Field RR Lyrae Stars. ApJ, 363, 159–167

LEONARD, P. J. T., 1989, Stellar Collisions in Globular Clusters and the Blue Straggler Problem. AJ, 98, 217–226

MATEO, M., HARRIS, H. C., NEMEC, J., OLSZEWSKI, E. W., 1990, Blue Stragglers as Remnants of Stellar Mergers: The Discovery of Short-Period Eclipsing Binaries in the Globular Cluster NGC 5466. AJ, 100, 469–484

McCREA, W. H., 1964, Extended Main-Sequence of Some Stellar Clusters. MNRAS, 128, 147–155

MILLIS, A. J., LILLER, M. H., 1980, The Variable Stars in NGC 6584. AJ, 85, 235–241

NEMEC, J. M., COHEN, J. G., 1989, Blue Straggler Stars in the Globular Cluster NGC 5053. ApJ, 336, 780–797

NEMEC, J. M., HARRIS, H. C., 1987, Blue Straggler Stars in the Globular Cluster NGC 5466. ApJ, 316, 172–188

PARESCE et al., 1991, Blue Stragglers in the Core of the Globular Cluster 47 Tucanae. Nature, 352, 297–301

RICHER, H. B., FAHLMAN, G. G., 1988, Deep CCD Photometry in Globular Clusters VI: White Dwarfs, Cataclysmic Variables, and Binary Stars in M71. ApJ, 325, 218–224

SAIO, H., WHEELER, J. C., 1980, The Evolution of Mixed Long-Lived Stars. ApJ, 242, 1176–1182

SANDAGE, A., 1953, The Color-Magnitude Diagram for the Globular Cluster M3. AJ, 58, 61–75

SANDAGE, A., HARTWICK, F. D. A., 1977, Remote Halo Globular Cluster Palomar 5. AJ, 82, 459–464

SARAJEDINI, A., 1992a, CCD Photometry of the Globular Cluster NGC 5897: Morphology of the Color-Magnitude Diagram. AJ, 104, 178–189

SARAJEDINI, A., 1992b, Globular Cluster Photometry Near the Turnoff: Blue Stragglers, Relative Ages, and the Horizontal Branch. PhD thesis, Yale University.

SARAJEDINI, A., 1993a, A CCD Color-Magnitude Diagram for the Globular Cluster IC 4499. AJ, 105, 2172-2181

SARAJEDINI, A., 1993b, A Survey of Galactic Globular Clusters for Blue Straggler Stars. in Saffer R, ed, Blue Stragglers, ASP Conference Series, in press

SARAJEDINI, A., 1994a, CCD Photometry of the Galactic Globular Cluster NGC 6535 in the B and V Passbands. PASP, in press

SARAJEDINI, A., 1994b, A Technique for the Simultaneous Determination of Globular Cluster Metallicity and Reddening Using $(V, V-I)$ Color-Magnitude Diagrams. AJ, February

SARAJEDINI, A., DA COSTA, G. S., 1991, CCD Photometry of NGC 6101: Another Globular Cluster With Blue Straggler Stars. AJ, 102, 628-641

SARAJEDINI, A., FORRESTER, W., 1994, in preparation

SMITH, G. H., McCLURE, R. D., STETSON, P. B., HESSER, J. E., BELL, R. A., 1986, CCD Photometry of the Globular Cluster Palomar 5. AJ, 91, 842-854

STETSON, P. B., 1987, DAOPHOT: A Computer Program For Crowded-Field Stellar Photometry. PASP, 99, 191-222

STETSON, P. B., HARRIS, W. E., 1988, CCD Photometry of the Globular Cluster M92. AJ, 96, 909-975

STETSON, P. B., VANDENBERG, D. A., BOLTE, M., HESSER, J. E., SMITH, G. H., 1989, CCD Photometry of the Anomalous Globular Cluster Palomar 12. AJ, 97, 1360-1396

STRYKER, L. L., 1993, Blue Stragglers. PASP, 105, 1081-1100

WEBBINK, R. F., 1985, Structure Parameters of Globular Clusters. in Goodman J., Hut P., eds, Dynamics of Star Clusters, IAU Symp. No. 113, Reidel, Dordrecht, pp. 541-577

WHEELER, J C., 1979, Blue Stragglers as Long-Lived Stars. ApJ, 234, 569-578

ZINN, R. J., 1985, The Globular Cluster System of the Galaxy. IV. The Halo and Disk Subsystems. ApJ, 293, 424-444

ZINN, R. J., WEST, M. J., 1984, The Globular Cluster System of the Galaxy. III. Measurements of Radial Velocity and Metallicity For 60 Clusters and a Compilation of Metallicities For 121 Clusters. ApJS, 55, 45-66

Discussion

PHILIP: Sometime ago Peter Stetson and Bill Harris (1977, AJ, 82, 954) did a UBV study of NGC 1904. There were a number of stars below the horizontal branch which could be blue stragglers. I have made a 4-color CCD study of NGC 1904. In my y, (b-y) diagram I recover the Stetson and Harris stars, but I also find new members. The CCD allows you to go closer to the cluster center and I am picking up new members there. This is in agreement with the statements in your paper.

CARNEY: One of the interesting implications of your work is that there will be quite a number of blue straggler "descendents", at least in the crowded regions where they seem to predominate. Flavio Fusi-Pecci and his colleagues argue that some red horizontal branch stars may be produced this way. Your data, with good coverage of both the blue straggler and horizontal branch sequences, should be able to test the idea. Any results yet?

SARAJEDINI: I have not done this kind of analysis mostly because I am concerned that the field stars severely contaminate the red HB stars of the cluster. The key is to obtain membership information for the candidate cluster red HB stars, and then proceed with the kind of analysis you suggest. Andy Layden and I have submitted an observing proposal to conduct such a study.

McNAMARA: It is increasing clear that a variety of objects occupy the blue straggler domain. Let me point out that standard evolutionary theory predicts the correct masses and absolute magnitudes of the SX Phe stars. All field SX Phe stars rotate with $v \sin i < 20$ km s^{-1}. It is in my view difficult to reconcile these facts with the merger hypothesis.

SARAJEDINI: As Ruth Peterson has pointed out, the pulsation mechanism may act to spin down these stars.

PETERSON: Field RR Lyrae stars also fail to show significant rotation though field BHB stars do. Perhaps pulsation itself precludes rapid rotation. There is some concern that crowding effects in the more concentrated clusters may have affected the detection of the relatively low-luminosity blue stragglers. Have you determined crowding effects from artificially introducing stars from the cluster luminosity function? Can you say whether there are radial trends in the luminosity function in your data set?

SARAJEDINI: Doing the artificial star experiments is the next step in the analysis. Concentrating on the effects of crowding on the luminosity function will be the main goal of these experiments. As for radial trends in the luminosity functions, I intend to pursue this question and try to include the results in my proceedings contribution.

LITTLE: Have you considered the infall of a condensed star into the core of a massive star, so extending the main sequence lifetime by mass accretion, as an explanation of blue stragglers?

SARAJEDINI: I have not considered this idea mostly because the binary star/merger/collision theories do a good job of explaining the properties of blue stragglers.

UV Observations with FAUST and the Galactic Model

By NOAH BROSCH

The Wise Observatory and the School of Physics and Astronomy, Beverly and Raymond Sackler Faculty of Exact Sciences, Tel Aviv University, Tel Aviv 69978, ISRAEL

The North Galactic Pole (NGP) region was observed in the space–UV by the FAUST experiment. The observations are analyzed and the UV sources are identified with optical counterparts. Most (50/81) sources are A–F stars, but a few evolved objects (HBB, sdO/B and WDs) were detected, along with seven galaxies and one AGN. The observations are compared with a model of the distribution of UV stars in the Galaxy. The method of deriving UV properties of stars from their optical properties is described and some outstanding sources are identified.

1. Introduction

Various populations of Galactic objects emit significant radiation in the far ultraviolet. Their understanding has increased significantly in the last two decades, mainly because of the all sky survey of TD-1 in the mid-70's. The TD-1 satellite was launched by ESRO and surveyed the sky from 1972 to 1974. Boksenberg et al. (1973) described the UV sky survey telescope on TD-1. For a point source, the S2/68 experiment on TD-1 provided ~300 Å broadband flux measurements at 2740 Å and ~330 Å wideband photometry at 1565 Å 1965 Å and 2365 Å. The absolute photometry of the S2/68 experiment for the short waveband was claimed to be accurate to ~ 10 % (Jamar et al. 1976). Gondhalekar (1991) mentions than the experiment is not linear for sources fainter than 10^{-12} erg/sec/cm^2/Å.

The results of the TD-1 survey were published as a catalog that contains ~31,000 stars detected with S/N $\geq$ 10 (Thompson et al. 1978). This S/N limit corresponds to a flux density limit of 10^{-12} erg/sec/cm^2/Å, or about m_{UV} ~9. An extended version of the TD-1 catalog contains 58,012 stars and is part of the PhD thesis of Landsman (1984). In the 1565 Å band, the TD-1 full catalog is complete to ~8.5 mag. Note that although 58,012 stars are included, only 47,039 objects are brighter than the non-linearity limit in at least one band.

The ANS satellite obtained photometric information on 3573 ultraviolet stellar objects and a catalog of sources was published by Wesselius et al. (1982). The International Ultraviolet Explorer satellite obtained some 85,000 spectra in the region 1200–3100 Å, of which about 55,000 were obtained at low dispersion (6 Å resolution) and can be used to derive broad-band UV properties of the objects.

2. FAUST observations

The North Galactic Pole (NGP) region is well suited for studies of the ultraviolet stellar population because of the very low expected amount of intervening extinction at high galactic latitudes. Typical H I column densities at the NGP are 1-2 x 10^{19}cm^2 (Elvis et al. 1989). Hence, at the NGP we can expect the highest detectability of galactic and extragalactic UV sources. Thus we observed it with the FAUST telescope (Bowyer et al. 1993).

2.1. The FAUST telescope

FAUST is a telescope with an 8 degree wide field of view, a sensitivity bandpass between 1400 and 1800 Å, and an angular resolution of 3.5 arcminutes. The detector is a microchannel plate intensifier with wedge-and-strip anode, which records the position of each detected photon. The details of the image construction from these data, and subsequent reductions, are given in Bowyer et al. (1993). The FAUST image of NGP region was taken on March 27, 1992. The maximal exposure of any sky pixel (1.11 arcmin) was 13.2 minutes. The final image consists of a photon flux (in counts per second), which can be converted to physical units with a calibration derived in–flight. The total sky coverage of the NGP frame is 69 square degrees, larger than the instantaneous field–of–view of FAUST, because of Shuttle drift during the exposure, which is compensated in the image construction process.

The identification process is based on catalog searches and on CCD photometry and spectrophotometry performed at Wise Observatory. The number versus magnitude of sources detected in the FAUST image is compared with the predictions of the distribution of stars in the UV (Brosch 1991).

2.2. Source Identification

We developed an automatic detection technique for impartial source detection based on the size of a typical stellar image of FAUST, which extends for more than a single pixel. A square box three pixels wide was translated across the FAUST image in single pixel steps. The typical count rate within the box was adopted as the median of the nine values. This was compared with the typical neighborhood background count rate, measured in eight square boxes 6 pixels wide distributed symmetrically around the location checked with the 3×3 box. In each *sky box* the mean and standard deviation of pixel values was calculated. The median of the means and the median of the box standard deviations were adopted as the most significant "sky" level and typical deviation from this level.

This procedure identified 81 sources in the NGP frame to a signal–to–noise ratio of ~ 4.4, but not to a uniform detection depth, because of the different exposure levels across the field. The list of objects was transformed from internal pixel coordinates to the celestial coordinate system, based on about 20 objects identified in the TD–1 and SAO catalogs. The FAUST objects whose coordinates showed a large deviation from the catalog were rejected and the transformation was recalculated using the more accurately registered objects. The final mean coordinate accuracy is ~ 0.6 arcmin, about 1/6th of the resolution in the image.

The list of detected objects was used to calculate the FAUST count rate from each. The estimated measurement error is in most cases smaller than 0.35 mag. The photometry integrated all count rates within a predefined circular aperture centered on the detected location of the object, subtracting the expected count rate from the immediate neighborhood. The faintest source has a net count rate of 0.044 cps and only ~ 35 counts were collected during the UV exposure. This is also the only source which we could not associate with an optical counterpart and it may be an artifact. In this case, the faintest detection would have a total of 126 counts (NGP–55 = AN Com).

The identification process is based on searches of the following catalogs, in the order presented here, to identify possible optical counterparts: SAO, TD–1, TD–1 extended (Landsman 1984), Hipparcos Input Catalog (HIC), and the Hubble Space Telescope Guide Star Catalog (GSC). The decision to accept a specific object as a possible identification is based on proximity and color; the bluer object is adopted when more than one object is within the acceptance radius (~ 3 arcmin, the size of a typical stellar image of FAUST).

2.3. FAUST photometric calibration

Bowyer et al. (1993) used 16 objects on FAUST images observed by IUE to establish an in-flight calibration from FAUST count rate to photons/s/cm^{-2}/Å shown as Figure 2 in Bowyer et al. The relation can be transformed to UV monochromatic magnitudes m_F assuming that **all** photons are monochromatic at 1650 Å, the wavelength of peak response for FAUST:

$$m_F = 11.62 - 2.5 log(\Sigma N) \qquad (2.1)$$

Here ΣN is the total count density rate within a predefined aperture.

We compared the FAUST results with those of TD–1 in the shortest band, despite the possibility (Gondhalekar 1990; Bowyer et al. 1993) that TD–1 photometry is not as accurate as originally thought (Jamar et al. 1976). The two bandpasses overlap, with that of FAUST being wider, peaking at a longer wavelength, and extending to somewhat shorter wavelengths. The comparison, including all 14 identified TD–1 objects with measurable flux in the short TD-1 band, shows significant scatter of the faint objects around a linear relation defined over five magnitudes. This is due mainly to measurement errors of TD-1 and its lack of linearity at the faint end. The non-linearity of TD–1, mentioned by Gondhalekar (1990), apparently sets in from $m_{1565} \simeq 10$ mag, with fainter stars measured brighter than they really are.

3. Wise Observatory data

All FAUST sources which could not be uniquely identified with known objects were observed with the Wise Observatory RCA or TI CCD cameras in imaging mode and through B, V and R filters. The objects were observed together with photometric standards from Landolt's catalog (Landolt 1992). The standards were observed a number of times during the nights of 14 and 15 July 1992, together with single observations of the FAUST objects. The reduction included a simultaneous solution for the extinction and the transformation to the standard bands, up to and including second order terms. The measurement errors were derived from the reconstruction of the magnitudes and colors of the standard stars and are $\sigma_V = 0.03$, $\sigma_{B-V} = 0.03$ and $\sigma_{V-R} = 0.03$.

A number of sources could be associated with HST Guide Star Catalog (GSC) objects (Lasker et al. 1990 et seq.) where the natural "quick–V" (QV) magnitudes are listed along with accurate coordinates for the object. The photometric accuracy of GSC QV magnitudes is estimated at 0.9 mag (3σ) for all fields and from comparisons with photometric surveys (Russell et al. 1990).

Some FAUST sources could not be identified with known SAO stars or RNGC galaxies but appeared bright enough on the Palomar Sky Survey (PSS) to warrant an attempt of ground-based spectroscopy using Wise Observatory resources. Seven objects were observed with the Faint Object Spectrograph Camera (FOSC) on May 11 and 12 and July 8 and 9, 1992, with additional spectroscopy obtained from February to May 1993. In total, spectra of 16 FAUST–NGP objects were analyzed, some observed more than once. The FOSC spectra, in the configuration chosen for observing the FAUST-NGP candidates, cover the region from ~4000 Å to ~8000 Å with ~4 Å per pixel.

Each object was observed in a slitless configuration together with spectrophotometric standard stars and a He-Ar arc with a 2" slit. Flat fields were obtained of the twilight sky and of an internal incandescent filament lamp. The spectra were extracted from the CCD images after debiasing and were flat fielded and divided pixel-by-pixel by a similarly processed spectrum of the standard star.

4. Results

Most detected sources can be identified with known stars and most spectra show only Balmer absorption features. The lines are pronounced, but **not** extremely wide, indicating that the objects are **not** DA white dwarfs. No He II features are observed. In a number of cases we could not remove completely instrumental artefacts and sky emission lines and cannot establish the presence or absence of He I $\lambda 5876$ Å or Na I $\lambda 5890$ Å.

The FOSC spectra were compared with "template" spectra from Jacoby et al. (1984) and from Silva & Cornell (1992). Most appear to be early-type stars, from mid-B to late-A or F. The resolution is not sufficient to distinguish between main sequence and giant luminosity classes, but supergiants appear to be ruled out because supergiant B stars have almost no H_α absorption while late A supergiants show reasonably deep He I $\lambda 5876$Å absorption. The effective temperature was estimated from the assigned spectral class and the calibration of Straizys & Kurilene (1981).

The luminosity classification is supported by comparing the equivalent widths (EWs) of H_β and H_γ lines measured in the spectra with those from the models of Kurucz (1979), where the luminosity class is represented by surface gravity g. EWs from the Kurucz models for some log g families were plotted together with the measured EWs of the FAUST NGP sources. We found that $\log g = 4.0$ or 4.5 fits best the objects and concluded that these stars are *not* giants or supergiants, but most likely dwarfs or subdwarfs. Assuming a main sequence classification and no extinction, which is a justified assumption given the galactic location, and considering the apparent and absolute magnitudes, puts the objects at a distance of 1–3 kpc.

NGP-11 was included in our program before it was identified as BD+21° 2417. The Wise Observatory spectroscopy classified it B6-A2V, whereas the catalog puts it at A0. The photometry yielded $V = 10.39$ whereas the catalogued magnitude is 10.3. We regard the agreement as further confirmation of our accuracy and method of spectral classification.

In Table 1 we show the distribution of different types of sources according to their UV magnitude:

UV mag	B0-B9	A0-A9	F0-F9	G0-K9	HB, sdO/B	WD	AGN/galaxies
≤ 6.9	0	3	0	0	1	0	0
7.0–7.9	0	1	0	0	1	0	0
8.0–8.9	0	2	1	0	0	1	0
9.0–9.9	1	3	3	1	0	0	0
10.0–10.9	0	4	5	0	1	2	2
11.0–11.9	1	5	5	1	1	1	3
12.0–12.9	1	2	7	3	1	1	2
13.0–13.9	1	3	5	3	0	0	1
Total	4	23	26	8	5	5	8

5. Discussion

5.1. The UV sky model

A model of the stellar distribution in the galaxy in the UV was described by Brosch (1991). The predictions of this model, which is an adaptation of the Bahcall–Soneira model of the Galaxy (Bahcall & Soneira 1980), were compared to the UV measurements of FAUST by calculating the projected stellar densities in the visible, and transforming these to the UV with color-color relations between B–V and any UV–V color.

The transformations from optical color to UV were derived using the spectral response curve of FAUST as given in Bowyer et al. (1993), convolved with the spectral distributions of 71 stars from the ESA and NASA IUE spectrophotometric standard star catalogs with very small reddening. This yielded *true* FAUST UV magnitudes for stars with accepted unique classification and known optical magnitude and colors. The transformation could be successfully modelled as two straight lines, one for luminosity class V (N=47, correlation coefficient=0.96) and another for classes I to IV (N=24, correlation coefficient=0.99), as follows:

- For luminosity class V

$$[UV - V] = (-0.08 \pm 0.13) + (12.01 \pm 0.55) * (B - V) \tag{5.1}$$

- For classes I to IV

$$[UV - V] = (-0.55 \pm 0.05) + (10.69 \pm 0.24) * (B - V) \tag{5.2}$$

Using this transformation and the UV galaxy model we predicted the expected stellar density at the NGP for a ~40 square degree wide area in the central region of the frame, with ~uniform exposure level, and the model predictions were scaled to this area. Using the "nominal" band of 1500 Å, as in Brosch (1991), yields an underprediction of star density, i.e., the NGP field as measured by FAUST appears to have many more UV-bright stars than expected, whereas using the proper spectral band the predictions were in good agreement with the measurements. The model is apparently very sensitive to the accurate spectral bandpass of the survey.

Interestingly, we managed to reproduce fairly well the measured distribution of UV sources, without invoking either a stellar halo or a thick galactic disk; these components are not included in the original Bahcall–Soneirs model. This indicates that, at least to the detection limit of FAUST, such components are not required to model the stellar distribution.

5.2. Aberrant sources

The FAUST sample allows the evaluation of the accuracy of inferring the UV brightness of stars while relying on optical colors only. For this purpose we used the identifications of the 81 FAUST sources and their apparent magnitudes. In cases where both B–V and spectral types were available, these were used. If no luminosity class was available, it was assumed to be V. Non–stellar objects were discarded, as were stars for which no B–V or spectral type were available. The V-magnitude was adopted mainly from the photometric catalogs, but for some sources was taken from the SAO catalog. Those SAO sources where no B–V was available, were assigned the B–V they would have had if no reddening was present; this is justified at the Galactic Pole.

The procedure yields reasonable (UV–V) colors even when the exact spectral type, or luminosity class, or both, are not known and even for late–type stars; NGP-56 (HD109519), classified K0 based on positional coincidence with a HIC object, is predicted within one magnitude of its observed (UV–V) color despite the lack of many late–type stars in the transformation relations. There are some exceptions to the rule, such as

NGP-52 (GSC1229+2440), predicted ~10 mag fainter than measured. This, however, is based on the QV magnitude of the GSC source we correlated with the FAUST position. The V magnitude we derive for our spectroscopy is fainter than the QV in the GSC, making the UV excess issue even more acute. It is possible that the excess is genuine and that the star is peculiar in some way. This should be determined by a more detailed study. Carnochan et al. (1975) found many more UV-bright sources than expected, from the TD-1 observations of a small fraction of the sky.

The same may be true for sources that are UV-faint with respect to their predicted UV magnitudes. Some that are more UV-deficient are NGP-57 (GSC1233+2057) and NGP-75 (SAO 82503), by about four magnitudes each.

5.3. Location of sources

Given the apparent magnitude of the sources and their absolute magnitudes from their spectral identification, it is possible to locate the stars in three-dimensional space. We have neglected reddening since this effect is clearly small in the direction of the North Galactic Pole. The distances for main-sequence stars range up to ~3.6 kpc (for NGP-14 = GSC1230+2139), fairly high above the galactic plane for a stellar population with low scale height. The presence of normal, Population I young stars at large distances from the plane has been discussed by Keenan (1992), who concluded that some are genuinely young population II objects.

In addition, Lu (1994) did not find any B stars at the South Galactic Pole, in contrast with the findings here. Note though that most B stars found here are late-B stars, which may be misidentified early-A stars. Much higher spectral resolution than achieved here is required to distinguish between normal B stars, early A stars and post-AGB objects. Note also that young, few Gyr old clusters are found above the galactic disk, albeit at not so high galactic latitudes (Janes 1994).

5.4. Evolved stars and non-stellar objects

We find among 81 FAUST sources four white dwarfs, of which two are ROSAT WFC bright sources, one of which may be a subdwarf, three previously known subdwarfs and fifty A and F stars, some of which may be subdwarfs. The WDs appear over-abundant; the distribution given in Boyle (1989) predicts that only ~1.4 WD should be present in the NGP area surveyed by FAUST. This is by assuming that only WDs brighter than B = 16 can be detected by FAUST (which is equivalent to assuming [1650]-B colors bluer than -4 for the faintest objects).

The objects identified as subdwarfs are under-represented; when comparing with the preliminary results of the Hamburg Schmidt survey (Dreizler et al. 1994) would expect that almost 50 % of the sources be subdwarfs, as claimed by Carnochan et al. (1975) from the preliminary analysis of the TD-1 results. However, if these claims are taken at face value, about 1300 objects similar to those found by Carnochan et al. (1975) should exist over the sky, or only two in the FAUST sample to the depth of the TD-1 survey. The numbers reported here are in the right order of magnitude. The tentative conclusion is that probably most subdwarfs are fainter that the UV magnitude limit of FAUST or TD-1, thus are not detected in our data set.

In addition to the stellar sources, six large galaxies, which are relatively nearby spirals, and one dwarf galaxy were detected, along with one AGN. These sources will not be discussed here.

6. Conclusions

We analyzed a UV image covering 69 square degrees of sky near the North Galactic Pole. We concentrated on 81 UV sources and identified them with known objects. Only one source could not be associated with an optical counterpart; this is the faintest in our sample and may be spurious. Among 62 "normal" stars, *i.e.*, not WDs or HB/sd stars, we found that 81 % belong to A or F spectral classes. A comparison of the star counts and the predictions of a model for the galactic UV stellar population shows good agreement. The entire sample of UV-measured objects from the FAUST flight, which comprises several thousand UV sources at various galactic latitudes, will allow firmer statistical conclusions on the distribution of UV-bright stars and possibly the identification of more peculiar UV sources.

Observations at the Wise Observatory are supported by grants from the Israel Academy of Sciences and the Ministry of Science and the Arts. The UV astronomy effort at Tel Aviv University is supported by special grants to develop a space UV astronomy experiment (TAUVEX) from the Ministry of Science and the Arts and the Austrian Friends of Tel Aviv University. The access to the FAUST data and continuing support of UV astronomy at Tel Aviv on the part of the Berkeley Space Astronomy Group (S. Bowyer and collaborators) is gratefully acknowledged.

REFERENCES

BAHCALL, J. N., SONEIRA, R. M., 1980, The Universe at Faint Magnitudes. I. Models for the Galaxy and the Predicted Star Counts. ApJS, 44, 73-110

BOKSENBERG, A., EVANS, R. G., FOWLER, R. G., GARDNER, I. S. K., HOUZIAUX, L., HUMPHRIES, C. M., JAMAR, C., MACAU, D., MACAU, J. P., MALAISE, D., MONFILS, A., NANDY, K., THOMPSON, G. I., WILSON, R., WROE, H., 1973, The Ultra-Violet Sky-Survey Telescope in the TD-1A Staellite. MNRAS, 163, 291-322

BOYLE, B. J., 1989, The Space Distribution of DA White Dwarfs. MNRAS, 240, 533-549

BOWYER, S., SASSEEN, T. P., LAMPTON, M., WU, X., 1993, In-flight Performance and Preliminary Results from the Far Ultraviolet Space Telescope (FAUST) Flown on ATLAS-1. ApJ, 415, 875-881

BROSCH, N., 1991, A model of the Galaxy in the Ultraviolet. MNRAS, 250, 780-785

CARNOCHAN, D. J., DWORETSKY, M. M., TODD, J. J., WILLIS, A. J., WILSON, R., 1975, A Search for Ultraviolet Objects. Phil. Trans. R. Soc. Lond. A., 279, 479-485

DREIZLER, S., HEBER, U., JORDAN, S., ENGELS, D., 1994, Faint Blue Stars from the Hamburg Schmidt Survey. in Adelman S. J., Upgren A. R., Adelman C. J., eds, Hot Stars in the Halo, Cambridge University Press, Cambridge, pp. 228-237

ELVIS, M., LOCKMAN, F. J., WILKES, B. J., 1989, Accurate Galactic N_H Values Towards Quasars and AGN. AJ, 97, 777-782

GONDHALEKAR, P. M., 1990 The Ultraviolet Starlight in the Galaxy. in Bowyer S., Leinert C., eds, The Galactic and Extragalactic Background Radiation, Kluwer Academic Publishers, Dordrecht, pp. 49-62

JACOBY, G. H., HUNTER, D. A., CHRISTIAN, C. A., 1984, A Library of stellar spectra. ApJS, 56, 257-281

JAMAR, C., MACAU-HERCOT, D., MONFILS, A., THOMPSON, G. I., HOUZIAUX, L., WILSON, R., 1976, Ultraviolet Bright-Star Spectrophotometric Catalogue. ESA SR-27

JANES, K. A., 1994, Can Stars Still Form in the Galactic Halo?, in Adelman S. J., Upgren A. R., Adelman C. J., eds, Hot Stars in the Halo, Cambridge University Press, Cambridge, pp. 330-339

KEENAN, F. P., 1992, Star formation in the Galactic halo. QJRAS, 33, 325-333

LANDOLT, A. U., 1992, UBVRI photometric standards in the magnitude range $11.5 \leq V \leq 16.0$ around the celestial equator. AJ, 104, 340–371

LANDSMAN, W. B., 1984, PhD Thesis, The Johns Hopkins University

LASKER, B. M., STURCH, C. R., MCLEAN, B. J., RUSSELL, J. L., JENKNER, H., SHARA, M. M., 1990, The Gude Star Catalog. I. Astronomical foundation and image processing. AJ, 99, 2019–2058

LU, P. K., 1994, Hot Stars at the South Galactic Pole. in Adelman S. J., Upgren A. R., Adelman C. J., eds, Hot Stars in the Halo, Cambridge University Press, Cambridge, p. 124

RUSSELL, J. L., LASKER, B. M., MCLEAN, B. J., STURCH, C. R., JENKNER, H., 1990, The Guide Star Catalog. II. Photometric and astrometric models and solutions. AJ, 99, 2059–2081

SILVA, D. R., CORNELL, M. E., 1992, A new library of stellar spectra. ApJS, 81, 865–881

STRAIZYS, V., KURILENE, G., 1981, Fundamental stellar parameters derived from evolutionary tracks. Ap&SS, 80, 353–368

THOMPSON, G. I., NANDY, K., JAMAR, C., MONFILS, A., HOUZIAUX, L., CARNOCHAN, D. J., WILSON, R., 1978, Catalogue of stellar ultraviolet fluxes. The Science Research Council

WESSELIUS, P. R., VAN DUINEN, R. J., DEJONSE, A. R. W., AALDERS, J. W. G., LUINSE, W., WILDEMAN, K. J., 1982, ANS ultraviolet photometry, catalogue of point sources., A&AS, 49, 427–474

Discussion

BIDELMAN: In most parts of the sky, the Henry Draper Catalogue types for early B stars fainter than about 8th magnitude are systematically considerably too late. This will affect your predictions of objects to be seen.

BROSCH: We have synthesized the entire TD-1 catalog using the apparent magnitudes, colors, spectral types and luminosity classes listed in the Hipparcos Input Catalog and in the SAO. We found that the prediction accuracy has a typical error of 0.5 mag in the longest TD-1 band, degrading to slightly more than one magnitude in the shortest TD-1 band. This shows that the method works in principle, and that it can be much improved if the stellar parameters would be known better. It would be extremely useful if unpublished data, such as revised spectral types could be made available to us for UV prediction purposes.

LANDSMAN: Your identifications include 8 G-K stars. The ROSAT WFC survey has shown that a significant fraction of white dwarfs could be optically hidden in late-type binaries. How certain are you of the late-type identifications?

BROSCH: The identifications rely, by definition, on coincidence with optical counterparts. Identifying a Faust source as a G-K star does not preclude it being a binary with a hot WD that produces the UV flux. In fact, by comparing the predicted UV flux with the measured flux we may have a way of finding candidates for further detailed studies. It would be a way to identify all 8 GK stars as mixed type binaries with hot WDs; then the number of WDs would then triple in the sample.

LU: Yesterday we found no B star brighter than 16.0 magnitude at SGP, yet the galaxy model predicted even more. Can you comment on that?

BROSCH: We are fairly confident of some B star identifications. In particular, one object was not found in the original catalog search and was observed at the Wise Observatory. The classification from rather low resolution spectroscopy. The colors were confirmed by subsequently finding the star in a publication, with just the right spectral type, magnitude and colors. In conclusion, there are some (a few) B stars in the NGP region. The model predictions do not refer specifically to spectral types, although they are available, but to numbers of stars per magnitude interval. Note also that most B stars we found are late-B. They may be misidentified early-A stars.

Hot Stars at the South Galactic Pole

By PHILLIP K. LU

Center for Galactic Astronomy, Western Connecticut State University, Danbury, CT 06810, USA

In the course of a new thin-prism survey for F and G-stars within a 40 sq-deg area centered on the SGP, a set of 3155 stars with spectral feature earlier than G5 were selected to a limiting magnitude of 16. The later F and early G-stars will be used as tracer objects for the K(z) study. A list of 155 stars was selected between A0 and F5 or mean color index of b-y < 0.288 for this study. This represents less than 5 % of the catalog 3155 stars between the types of A, F and early G-stars that now have MK spectral types (1274 stars) or uvby, H-β photometry (1403 stars). The results show that there are: a) about one star per sq-deg of the stars between A0 to F0, b) no O and B-stars brighter than 16 magnitude was found within this 40 sq-deg, c) nearly 30 % of the listed stars with large c_1 index are probable Population II types, and d) seven objects with small m_1 and large c_1 are suspected to be candidates for blue horizontal-branch stars.

1. Introduction

Studies are now in progress to determine the galactic gravitational force law K(z), the local mass density in the Milky Way galaxy using faint main-sequence F and G-stars as tracer objects to about $z = 1.5$ kiloparsec. The principal objectives of the study are to investigate the velocity distribution, to search for evidence of a thick disk, and to conduct a dynamical study of the dark matter in the galactic disk. The description, planning and some preliminary results for this study were reported elsewhere (Lu, 1988, 1989, 1990, 1991a, 1991b; Lu et al. 1992 [Paper I], 1993). Some of the most comprehensive reports by other investigators are included in Philip & Lu (1989). The earliest investigations of the velocity dispersion and mass density were made by Oort (1932, 1960) and led to the first suggestion of missing mass in the Galaxy. The mean mass density value in the solar neighborhood is now called the "Oort Limit".

The hot stars selected near the South Galactic Pole in this study are only a small sample of those stars that now have 2D-Frutti spectral type or four color and H-β photometry. Although A and early F stars are too young to be in dynamic equilibrium for the study of velocity dispersion (Bahcall 1984; King 1989), they are still useful for providing data for:

1) detecting blue horizontal-branch stars when their m_1 index is small and their c_1 index is large,

2) segregating individual objects of the halo population with large c_1 indices, and

3) the statistical study of galactic structure as function of z-scale height distance above the plane.

Early studies and survey for early-type stars near the South Galactic Pole include those of Haro & Luyten (1962), Westerlund (1963), Bok & Basinski (1964), Luyten (1966), Philip & Sanduleak (1968), MacConnell et al. (1971), and Slettebak & Brundage (1971). The more recent survey include those by Preston et al. (1991), Hauck (1993), and Kilkenny et al. (1994).

2. Observations and results

2.1. The selection of hot stars

The criterion for selecting A and early F stars in this paper is based on the 2DF spectral types being earlier than F5 or color index of b-y being less than 0.288 (Crawford 1979). These stars all belong to the catalog of 3155 stars that resulted from the 1.5° thin prism spectral survey covering about 40 sq-deg centered at the SGP. Detailed planning and designs are described by Lu et al. (1992, Paper I) while its outline is recapitulated here. A direct film copy covering the same area provided by the UK Schmidt Telescope Unit (UKSTU) at the Royal Observatory, Edinburgh, Scotland, was scanned using the Yale PDS 2020s microdensitometer with a 40 x 40 micron aperture which provide the positions of the catalog. The accuracy in the position is about 2.6" pixel^{-1}. The overall catalogue position accuracy is about ± 1.5" which is constrained by the half size of a pixel.

This 40 sq-deg. area was divided into 6 zones (named as SGP1 to SGP6), each containing 18 frames due to the size of the AED (Advanced Electronic Display) dimension of 750 x 1000 pixels. Therefore, the 18 frames per SGP zone were delineated as SAxxxyy, the name of the star; the x's are pixel numbers in right ascension and the y's are pixel numbers in declination.

Table 1 consists four types of databases which including a) name and positions for 155 A and early F stars at the South Galactic Pole, b) MK spectral type, radial velocity, c). Strömgren 4-color and H-β photometry, and d). Cross references for HD, SAO, Bok & Basinski (1964) and Slettebak & Grundage numbers (1971).

2.2. 2D-Frutti spectra

One hundred and fifteen of the listed hot stars now have 2DF spectral types. These spectra have been obtained with the 2D-Frutti spectrograph at CTIO. The system is a combined image-tube and CCD system with 132 pixels in y and 3080 pixels in x and with a dispersion of 45 Å mm^{-1} at second order. Spectral and luminosity standards for A, F and G-stars between various luminosity classes were obtained using the identical 2DF settings for the calibration of classification (Lu 1994). The accuracy in classifying in spectral type largely follows the signal-to-noise (S/N) ratio in the 2DF spectra. The mean difference is 0.81 ± 1.93 subclass for 71 HD stars (Paper I) in all luminosity classes compared to the Michigan Catalogue of Spectral Types for HD Stars (Houck 1982) which have both spectral and luminosity types. The mean 2DF spectral distribution for these hot stars equals to F1.2 ± 4.9 subclass (Figure 1) which agrees well with the mean photometric (b-y) color index of 0.232 or F2 (Figure 2). Spectral comparisons between the SGP hot stars and SB stars are shown in Figure 3. Crosses are those stars with peculiar spectra or spectral types without subclasses listed in the SB catalog (Slettebak & Brundage 1971).

2.3. Direct CCD and ASCAP photometry

Direct prime focus CCD images using the 0.9-m telescope with a TI chip and ASCAP (Automated Single Channel Aperture Photometry) measurements with the Yale 1-m telescope at CTIO were obtained for 1403 stars in the uvby, H-β system. Distinguishing a subdwarf from a normal dwarf is very essential in the K(z) study since about 7 % metal deficient or intermediate Population II F-stars have been detected in the 2DF spectra (Paper I). In this study, nearly 30 % of the hot stars have c_1 indices great or equal to 0.8 and are presumably Population II stars (Figure 4). About 11 stars in Table 1 with small m_1 indices and large c_1 values probably could be considered as the candidates for blue horizontal-branch (BHB) stars (Figure 4).

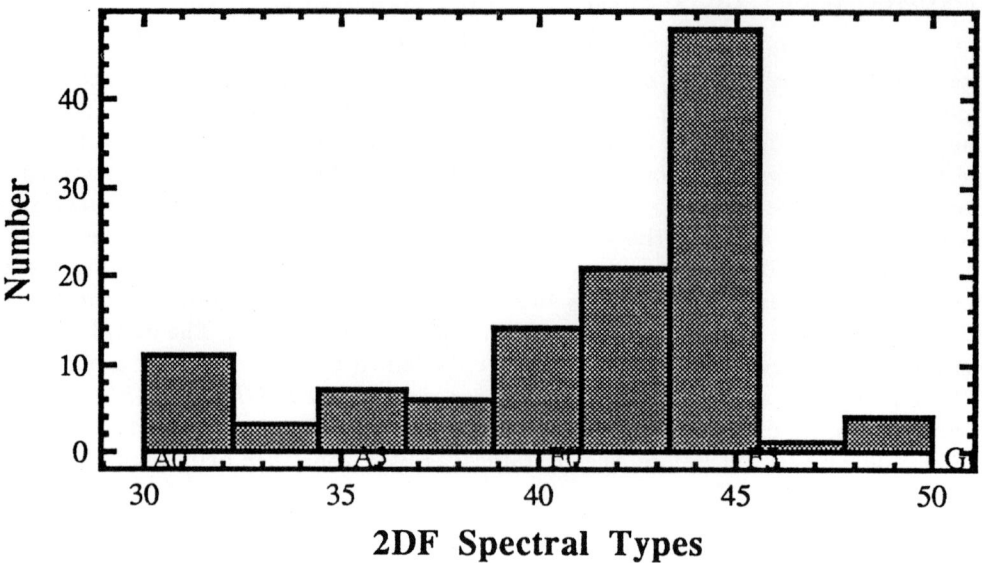

FIGURE 1. 2D-Frutti spectral types distribution for the hot SGP stars

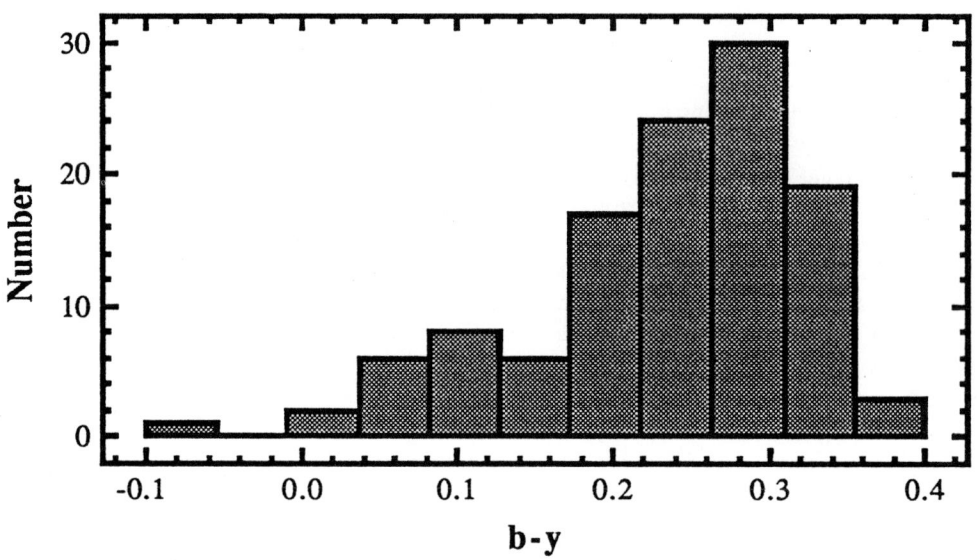

FIGURE 2. Distribution of (b-y) color index for the hot SGP stars

The observations normally require 4 to 5 minutes' integration time for the TI chip to reach a 16 th magnitude star in the vby bandpasses. However, the very low quantum efficiency at the u-band of the CCD chip required quadruple integration times to achieve the same intensity. The ASCAP system is generally more suitable and efficient than CCD imaging using the Strömgren system since the u-bandpass in a photon counting system

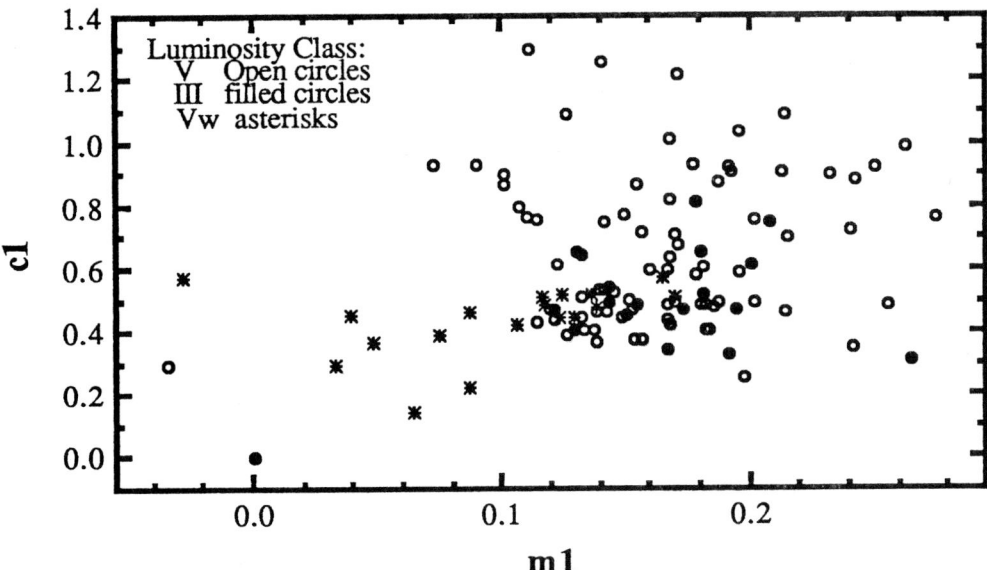

FIGURE 3. Metallicity index m_1 vs luminosity index c_1 plane for the hot SGP stars. Open circles for dwarf stars, filled circles for the giants and asterisks for the Population II weak-lined objects.

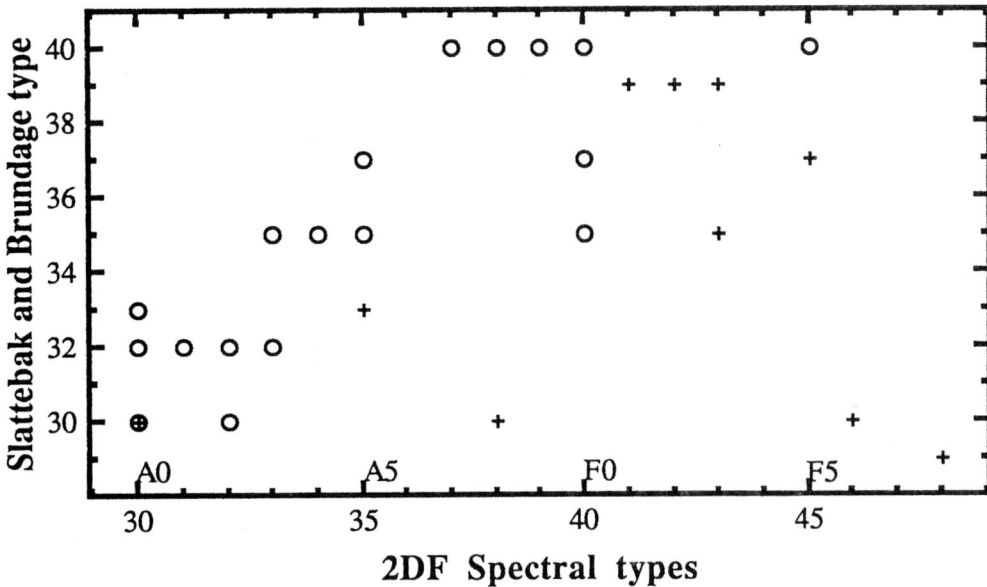

FIGURE 4. Comparison of spectral types between 2DF and Slettebak-Brundage spectral types. Plus signs are for the peculiar spectra in SB.

is about a factor of three faster than in the CCD system with comparable accuracy and it is designed for single object photometry.

Neutral density filters of 2.5 to 7.5 magnitudes which were used for Crawford's bright standards (Crawford 1975) caused some calibration problems because the transmission curves of the combined neutral density filter set are not well known. A systematic difference appears to exist in Twarog's secondary standards (Twarog 1984) and data from Olsen (1983) and McFadzean et al. (1982). A new set of HD stars fainter than 7 th magnitude selected from McFadzean et al. (1982) were observed and calibrated as secondary standard stars within the 40 sq-deg of this catalog. Results and the discussion of their residuals were reported earlier (Lu 1993).

The mean color distribution of (b-y) for 155 hot stars equals to 0.232 ± 0.088 compares to 0.359 ± 0.080 for 1403 F and G stars which corresponds spectral type of F9 V. Spectral, color distribution and a typical (b-y) vs. m_1 diagrams are shown in Paper I.

2.4. Cross-correlation radial velocities

Analyses to obtain the radial velocity with an accuracy of about ± 5 km s^{-1} at 0.5 Å pixel^{-1} using 2D-Frutti spectra are now in progress. Preliminary results were reported by Lu (1991a, b). Spectra with a very low signal-to-noise ratio will generally affect the accuracy and the determination of velocity using cross-correlation and the new technique reported by Carney et al. (1987) for extracting metallicities from high resolution and low S/N spectra. Radial velocity reductions using cross correlation techniques were first done with the CCPROG developed at Yale after considerable time and effort was spent in trying to use the early version of FXCOR in IRAF. The revised FXCOR seems working well in comparison with the RVSAO except that the RVSAO is more efficient and conducts reduction non-interactively.

The radial velocity distribution for the stars with spectral class A0 to F5 peaks at 13.6 ± 35.5 km s^{-1} (Figure 5). This compares with 13.7 ± 31.4 km s^{-1} for the weighted mean velocity of about 700 F and G-stars. The z-scale height distance distribution has a mean of 499 ± 335 pc which is about 150 pc greater than to the sample of 850 stars (Figure 6).

3. Discussion

The spatial distribution of the entire SGP catalogue and the hot stars appears to be random. The small clumping near and surrounding NGC288 can be suggested due to cluster members and high density of stars per square degree in that area. At least 4 stars are suspected to be cluster members, either RR Lyrae stars or blue stragglers. Of the 155 stars listed in Table 1, there are about 1 star per sq-deg for stars early than F0. This finding is identical with that of Slettebak & Brundage's survey (1971) of 957 stars in an 840 sq-deg area to a limiting magnitude of about 14 to 15. This thin prism survey essentially identified every star listed in SB catalog within this 40 sq-deg. For the later types (F1 to F5), there are about 4 stars per sq-deg with magnitude range brighter than 15; no bright B-star is found within this 40 sq-deg area; and 4 stars with a 'cm' in the remark column are probably the cluster members of NGC288.

On the basis of appropriate spectral type at maximum light, three stars or 2 % of the listed hot stars should be pulsating variable either RR Lyrae or Cepheid variables. The star SGP7552 30 has been suspected as variable based on the pixel-magnitude determined using UK Schmidt direct film copies and its recent 4-color photometry. SGP28537 2, or SB 405, also CD-28° 307 is a Cepheid variable and SGP09515 36 has been also suggested

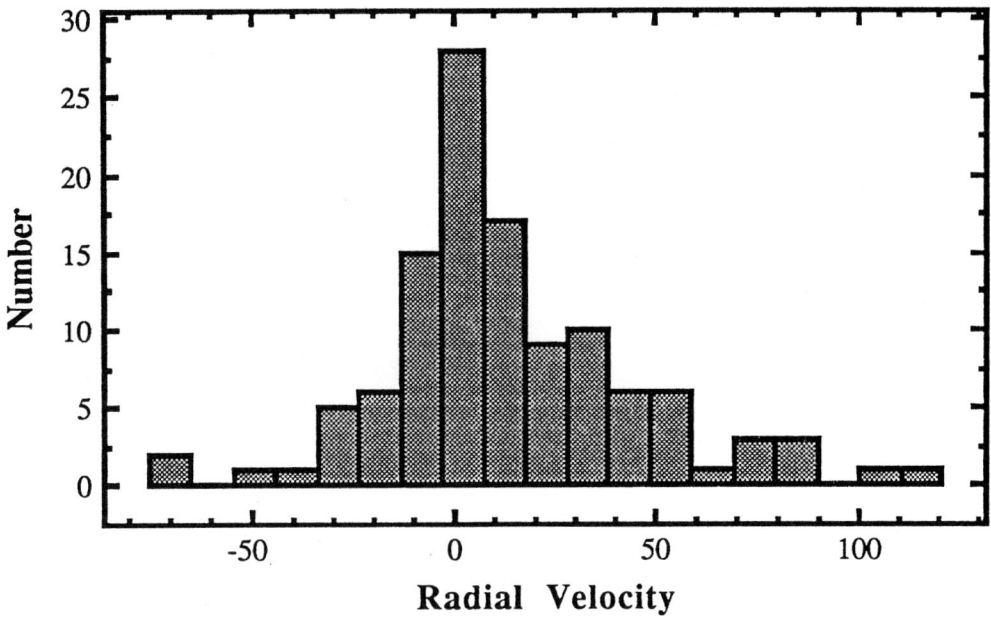

FIGURE 5. Radial velocity distribution for the hot SGP stars

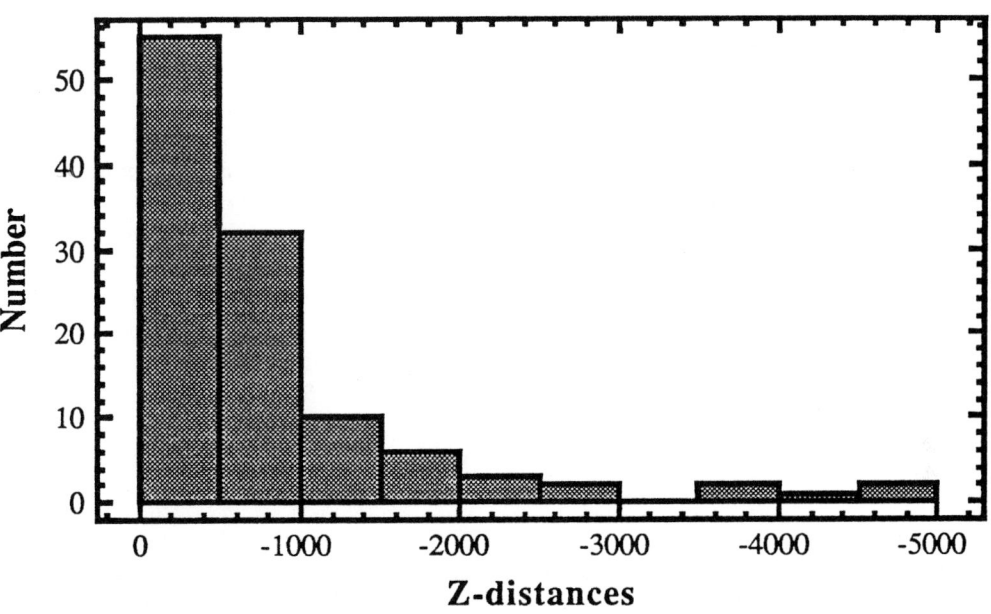

FIGURE 6. The z-scale height distance distribution for the hot SGP stars

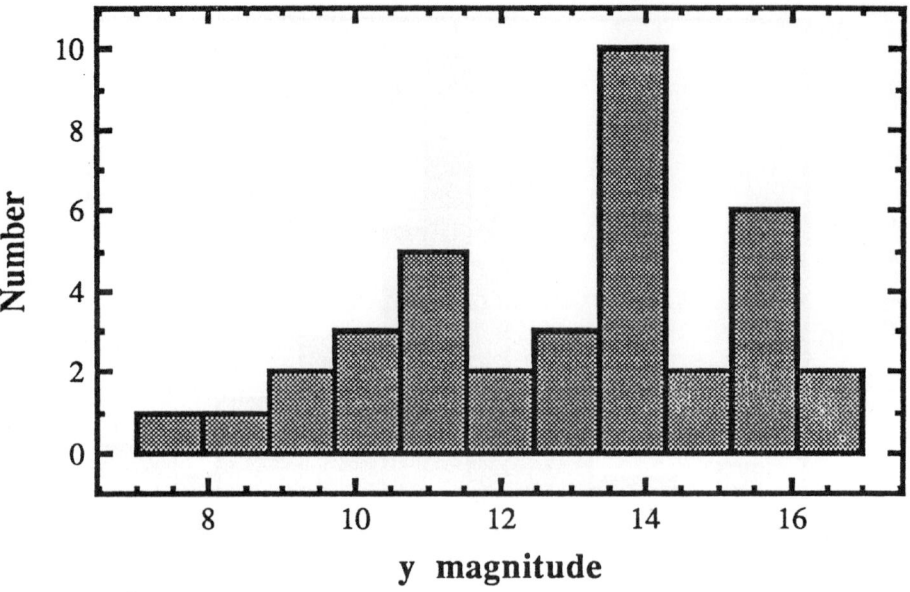

FIGURE 7. Visual magnitude distribution for A0 to A5 stars

as a variable by Bok & Basinski (1964). In addition, SGP38060 26, SB 371, and UV Scl are known as a RR Lyrae stars.

The magnitude distributions in the range A0 to A5 and A6 to F5 as function of visual magnitude are shown in Figures 7 and 8, respectively. Although, the sample is incomplete since both spectral and photometric observations are still in progress. Nevertheless, the magnitude distribution for A0 to A5, which would include hot subdwarfs, white dwarfs and horizontal-branch stars peaks at faint magnitudes and indicates that the sample is composed largely of normal Population I stars. This is clearly shown from spectral classification and photometric analyses in section II. The distribution for the late-type hot stars (A6-F5) reaches a rather flat maximum. Similar finding has been obtained by Philip & Sanduleak (1968) near South Galactic Pole. These results suggest that stellar population gradual change from the old disk population into mostly halo population for stars fainter than 11 or 12 apparent magnitude.

In the course of this work I have been a Visiting Astronomer at Cerro Tololo Inter-American Observatory, which is operated by the Association of Universities for Research in Astronomy, Inc., under contract with the National Science Foundation. I thank Drs. Robert Williams (former director) and Mark Phillips and their staff at CTIO for all their support, and the staff of UKSTU, Royal Observatory, Edinburgh, Scotland, who provided the needed film copies to conduct our earlier search for F-stars. Special thanks is due to Donald Platt, who has done all the radial velocity reductions. This research has been supported in part by grants from the National Science Foundation (AST 8713183) and from the Connecticut State University system.

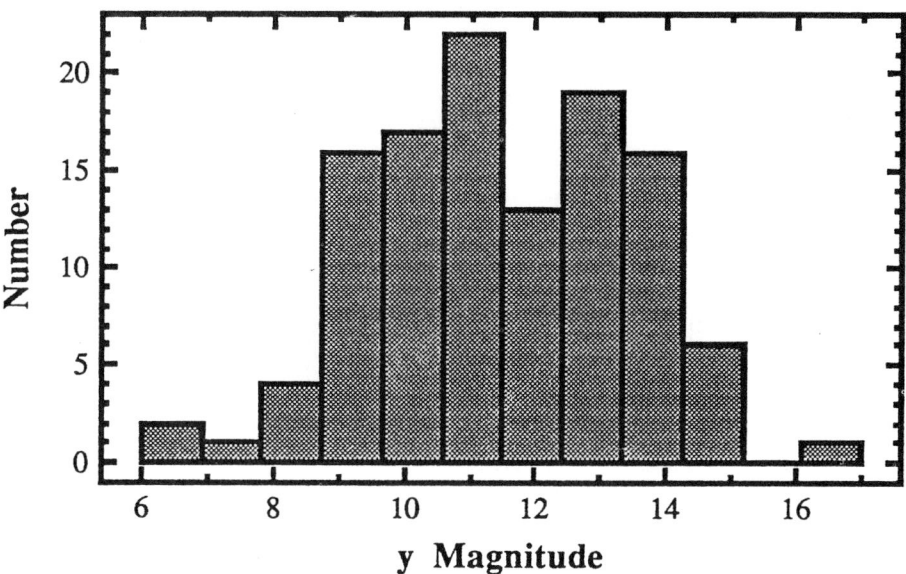

FIGURE 8. Visual magnitude distribution for A6 to F5 stars

REFERENCES

BAHCALL, J. N., 1984, ApJ, 276, 156
BOK, B. J., BASINSKI, J., 1964, Memoirs of Mt. Stromlo Obs., no. 16
CARNEY, B. W., LAIRD, J. B., LATHAM, D. W., KURUCZ, R. L., 1987, AJ, 94, 1066
CRAWFORD, D. L., 1975, AJ, 80, 955
CRAWFORD, D. L., 1979, in Philip A. G. D., ed, Problems of Calibration of Multicolor Photometric Systems, Dudley Obs., Schenectady, N.Y., p 23
HARO, G., LUYTEN, W., 1962, Bol. Obs. Tonantzintla y Tacubaya, 22
HOUK, N., 1982, Michigan Catalogue of Two-Dimensional Spectral Types for the HD stars, Vol. 3, Univ. of Michigan, Ann Arbor, MI
KILKENNY, D., O'DONOGHUE, D., STOBIE, R. S, CHEN, A., KOEN, C., SAVAGE, A., 1994, in Adelman S. J., Upgren A. R., Adelman C. J., eds, Hot Stars in the Halos, Cambridge University Press, Cambridge, p. 70
KING, I. R., 1989, in Philip A. G. D., Lu P. K., eds, The Gravitational Force Perpendicular to the Galactic Plane, L. Davis Press, Schenectady, N.Y. p 147
LU, P. K., 1988, in Philip A. G. D., ed, The Workshop on the Calibration of Stellar Ages, L. Davis Press, Schenectady, N.Y., p 217
LU, P. K., 1989, in Philip A. G. D., Lu P. K., eds, The Gravitational Force Perpendicular to the Galactic Plane, L. Davis Press, Schenectady, N.Y. p 95
LU, P. K., 1990, in Jarvis, Terndrup, eds, ESO/CTIO Workshop on Bulges of Galaxies, La Serena, Chile, p. 351
LU, P. K., 1991a in Janes K. A., ed, Formation and Eolution of Stars Clusters, ASP Conf. Ser., 13, 581
LU, P. K., 1991b, in Philip A. G. D., Upgren, A. R., eds, Objective-prism and Other Surveys, Van Vleck Obs Contr., 12, p. 33
LU, P. K., 1994 in Philip A. G. D., Hauck B., Upgren A. R., eds, "Databases for Galactic Structure", L. Davis Press, Schenectady, p. 25

Lu, P. K., Miller, J., Platt, D., 1992, ApJS, 83, 203

Luyten, W. J., 1966, A Search for Faint Blue Stars, The SGP, number 34 University of Minn. Obs. Publ.

MacConnell, J., Frye, R .L., Bidelman, W. P., Bond, H. E., 1971, PASP, 83, 98

McFadzean, A. D., Hilditch, R. W., Hill, G., 1982, MNRAS, 205, 525

Olsen, E. H., 1983, A&AS, 54, 55

Oort, J. H., 1932, Bull. Astron. Inst. Netherlands, 6, 249

Oort, J. H., 1960, Bull. Astron. Inst. Netherlands, 15, 45

Philip, A. G. D., Lu, P. K., 1989, eds, The Gravitational Force Perpendicular to the Galactic Plane, L. Davis Press, Schenectady, N.Y.

Philip, A. G. D., Sanduleak, H., 1968 Bol. Obs. Tonantzinla y Tacubaya, 4, number 30, 253

Preston, G. W., Schectman, S. A., Beers, T. C., 1991, ApJ, 375, 121

Slettebak, A., Brundage, R. K., 1971, AJ, 76, 338

Twarog, B. A., 1984, AJ, 89, 523

Westerlund, B. E., 1963, MNRAS, 127, 83

Discussion

PHILIP: What was the breakdown of your classification relative to Population I and Population II in the A star range? You did obtain 4-color photometry for these stars, I believe.

LU: About 30 % of the hot stars presumably are Population II stars between A0 - F5. About 7 % are Population II metal weak dwarf stars between A0 - G5. There are only 1 to 2 giants among the 43 hot stars. Detailed analysis of the 4-color photometry is in progress.

PETERSON: Have you observed HB stars within the globular cluster NGC 288?

LU: No, I have not. The only 2048 pixel by 2048 pixel CCD frames of NGC 288 I obtained have not yet been reduced. However, about a half dozen blue stars in the Oort region of NGC 288 that have b-y < 0.1 might be blue HB stars.

Editorial note: Lu upon reducing his data from his most recent run at CTIO discovered two new B-stars.

SA	Frm	#	Ra (2000)	Dec	Spt	Rv	err	#	y/V	b-y	m1	c1	Hb	#	HD/Bok	SAO	SB#	Spt
190	0	14	10120.1	-303641	F5V	74.8	1.8	4	10.641	0.281	0.242	0.370	2.706	2	3161b	192868		
190	0	22	10259.6	-304203	F5V	2.5	4.6	8	11.473	0.194	0.262	0.399	2.774	1	3100b			
190	0	31	10137.1	-305615	F5V	-12.5	5.1	6	9.676	0.255	0.187	0.500	2.745	1	6068	192872		
285	0	5	5924.0	-303458	F3III	44.7	7.9	6	9.449	0.294	0.173	0.471	2.709	14	5815	192848		
380	0	18	5517.6	-303131	F5III	18.5	2.0	9	10.804	0.310	0.155	0.492	2.682	1		192816		
380	0	24	5549.5	-305655	F5V	24.9	5.0	9	12.780	0.325	0.151	0.458	2.650	1				
475	0	18	5153.4	-305733	F0V	-5.2	2.2	5	9.336	0.252	0.180	0.655	2.752	1	5024	192783		
570	0	3	5012.6	-303600	F5:				11.458	0.293	0.142	0.456	2.655	1				
570	0	27	4958.4	-303533	F5V	18.4	5.6	8	10.738	0.269	0.187	0.494	2.711	1		192767		
665	0	15	4507.8	-304352	F4:				13.166	0.275	0.311	0.288	2.696	1				
0	7	16	11039.2	-295439	F2III	81.8	22.1	2	14.103	0.298	0.167	0.346	2.594	1		60b		
95	7	3	10649.5	-301813	F3Vw	-33.0	5.5	5	13.032	0.329	0.106	0.428	2.644	1				
95	7	28	10416.7	-300116	A3III	23.4	3.4	5	9.769	0.151	0.178	0.812	2.844	9	6365	192903	431	A5
475	7	1	5303.0	-302411	F5Vw	-75.0	5.9	10	11.070	0.287	0.165	0.572	2.703	1				
475	7	15	5206.8	-301711	G0V	12.4	7.0	5	13.872	0.272	0.180	0.485	2.680	1				
475	7	20	5116.7	-300122	F4:				11.524	0.262	0.150	0.515	2.724	1				
475	7	31	5048.6	-300309	F2:				13.066	0.242	0.141	0.577	2.730	1				
570	7	4	5034.3	-295804	A3V	-13.2	10.5	8	11.530	0.101	0.232	0.900	2.924	1			344	A2
665	7	17	4658.1	-301050	F2V	5.5	2.5	8	10.308	0.189	0.201	0.756	2.794	1		192736		
015	12		10851.6	-293720	F5V IV	38.4	3.3	6	8.194	0.313	0.183	0.411	2.645	16	6868	166834		
9515	13		10405.3	-292352	F5IV V	4.6	3.3	4	9.899	0.305	0.181	0.517	2.692	1	248b			
9515	25		10622.1	-294702	F5Vw	-32.3	6.0	4	12.175	0.321	0.123	0.449	2.639	1	199b			
9515	36		10700.9	-293654	A9V	40.1	3.4	6	9.383	0.199	0.168	0.640	2.802	2	6670	166811	451	F0
19015	1		10314.1	-294351	A0V	70.1	6.1	2	11.141	0.010	0.171	0.677	2.974	1				
19015	4B		10310.7	-293400	F3:				13.929	0.258	0.260	0.176	2.681	1				
19015	4A		10310.7	-293400	F3:				13.929	0.258	0.260	0.176	2.681	1				
19015	8		10201.1	-293116	F3V	42.5	9.3	5	12.876	0.192	0.114	0.762	2.762	1			411	A5p
19015	35		10209.0	-295146	F4:				12.923	0.284	0.165	0.580	2.666	1				
38015	25		5654.5	-294538	A5V IV	11.4	19.1	8	10.101	0.102	0.101	0.900	2.894	1		192831	379	A7
47515	6		5257.4	-295433	A2:				13.738	0.116	0.103	1.226	2.776	1			355	A
47515	12		5339.9	-293412	F5V	18.6	8.9	7	13.660	0.244	0.119	0.469	2.758	1				
57015	1		5034.4	-295456	F5V IV	-21.9	6.8	10	12.147	0.250	0.139	0.539	2.758	1				
57015	13		4845.9	-293900	F5III	-3.3	8.3	10	12.918	0.251	0.264	0.316	2.754	1				
57015	20		4914.5	-292909	F5V	28.6	5.6	6	10.639	0.192	0.215	0.705	2.829	1			335	A7
022	25		10734.5	-291712	F2V	2.8	2.9	10	9.370	0.263	0.122	0.613	2.736	1	6724	166816		
19022	3		10315.4	-291131	F5V	58.2	11.6	4	13.310	0.233	0.107	0.795	2.714	1			418	A
28522	21		5901.1	-292406	A4V	30.6	6.2	5	9.303	0.112	0.177	0.936	2.891	15	5769	166719	393	A5V
57022	29		4756.6	-292039	F0V	4.0	5.0	8	7.536	0.226	0.141	0.754	2.802	2	4623	166584	327	F0III
66522	6		4616.0	-290703	A8Vp	0.4	10.3	4	13.492	0.209	0.110	0.769	2.736	1			315	A0:
66522	14		4453.0	-291415	A0				13.858	0.172	0.101	0.873	2.719	1			302	A0
030	4		10846.8	-284120	A8IVIII	108.3	70.0	4	14.600	0.211	0.130	0.658	2.716	1	1125b			
030	20		10736.1	-284218	A8V	46.2	5.2	6	9.100	0.200	0.149	0.774	2.803	2	6723	166817	455	F0III
9530	1		10653.0	-284642	F5Vw	-120.7	0.0	1	13.495	0.327	0.074	0.395	2.615	1	154b			
9530	18		10540.8	-283610	A3:				15.857	0.129	0.105	1.336	2.770	1				
19030	23		10149.2	-283723	F5:				10.956	0.250	0.160	0.536	2.682	1				
28530	8		5745.9	-283323	F5III	20.7	4.6	9	12.043	0.292	0.194	0.469	2.700	1				
28530	18		5910.8	-285128	F5V				14.446	0.234	0.201	0.499	2.696	1				
47530	1		5406.6	-284505	A5V	12.2	3.4	8	11.079	0.093	0.192	0.906	2.883	1			361	A5
47530	10		5313.3	-284056	F3:				14.149	0.245	0.125	0.563	2.689	1				
57030	33		4841.1	-282942	F5V	4.0	20.9	8	6.635	0.216	0.178	0.583	2.831	2	4691	166593	333	F0
66530	5		4640.1	-284138	F5V	5.0	48.5	5	13.955	0.249	0.153	0.475	2.685	1				
66530	7		4617.0	-284415	A2:				13.839	0.075	0.454	0.546	2.861	1				
66530	36		4532.0	-284106	A0V	-4.0	17.7	9	10.098	0.067	0.195	1.039	2.925	21	4329	166540	306	A2
66530	37		4604.4	-284933	A8V	8.6	4.8	9	9.648	0.166	0.168	0.818	2.830	11	4399	166550	312	F0
76030	19		4224.8	-283422	A0V	-0.2	10.7	9	13.627	0.105	0.191	0.929	2.838	1			285	A3
76030	20		4230.3	-283433	A1V				11.218	0.077	0.168	1.013	2.842	1			287	A2
037	2		10959.2	-275833	F2Vw	34.4	10.0	6	13.039	0.208	0.039	0.460	2.742	1			469	A p
037	21		10742.4	-281551	F0III	35.1	3.7	6	12.409	0.206	0.191	0.333	2.753	1				
9537	31		10658.2	-280745	A5:				12.076	0.115	0.187	0.881	2.787	1			449	A0
19037	8		10320.9	-275731	F5V	9.6	5.9	7	10.338	0.314	0.133	0.408	2.672	1	2066b			
19037	21		10052.2	-280528	A0	39.9	72.8	6	11.417	0.054	0.242	0.882	2.881	1	2242b		408	A
28537	2		10027.8	-281221	F3V	112.7	6.8	9	13.137	0.219	0.089	0.937	2.729	1	2248b		405	A p
28537	21		5749.6	-281306	F3V IV	74.5	0.9	3	9.509	0.267	0.154	0.376	2.735	1	5645	166706		
28537	22		5747.3	-280850	F5V	75.2	3.0	3	10.977	0.280	0.126	0.394	2.701	1	2310b			
28537	29		5725.7	-280618	F5:				12.931	0.279	0.139	0.489	2.660	1				
38037	26		5440.5	-281353	:A5				13.532	0.128	0.126	1.096	2.768	1			363	A3
57037	6		4945.3	-280520	F0:				14.271	0.191	0.255	0.451	2.710	1				
57037	32		4941.4	-280342	F4:				14.796	0.256	0.255	0.476	2.667	1				
66537	5		4641.3	-275612	F6V	-18.4	7.2	9	13.368	0.182	0.240	0.729	2.789	1			317	A
66537	19		4720.7	-281809	F2Vw	9.0	2.3	4	10.082	0.277	0.124	0.518	2.718	1		166576		
76037	2		4338.0	-281559	F5V	12.3	4.0	6	10.635	0.323	0.138	0.366	2.670	1				
76037	14		4249.2	-281916	F5Vw	-7.0	3.5	8	11.077	0.295	0.086	0.468	2.694	1				
045	4		10845.8	-274149	F5Vw	-2.9	5.0	5	10.676	0.278	0.136	0.518	2.713	1				
045	20		10738.0	-271632	F5Vw	10.9	4.8	5	10.467	0.277	0.170	0.511	2.715	1				
9545	19		10556.9	-273542	F4V	12.9	2.4	8	9.782	0.291	0.148	0.446	2.705	10	6548	166797		
9545	22		10600.5	-273211	F5Vw	-65.1	7.8	5	11.896	0.364	0.048	0.365	2.604	1			442	A7p

19045	1	10340.9	-274411	F8Vw				12.968	0.323	0.032	0.296	2.629	1	2049b	421 B:
38045	13	5636.6	-272359	A0:				14.858	0.017	0.250	0.923	2.955	1		
47545	9	5250.5	-272547	F4Vw?				9.966	0.297	0.129	0.414	2.699	1		166648
57045	13	4914.4	-273142	F0V	14.5	8.4	4	11.419	0.182	0.155	0.873	2.792	2		337 A5
57045	14	4919.5	-272500	A9:	15.5	3.0	8	13.390	0.210	0.554	0.269	2.319	1		
57045	15	4929.2	-272315	A5V	10.7	8.1	3	13.390	0.210	0.072	0.934	2.728	1		338 A3p
57045	23	5036.5	-272606	F0III	21.5	3.1	6	9.454	0.224	0.200	0.615	2.775	10	4876 166620	343 F0
57045	24	4923.0	-273616	F5V	-21.8	5.7	8	8.538	0.287	0.167	0.442	2.725	9	4763 166606	
66545	9	4734.4	-273822	F5Vw	25.5	10.0	6	12.311	0.319	0.086	0.223	2.641	1		
66545	24	4510.2	-274601	F5V	-4.9	7.5	8	8.933	0.252	0.185	0.481	2.760	1	4289 166536	
76045	14	4316.7	-274715	F5V	58.1	36.3	6	13.786	0.337	0.197	0.256	2.601	1		
052	7	10837.4	-265312	A0V	31.7	10.0	4	12.280	-0.101	0.195	0.594	3.005	1		460 A0p
052	16	10939.3	-265256	F5Vw	-6.8	2.6	8	9.424	0.302	0.117	0.486	2.701	1	6957 166842	
9552	17	10418.0	-270918	A5III	13.4	4.6	5	9.626	0.177	0.207	0.753	2.822	8	6364 166778	430 A5
9552	19	10534.5	-265720	F0V	5.4	5.4	5	8.123	0.242	0.181	0.605	2.777	8	6491 166790	
19052	18	10211.0	-270205	F5V	-10.3	3.2	3	11.887	0.333	0.121	0.439	2.634	1	2098b	
19052	37	10245.7	-265941	F5V	4.0	2.1	5	10.061	0.305	0.157	0.373	2.688	14	6196 166765	
28552	11	5941.1	-264539	A5V	-4.8	0.8	2	11.157	0.072	0.262	0.985	2.910	1		399 A5
38052	19	5420.3	-265534	A0V				12.827	0.046	0.171	1.219	2.855	1		362 A3
47552	15	5148.2	-271412	A0:				13.884	-0.090	0.096	0.259	2.976	1		
47552	23	5212.1	-270233	A5				15.516	0.073	0.111	1.302	2.828	1		
47552	30	5058.4	-271018	A8:				13.864	0.207	0.373	0.296	2.750	1		
57052	4	4951.3	-271319	F2V	17.1	2.5	6	11.259	0.287	0.138	0.466	2.690	1		
57052	24	4835.2	-271044	F1V	-39.3	5.2	4	14.603	0.141	0.275	0.766	2.739	1		332 A
66552	17	4533.4	-271349	F5V	12.8	3.5	4	10.997	0.304	0.167	0.490	2.694	1		
66552	18	4534.8	-270747	F5Vw	-31.3	1.7	4	12.264	0.289	0.129	0.446	2.673	1		
66552	21	4445.9	-271209	A2V				12.731	0.116	0.140	1.260	2.865	1		298 A2
66552	22	4740.8	-265832	F0V	23.3	5.0	8	9.156	0.240	0.167	0.603	2.764	10	4586 166581	
76052	5	4406.5	-270133	F5V	8.5	13.7	6	12.193	0.276	0.129	0.407	2.686	1		
060	24	10743.5	-263954	F5Vw	50.3	22.6	4	10.858	0.321	0.138	0.478	2.661	1		
9560	17	10520.5	-262116	F5IV	-27.4	1.8	4	11.086	0.322	0.121	0.469	2.656	1		
9560	28	10434.9	-262729	F7:				14.225	0.366	0.166	0.345	2.528	1		432 A:
9560	42	10555.5	-264343	A2Vp	3.3	2.1	3	8.425	0.094	0.213	0.907	2.921	9	6532 166796	
28560	28	5901.2	-263137	F3IV V	-33.0	3.7	6	9.769	0.151					5768 166718	
38060	26	5559.0	-262255	A5?				13.231	0.039	0.333	0.984	2.934	1		371 A3p
47560	2	5327.9	-263929	F3:				16.087	0.253	0.188	0.460	2.695	1		
47560	5	5304.2	-263829	A3:				15.624	0.046	0.098	0.958	2.917	1		
47560	6	5259.3	-263859	A0:				16.289	-0.018	0.236	0.475	2.977	1		
47560	7	5320.4	-263603	A0:				15.573	-0.014	0.169	1.286	2.910	1		
47560	12	5346.7	-262213	F5V	8.6	7.5	4	12.304	0.250	0.255	0.490	2.734	1		
47560	28	5149.6	-263043	F3III	53.0	4.6	11	9.824	0.257	0.143	0.546	2.739	12	4999 166634	
47560	36	5229.7	-264315	A4:				15.607	-0.132	-0.036	1.174	2.826	1		
47560	40	5135.1	-264157	A0:				16.741	-0.049	0.093	-0.133	3.007	1		
57060	11	4846.0	-264519	F5V	-0.3	6.5	4	13.009	0.280	0.182	0.406	2.707	1		
66560	23	4438.2	-261859	F0IVIII	-14.3	4.9	7	10.768	0.252	0.132	0.646	2.749	1		
76060	17	4257.2	-263741	F5V	31.7	2.4	6	9.654	0.263	0.182	0.491	2.746	7	4073 166511	
067	23	10750.2	-255918	A9:				12.940	0.205	0.297	0.446	2.768	1		
9567	1	10647.4	-260812	F1:				14.858	0.255	0.324	0.261	2.703	1		
9567	21	10513.8	-250529	F5Vw	-9.3	6.1	4	11.288	0.349	0.116	0.508	2.638	1		
19067	20	10142.2	-254527	A8:				14.082	0.239	0.076	0.635	2.729	1		
19067	21	10040.9	-254438	A4:				14.026	0.162	0.113	0.088	2.804	1		
19067	25	10149.2	-255311	F0V	44.9	2.2	5	9.741	0.219	0.160	0.600	2.776	1	6088 166758	
28567	32	5743.6	-261322	F0Vw	-52.4	15.2	6	9.977	0.237	0.170	0.714	2.752	1	5630 166705	386 F0
38067	17	5409.7	-260540	F5V	79.6	2.0	4	11.915	0.331	0.114	0.436	2.634	1		
47567	4	5256.1	-260023	F3V IV	67.2	4.4	4	11.569	0.318	0.137	0.407	2.652	1		
47567	7	5239.4	-255423	A4:				14.470	0.173	0.010	1.116	2.755	1		353 A:
47567	8	5226.1	-255008	F2Vw	-20.2	7.5	4	12.500	0.379	0.064	0.147	2.579	1		
47567	19	5107.6	-254448	F2V	56.2	33.4	4	13.275	0.266	0.168	0.422	2.678	1		
57067	8	4958.6	-255815	F5III	37.1	3.1	4	10.789	0.292	0.143	0.496	2.688	1		166612
075	21	10834.6	-252704	F3:				12.479	0.257	0.203	0.468	2.720	1		
075	23	10959.8	-252558	F5V	-4.6	3.8	8	9.906	0.295	0.142	0.464	2.707	1	6993 166845	
9575	30	10552.5	-252853	F0V	2.9	5.0	8	9.234	0.235	0.145	0.528	2.770	9	6516 166795	440 F0
9575	36	10344.3	-251900	A5:				11.622	0.137	0.128	0.735	2.706	1		
28575	31	10025.7	-251753	A4::				13.386	0.148	0.187	0.878	2.751	1		404 A p
38075	15	5649.0	-252146	A5V	54.0	6.1	4	7.176	0.071	0.214	1.094	2.934	4	5524 166699	377 A5V
47575	10	5302.7	-252514	F3V	-18.7	14.9	10	12.676	0.387	-0.035	0.301	2.584	1		
47575	19	5122.2	-251308	F2III	9.8	15.2	7	13.513	0.213	0.180	0.652	2.758	1		
66575	25	4615.6	-253215	A7V	-4.0	2.7	5	9.049	0.193	0.157	0.719	2.811	8	4414 166557	314 F0V
9582	16	10508.0	-245242	A8:				11.348	0.287	0.129	0.536	2.598	1		
19082	1	10250.9	-250626	F8V IV	30.0	8.6	7	11.446	0.260	0.214	0.466	2.655	1		
28582	23	5954.9	-244351	F5Vw	-12.7	5.4	10	10.479	0.232	-0.009	0.572	2.755	1		166732
47582	4	5331.7	-250528	A5:				15.230	0.132	0.553	0.832	2.668	1		
47582	25	5359.1	-250230	F2V	33.0	4.4	8	8.743	0.261	0.241	0.351	2.746	2	5250 166661	
47582	26	5359.9	-250238	F3V				8.907	0.278	0.132	0.513	2.728	2	5251 166662	
47582	27	5312.3	-244905	F5V	-2.5	5.9	8	6.203	0.271	0.170	0.499	2.799	1	5156 166651	
47582	29	5304.9	-245702	F3V IV	33.3	4.0	9	9.300	0.291	0.152	0.503	2.711	10	5145 166649	
66582	1	4650.6	-245113	F5V	-12.4	5.6	7	10.539	0.294	0.141	0.534	2.697	3		
66582	21	4506.8	-245509	F5V	83.9	7.6	5	9.884	0.310	0.132	0.451	2.685	6	4288 166535	

TABLE 1. Spectral and photometric data for the hot SGP stars

Clusters

Population II Horizontal Branches: A Photometric Study of Globular Clusters

By KENT A. MONTGOMERY
AND KENNETH A. JANES

Department of Astronomy, Boston University, Boston, MA 02215, USA

We present here the first results of a V, B-V photometric survey of 19 globular clusters, with the goal of obtaining statistically interesting numbers of stars on the red giant and horizontal branch sequences. Using the 0.9-meter telescopes at KPNO and CTIO, equipped with 2048×2048 pixel CCD's, we were able to image the entire area of all the clusters, with the exception of 47 Tuc, on either a single frame or a 2×2 array of overlapping fields. We specifically chose clusters that are bright, populous, minimally affected by reddening and which span a wide range of metallicities from -0.7 to -2.26. Although many of them have been extensively observed for deep photometry of their main sequences, typically the evolved stars have been studied only photographically and their red giant and horizontal branch sequences are poorly defined. For most of the clusters we measured over 10,000 stars, typically reaching V magnitude 21. The color-magnitude diagrams show well-populated and well-defined horizontal and red giant branches, and a wide range of horizontal branch types. We have developed computer software to determine the fiducial sequences of the clusters in a systematic way, which can be compared with each other and with theoretical isochrones. Our fiducial sequences agree well with other recently published sequences. The eventual goal of the project is to do detailed comparisons of the cluster color-luminosity distributions with theoretical models which may, among other things, help determine if age is the second parameter affecting horizontal branch type.

1. Color-Magnitude Diagrams

We obtained images of 19 globular clusters over 2 observing runs, one each at KPNO and CTIO, using 0.9 meter telescopes equipped with 2048×2048 CCDs, and the standard B and V filters. Thirteen of the clusters were observed from KPNO over 6 nights and the other 6 were observed over 2 nights at CTIO. Exposures were typically 300 seconds in V and 540 seconds in B. In addition to the cluster images, images of standard stars from Landolt (1983) were taken to transform the photometry to the standard system. The standard deviations (in magnitudes) for the stars used in transforming to the standard system were 0.018 in V and 0.023 in B, for the KPNO observing run and for the CTIO observing run they were 0.019 in V and 0.026 in B.

The color-magnitude diagrams were produced by taking an annulus of stars about the center of the cluster. This technique typically produced 2000-3000 stars per diagram, removes stars which have large photometric errors due to crowding in the center, and reduces the number of field stars in the diagram.

2. Fiducial Sequences

Also presented in this paper is a fiducial sequence for the globular cluster M92. The fiducial sequence was created by selecting points along the cluster sequences and connecting them with cubic splines. A χ^2 evaluator was created by summing the distance of each star from the fiducial line and weighting it by the photometric errors. A gradient search was then used to determine the best fit for the fiducial line to the data. The gradient search involved finding the change in χ^2 as each selected point was slightly varied

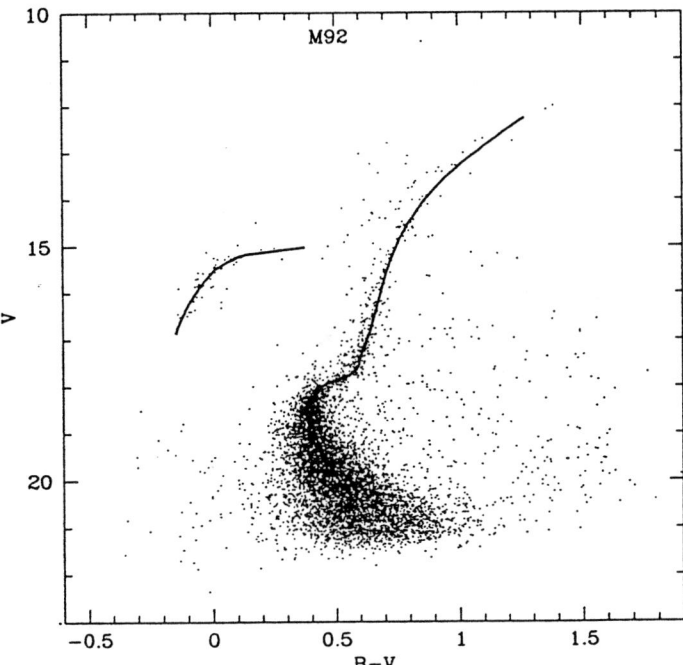

FIGURE 1. M92 color-magnitude diagram for all stars between 2.8 and 7.9 arcminutes from the cluster center, and the derived fiducial sequence.

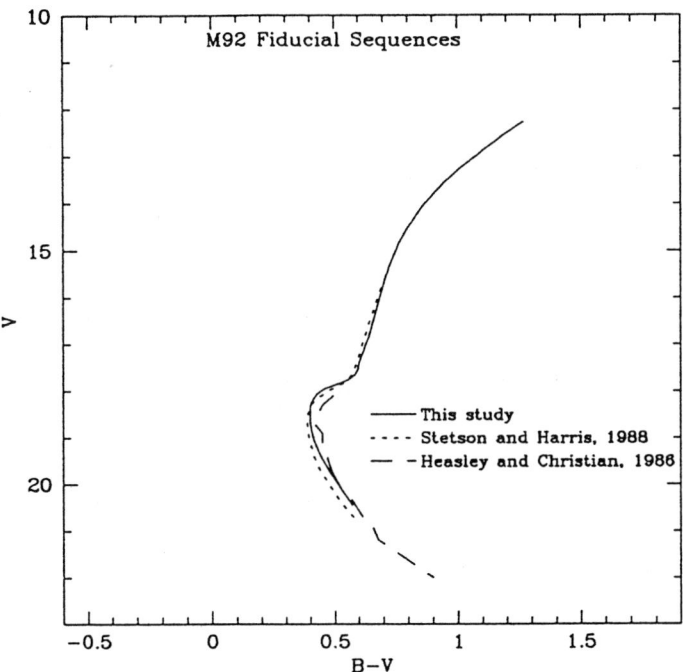

FIGURE 2. Comparison between our fiducial sequence and the fiducial sequences of Stetson & Harris (1988), and Heasley & Christian (1986).

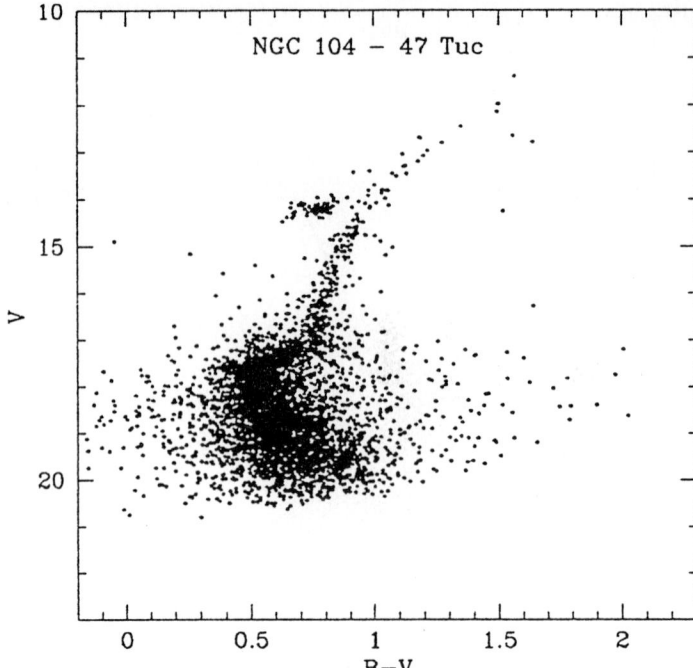

FIGURE 3. Color-Magnitude diagram of NGC 104, 47 Tuc.

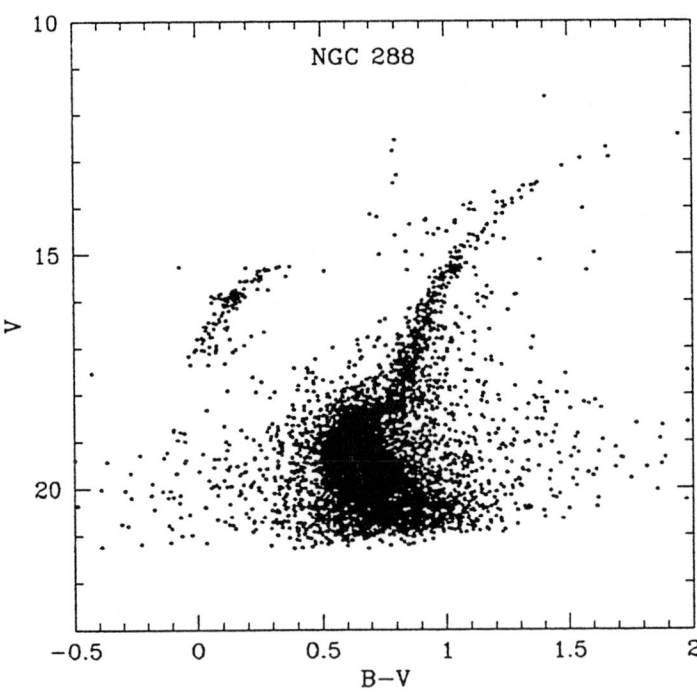

FIGURE 4. Color-Magnitude diagram of NGC 288.

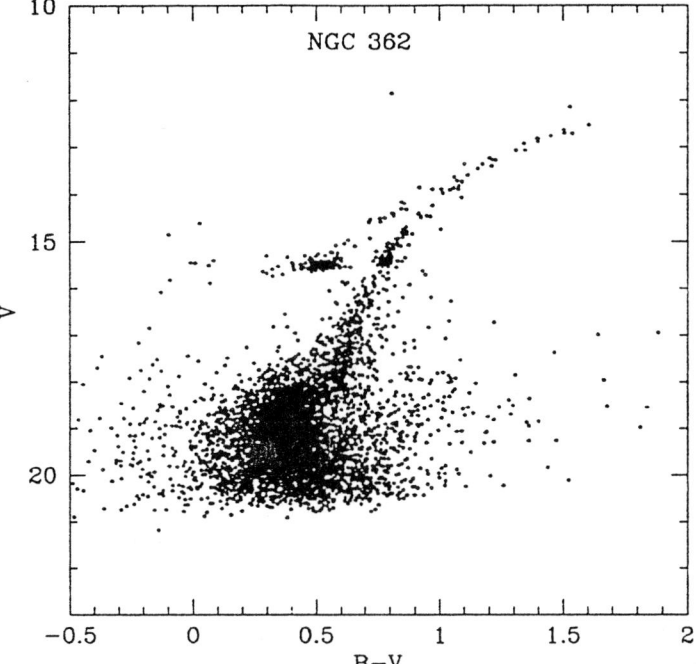

FIGURE 5. Color-Magnitude diagram of NGC 362.

in color and then simultaneously moving all the points and determining a new χ^2. This process was repeated until a minimum χ^2 was determined or the changes in color for each selected point were less than 0.00001 magnitudes. This technique will be applied to all 19 clusters.

3. Future Plans

This paper represents a preliminary analysis of the dataset. After comparisons with already published photometry, and a determination of internal errors, the ultimate goal is to determine specific stellar evolution parameters such as the ratio of red giant branch stars to horizontal branch stars, or width of the horizontal branch, both of which are functions of the helium abundance of the cluster. In addition, the fiducial sequences of the clusters can be used to determine relative ages, and comparisons with published isochrones will give absolute ages. Both the relative ages and absolute ages are critical for understanding the type of formation of the Galaxy.

REFERENCES

LANDOLT, A. U., 1983, AJ, 104, 340
STETSON, P. B., HARRIS, W. E., 1988, AJ, 96, 909
HEASLEY, J. N., CHRISTIAN, C. A., 1986, ApJ, 307, 738

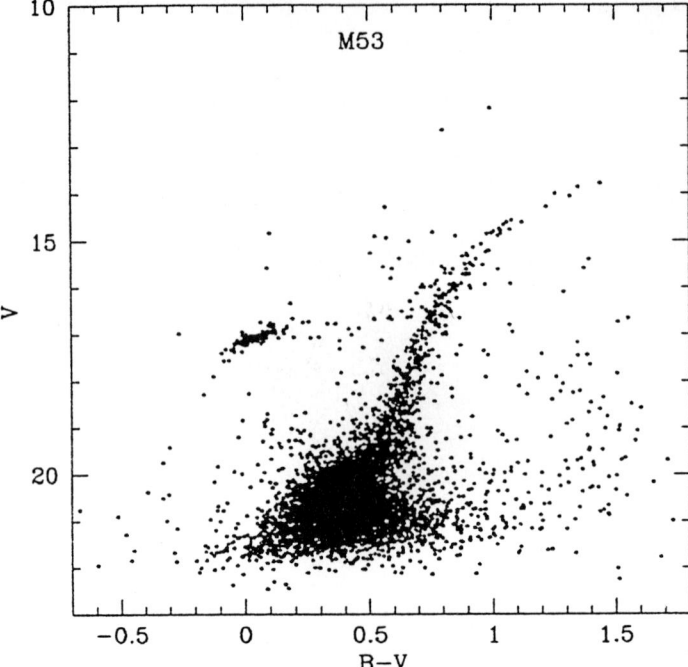

FIGURE 6. Color-Magnitude diagram of M53.

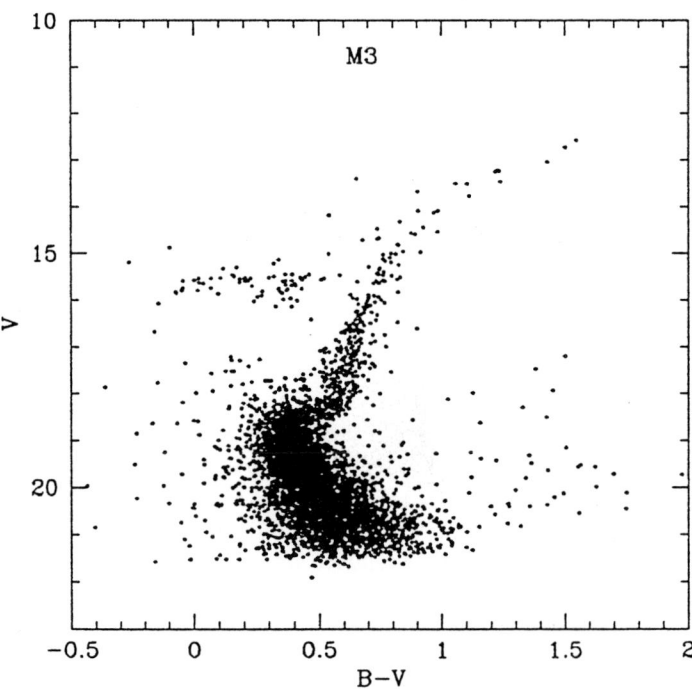

FIGURE 7. Color-Magnitude diagram of M3.

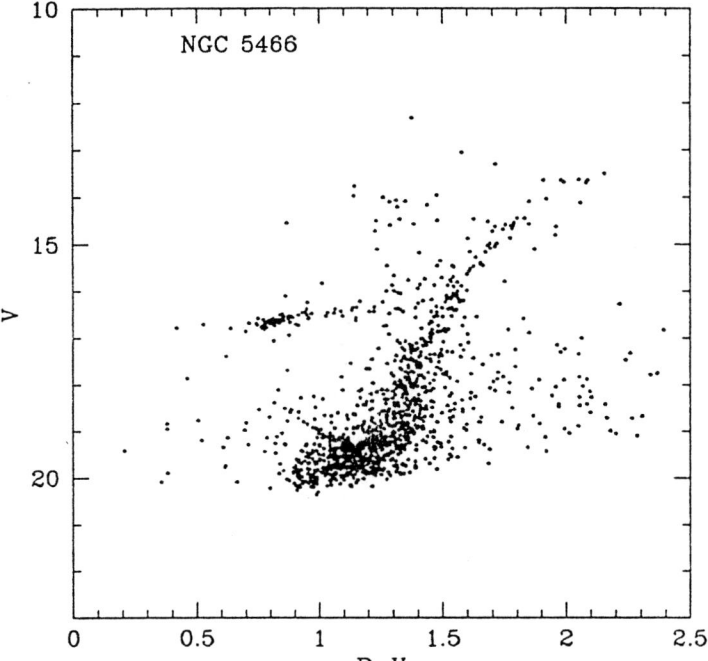

FIGURE 8. Color-Magnitude diagram of NGC 5466.

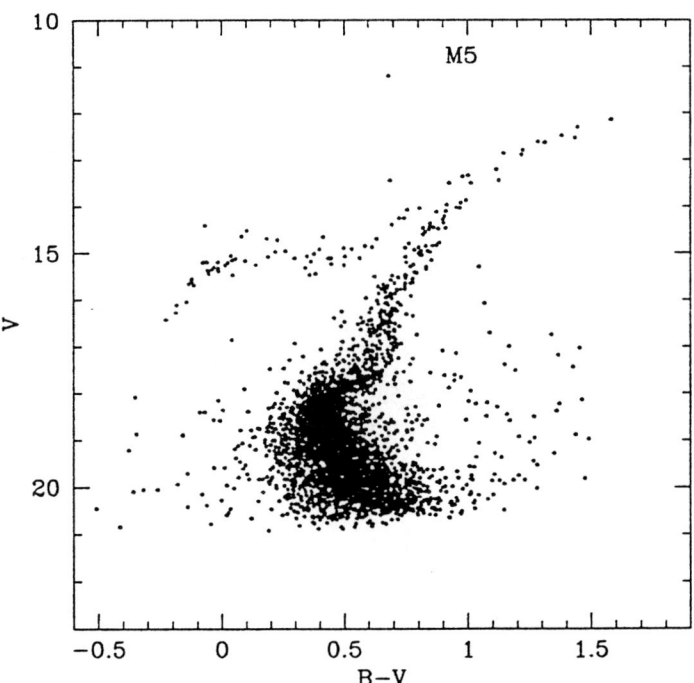

FIGURE 9. Color-Magnitude diagram of M5.

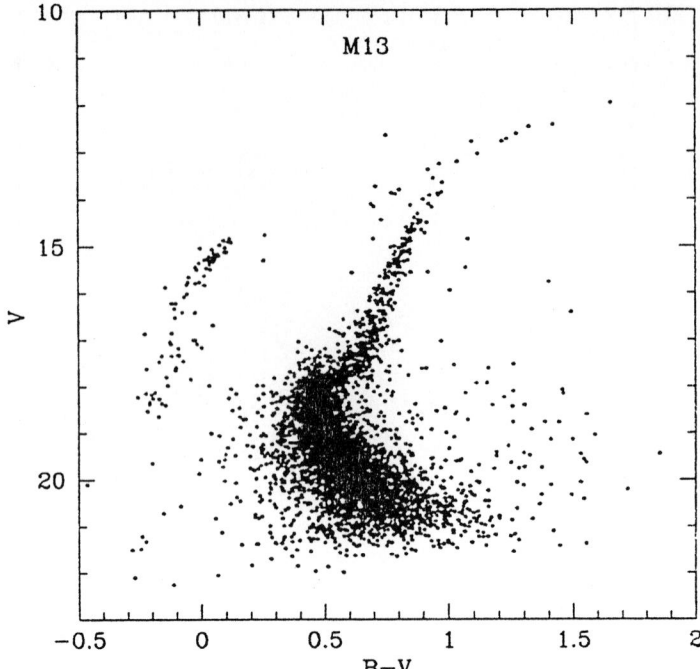

FIGURE 10. Color-Magnitude diagram of M13.

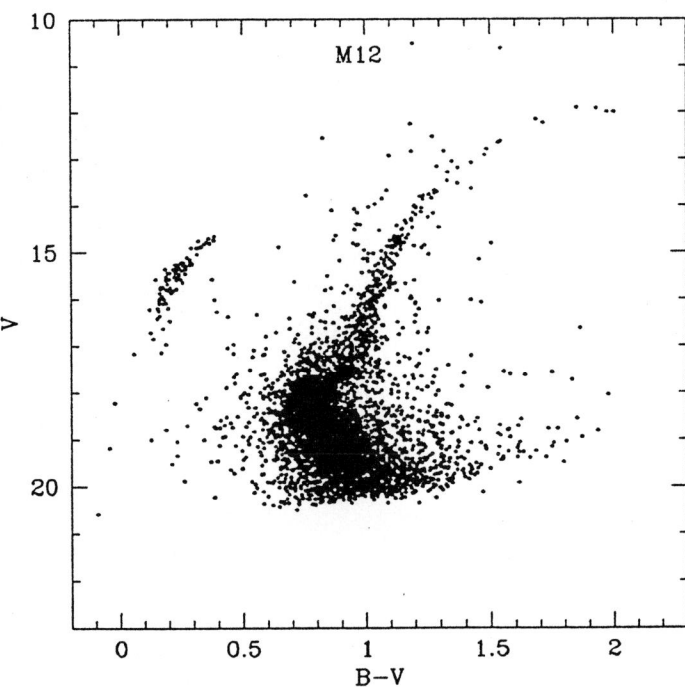

FIGURE 11. Color-Magnitude diagram of M12.

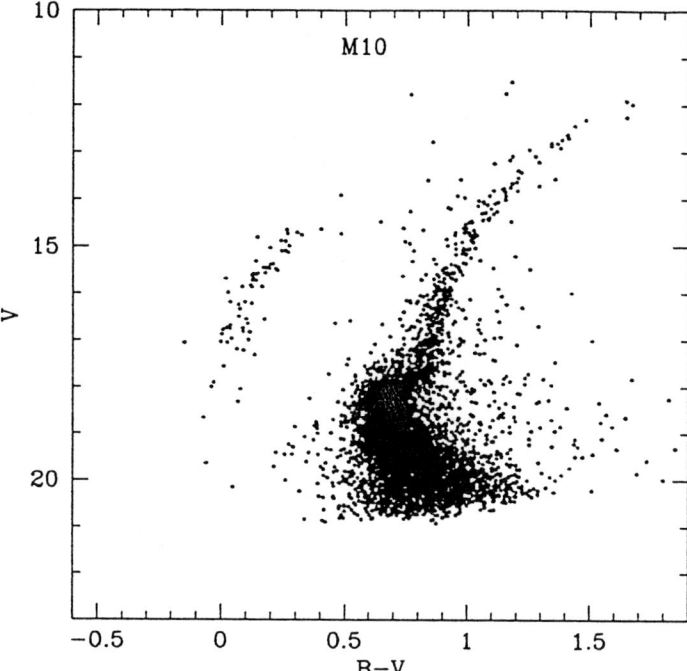

FIGURE 12. Color-Magnitude diagram of M10.

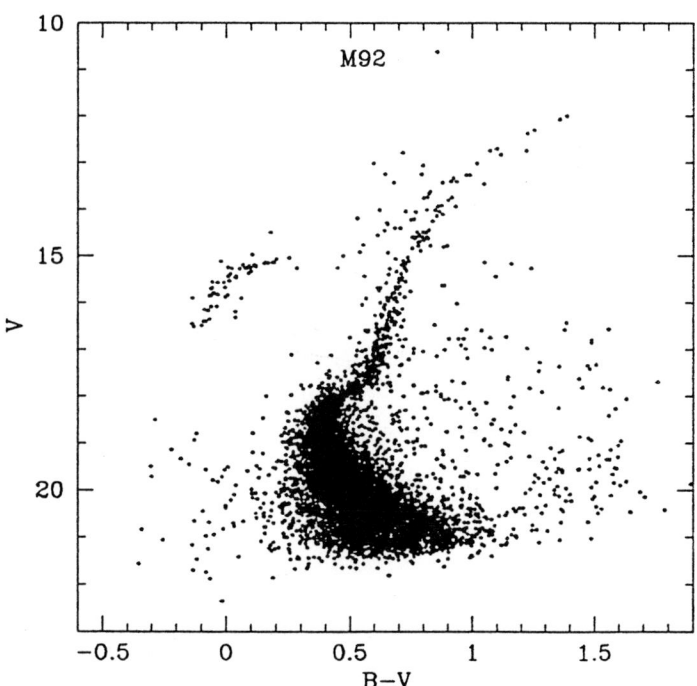

FIGURE 13. Color-Magnitude diagram of M92.

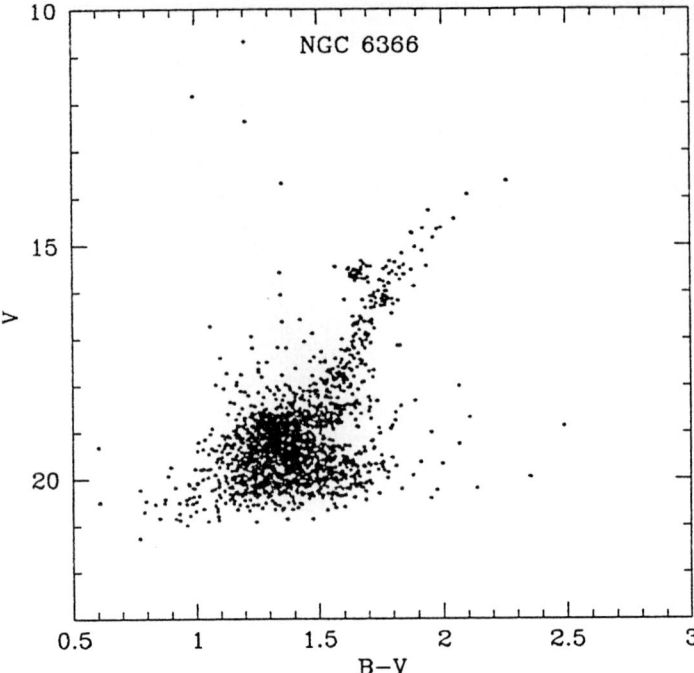

FIGURE 14. Color-Magnitude diagram of NGC 6366.

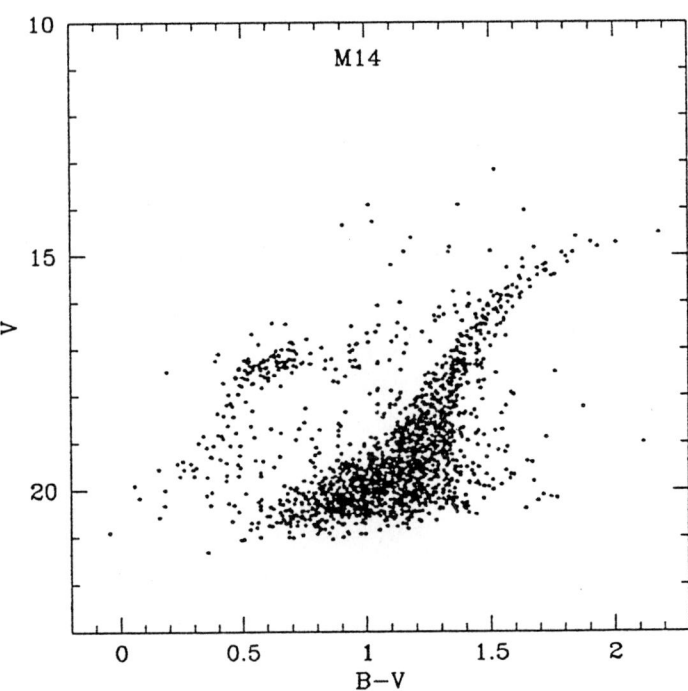

FIGURE 15. Color-Magnitude diagram of M14.

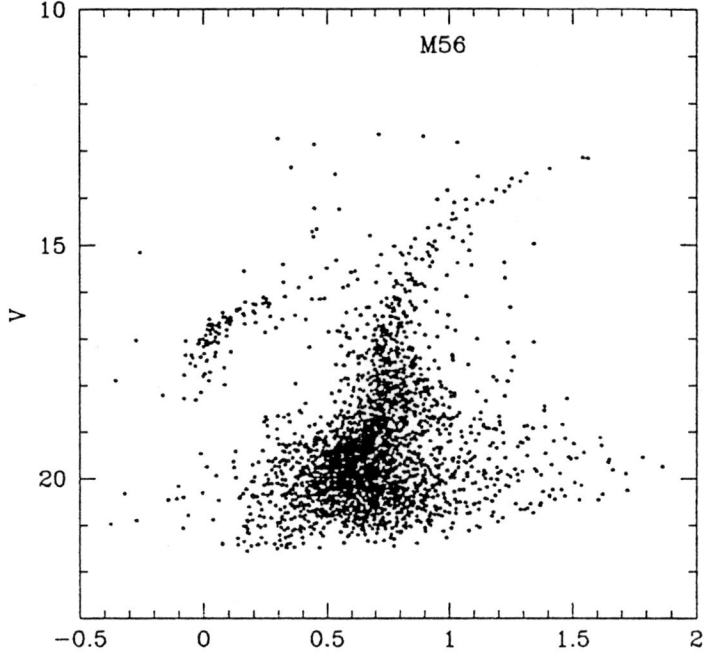

FIGURE 16. Color-Magnitude diagram of M56.

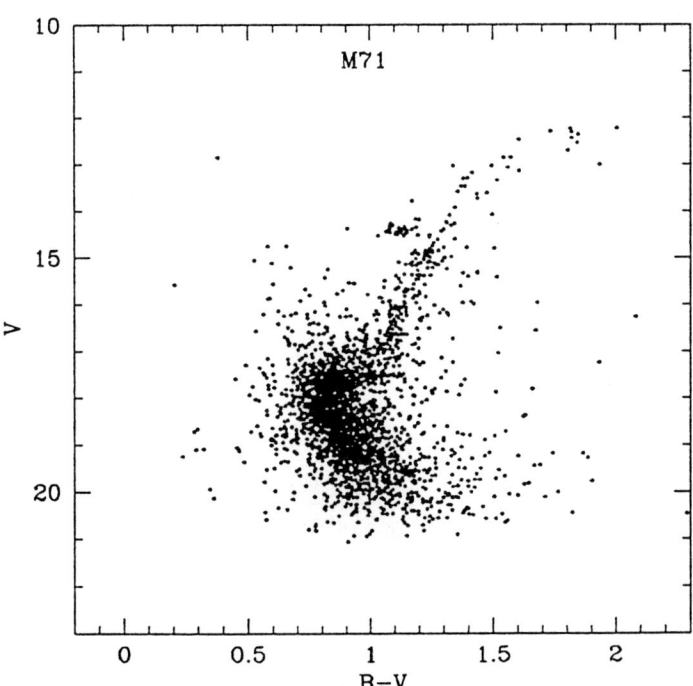

FIGURE 17. Color-Magnitude diagram of M71.

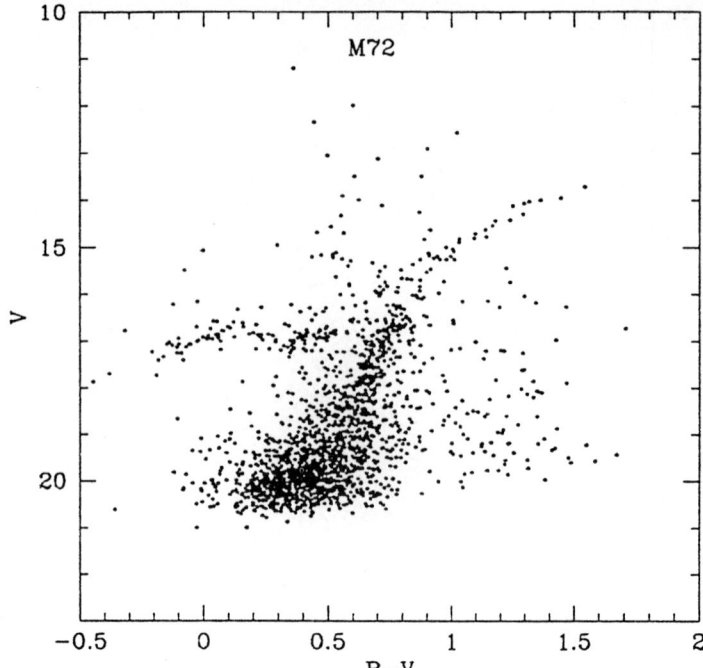

FIGURE 18. Color-Magnitude diagram of M72.

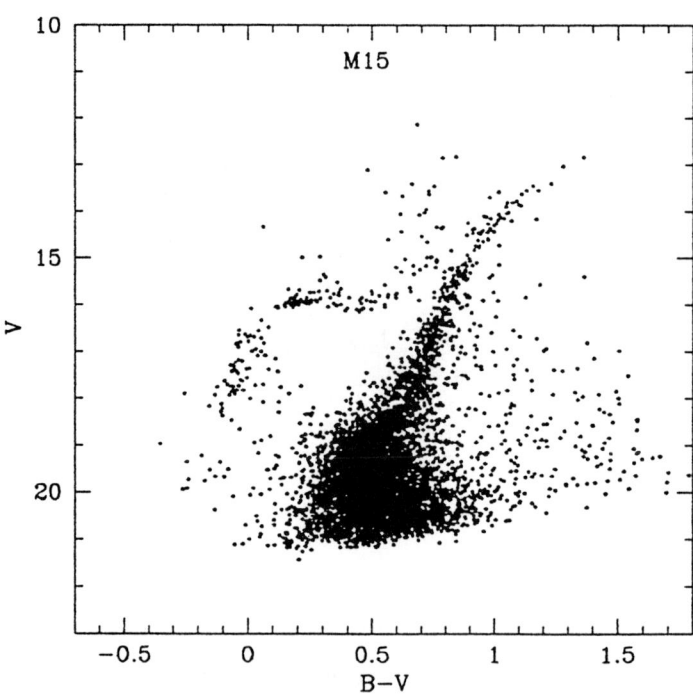

FIGURE 19. Color-Magnitude diagram of M15.

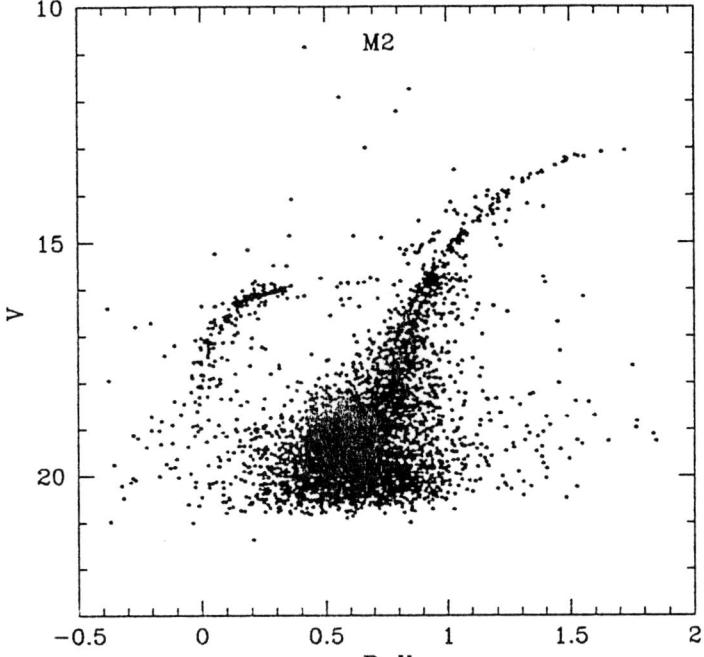

FIGURE 20. Color-Magnitude diagram of M2.

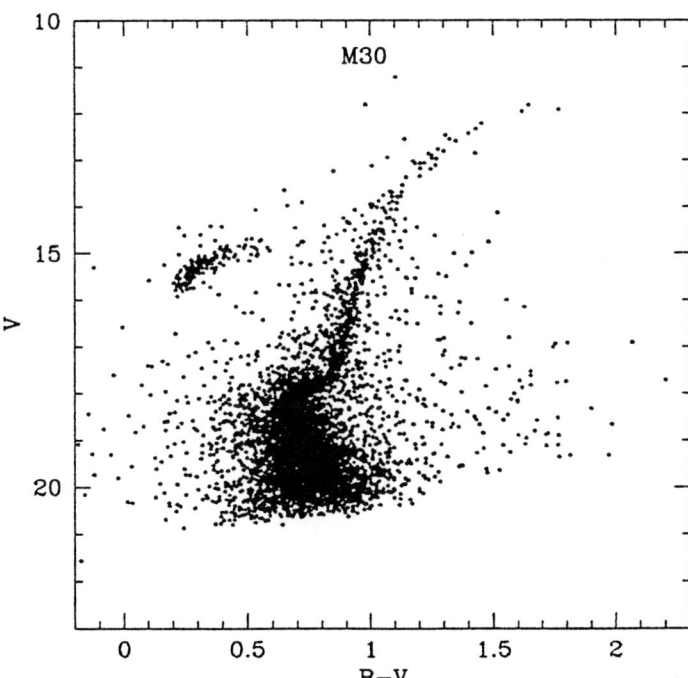

FIGURE 21. Color-Magnitude diagram of M30.

Discussion

SARAJEDINI: What kind of astrophysically interesting quantities will these data put constraints on?

MONTGOMERY: By looking at ratios of stars on the RGB to HB and HB to AGB we can put limits on the relative lifetimes at these positions in the color magnitude diagrams. We can also look for variations in the abundance and metallicity from the width and luminosity of the HB and the luminosity of the tip of the RGB.

PETERSON: What plans do you have to establish cluster membership of potentially rare species?

MONTGOMERY: We observed well defined halo clusters to decrease the number of field stars present in the sample, but other than a couple of clusters which have proper motion studies we have no other means of determining membership for the cluster.

BEERS: Now that the ground-based data has improved so dramatically, I suspect that problems due to completeness of counts of various subtypes such as FHB, Red Giants, etc. as a result of crowding in the central region, may begin to dominate over other errors.

The Period-Shift Effect in Oosterhoff Type II Globular Clusters

By MÁRCIO CATELAN

Instituto Astronômico e Geofísico, Universidade de São Paulo, Caixa Postal 9638,
CEP 01065-970 São Paulo, SP, BRAZIL

The mean "fundamentalized" periods of RR Lyrae variables in Oosterhoff type II Galactic globular clusters with "blue" horizontal branches are shown not to present statistically significant correlations with the horizontal-branch (HB) morphology, as represented by the parameter $(B-R)/(B+V+R)$. This result is compared with the predictions of the synthetic HB models of Catelan (1993), which suggest instead a strong correlation between the periods of RR Lyrae variables and the HB morphology. Possible reasons for this disagreement are explored. Viable candidates include variations among the clusters in stellar evolution-controlling parameters *other than age and/or mass loss* on the red-giant branch, and/or inadequacies of the employed evolutionary tracks for HB stars. This is expected to have a strong impact upon the interpretation of the Oosterhoff dichotomy and second-parameter problems.

1. Introduction

Oosterhoff (1939) was the first to note that Galactic globular clusters (GGCs) are divided into two distinct classes, according to the mean periods of their ab-type RR Lyrae variables. Arp (1955) discovered that Oosterhoff type II (OoII) clusters ($\langle P_{ab} \rangle \approx$ 0.65 d) are systematically more metal-poor than their OoI ($\langle P_{ab} \rangle \approx 0.55$ d) counterparts. Castellani (1983) and Renzini (1983) later suggested that the effect was associated with the non-monotonic variation of the horizontal-branch (HB) morphology with metallicity which was found to exist in the halo of the Galaxy. The reason for this particular behavior is still unknown; a popular idea is that it is related with a possible non-linear variation of mass loss on the red-giant branch (RGB) with metallicity [Fe/H]. On the other hand, Sandage (1990 and references therein) has claimed that the Oosterhoff-Arp dichotomy actually involves a star-by-star effect at any given (fixed) effective temperature; this became known as the "Sandage period-shift effect," whose reality has often been questioned (Caputo 1988; Lee et al. 1990). Sandage (1993) has recently reviewed the arguments in favor and against the Sandage period-shift effect. At any rate, the ability of the models to explain these phenomena still remains to be proved (Sweigart et al. 1987; Rood 1990; Fernley 1993; Catelan 1992, 1994b).

In the present paper, the mean "fundamentalized" periods of RR Lyrae variables in 8 OoII GGCs spanning a considerable range in (blue) HB morphologies are analyzed, on the basis of data available in the literature, in an attempt to provide further insight into the evolutionary interpretation of the above problems.

2. Periods of RR Lyrae variables and the HB morphology: observations and theory

2.1. The data

Attention has been focused on the OoII clusters with the bluest HB types $[(B-R)/(B+V+R) \equiv L_{BVR} > 0.65]$ with significant numbers of RR Lyrae variables ($N_{RR} \geq 11$ and/or $N_{ab} \geq 7$). These limits were set forth in such a way as to bring the effect of statistical fluctuations to an acceptable minimum, while preserving a reasonable number

GGC	$\frac{(B-R)}{(B+V+R)}$	[Fe/H]	N_{RR}	f_c	$\Delta\langle P_f\rangle^\circ$	$\Delta\langle P_f\rangle^{p1}$	$\Delta\langle P_f\rangle^{p2}$	$\Delta\langle P_f\rangle^{p3}$
M53	0.76	−2.04	33	0.455	+0.026	−0.004	−0.069	−0.056
N5286	0.91	−1.79	16	0.375	+0.009	+0.049	−0.035	−0.039
N5466	0.68	−2.22	20	0.450	+0.006	+0.003	−0.060	−0.067
N5986	0.95	−1.67	9	0.222	+0.052	+0.063	+0.022	−0.000
M92	0.88	−2.24	12	0.333	+0.038	+0.029	+0.004	−0.010
M22	0.94	−1.75	19	0.526	+0.014	+0.051	+0.005	−0.016
M15	0.72	−2.15	67	0.567	+0.026	+0.009	−0.054	−0.069
M2	0.96	−1.62	17	0.294	+0.033	+0.032	+0.024	−0.003

TABLE 1. Relevant properties for 8 OoII GGCs

Case	a	b	r
Observations	−0.023 ± 0.004	0.057 ± 0.052	0.410
Predictions 1	−0.133 ± 0.037	0.191 ± 0.043	0.875
Predictions 2	−0.277 ± 0.055	0.302 ± 0.065	0.886
Predictions 3	−0.234 ± 0.034	0.237 ± 0.040	0.924
Predictions 4	−0.090 ± 0.032	0.141 ± 0.037	0.866

TABLE 2. Regression and correlation coefficients for Eq. (2.1)

of GGCs in the sample. Among the excluded objects, one finds M55, NGC 4833, NGC 6397, and NGC 5053. Pulsational information for the resulting sample of OoII clusters was drawn from Castellani & Quarta (1987), whereas L_{BVR} values were taken from Lee (1990). Relevant quantities for the resulting sample of 8 OoII GGCs are provided in Table 1, where *logarithmic* period shifts are given, which have been computed with respect to M3.

2.2. The models

Synthetic HB models have been computed as described in detail by Catelan (1993). In the present work, the models for the clusters in Table 1 are those which match the observed L_{BVR} values very accurately for the adopted chemical composition. The Lee & Demarque (1990) tracks have been employed. The effect of an enhancement of the α-elements by a factor of 3 has also been studied, following the prescriptions of Chieffi et al. (1991). The analysis is restricted to fundamentalized mean periods, to avoid the theoretical uncertainties related with the RR Lyrae mode transition in the HR diagram and with the temperature definition and determination for these variables. See Catelan (1994a) for a more general discussion. The predicted period shifts for the three theoretical situations analyzed, $(Y_{MS}, [\alpha/Fe]) = (0.20, 0)$, $(0.23, 0)$, and $(0.23, 0.48)$, are provided in the last three columns of Table 1 (entries labeled p1, p2, and p3, respectively).

2.3. Comparison between the models and the observations

In Table 2 are listed the least-squares coefficients for relations of the type

$$\Delta\langle\log P_f\rangle = a + b\,\frac{(B-R)}{(B+V+R)}. \qquad (2.1)$$

The associated standard errors are also given, as are the Pearson correlation coefficients r. In Figure 1, the relations obtained in this way are also displayed. In Table 3, the confidence levels for the rejection of the hypothesis $b_p = b_{obs}$ are provided, having been obtained from a Student's t-test.

Analysis of these tables discloses that the present models are inadequate representa-

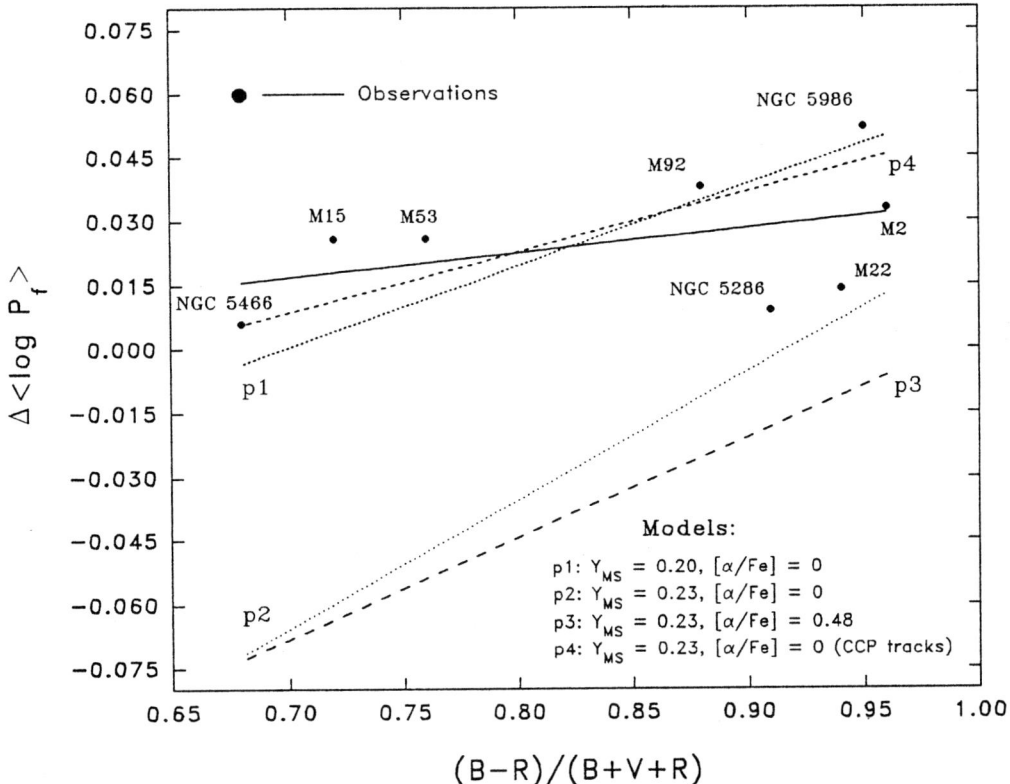

FIGURE 1. Confrontation between predicted and observed period shifts $\Delta \langle \log P_f \rangle$

Case	t	c.l.
Predictions 1	−1.992	93.0%
Predictions 2	−2.967	98.8%
Predictions 3	−2.760	98.3%
Predictions 4	−1.113	71.1%

TABLE 3. Student's test t-values and confidence levels

tions of the observations for OoII clusters (see also Figure 1). It may be noted, in this regard, that whereas discrepancies in zero points also exist for the $\Delta \langle \log P_f \rangle - L_{BVR}$ (cf. Figure 1) and $\langle \log P_f \rangle - L_{BVR}$ relations in general, these have not been considered for the computation of the c.l. entries in Table 3. This is because the basic prediction of the evolutionary models of the HB is that the periods of RR Lyrae variables should increase strongly with L_{BVR} (Lee 1990; Lee et al. 1990; Catelan 1992, 1993) for blue HB morphologies – and this is the effect we are primarily interested in testing. Figure 2 illustrates this point very clearly.

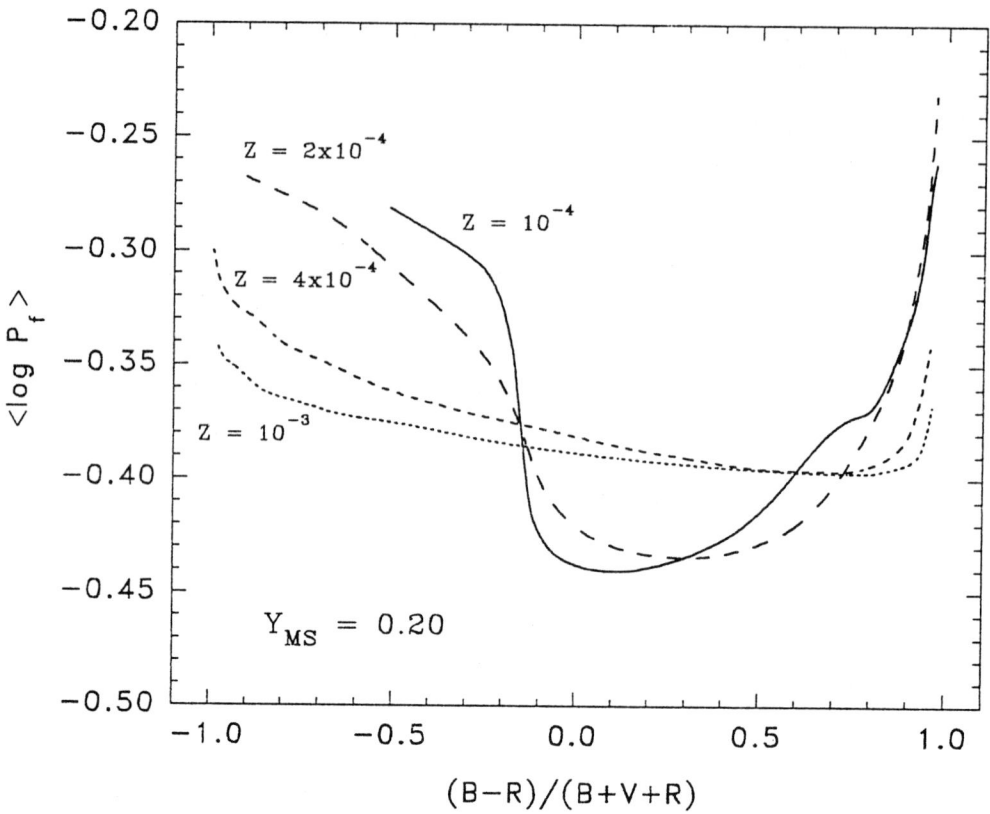

FIGURE 2. Predicted period shifts as functions of HB morphology and metallicity for $Y_{MS} = 0.20$. Note the sharp increase in $\Delta \langle \log P_f \rangle$ with L_{BVR} for blue HB morphologies, especially at very low metallicities

3. Searching for a solution

3.1. HB morphology parameters

A solution to the problem at hand could be that our models with $L_{BVR} \lesssim 0.85$ are too red, and vice-versa for $L_{BVR} \gtrsim 0.85$. However, Rood (1990) argues that for M15 ($L_{BVR}^{\rm Lee} = 0.72$) a more adequate value should be $L_{BVR} = 0.45$. Preston et al. (1991) independently report $L_{BVR} = 0.63$ for this cluster. Similarly, comparison of the values of $B/(B+R)$ [Mironov's (1973) HB morphology parameter] compiled by Zinn (1980) with those predicted by the present models indicate that the latter may be too *blue* for $L_{BVR} \lesssim 0.85$. It may be noted as well that, for clusters with $L_{BVR} \lesssim 0.85$, one finds $\langle L_{BVR}^{\rm Lee} - L_{BVR}^{\rm Preston} \rangle \simeq +0.07$, whereas $\langle L_{BVR}^{\rm Lee} - L_{BVR}^{\rm Preston} \rangle \simeq -0.036 \pm 0.04$ for bluer objects.

3.2. Metallicities

The adopted metallicity scale is that of Zinn & West (1984). Detailed comparison between the implied Z values and those adopted herein (in the solar-scaled and α-enhanced cases alike) discloses that, for M53, NGC 5466, M2, NGC 5986, and M92, they are compatible to within the 1σ level. Small discrepancies exist for M15, NGC 5286, and M22 – these being in the opposite sense to what is needed to explain the present results.

3.3. Mass dispersion on the HB

The adopted σ_M values for the present sample of GGCs are based on (but not identical to) the values reported by Lee (1990). The latter may require a downward shift (cf. Catelan 1993). This being the case, the predicted period shifts would increase much more significantly for the $L_{BVR} \gtrsim 0.85$ clusters than for the redder objects – again, the opposite of what is needed here.

3.4. Statistical fluctuations

Evidence that statistical fluctuations are affecting some of the clusters in the present sample may be provided by the fact that three of these objects present smaller fractions of c-type variables f_c than usually found in OoII GGCs: M92, NGC 5986, and M2 (cf. Table 1). It is interesting to note that these objects are precisely those which have the largest $\Delta \langle \log P_{\rm f} \rangle^\circ$ in Table 1, being also objects with very blue HBs ($L_{BVR} > 0.85$). This means that they are responsible for pushing b_{obs} toward a positive value. First-overtone variables lie at higher temperatures than fundamental variables; thus, their fundamentalized periods are expected to be smaller, since $\log P_{\rm f} \propto -3.48 \log T_{eff}$ (van Albada & Baker 1971). It follows that, were it not for these fluctuations, these clusters would probably have smaller $\Delta \langle \log P_{\rm f} \rangle^\circ$ values, and c.l. in Table 3 would have been even closer to unity.

In addition, based on Figure 2 of Castellani & Quarta (1987) it is quite straightforward to estimate that the probability that the difference between the observed and predicted b values stems from statistical fluctuations affecting $\langle \log P_{\rm f} \rangle$ for the clusters with the bluest HB morphologies ($L_{BVR} > 0.80$) is exceedingly low: for the case $(Y_{MS}, [\alpha/{\rm Fe}]) = (0.23, 0.48)$, for example, it is $\lesssim (16\%)^5 \sim 0.01\%$.

3.5. Instability strip topology

The zero point and functional dependencies of the temperature of the first harmonic blue edge on mass and luminosity have been explored, together with the width of the instability strip, in a wide range of situations. Details are provided in Catelan (1994a). Under no circumstance analyzed did the agreement between the models and the observations improve significantly.

3.6. The evolutionary tracks

The tracks of Castellani et al. (1991) have been used for the construction of an alternative set of synthetic HB models. For the case $(Y_{MS}, [\alpha/{\rm Fe}]) = (0.23, 0)$, the models based on these tracks appear more satisfactory at reproducing the observations than are those based on the Lee & Demarque (1990) tracks, though significant discrepancies are still present. This is shown by the entries labeled "Predictions 4" in Tables 2 and 3. It may be noted that the Castellani et al. (1991) tracks present less extended blueward loops than the Lee & Demarque (1990) ones for the same chemical composition, which is the reason why the "Predictions 4" entries are similar to the "Predictions 1" entries in these tables (see also Figure 1).

3.7. Variations in stellar evolution-controlling parameters

Another appealing possibility is that some parameters which have been assumed constant in the present study in reality are not. Examples are ΔM_c (increase in helium core mass with respect to the "standard" value; we have here assumed $\Delta M_c \equiv 0$ throughout), Y_{MS}, and $[\alpha/{\rm Fe}]$. It cannot be overemphasized that this is *not* the case of age and/or mass loss alone, since, for a given chemical composition, the particular combination of age and overall mass loss on the RGB which leads to the mean evolutionary mass on the HB that is needed to account for a cluster's L_{BVR} value is totally irrelevant, as far as the predicted $\langle \log P_f \rangle$ value is concerned. Additional evidence in favor of such variations may be provided by the study of Catelan & de Freitas Pacheco (1993), where the difficulty in accounting for the differences in HB morphology between prototypical "second-parameter" GGCs in terms of age differences alone has been pointed out. In this regard, it may be interesting to note that the clusters in the present sample with $L_{BVR} \gtrsim 0.85$ lie at systematically smaller distances from the center of the Galaxy than those with redder HB types.

4. Conclusions

The present synthetic HB models are unable to account for the observed dependence of RR Lyrae mean fundamentalized periods on the GGC HB morphology types for OoII clusters, as indicated by Tables 1–3 and Figure 1. Possible solutions to the problem are the inadequacy of the Lee & Demarque (1990) tracks and variations in second-parameter candidates other than age and/or mass loss on the RGB alone among the GGCs. The Castellani et al. (1991) tracks, based upon improved input physics with respect to the former, are found to give a better, though not fully satisfactory, agreement with the observations, but maintaining the need for "third" parameters among the GGCs (given particularly the comments in Sections 3.1 through 3.4). Therefore, an explanation to the problem that has been pointed out in the present study is expected to have a strong impact upon the interpretation of the second-parameter and Oosterhoff dichotomy problems.

I am indebted to Dr. Alvio Renzini for his comments and suggestions. This work has been supported by FAPESP (grant 92/2747-8).

REFERENCES

ARP, H. C., 1955, Color-magnitude diagrams for seven globular clusters. AJ, 60, 317–337

CAPUTO, F., 1988, The galactic globular cluster system: period-shift versus metallicity and the case of ω Centauri. A&A, 189, 70–73

CASTELLANI, V., 1983, Old and new problems in the first stellar generations. Mem. Soc. Astron. Ital., 54, 141–152

CASTELLANI, V., CHIEFFI, A., PULONE, L., 1991, The evolution of He-burning stars: horizontal and asymptotic branches in galactic globulars. ApJS, 76, 911–977

CASTELLANI, V., QUARTA, M. L., 1987, The Oosterhoff dichotomy revisited. I. The ranking of RR Lyrae periods versus metallicity. A&AS, 71, 1–10

CATELAN, M., 1992, The Sandage effect revisited. A&A, 261, 457–471

CATELAN, M., 1993, Synthetic horizontal-branch models for Galactic globular clusters. A&AS, 98, 547–582

CATELAN, M., 1994a, The period-shift effect in Oosterhoff type II globular clusters with blue horizontal-branch morphologies. AJ, in press

CATELAN, M., 1994b, On the Oosterhoff-Arp-Sandage period-shift effect and its evolutionary interpretation. A&A, in press

CATELAN, M., DE FREITAS PACHECO, J. A., 1993, Age differences between old stellar populations from the HB morphology-metallicity diagram. AJ, 106, 1858–1869

CHIEFFI, A., STRANIERO, O., SALARIS, M., 1991, Do we really need α-enhanced isochrones? in Janes K. A., ed, The Formation and Evolution of Star Clusters, ASP Conf. Series, 13, 219–222

FERNELY, J., 1993, A re-analysis of the period shifts in RR Lyrae stars. A&A. 268, 591-606.

Lee, Y.-W., 1990, On the Sandage period shift effect among field RR Lyrae stars. ApJ, 363, 159–167

Lee, Y.-W., Demarque, P., 1990, The evolution of horizontal-branch stars: theoretical sequences. ApJS, 73, 709–746

Lee, Y.-W., Demarque, P., Zinn, R., 1990, The horizontal-branch stars in globular clusters. I. The period-shift effect, the luminosity of the horizontal branch, and the age-metallicity relation. ApJ, 350, 155–172

Mironov, A. V., 1973, Chemical composition of globular-cluster stars and the form of the horizontal branch. Soviet Astron., 17, 16–23

Oosterhoff, P. Th., 1939, Some remarks on the variable stars in globular clusters. Observatory, 62, 104–109

Preston, G. W., Schectman, S. A., Beers, T. C., 1991, Detection of a Galactic color gradient for blue horizontal-branch stars of the halo field and implications for the halo age and density distributions. ApJ, 375, 121–147

Renzini, A., 1983, Current problems in the interpretation of the characteristics of globular clusters. Mem. Soc. Astron. Ital., 54, 335–357

Rood, R. T., 1990, The RR Lyrae. in Cacciari C., Clementini G., eds, Confrontation Between Stellar Pulsation and Evolution, ASP Conf. Series, 11, 11–21

Sandage, A. R., 1990, The Oosterhoff period effect: luminosities of globular cluster zero-age horizontal branches and field RR Lyrae stars as a function of metallicity. ApJ, 350, 631–644

Sandage, A. R., 1993, Temperature, mass, and luminosity of RR Lyrae stars as functions of metallicity at the blue fundamental edge. II. AJ, 106, 703–718

Sweigart, A. V., Renzini, A., Tornambè, A., 1987, On the interpretation of the Sandage period-shift effect among globular-cluster RR Lyrae variables. ApJ, 312, 762–777

van Albada, T. S., Baker, N., 1971, On the masses, luminosities, and compositions of horizontal-branch stars. ApJ, 169, 311–326

Zinn, R., 1980, The globular cluster system of the Galaxy. II. The spatial and metallicity distributions, the second parameter phenomenon, and the formation of the cluster system. ApJ, 241, 602–617

Zinn, R., West, M. J., 1984, The globular cluster system of the Galaxy. III. Measurements of radial velocity and metallicity for 60 clusters and a compilation of metallicities for 121 clusters. ApJS, 55, 45–66

Ultraviolet Observations of Globular Clusters

By WAYNE B. LANDSMAN

Hughes STX, Code 681, NASA/GSFC, Greenbelt, MD 20771, USA

I highlight some recent work based on ultraviolet observations of globular clusters. Images of the globular clusters M79, NGC 1851, ω Cen, M3, and M13 were obtained with the Ultraviolet Imaging Telescope (UIT) in 1990 December. The blue and extreme horizontal branch (HB) stars in M79, ω Cen, and NGC 1851 are found to be 0.2 – 0.5 mag fainter than predicted by HB models using nominal distances and reddenings. The origin of this discrepancy is not yet understood. In NGC 1851, two hot ($T_{eff} >$ 20,000 K) subdwarfs have been identified, and one has been confirmed as a radial velocity member by followup ground-based spectroscopy. Since NGC 1851 does not have an extended horizontal branch, these stars are unlikely to be a product of single star evolution. Several candidate UV-bright stars in M79 and ω Cen are identified on the UIT images, and a program of IUE and optical followup spectroscopy is in progress. Of particular interest is ROA 5342 in ω Cen, which is identified as a sdO cluster member from ground-based spectroscopy.

Observations of NGC 6397 were obtained with the Ultraviolet Spectrometer on the Voyager 2 spacecraft. Although a hot stellar-like spectrum is detected, the dominant source of this emission is probably dust scattered light from hot foreground stars, unrelated to NGC 6397.

1. Introduction

Ultraviolet images of globular clusters have a strikingly different appearance from the familiar ground-based images. At 1500 Å, all stars cooler than $\sim$ 7500 K are suppressed, including the main sequence stars, the red giant and asymptotic giant branch (AGB) stars, and the RR Lyrae and red horizontal branch (HB) stars. Even the contamination due to foreground stars is strongly suppressed. What remains on an ultraviolet image are the hot HB stars, and an occasional luminous hot post-AGB star. Ultraviolet imagery is thus an excellent tool for the discovery of hot stars in globular clusters, and for study of their radial distribution and fundamental parameters.

Ultraviolet images of five globular clusters were obtained with the Ultraviolet Imaging Telescope (UIT) during the first flight of the *Astro* observatory in 1990 December (Stecher et al. 1992). The UIT images are recorded on image-intensified film with about 3″ spatial resolution. The observed clusters span a narrow range in metallicity (Table 1), but are useful for second-parameter studies, since ω Cen, M79, and M13 contain an extended tail of hot HB stars, while NGC 1851 and M3 do not. The images of ω Cen, M3, and M13 were obtained on the daytime side of the Shuttle orbit, and were successfully observed only with a single anti-dayglow filter centered at 1620 Å, while M79 and NGC 1851 were observed with two broad-band filters centered at 1520 Å and 2490 Å.

This paper highlights globular cluster studies with UIT in three areas: (1) the discrepancy between observed ultraviolet HB magnitudes and predictions of theoretical HB models, (2) the discovery of two hot subdwarfs in NGC 1851, a globular cluster not previously known to contain such stars, and (3) spectroscopic followup of newly identified UV-bright stars in M79 and ω Cen. I also present results of a recent observation of NGC 6397 with the Voyager ultraviolet spectrometer.

Name	(m-M)₀	E(B-V)	[Fe/H]	Exp Time (s)	Comment
M79	15.57	0.01	−1.69	1116	
NGC 1851	15.49	0.02	−1.29	543	
ω Cen	13.45	0.15	−1.59	291	Daytime
M3	15.02	0.01	−1.66	199	Daytime
M13	14.29	0.02	−1.65	46	Daytime

TABLE 1. Globular clusters observed with UIT

2. Absolute luminosity of the ZAHB

Based on the LTE stellar atmospheres of Kurucz (1992), the ultraviolet fluxes of the hot ($T_{eff} >$ 10,000 K), high-gravity ($\log g >$ 4) stars on the blue horizontal branch are predicted to have only a very weak dependence on metallicity and gravity. For example, the UIT 1620 Å flux of a solar metallicity Kurucz model with $T_{eff} =$ 20,000 K differs by only 0.03 mag from a model with the same T_{eff} but with $\log Z/Z_\odot =$ -3. Therefore, the transformation of a UV color-magnitude diagram to a theoretical (T_{eff}, log L) diagram should be especially secure. Thus, we were surprised to discover that for both NGC 1851 (Parise et al. 1994) and ω Cen (Whitney et al. 1994) the blue HB stars observed by UIT are approximately 0.5 mag fainter than predicted by current zero-age horizontal branch (ZAHB) models, using the nominal cluster distances, reddenings and metallicities given in Table 1. A somewhat smaller discrepancy of 0.2–0.3 mag is found for the M79 ultraviolet color-magnitude diagram (Hill et al. 1992).

The discrepancy of the UIT fluxes with the theoretical ZAHB in NGC 1851 and ω Cen could be removed by increasing the assumed cluster reddening E(B-V) values by 0.1 mag, by assuming extremely steep ultraviolet extinction properties of the interstellar dust, such as that found in the Small Magellanic Clouds, or by increasing the cluster distance moduli by 0.5 mag. Alternatively, the ZAHB models could be made 0.5 mag fainter if the adopted helium core mass were decreased by ~ 0.05 M$_\odot$ from its canonical value (Sweigart 1994). Another way to remove the discrepancy is to suppose that most of the hot stars seen by UIT are not the product of single star evolution, so that standard HB models do not apply. Strong criticisms can be made against each of these scenarios, which appear to conflict with a variety of measurements at other wavelengths. Note, however, that Noble et al. (1991) and Nemec et al. (1993) argue for a larger distance modulus for ω Cen, and that there have been several claims in the literature of a larger reddening toward NGC 1851, as reviewed – but not endorsed – by Walker (1992).

Could there be an offset of 0.5 magnitudes in the UIT calibration? The absolute calibration of UIT is mainly based on comparison of $\sim$ 20 stars observed in common with IUE. However, the linearity of UIT must also be verified, since the HB stars are typically $\sim$ 5 magnitudes fainter than the IUE calibration stars. Extensive tests on UIT images of differing exposure times have revealed some linearity problems, but at a less than 10 % level. The current UIT calibration is especially convincing for ω Cen, where there are several long IUE exposures of UV-bright stars in the cluster. Verification of the UIT fluxes may be possible with observations of blue HB stars with the Faint Object Spectrograph on HST, or with the reflight of the *Astro* mission in December 1994, when several additional UIT globular cluster pointings are planned.

Name	m_{1520}	m_{2490}	V	B–V	Comment
UIT-44	15.9	17.3	18.6	0.11	RV member
UIT-31	16.1	17.1	18.6	–0.22	=Walker 1243

TABLE 2. Hot Subdwarfs in NGC 1851

3. Hot Subdwarfs in NGC 1851

Parise et al. (1994) identified 46 blue horizontal branch stars outside of the central 70" in NGC 1851 on UIT images at 2490 Å and 1520 Å. (The core of the cluster could not be studied since it is heavily saturated by the presence of the luminous hot post-AGB star UV-5.) Two of these stars, UIT-44 and UIT-31, showed $m_{2490} - m_{1520}$ colors indicating effective temperatures greater than 20,000 K. This result was somewhat surprising because ground-based studies of NGC 1851 had not shown the presence of a hot extended HB. The recent study of Walker (1992) confirmed a bimodal temperature distribution of HB stars, but with only 27 % of the HB stars blueward of the RR Lyrae gap, and none hotter than 10,000 K. Landsman et al. (1994a) obtained CTIO spectrophotometry of UIT-44 and confirmed it as a radial velocity cluster member. The spectrum showed only weak Balmer lines typical of helium-poor hot subdwarfs. Simultaneous model fitting of the Balmer lines yielded $T_{eff} = 29,700 \pm 1630$ K and $\log g = 5.39 \pm 0.27$.

Table 2 gives the visible and ultraviolet photometry for the two hot subdwarfs in NGC 1851. The V and B-V magnitudes for UIT-44 are from the CCD photometry in Landsman et al. (1994a), while the values for UIT-31 are from the Walker (1992) catalog. While UIT-31 is not a confirmed cluster member, the similarity of its ultraviolet magnitudes and color to UIT-44 strongly suggest membership. The presence of these two stars is not easily explained by models of single star evolution, since a mechanism must be found to explain their large separation in effective temperature from other HB stars in the cluster. Possible evidence for the binarity of UIT-44 is its observed value of B–V = 0.11, which is too red to be compatible with the ultraviolet photometry or the temperature derived from Balmer line fitting.

4. UV-bright Stars

The term "UV-bright star" is used here to refer to a star more than ~ 1 magnitude brighter the observed ZAHB luminosity. This definition is motivated by theoretical tracks which predict that hot HB stars will be ~ 1 magnitude brighter than the ZAHB at the time of their core helium exhaustion (Dorman, Rood & O'Connell 1993). The scenario where most hot UV-bright stars are evolved off of the extended HB (i.e., AGB-manque stars) is consistent with the absence of UV-bright stars found on the UIT image of NGC 1851 (except for the known post-AGB star UV5). On the other hand, about 30 UV-bright stars are visible on the UIT image of ω Cen, and Hill et al. (1992) identified two UV-bright stars on the M79 images. The tracks of Dorman et al. (1993) predict that the number of UV-bright stars should be 15–20 % of the number of core helium burning stars on the extended HB, and this is roughly what is observed in the ultraviolet color-magnitude diagram of ω Cen (Whitney et al. 1994).

We have begun a program of ground-based and IUE spectroscopy of the UV-bright stars discovered on the UIT images of M79 and ω Cen. Tables 3 and 4 list the stars for which either optical or IUE observations are already available, although in most cases, our analysis is still preliminary. The listed luminosities and temperatures are estimated

Name	m_{1620}	m_{2490}	V	B-V	T_{eff}	lbol
UIT-1(F1179)	13.4	14.4	18.61	-0.28	> 40000	> 2.4
UIT-2(F2804)	14.2	14.5	16.83	-0.09	13000	2.0

TABLE 3. Newly identified UV-bright stars in M79

Name	m_{1620}	V	B-V	Sp.T.	T_{eff}	lbol
UIT-1	10.8			B-type	18000	2.9
UIT-2	11.9			B-type	19000	2.5
ROA5342	12.2	15.89	-0.15	sdO	> 40000	> 2.6
ROA5857	12.9	15.37	0.04	HBB	16000	2.1
Dk 3873	12.9	16.00	0.04	sdO?	> 20000	> 2.1

TABLE 4. Newly identified UV-bright stars in Omega Cen

from the UV and visible photometry, and the adopted cluster distances and reddenings in Table 1. For the hotter stars, the value of T_{eff} derived this way is very sensitive to errors in the photometry or the assumed reddening, and only lower limits are given. In M79, the optical identifications and V and B-V magnitudes are taken from the catalog of Ferraro et al. (1992). Further discussion of the IUE and optical spectra of the M79 stars is given in Cheng et al. (1993)

IUE spectra of the stars UIT-1 and UIT-2 in ω Cen were discussed by Landsman et al. (1992). Both stars show low-resolution IUE spectra similar to those of main-sequence B-type stars, but with weaker absorption lines. Unfortunately, ground-based spectroscopy of these stars may not be possible, because they are within 2′ of the dense cluster center. The identifications and V and B–V magnitudes for the other three UV-bright stars in ω Cen are taken from Dickens et al. (1988). Low-resolution spectroscopy of ROA 5342 and ROA 5857 at CTIO confirm both stars as radial velocity cluster members. ROA 5857 has a luminous BHB-type spectrum with strong Balmer lines visible up to n = 12. ROA 5342 has an sdO spectrum with Balmer lines accompanied by He II absorption (Figure 1). The equivalent widths of the He II lines closely match those of the sdO star ROB 162 in NGC 6397 (Heber & Kudritzki 1986). This suggests that ROA 5342 that might be similar to ROB 162 in having a normal helium abundance and a post-AGB evolutionary status. The IUE spectrum of ROA 5342 (Figure 2) shows weak emission features near He II 1640 Å and C III 1909 Å, which might indicate an associated planetary nebula. We are preparing a more detailed investigation of ROA 5342 (Landsman et al. 1994b), and have recently obtained an echelle spectrum at the CTIO 4-m.

No optical spectra were obtained of Dk 3873. However, its IUE spectrum matches quite closely to that of ROA 5342 (Figure 2), which is suggestive both of its cluster membership and of its high effective temperature.

5. Far-UV Observations

The 912–1216 Å region can provide a sensitive measure of the effective temperature of stars with T_{eff} > 25,000 K, for which the flux longward of Lyα is a poor discriminator. There have been successful programs with both the HUT and Voyager ultraviolet spectrometers to detect globular clusters shortward of 1216 Å. The Hopkins Ultraviolet Telescope (HUT) was co-pointed with UIT during the *Astro-1* mission and obtained 912 – 1860 Å spectra of hot stars in M79, M3, and NGC 1851 (Dixon, Davidsen & Ferguson

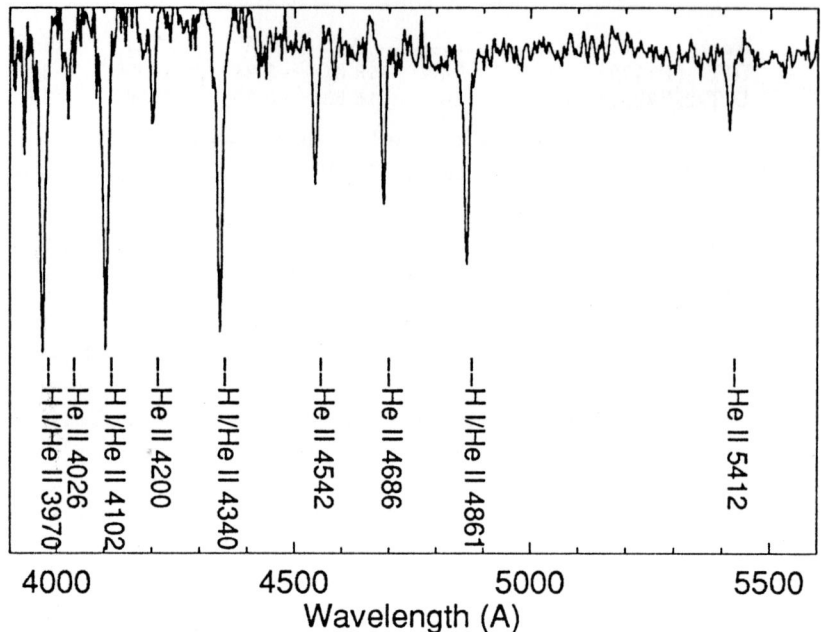

FIGURE 1. A rectified CTIO spectrum of ROA 5342 in ω Cen

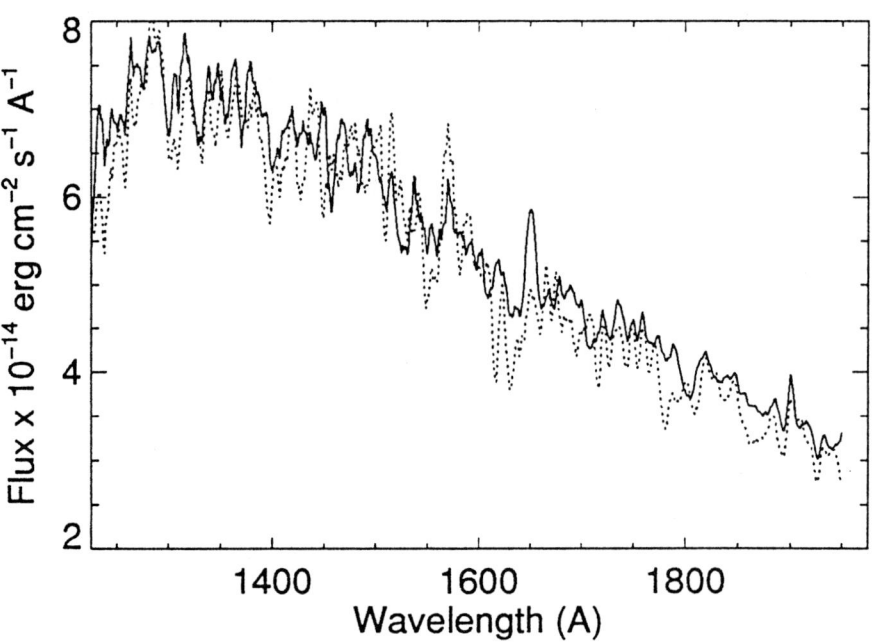

FIGURE 2. IUE spectrum SWP 48266 of ROA 5342 (solid line), and SWP 48271 of Dk 3873 (dotted line). The Dk 3873 spectrum has been multiplied by a factor of 1.6.

1994). In M79, the HUT large (18″ × 116″) aperture was used to integrate the flux of numerous blue and extended horizontal branch stars, while in M3 and NGC 1851, the HUT aperture was centered, respectively, on the hot post-AGB stars, vZ1128 and NGC 1851-UV5. The Voyager ultraviolet spectrometer was used to obtain 912 – 1700 Å spectra of the clusters NGC 6752, M13 and M92 (Holberg 1990a). The 0.1° × 0.87° Voyager FOV is well-matched to globular clusters, and partially compensates for its limited sensitivity ($\sim 5 \times 10^{-13}$ erg cm^{-2} s^{-1} Å^{-1}).

Only one additional globular cluster was observed with Voyager before the guest observer program was canceled in early 1993. Figure 3 shows a Voyager 2 spectrum of NGC 6397 obtained during 5 days of integration in February 1993. The horizontal branch of NGC 6397 does not contain an extended blue tail (Alcaino et al. 1987), and so the dominant contribution to the Voyager spectrum was expected to be from the well-studied hot post-AGB star ROB 162 (Heber & Kudritzki 1986). However, there are three reasons to believe that ROB 162 was *not* the main contributor to the Voyager spectrum. First, the Voyager flux longward of Lyα is $\sim$ 3 times larger than the IUE flux of ROB 162. Second, there was no evidence of a point-source response during the limit cycle motion of the spacecraft. Finally, a Voyager "background" observation at an offset position 26′ from the cluster center shows a similarly strong spectrum.

The source of the diffuse emission detected by Voyager is probably dust scattered starlight unrelated to NGC 6397. Such diffuse galactic light has been detected by Voyager at similar intensities at other locations at low galactic latitude (Holberg 1990b). (The galactic latitude of NGC 6397 is -12°.) Thus, the measurement of the 912-1200 Å spectrum of ROB 162 must wait for future spectrographs with smaller apertures, such as the reflight of HUT in December 1994 on the *Astro*-2 mission.

This paper summarizes work done with various collaborators including K.-P. Cheng, Van Dixon, Bob Hill, Paul Hintzen, Bob O'Connell, Ron Parise, Rex Saffer, Ted Stecher, and Jon Whitney. I thank Jay Holberg for assistance with the Voyager data reduction. This work is supported by NASA Voyager grant S-97230E to Hughes STX.

REFERENCES

ALCAINO, G., BUONANNO, R., CALOI, V., CASTELLANI, V., CORSI, C., E., IANNICOLA, G., LILLER, W., 1987, AJ, 94, 917

CHENG, K.-P., HILL, R., HINTZEN, P., STECHER, T. P., BOHLIN, R. C., O'CONNELL, R. W., ROBERTS, M. S., SMITH, A. M., 1993, BAAS, 25, 1407

DICKENS, R. J. BRODIE, I. R., BINGHAM, E. A., CALDWELL, S. P., 1988, Rutherford Appleton Laboratory, RAL 88-04

DIXON, W. V., DAVIDSEN, A. F., FERGUSON, H. C., 1994, AJ, in press

DORMAN, B., ROOD, R. T., O'CONNELL, R. W., 1993, ApJ, 419, 596

FERRARO, F. R., CLEMENTINI, G., FUSI PECCI, F., SORTINO, R., BUONANNO, R., 1992, A&A, 256, 391

HEBER, U., KUDRITZKI, R. P, 1986, A&A, 169, 244

HILL, R. S., HILL, J. K., LANDSMAN, W. B., et al., 1992, ApJ, 395, L17

HOLBERG, J., 1990a, in Kondo Y., ed, Observatories in Earth Orbit and Beyond, IAU Symp. 123, p. 49

HOLBERG, J., 1990b, in Bowyer S., Leinert C., eds, The Galactic and Extragalactic Background Radiation, IAU Symp. 139, p. 220

KURUCZ, R. L., 1992, in Barbuy B., Renzini A., eds, The Stellar Populations of Galaxies, IAU Symp. 149, p. 225

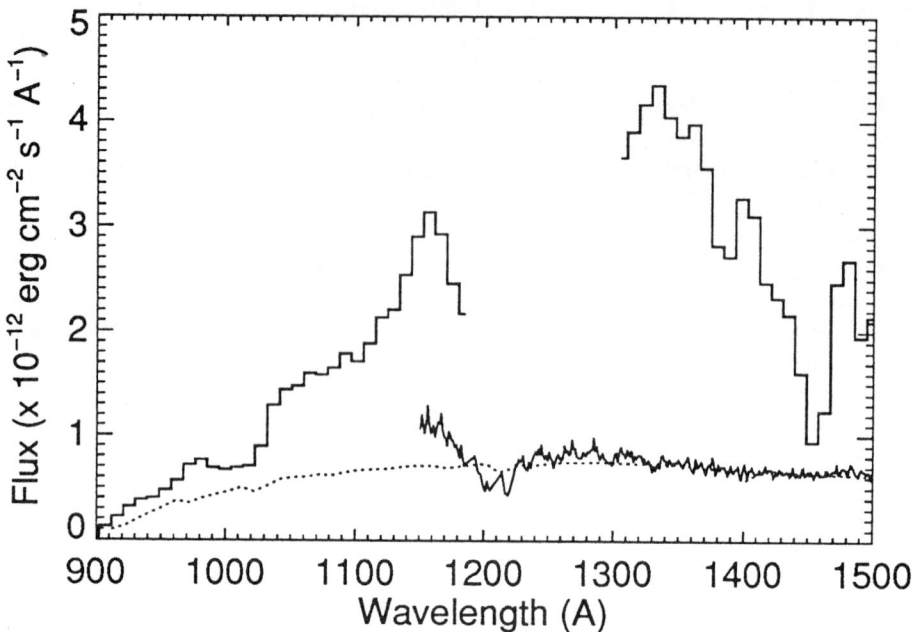

FIGURE 3. Voyager spectrum of NGC 6397; the spectral region contaminated by diffuse interplanetary Lyα has been deleted. Also shown is an IUE spectrum (SWP 15322) of the hot post-AGB star ROB 162. The dotted line shows a Kurucz (1992) model with $T_{eff} = 50{,}000$ K, $\log g = 4.5$, and $\log Z/Z_\odot = -1$ which has been normalized to the V = 13.23 magnitude of ROB 162 and reddened with E(B-V) = 0.18. The parameters used for the Kurucz model are those determined spectroscopically for ROB 162 by Heber & Kudritzki (1986).

LANDSMAN, W. B., O'CONNELL, R. W., WHITNEY, J. H., et al., 1992, ApJ, 395, L21.

LANDSMAN, W. B, SAFFER, R., HINTZEN, P., 1994a, in preparation

LANDSMAN, W. B, WHITNEY, J., CHENG, K. -P., HINTZEN, P., O'CONNELL, R. W., STECHER, T. P., 1994b, in preparation

NEMEC, J. M., NEMEC, A. F., LUTZ, T. E., 1993, in Saffer R, ed, Blue Stragglers, ASP Conference Series 53, p. 145

NOBLE, R. G., DICKENS, R. J., BUTRESS, J., GRIFFITHS, W. K., PENNY, A. J., 1991, MNRAS, 250, 314

PARISE, R. A. et al., 1994, ApJ, in press

STECHER, T. P., BAKER, G. R., BARTOE, D. D., et al., 1992, ApJ, 395, L1

SWEIGART, A. V., 1994, ApJ, in press

WALKER, A. R., 1992, PASP, 104, 1063

WHITNEY, J. N., O'CONNELL, R. W., ROOD, R. T., et al., 1994, in Adelman S. J., Upgren A. R., Adelman C. J., eds, Hot Stars in the Halo, Cambridge University Press, Cambridge, p. 163

UV Photometry of Hot Stars in Omega Centauri

By JONATHAN H. WHITNEY[1], R. W. O'CONNELL[1],
R. T. ROOD[1], B. DORMAN[1], R. C. BOHLIN[2],
K. P. CHENG[3], P. M. N. HINTZEN[3], W. B. LANDSMAN[4],
M. S. ROBERTS[5], A. M. SMITH[3], E. P. SMITH[3]
AND T. P. STECHER[3]

[1]University of Virginia, P.O. Box 3818, Charlottesville, VA 22903, USA

[2]Space Telescope Science Institute, 3700 San Martin Drive, Baltimore, MD 21218, USA

[3]NASA/GSFC, Code 681, Greenbelt, MD 20771, USA

[4]Hughes STX, 4400 Forbes Boulevard, Lanham, MD 20706, USA

[5]National Radio Astronomy Observatory, Edgemont Road, Charlottesville, VA 22903, USA

Far-ultraviolet exposures (center 1620 Å, bandwidth 225 Å, field of view 40' diameter) were obtained of the globular cluster ω Cen with the Ultraviolet Imaging Telescope during the ASTRO-1 mission of December 1990. In the FUV we detect 1957 sources, down to an apparent FUV magnitude of 16.4. We also obtained CCD observations in the Strömgren u band at CTIO covering a field 15' square. Using these observations, we construct a FUV, u color-magnitude diagram to study the luminosity function and HB structure and to compare with theoretical evolutionary tracks. Of special interest are the hot stars with small envelopes, which lie on the extreme HB (with $T_{eff} > 20,000$ K) and evolved stars above the HB.

1. Introduction

Far-ultraviolet exposures were obtained of the globular cluster ω Cen with the Ultraviolet Imaging Telescope during the Astro-1 mission in December 1990. Combining these with CCD observations in the Strömgren u band allows us to construct an FUV, u color-magnitude diagram for the cluster.

The FUV observations give us a complete sample of the hot horizontal branch stars and their progeny even in the dense core of the cluster, free of the cool star background which dominates at longer wavelengths. This allows us to study the HB structure and compare with theoretical evolutionary tracks transformed into color-magnitude space by use of model stellar atmospheres. Of special interest are the hot stars with small envelopes, which lie on the extreme HB, and evolved stars above the HB.

2. Observations and Data Reduction

The globular cluster ω Cen was observed in the Far UV by UIT using the B5 filter ($\lambda_0 = 1620$ Å, $\Delta\lambda = 225$ Å) for 291 seconds. The field of view is 40' in diameter, with a plate scale of 56.8 arcsec mm^{-1}. The image was recorded on Kodak IIa-O film and digitized by scanning with a microdensitometer (Stecher et al. 1992). In February 1992, we also obtained 900 second exposure with the 0.9m telescope at CTIO in the Strömgren u band ($\lambda_0 = 3515$ Å, $\Delta\lambda = 300$ Å) on a Tek2048 CCD. The field of view is 15' square, centered on the cluster.

The digitized, flat-fielded FUV image was reduced using specifically adapted IDL versions of the DAOPHOT routines (Stetson 1987) to find stars and determine magnitudes with point-spread function (psf) fitting. The FUV fluxes obtained were then directly

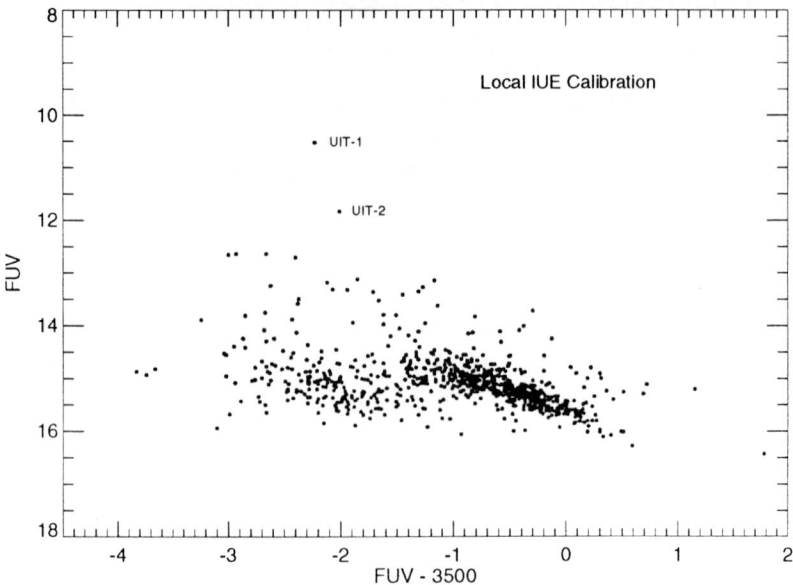

FIGURE 1. The FUV, m3500 CMD for ω Cen

calibrated to two isolated stars in the frame which had also been observed with IUE. The optical CCD data was reduced using the standard IRAF/DAOPHOT routines, and calibrated with standards observed the same night.

An astrometric solution was determined for each frame by matching detected sources with those of Dickens et al. (1988). Using these solutions the sources in the FUV image and the optical image were matched to produce an FUV, m3500 color-magnitude diagram. The magnitude distribution for the 1957 sources detected in the FUV ranges from 9.9 to 16.5, with a peak at 15.3 and a FWHM of 0.8 magnitudes. A total of 64 stars are brigher than magnitude 14.0, placing them well above the HB. The 789 stars which have the highest confidence in matching and photometic quality make up the CMD shown in Figure 1.

3. Discussion

For comparison, theoretical models of horizontal branch evolution (Dorman et al. 1993), were transformed to FUV, m3500 colors and magnitudes using the models of Kurucz (1991). The data are dereddened and the tracks shifted by the distance modulus, derived from fitting HB evolutionary models of Dickens et al. (1988) optical data. The evolutionary models have ZAHB envelope masses ranging from 0.003 to 0.145; the 3 hottest points on the ZAHB correspond to $M_{env} = 0.003, 0.005, 0.010\ M_\odot$. The tracks overlaid on the CMD are shown in Figure 2.

The theoretical tracks match the observed HB quite well for the redder regions, but are too bright on the blue end. This same effect is observed in ultraviolet observations of M79 (Hill et al. 1992) and NGC 1851 (Parise et al. 1994). The origin of the offset is unclear, but there are several possibilities: (1) theoretical atmospheres may not model blue HB stars well, since HB abundances may be unusual and non-LTE effects may be important; (2) reddening (and the reddening law) are not uniquely determined for ω Cen.

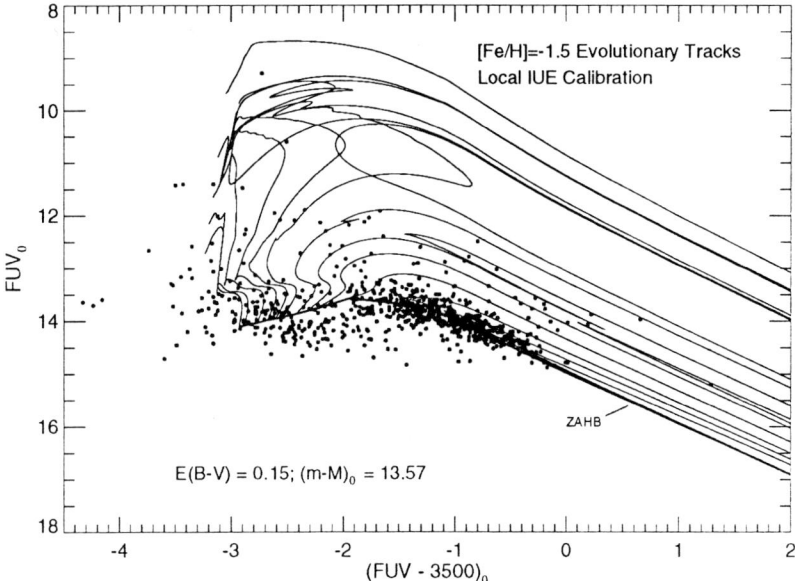

FIGURE 2. The FUV, m3500 CMD for ω Cen with theoretical evolutionary tracks overlaid. The models have [Fe/H] = −1.5 and $Y = 0.25$.

(3) the discrepancy may be genuine, and the stars here may be a population distinct from the HB.

The bluest stars are also unusual, being bluer than the lowest envelope mass model plotted here ($M_{env} = 0.003\ M_\odot$). These objects approach the bluest colors possible from blackbody emission. Possibly they are stars evolving towards the white dwarf cooling curve.

The CMD of Figure 1 shows an obvious break in the HB population density at $FUV - 3500 = -1.5$ ($T_{eff} \sim 15,000$ K). The gap has also been observed in optical CMDs; the bimodal population of the hot stars is made clear in Figure 3. The hotter group is also somewhat less luminous than the models predict. If this difference is real, the hot blue stars may in fact represent a distinct population, such as interacting binaries (Bailyn et al. 1992) or late He ignition objects (Castellani & Castellani 1993).

Two other areas in the CMD also contain interesting objects. First, a number of bright ($FUV < 14.0$), blue ($FUV - u < -2.0$) stars above the HB are most likely AGB-Manqué stars, which have such low envelope mass that they never return to the AGB. Second are the bright stars UIT-1 and UIT-2 identified from the UIT frame (Landsman et al. 1992) stand out in the CMD. Both could be AGB-Manqué stars. The only candidates bright enough to be post-AGB stars are ROA-5701, which was outside the CCD field of view, and (possibly) UIT-1.

4. Conclusions

Exploring the globular cluster ω Cen in the ultraviolet brings out the hot stars especially well, providing a look at a number of interesting hot HB and post-HB stars difficult to observe optically. The richness of the cluster offers a significant sample of the relatively short lived hot post-HB evolutionary states.

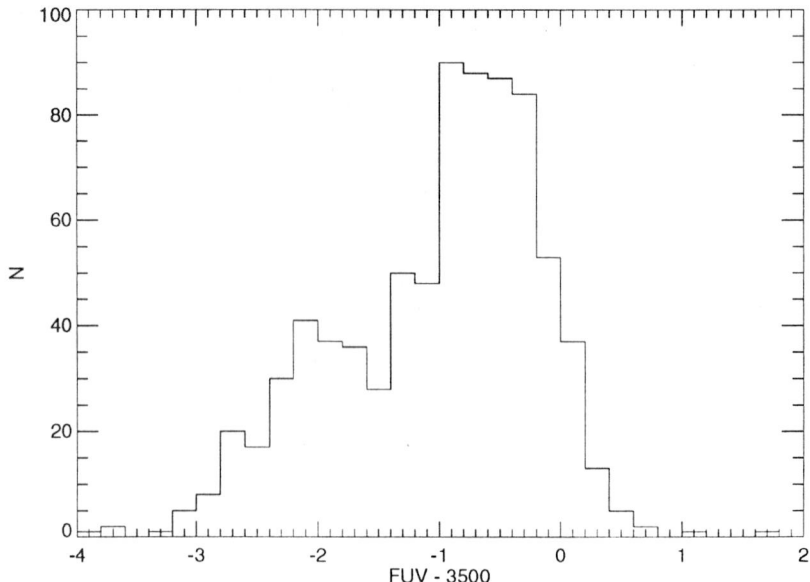

FIGURE 3. The distribution of stars in ω Cen by color

This work was supported financially by NASA grant NAGW-2596.

REFERENCES

BAILYN, C. D., SARAJEDINI, A., COHN, H., LUGGER, P. M., GRINDLAY, J. E., 1992, AJ, 103, 1564

CASTELLANI, M., CASTELLANI, V., 1993, ApJ, 407, 649

DICKENS, R. J., BRODIE, I. R., BINGHAM, E. A., CALDWELL, S. P., 1988, Rutherford Appleton Laboratory, RAL 88-04

DORMAN, B., ROOD, R. T., O'CONNELL, R. W., 1993, ApJ, 419, 596

HILL, R., HILL, J. K., LANDSMAN, W. B., et al., 1992, ApJ, 395, L17

KURUCZ, R. L., 1991, in Philip A. G. D., Upgren A. R., Janes K. A., eds, Precision Photometry: Astrophysics of the Galaxy, L. Davis Press, Schenectady, p. 27

LANDSMAN, W. B., O'CONNELL, R. S., WHITNEY, J. H., et al., 1992, ApJ, 395, L21

PARISE, R., et al., 1994, in preparation

STECHER, T. P., BAKER, G. R., BARTOE, D. D., et al., 1992, ApJ, 395, L1

STETSON, P. B., 1987, PASP, 99, 191

Discussion

SARAJEDINI: Bailyn et al. (1992, AJ, 103, 1564) claim that the extreme blue HB stars in Omega Cen are more centrally concentrated than the stars on the flat part of the HB and the subgiant stars. Those data extended from 3 to 14 arcminutes. Does this finding disagree with your data?

CARNEY: Is there a potential explanation for the discrepancy between observed and model luminosities for the hottest stars?

DORMAN: Assuming that the problem is real, the sources of uncertainty lie in the models or the color transformations (or both). As far as the models are concerned their Bolometric luminosity is determined by the core mass: this is well-constrained (see Sweigart's paper on

p. 17) but could be lower if the stars are actually examples of the models of Castellani & Castellani (1993) who found that stars that lose sufficient mass on RGB to fail to reach the helium flash may ignite helium later (if the mass loss is not too great) thus becoming similar to the EHB stars, but with smaller core mass. The difficulty with this is, that we expect all of them to be very blue, because they would have very thin envelopes. So, my bets would be on a still-incomplete line list in the Kurucz database at wavelengths < 2000 Å.

Spectroscopic and UBV Observations of Blue Stars at the NGP

By DAVID J. BELL[1], H. L. DETWEILER[2],
KENNETH M. YOSS[1], STEFANO CASERTANO[3],
GRANT BAZAN[4], ANURAG SHANKAR[4],
ROSA MURPHY[1] AND SEAN POINTS[1]

[1]Department of Astronomy, University of Illinois at Urbana-Champaign, Urbana, IL 61801, USA

[2]Department of Physics, Illinois Wesleyan University, Bloomington, IL 61702, USA

[3]Physics and Astronomy Department, The Johns Hopkins University, Baltimore MD 21218, USA

[4]Steward Observatory, University of Arizona, Tucson, AZ 85721, USA

Spectra have been obtained of 34 stars near the NGP with photographic $V < 17$, $(B-V) < 0.5$. Spectral types and luminosity classes have been estimated from line-ratios and Balmer line-widths, based on criteria calibrated from observations of standard stars. Eighteen of the candidate stars fall within the spectral range from B9 to A5; one has broad shallow Balmer lines characteristic of a white dwarf; two appear to be B subdwarfs; three horizontal-branch stars; and twelve main-sequence stars. UBV photometry has been performed for about half of the program stars, and the observed colors are consistent with the derived spectral types. The presence of Mg II 4481 Å in the main-sequence candidates and lack in the horizontal-branch candidates is further confirmation of the luminosity classifications. Nine of the main-sequence candidates have estimated z-distances of 1 kpc or greater, confirming that a significant fraction of halo blue stars appear unevolved.

The radial-velocity dispersion for the B9 to A5 program stars with $V > 12$ for both horizontal-branch and main-sequence stars combined is $\sim 80 \pm 20$ km s^{-1}, consistent with previous results for blue halo stars.

1. Introduction

The existence of hot stars well above the Galactic plane has long been known (cf. extensive references, Philip 1994). However, the question of the percentage of blue stars in the total halo population has not yet been rigorously addressed, as it requires analysis of a complete unbiased sample. Of particular interest is the dividing line between hot blue stars and the remainder of the halo population. Our ultimate objective is to determine the fraction of the total population which is blue by establishing the frequency distribution of $(B-V)$ as a function of V.

As can be seen in Figure 1, the percentage of hot stars in an unbiased halo sample is relatively small. The histogram represents all stars with $15 < V < 16$ in the Weistrop NGP sample (Weistrop 1972), with the photometric corrections of Yoss et al. (1989). Although the Weistrop (photographically determined) colors are of relatively low accuracy (± 0.1 mag.), the corrected values are quite consistent with our photoelectric $(B-V)$ colors. In Figure 1, the continuous curve is the result of Galactic modeling; the poor fit at the red end is partially due to the uncertainties in the colors, and illustrates the need for better colors. At this time we can not accurately predict the actual percentage of hot stars, but based on the present colors it seems to be ~ 1 %.

Galactic modeling eventually will help establish the line of demarcation between the

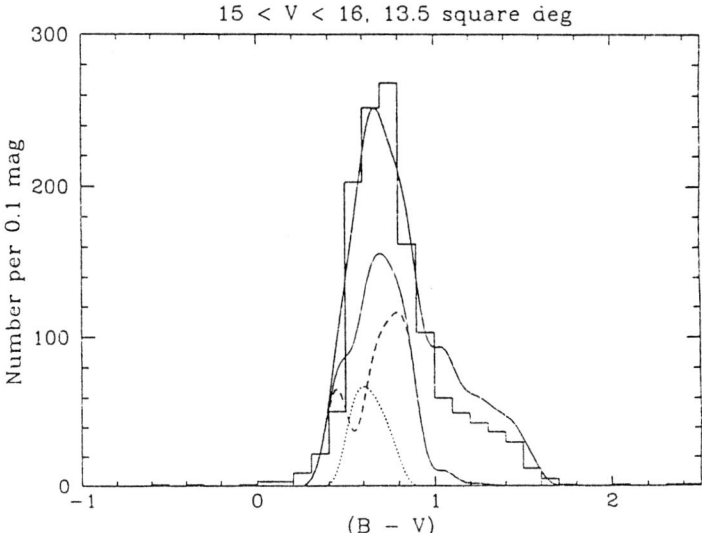

FIGURE 1. A comparison between the observed $(B-V)$ frequency distribution (histogram) and that expected by modeling (thick solid line) for Weistrop NGP stars with $15 < V < 16$. The expected distribution of giants (thin solid line) is also shown, and is subdivided into those due to the Galactic halo (dashed line) and the thick disk (dotted line).

"normal" halo population and the hot blue-star component, the origin of which presumably is unusual relative to the bulk of halo stars. We additionally want to determine from our blue-star component the numbers of main-sequence (young, high mass) and horizontal-branch (old, low mass) stars. Our initial results are presented here in this interim report.

2. Spectroscopic Observations

Spectra have been obtained of 34 blue NGP stars with the Steward Observatory 90-inch reflector and the Mount Laguna 1-meter, and CCD spectrographs. We also obtained spectra of standard stars, including three blue horizontal-branch stars in M13 and M92. Observations were obtained at Steward on one night in April 1992, with a resolution of ~ 1.5 Å, and on three nights in April 1993, with a resolution of ~ 3.0 Å. The Mount Laguna observations were obtained on one night in April 1992, with a resolution of ~ 8 Å.

The program stars were selected from the Weistrop catalog with $V < 17$, and $(B-V) < 0.5$. As expected, several turned out to be late-type stars with inaccurate colors, but eighteen of the observed stars showed strong Balmer lines. Figure 2a shows one of the strongest Balmer-line stars, W 35802, with $V \sim 16.0$, $(B-V) \sim +0.1$ and a derived spectral type of A0, and Figure 2b shows one of the weakest Balmer-line stars, W 44475, at $V = 14.74$, $(B-V) = 0.26$, and a derived spectral type of A3.

We determined spectral types of the program stars through spectrographic line ratios, calibrated through the standard-star spectra. We relied primarily on the ratios FeI 4325/Hγ, and K-line/Hδ, but also monitored the He lines and the Mg II 4481 Å line. The spectral types are accurate to $\sim 1-2$ subclasses, and range from B9 to A5 for the Balmer-line stars.

The Balmer lines were fitted empirically using the Lorentz profile in the Minnaert-

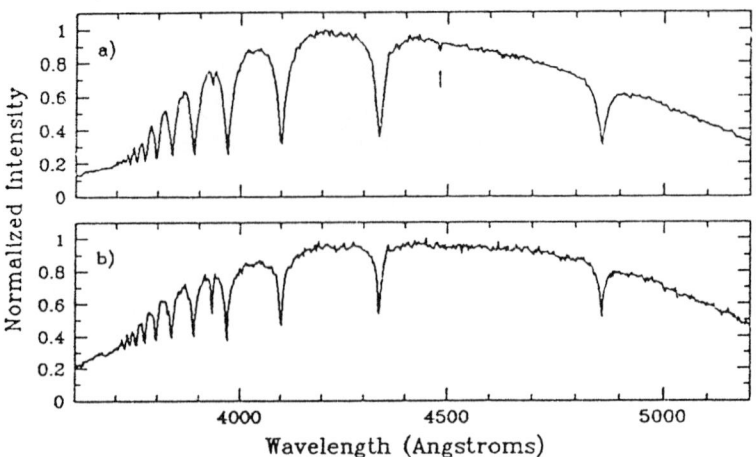

FIGURE 2. Spectra of (a) one of the strongest Balmer-line stars, and (b) one of the weakest. The Mg II 4481 Å line, marked on the MS candidate (a), is not detected for the BHB candidate star (b).

Unsold formula, and the strengths measured in three ways: central depth, width at the 20 % level (measured from the continuum), and the equivalent width. Each can be used to separate horizontal-branch from main-sequence stars, but we relied primarily on the equivalent width, as it is the most accurately measured and is not affected by resolution. The measurements were obtained from spectra of three different resolutions, with some systematic differences. However, since the overlap in observations is not extensive (three program stars and three standard stars), and since the differences are small compared to the differences between main-sequence and horizontal-branch stars, for this interim report we have applied no corrections. Figure 3 shows the fitted Hδ profile for W 35802, with the 20 % width indicated.

Although profiles were fitted to Hβ, Hγ, and Hδ, only the results for Hδ are given in Table 1 for the eighteen stars with derived spectral type B9 to A5. The Table 1 entries begin with the Weistrop name; followed by V, $(B - V)$, $(U - B)$, and the number of photoelectric observations. When no photoelectric UBV is available, the Weistrop values (indicated with asterisks) are given with the corrections of Yoss et al. (1989) applied for V and $(B - V)$. The next three columns list the central depth, equivalent width, and 20 % width as measured from the Hδ profile fit. Next is the derived spectral type and an estimated luminosity category, white dwarf, subdwarf, horizontal-branch, marginal, or main-sequence.

Figure 4 shows Hδ equivalent width versus spectral type for standard stars, with the three globular-cluster horizontal-branch and two supergiant values marked. Figure 5 shows a similar diagram for our program stars, with white dwarf, subdwarf, horizontal-branch, main-sequence, and marginal candidates marked.

The eighteen stars within the spectral range B9 to A5 fall into three groups: 1) one program star has broad shallow Balmer lines characteristic of a white dwarf; 2) ten stars with Balmer-lines stronger than the three globular-cluster horizontal-branch stars, and therefore likely main-sequence stars; and 3) seven with strengths comparable to the horizontal-branch standards. Of the stars in the third group, two appear to be sub-

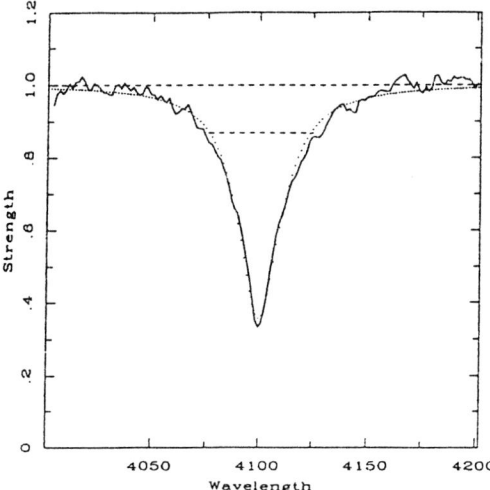

FIGURE 3. An empirical line-profile fit to Hδ of the star in Figure 2a.

ID	V	B − V	U − B	N	cd	EqW	D20%	Sp	class
W 7101	12.65	-0.11	-1.25	2	0.13	8.0	90.0	B9	wd
W 7324	15.7*	-0.1*	-1.1*		0.35	6.9	26.8	B9	sd
W 11335	15.16	0.23	0.05	2	0.46	10.2	30.2	A3	hb
W 11594	14.15	0.06	0.17	2	0.74	16.9	30.2	A1	ms
W 14771	15.6*	0.3*	0.3*		0.55	18.2	45.2	A2	ms
W 20836	15.2*	0.3*	0.0*		0.45	10.4	31.0	A4	ms?
W 22210	10.53	0.22	0.15	3	0.60	14.4	34.1	A5	ms
W 34584	10.63	-0.04	-0.07	3	0.59	12.0	27.0	A0	ms?
W 35802	16.1*	0.1*	0.2*		0.64	22.6	49.3	A0	ms
W 35842	15.1*	0.3*	0.0*		0.36	5.9	20.6	A5	hb
W 42576	8.79	0.31	0.18	2	0.44	12.8	41.2	A5	ms
W 44475	14.74	0.26	-0.14	1	0.45	9.7	28.6	A3	hb
W 50462	6.52	0.11	0.11	3	0.61	22.1	47.6	A2	ms
W 52540	15.6*	0.1*	0.5*		0.68	17.9	37.1	A4	ms
W 52762	16.8*	-0.1*	0.8*		0.74	22.9	41.7	A1	ms
W 60466	15.0*	0.3*	0.2*		0.58	16.0	38.1	A2	ms
W 61442	15.3*	0.3*	0.3*		0.58	14.9	34.6	A0	ms
W 64217	14.26	-0.26	-1.07	2	0.39	9.3	32.8	B9	sd

TABLE 1. Blue-Stars at the NGP

dwarfs, based on the ratio of 20 % width to central depth; similarly, three others appear to be horizontal-branch stars, based on the same ratio. The remaining two are discussed in the next paragraph.

The presence or absence of the Mg II 4481 Å line is a further criterion for classification. The ten stars of group 2) as well as two from group 3) all have the Mg II line, and we conclude that these twelve stars in fact are normal main-sequence A-type stars. There is also the possibility that some of these young-looking blue stars might arise from conditions similar to those that form blue stragglers in globular clusters, although most of the suggested mechanisms would be much less likely to work for field stars. In contrast, the

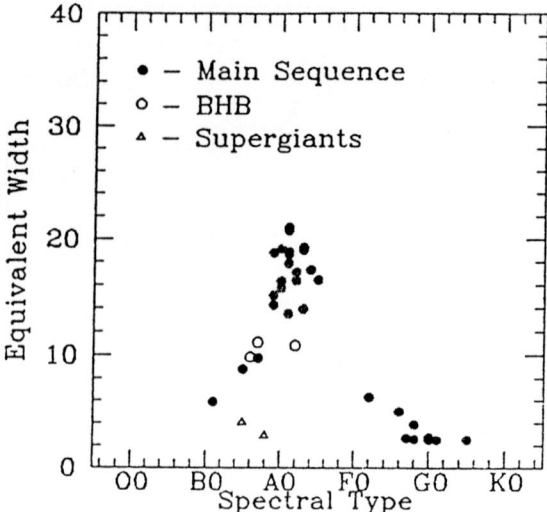

FIGURE 4. Distribution of Hδ equivalent widths of standard stars versus their spectral types. The main-sequence stars are shown as filled circles, three globular-cluster BHB stars as open circles, and two supergiants as triangles.

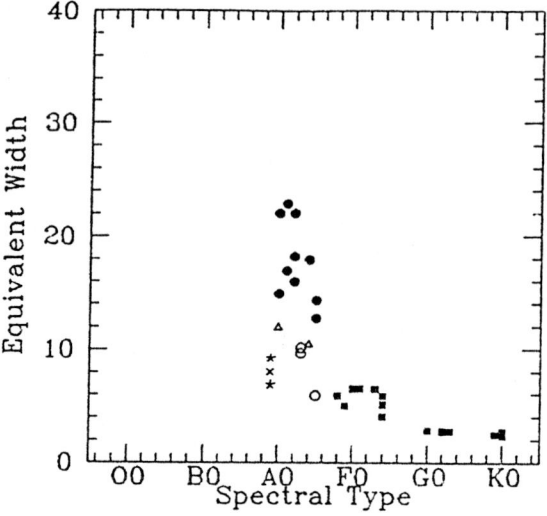

FIGURE 5. Distribution of the Hδ equivalent widths versus derived spectral types for the program stars. The blue main-sequence candidates are shown as filled circles, BHB as open circles, subdwarfs as stars, the white dwarf as an x, marginal main-sequence candidates as triangles, and stars with spectra later that type A5 as squares.

virtual absence of Mg II absorption in the horizontal-branch and subdwarf candidates is added support to those classifications.

The approximate z-distances, based on the assigned classifications, for three of the stars assigned main-sequence are substantially below 1 kpc, and thus these three can be ruled disk stars. The remainder of the nine stars assigned as main-sequence have a mean

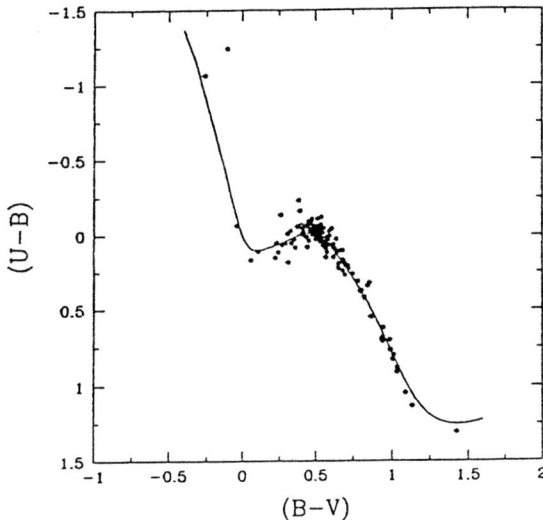

FIGURE 6. Two-color plot for NGP stars to $V = 15.3$, including about one-half of the blue-star candidates. The distribution of nearby main-sequence stars is shown as the solid line.

z of $\sim$ 8 kpc, and for the five horizontal-branch and subdwarf stars mean z $\sim$ 10 kpc. We therefore conclude, based on this initial study, that in the neighborhood of 50 % of the halo blue stars are young main-sequence A-type stars. Previous estimates of that fraction range from about one half (Philip 1968) to 5 % (Pier 1983).

Radial velocities were derived through cross-correlation of the program star spectra against those of nearby early-type radial velocity standards, with an accuracy of $\sim$ 10 km s^{-1}. The radial-velocity dispersion for the B9 to A5 program stars for both horizontal-branch and main-sequence stars combined is $\sim 80 \pm 20$ km s^{-1}, consistent with previous results for blue halo stars. The result is essentially the same for the two groups separately.

3. UBV Photometry

UBV photometry was acquired at McDonald and Mount Laguna Observatories for about half of the program stars in March and April 1993, as well as a number of cooler stars, for $V < 15.3$. The McDonald observations were obtained over 8 nights with the 36-inch, and the Mount Laguna observations on one night with the 1-meter and blue-sensitive phototubes, using standard UBV and red-leak filters. For a number of overlaps between Mount Laguna and McDonald no systematic differences are evident, and the derived mean errors are $\sim \pm 0.01$ mag for both V and $(B - V)$. The colors are consistent with the derived spectral types. The results are in the two-color plot of Figure 6, which includes stars over the entire color range, and in Table 1 for the blue-star candidates. The white dwarf and one suspected subdwarf were observed, and appear in the upper left of the plot, at $(B - V) = -0.11$ and -0.26 respectively.

4. Summary

From a sample of 34 NGP stars, eighteen have derived spectral types of A5 or earlier. Of these, twelve appear to be main-sequence stars, based on their Hydrogen line strengths as well as the presence of the Mg II 4481 Å line (but three are ruled out as halo occupants, based on their nearby z-distances); and five appear to be expected denizens of the halo, either horizontal-branch or sub-dwarf stars. One star clearly is a nearby white dwarf. The velocity dispersion for both main-sequence and horizontal-branch groups is $\sim 80 \pm 20$ km s^{-1}, consistent with previous determinations of halo A-type stars.

UBV colors were obtained for about half of the sample, and are consistent with the derived spectral types. However, due to the lack of extensively available accurate colors, we are not yet able to establish the frequency distribution of $(B-V)$ as a function of V, which is critical to good modeling of the Galaxy. Our future primary effort will be to obtain UBV (both photoelectric and CCD) photometry at Mount Laguna Observatory for all stars to a limiting magnitude of $V = 16$ within a 13.5 square-degree region. Because of their intrinsic interest, the limiting magnitude for the stars identified as blue by the current less-accurate photometry will be extended to $V = 17$. Additionally we will obtain spectra of all stars with accurate $(B-V) < 0.3$.

REFERENCES

PHILIP, A. G. D., 1968, AJ, 73, 1000

PHILIP, A. G. D., 1994, in Adelman S. J., Upgren A. R., Adelman C. J., eds, Hot Stars in the Halo, Cambridge University Press, Cambridge, p. 41

PIER, J. R., 1983, ApJS, 53, 791

WEISTROP D., 1972, AJ, 77, 366

YOSS, K. M., SHANKAR, A., BELL, D. J., NEESE, C. N., DETWEILER, H. L., 1989, PASP, 101, 653

Discussion

BEERS: What is the B-V color of the main sequence star whose spectrum you showed? I ask because the Ca II K line appears weak, and I was curious about the metallicity of this star.

BELL: It was too faint for us to observe it at McDonald Observatory. Our only estimate is Weistrop's (corrected) photographic value of (+0.1 0.1) consistent with an A0 star. The K-line is too temperature sensitive for us to use it for a metallicity estimate until we get accurate photometry, but the strong Mg II 4481 Å line suggests a Population I metallicity.

KINMAN: We can see the 4481 Å line quite clearly in M3 horizontal branch stars spectra and less well in M92 horizontal branch stars.

BELL: It is very weak or undetectable in our three M13/M92 spectra, which are of comparable S/N (100 per resolution element). We need to get many more BHB spectra for these and other clusters.

PETERSON: Mg II is seen at 3 Å FWHM resolution in some M13 BHB's, with high S/N. With some knowledge of effective temperature perhaps from both B-V photometry and Balmer line profiles, this line may provide an abundance. The Ca II line would provide a good check, provided interstellar absorption is minimal; at least an upper limit should be obtainable.

Population I Horizontal Branches: Probing the Halo-to-Disk Transition

By RANDY L. PHELPS[1] KENNETH A. JANES[2] AND KENT A. MONTGOMERY[2]

[1]Phillips Laboratory / GPOB, Hanscom AFB, MA 01731-3010, USA
[2]Department of Astronomy, Boston University, Boston, MA 02215, USA

The oldest Population I horizontal branch stars (also known as red giant "clump" stars) are not very hot, and are not quite in the halo, but they are useful nevertheless to probe the transition from the halo to the disk. Old clusters provide the best hope for establishing the chronology of the Galaxy's formation, and a key to estimating the distances, reddenings and ages of these clusters are the narrow range of color and luminosity of HB stars, including the Population I clump stars. In an extensive CCD photometric survey of potential old open clusters, we have identified a number of systems that are indeed old; some of them are among the oldest of the open clusters. We used a variation of two well-known morphological age indices, one based on the luminosity difference between the main sequence turnoff and the horizontal branch and the other on the color difference between the turnoff and the giant branch to rank the clusters in approximate order of age. The small range in color and luminosity of clump stars makes it possible to estimate the distances and reddenings of clusters when this information is not otherwise available.

Our data, together with previously published photometry, yields a list of 72 clusters of the age of the Hyades or greater, with 19 clusters being as old or older than M 67. We find that the age distribution of the open clusters overlaps, or nearly overlaps, that of the globular clusters.

1. Introduction

The only old objects in the Galaxy whose ages we can reliably date are the star clusters, and hence they constitute a powerful tool for probing Galactic structure and the formation and development of the Galactic halo and disk. In addition to the 135 or so known globular clusters, which define the halo, over 1200 open clusters are known to exist. However, only about one-third of the open clusters have sufficiently well-determined photometry to allow estimates of their ages and distances to be made. Since most open clusters are destroyed by interactions with molecular clouds on timescales of a few hundred million years or less (Spitzer 1958), the vast majority of those which we now observe are much younger than the Galaxy. Janes *et al.* (1988) found that the median age of clusters whose ages have been estimated is about 100 million years.

Some open clusters, however, survive for billions of years, and a few may be as old as the disk. Until recently only a handful of such clusters were known, with most of them located in the outer disk of the Galaxy and far from the Galactic plane, compared to most open clusters. Because the old clusters are of such importance for an understanding of Galactic evolution, several searches for additional clusters have been made. King (1964) published a list of clusters that appeared to be old on the basis of their appearance on the Palomar Observatory Sky Survey (POSS) prints. He recognized that a young cluster will appear to contain only a few bright stars, while in an old cluster there will be a substantial number of stars of about the same magnitude. "Old" in this context means a cluster with a significant population of giant stars which typically means clusters as old or older than the Hyades.

Janes & Adler (1982) presented a somewhat expanded list of unstudied, rich open clusters using the King prescription and Janes & Phelps (1990) derived a more extensive list based on a systematic examination of clusters in the Lund "Catalog of Open Clusters" (Lyngå 1987).

2. Data

Over the past 6 years we conducted a CCD photometric survey of candidate old open clusters as listed by Janes & Phelps (1990) and Janes & Adler (1982), and we also observed several other well known old open clusters to bring the photometry up to modern standards. New photometric data was obtained for 37 open clusters using the KPNO 0.9m and 2.1m telescopes and the CTIO 0.9m telescope. Data from the literature for an additional 58 clusters were also examined to increase the sample.

3. Analysis

The horizontal branch is a distinctive feature of globular cluster CMDs but the analogous structure in an open cluster CMD generally has the appearance of a "clump" just on the blue side of the red giant branch. We have adopted as our primary age parameter, an index δv, defined as the magnitude difference between the main-sequence turnoff and the clump. Observationally, when the luminosity of the main-sequence turnoff is defined as the magnitude at the bluest point on the CMD, it is intrinsically a poorly-defined quantity, as the cluster sequence lies along a line of nearly constant color for almost a magnitude in luminosity. The problem is compounded in the open clusters, first because they contain far fewer stars than the globular clusters, and second because many open clusters have a substantial binary star sequence. A better measure of the main sequence turnoff luminosity can be found at the inflection point on the CMD between the turnoff and the base of the giant branch. This parameter is well-defined even when the photometry is of marginal quality and even in the presence of a binary sequence or substantial numbers of field stars. We will call the magnitude at this inflection point the "turnoff magnitude". We have chosen the notation δV to avoid confusion with the slightly different index, ΔV, that has been used by others.

We have also defined another CMD age parameter, independent of δv, from color differences. Our color index parameter, $\delta 1$, is the difference in color index between the bluest point on the main sequence at the luminosity of the turnoff and the color of the giant branch one magnitude brighter than the turnoff luminosity. This parameter is especially useful since it can be measured in clusters with no noticeable clump. Since the giant branch of many of the younger clusters does not extend more than one magnitude above the turnoff, the $\delta 1$ parameter cannot be measured from the CMD's of clusters younger than about 1.5 - 2.0 Gyr.

We have estimated, wherever possible, both age parameters for all clusters that appear to be old, based on the available data. We were able to measure δV in most clusters, $\delta 1$ in many clusters, and both parameters in some clusters. Furthermore, some of the clusters were observed with B and V filters and others with V and I filters. Since all these variants are presumed to be measures of the single parameter, age, there is a need to combine them into one index. We first averaged multiple determinations of δV and $\delta 1$ made in a single color (BV or VI), weighting the uncertain values one-half the other measures. Next, since the δV measures from BV photometry and those from VI photometry agree, on average, to within 0.1 mag a straight average was taken when both were available for a cluster. Finally, while we have chosen δV to be our primary age indicator, some clusters do not have a visible red giant clump making it impossible to measure δv. There are, however, enough clusters for which both δV and $\delta 1$ can be measured so that correlations

Name	α(1950) (h:m)	δ(1950) (°:')	l (°)	b (°)	δV	q	Name	α(1950) (h:m)	δ(1950) (°:')	l (°)	b (°)	δV	q
NGC 188	00:39.4	+85:04	122.78	+22.46	2.4	b	Melotte 71	07:35.2	−11:57	228.95	+4.51	0.5	b
King 2	00:48.1	+57:55	122.88	−4.67	2.2	b	NGC 2420	07:35.5	+21:41	198.11	+19.65	1.6	a
IC 166	01:49.0	+61:35	130.08	−0.19	1.0	c	AM 2	07:37.3	−33:44	246.89	−5.09	2.5	b
NGC 752	01:54.8	+37:26	137.17	−23.36	0.9	a	Berkeley 39	07:44.2	−04:29	223.47	+10.09	2.4	a
NGC 1193	03:02.5	+44:11	146.81	−12.18	2.1	b	NGC 2477	07:50.5	−38:25	253.58	−5.83	0.5	b
King 5	03:11.0	+52:32	143.75	−4.27	0.4	b	NGC 2506	07:57.8	−10:39	230.57	+9.91	1.5	a
NGC 1245	03:11.2	+47:04	146.64	−8.93	0.7	b	Pismis 2	08:16.3	−41:28	258.83	−3.29	1.1	c
Hyades	04:24.0	+15:45	180.05	−22.40	0.4	a	NGC 2627	08:35.2	−29:46	251.58	+6.65	1.6	b
NGC 1798	05:08.1	+47:34	160.76	+4.85	1.0	b	Praesepe	08:37.2	+20:10	205.54	+32.52	0.3	a
NGC 1817	05:09.2	+16:38	186.13	−13.13	0.8	a	NGC 2660	08:40.6	−46:58	265.86	−3.03	0.4	b
Berkeley 17	05:17.4	+30:33	175.65	−3.65	2.8	b	M 67	08:47.7	+12:00	215.58	+31.72	2.3	a
Berkeley 18	05:18.5	+45:21	163.63	+5.01	2.3	b	NGC 2849	09:17.4	−40:20	265.27	+6.33	0.5	c
Berkeley 20	05:30.4	+00:11	203.50	−17.28	2.1	b	092-SC18	10:13.4	−64:22	287.1	−6.7	2.2	c
King 8	05:46.1	+33:37	176.40	+3.12	0.2	c	NGC 3680	11:23.3	−42:58	286.77	+16.93	1.0	b
Berkeley 21	05:48.7	+21:46	186.83	−2.50	1.6	c	NGC 3960	11:48.4	−55:25	294.41	+6.18	0.2	c
Berkeley 22	05:55.7	+07:50	199.80	−8.05	2.1	c	Cr 261	12:34.9	−68:12	301.69	−5.64	2.6	b
NGC 2141	06:00.3	+10:26	198.07	−5.79	1.6	c	NGC 4815	12:54.9	−64:41	303.63	−2.09	1.1	c
NGC 2158	06:04.4	+24:06	186.64	+1.76	1.4	b	096-SC04	13:11.8	−65:40	305.35	−3.17	0.2	c
NGC 2194	06:11.0	+12:49	197.26	−2.33	0.5	c	NGC 5822	15:01.5	−54:09	321.71	+3.58	0.8	c
NGC 2192	06:11.7	+39:52	173.41	+10.64	0.6	b	IC 4651	17:20.8	−49:54	340.07	−7.88	1.2	b
NGC 2204	06:13.5	−18:38	226.01	−16.07	1.4	b	IC 4756	18:36.5	+05:24	36.37	+5.26	0.4	b
NGC 2236	06:27.0	+06:52	204.37	−1.69	0.4	c	Berkeley 42	19:02.6	+01:48	36.17	−2.19	0.4	c
NGC 2243	06:27.9	−31:15	239.50	−17.97	2.2	b	NGC 6791	19:19.0	+37:45	70.01	+10.96	2.6	a
Tr 5	06:34.0	+09:29	202.86	+1.05	2.3	c	NGC 6802	19:28.4	+20:10	55.34	−0.93	0.4	c
NGC 2266	06:40.1	+27:01	187.78	+10.28	0.5	c	NGC 6819	19:39.6	+40:04	73.98	+8.47	1.7	b
Berkeley 29	06:50.4	+16:59	197.98	+8.03	2.1	a	NGC 6827	19:46.7	+21:04	58.24	−2.35	0.5	c
Berkeley 31	06:54.9	+08:20	206.26	+5.12	2.3	c	IC 1311	20:08.6	+41:04	77.70	+4.25	0.2	c
Berkeley 30	06:55.1	+03:17	210.80	+2.89	0.3	b	NGC 6939	20:30.4	+60:28	95.88	+12.30	1.4	c
Berkeley 32	06:55.4	+06:30	207.95	+4.40	2.4	b	NGC 6940	20:32.5	+28:08	69.90	−7.16	0.2	b
Tombaugh 2	07:01.2	−20:47	232.90	−6.84	1.5	a	Berkeley 54	21:01.3	+40:16	83.13	−4.14	2.5	c
NGC 2324	07:01.6	+01:08	213.45	+3.31	0.3	c	NGC 7044	21:11.1	+42:17	85.87	−4.13	0.7	b
NGC 2354	07:12.2	−25:39	238.42	−6.80	0.8	c	Berkeley 56	21:15.8	+41:41	86.04	−5.18	2.3	c
NGC 2355	07:14.1	+13:52	203.36	+11.80	0.4	b	NGC 7142	21:44.7	+65:34	105.42	+9.45	2.0	b
NGC 2360	07:15.5	−15:32	229.80	−1.42	0.5	a	King 9	22:13.6	+54:09	101.45	−1.84	2.0	c
Melotte 66	07:24.9	−47:38	259.61	−14.29	2.3	b	King 11	23:45.4	+68:21	117.16	+6.47	2.3	b
NGC 2423	07:34.8	−13:45	230.47	+3.55	0.1	c	NGC 7789	23:54.5	+56:27	115.49	−5.36	1.1	a

TABLE 1. Old Open Clusters

between the indices can be established, allowing a conversion between $\delta 1$ and δV to be made.

In this way it was possible to derive up to three versions of δV; one an average of the measured δV from the BV and/or VI photometry, one a conversion of $\delta 1(BV)$ into δV, and one a conversion of $\delta 1(VI)$ into δV. The adopted δV, taken to be the average of the directly measured value and the computed value(s), is listed in Table 1 for each of the clusters found to be old. The index "q" is a qualitative estimate of the reliability of δV, with clusters with the best-defined CMDs assigned a value of "a", clusters with moderately well-defined CMDs assigned a "b", and the most difficult clusters to evaluate given a "c".

4. Results

In our survey of old open clusters we identified 72 clusters with δV similar to or larger than the Hyades, which has $\delta v = 0.4$ (Table 1). Allowing for the uncertainty in δv (∼0.1 mag), there are 19 clusters as old or older than M67 (Table 2).

Name	δV	Name	δV
Berkeley 17	2.8	Melotte 66	2.3
Cr 261	2.6	M 67	2.3
NGC 6791	2.6	Tr 5	2.3
AM 2	2.5	King 2	2.2
Berkeley 54	2.5	NGC 2243	2.2
Berkeley 32	2.4	092-SC18	2.2
Berkeley 39	2.4	Berkeley 20	2.1
NGC 188	2.4	Berkeley 22	2.1
Berkeley 18	2.3	Berkeley 29	2.1
Berkeley 31	2.3	NGC 1193	2.1
Berkeley 56	2.3	King 9	2.0
King 11	2.3	NGC 7142	2.0

TABLE 2. The Oldest Open Clusters

Among the oldest open clusters are Be 17, Cr 261, NGC 6791, Be 54 and AM 2. Our value of $\delta v = 2.6$ for NGC 6791 confirms that it is among the very oldest of open clusters, but NGC 188 ($\delta v = 2.4$) is only a little older than M67 ($\delta v = 2.3$), which has an age of 3.5-6 Gyr (Montgomery, Marschall & Janes 1993). Be 17 appears to be the oldest cluster thus far detected and is likely as old as the youngest globular clusters.

We have also calibrated the clump luminosities and colors using 47 clusters with published reddenings and distances and found $<M_V(HB)> = 0.6 \pm 0.5$ and $<(B-V)(HB)> = 0.87 \pm -0.12$ for 24 clusters with $\delta V < 1.0$ and $<M_V(HB)> = 0.9 \pm 0.4$ and $<(B-V)(HB)> = 0.95 \pm 0.10$ for the older clusters.

Figure 1 shows the number of old open clusters as a function of δV. This diagram shows first that there was substantial cluster formation back to at least 10 Gyr ago, assuming the Garnavich et al. (1993) value of 9 Gyr for the age of NGC 6791 ($\delta v = 2.6$), and also that a significant number of clusters have also survived from that time. Since the typical lifetime of open clusters is about 100 Myr (Janes et al. 1988), it is particularly striking to see a distinct peak in the number of clusters near $\delta v = 2.3$, corresponding to an age near 5 Gyr, or 50 times the typical open cluster lifetime. The peak in the distribution, combined with the paucity of clusters from $1.6 < \delta V < 2$ suggests that there was a higher formation rate of clusters early in the history of the disk compared to intermediate times; for one would expect to see far fewer old clusters relative to intermediate age ones because of the typical short lifetimes of clusters. The age versus δv relation is not linear at large ages, however, and a more detailed interpretation of this distribution is currently in progress (Janes & Phelps 1994). The results, however, suggest that the Galactic disk was beginning to develop during the epoch when the last globular clusters were forming

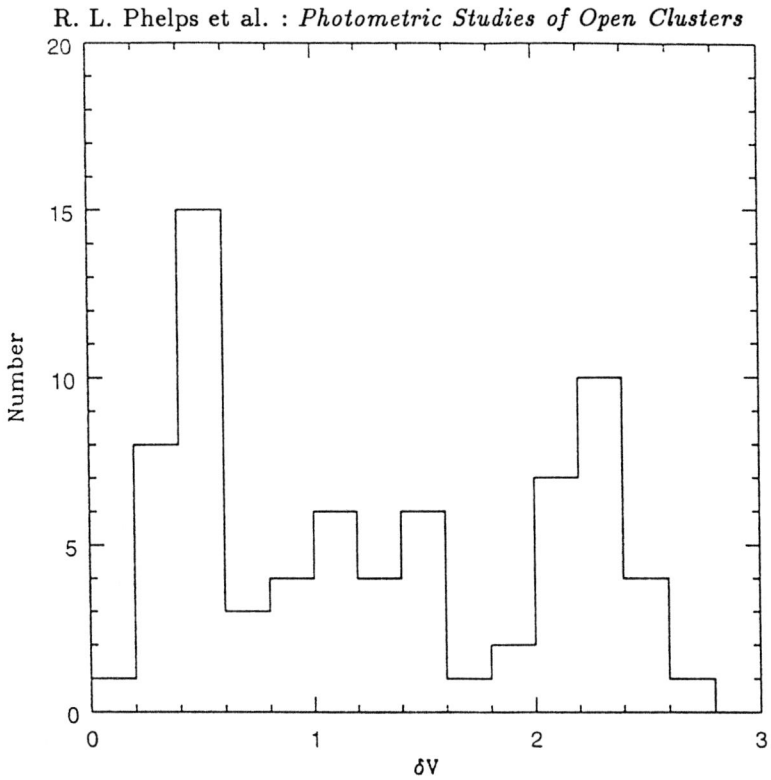
FIGURE 1. Histogram of adopted δVs

since Be 17, and perhaps the cluster Lyngå 7 (Ortolani, Bica & Barbuy 1993), "fill in the gap" between the oldest open clusters and the youngest globular clusters.

REFERENCES

CHABOYER, B., SARAJEDINI, A., DEMARQUE, P., 1992, AJ, 394, 515
GARNAVICH, P. M., VANDENBERG, D. A., ZUREK, D. R., HESSER, J. E., 1993, preprint
JANES, K. A., ADLER, D., 1982, ApJS, 49, 425
JANES, K. A., PHELPS, R. L., 1994, in preparation.
JANES, K. A., PHELPS, R. L., 1990, in Philip A. G. D., Hayes D. S., Adelman S. J., eds, CCDs in Astronomy II., L. Davis Press, Schenectady, p. 117
JANES, K. A., TILLEY, C., LYNGÅ G., 1988, AJ, 95, 771
LYNGÅ G., 1987, Catalog of Open Clusters, Centre de Donnees Stellaires, Strasborg
MONTGOMERY, K. A., MARSCHALL, L. A., JANES, K. A., 1993, AJ,106, 181
ORTOLANI, S., BICA, E., BARBUY, B., 1993, A&A, 273, 415
PHELPS, R. L., JANES, K. A., MONTGOMERY, K. A., 1994, AJ, in press
SPITZER, L., 1958, ApJ, 127, 17

Discussion

SARAJEDINI: Are there any blue straggler candidates in these clusters? If so, about how faint are they in V mag?

PHELPS: The field star contamination in the color magnitude diagram (CMD) is often quite high so it is difficult to tell. Typically, the stars in the portion of the CMD where blue stragglers would be expected are 17th magnitude. Without spectroscopy, which is clearly quite difficult, it is difficult to establish whether or not blue stragglers are present.

SWEIGART: Why do some of these clusters survive to 10^{10} years? Can you detect a metallicity gradient with age?

PHELPS: There are several reasons: The clusters generally are located far from the Galactic plane are generally richer (and, hence, less prone to disruption), and are generally located in the outer Galaxy where interactions with molecular clouds (which disrupt them) are far less frequent.

We are working with Eileen Friel, as part of a larger project, to obtain ages, distances, reddenings, and metallicities for the old open cluster system to address this, and other questions pertaining to the development of the Galactic disk.

Stars

Very Hot Subdwarf O Stars

By J. S. DRILLING[1], T. C. BEERS[2] AND U. HEBER[3]

[1]Department of Physics and Astronomy, Louisiana State University, Baton Rouge, LA 70803, USA

[2]Department of Physics and Astronomy, Michigan State University, E. Lansing, MI 48824, USA

[3]Dr. Remeis-Sternwarte, Universität Erlangen-Nürnberg, D96049 Bamberg, GERMANY

The following is a review and progress report on studies of very hot sudwarf O stars discovered in follow-up spectroscopy of the Case-Hamburg-LSU and Beers-Preston-Shectman objective-prism surveys.

1. Introduction

The OB$^+$ stars are stars which show continuous or nearly continuous spectra at a resolution of 10 Å. Drilling (1983) and Beers *et al.* (1992) have shown that follow-up spectroscopy of such stars found in objective-prism surveys can be used to obtain spatially complete samples of the hottest and/or most helium-rich subdwarf O stars in the Galaxy. Today, we would like to discuss two such samples of very hot sdO stars. The first was obtained by follow-up spectroscopy of the Case-Hamburg-LSU surveys, which are complete to photographic magnitude 12 for the entire Milky Way (see Stephenson & Sanduleak 1971), and for b = ±30° and l = ±60° (see Drilling 1987). The second was obtained from a partial spectroscopic follow-up of 89 plates from the Beers-Preston-Shectman survey, which cover 2225 square degrees at high galactic latitude, mostly in the southern hemisphere. These plates were taken with the Curtis Schmidt telescope and 4° objective prism at Cerro Tololo, which is the same instrumental setup as that used by Drilling (1987) in his extension of the Case-Hamburg surveys. In the Beers-Preston-Shectman survey, however, a 150 Å wide interference filter is used to isolate the spectral region centered on the Ca II H and K lines, which enables 180 Å mm^{-1} spectra widened to 0.2 mm to be obtained for all stars brighter than photographic magnitude 16 in a 25 square degree field in 90 minutes. This is a very powerful combination, indeed! The subdwarf O stars from the Beers-Preston-Shectman sample are spectroscopically very similar to those those from the Case-Hamburg-LSU surveys, but have much higher radial velocities (Drilling 1983; Drilling & Heber 1986; Beers *et al.* 1992).

2. The Case-Hamburg-LSU Sample

Twenty-two of the stars classified as OB$^+$ in the Case-Hamburg-LSU surveys are sdO stars, and these are listed in Table I of Drilling (1987). Effective temperatures, Bolometric corrections, surface gravities, and abundances for 11 of these stars have been determined from the fine analysis of high-resolution spectra covering the wavelength ranges 3950 - 4950 Å and 1150 - 2100 Å (Hunger, Gruschinske, Kudritzki & Simon 1981; Heber & Hunger 1987; Mendez *et al.* 1988; Heber, Werner & Drilling 1988; Husfeld, Butler, Heber & Drilling 1989; Rauch *et al.* 1991; Dreizler 1993). Schönberner & Drilling (1984) showed that the effective temperatures of these stars are strongly correlated with the quantity

$$R = \int_{1240}^{1945} F_\lambda^0 d\lambda / \int_{1945}^{3120} F_\lambda^0 d\lambda, \qquad (2.1)$$

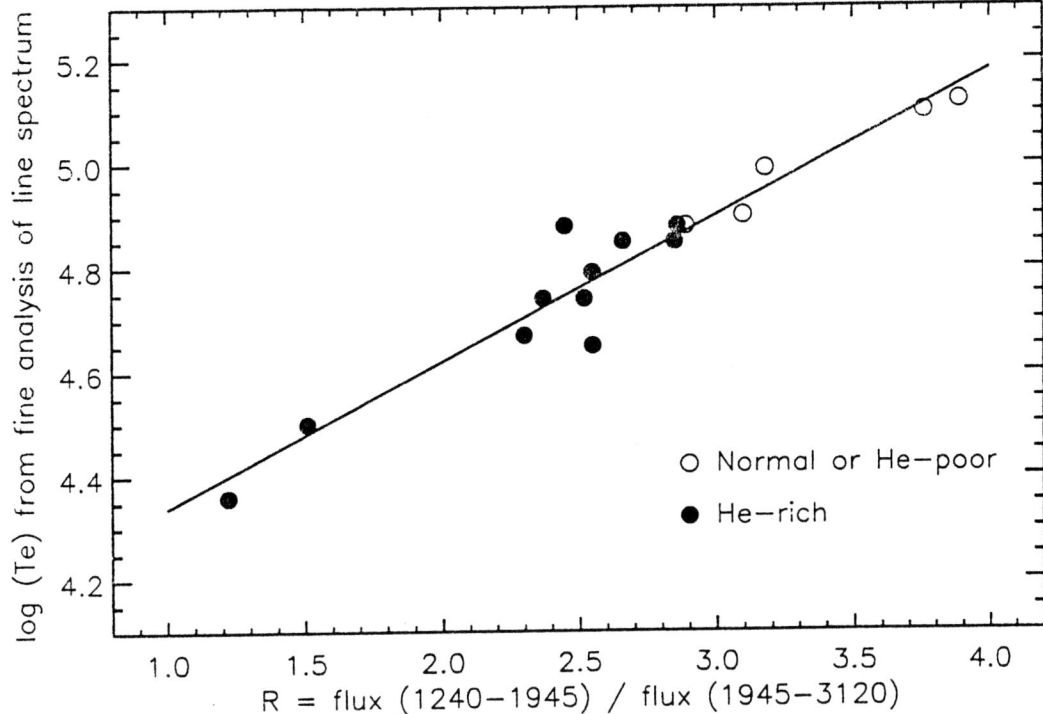

FIGURE 1. Effective temperature versus R for 15 very hot sdO stars and the two hottest known extreme helium stars.

the ratio of the integrated, dereddened fluxes in the two IUE cameras. In Figure 1, the effective temperatures determined from the fine analysis of high resolution spectra of stars from the Case-Hamburg-LSU sample, four other very hot sdO stars (Kudritzki & Simon 1978; Kudritzki et al. 1980; Auer & Shipman 1977; Mendez, Kudritzki & Simon 1983), and the two hottest extreme helium stars (Hamann, Schönberner & Heber 1982; Jeffery & Heber 1992), are plotted against R. The relation shown is for helium-rich stars, but stars which are not helium-rich have also been included by increasing T_{eff} by a factor of 1.25 following the model atmosphere result of Schönberner & Drilling (1984) that for a given value of R, the effective temperatures of helium-rich stars are 25 % higher than those of normal or helium-poor stars. The calibration can therefore be used for stars which are not helium-rich by dividing the resulting effective temperatures by 1.25. The straight line is the relation

$$log T_{eff} = 4.06 + 0.28R \pm 0.04, \qquad (2.2)$$

which was derived by Drilling (1992) from the same data *without using the extreme helium stars* (the two points farthest to the left in Figure 1)! We have now observed all but one of the 22 stars in the Case-Hamburg-LSU sample at low resolution with both IUE cameras, and have determined effective temperatures (and Bolometric corrections) for these 21 stars using the above calibration. The IUE fluxes were dereddened using the Seaton (1979) reddening law and the known color excesses. The color excesses were determined from the strength of the 2200 Å interstellar feature and from B-V by assuming $(B-V)_0$ = -0.35. The probable error in E_{B-V} using either of these methods alone is estimated to be ± 0.03. The resulting effective temperatures ranged from 42,000 K to 115,000 K.

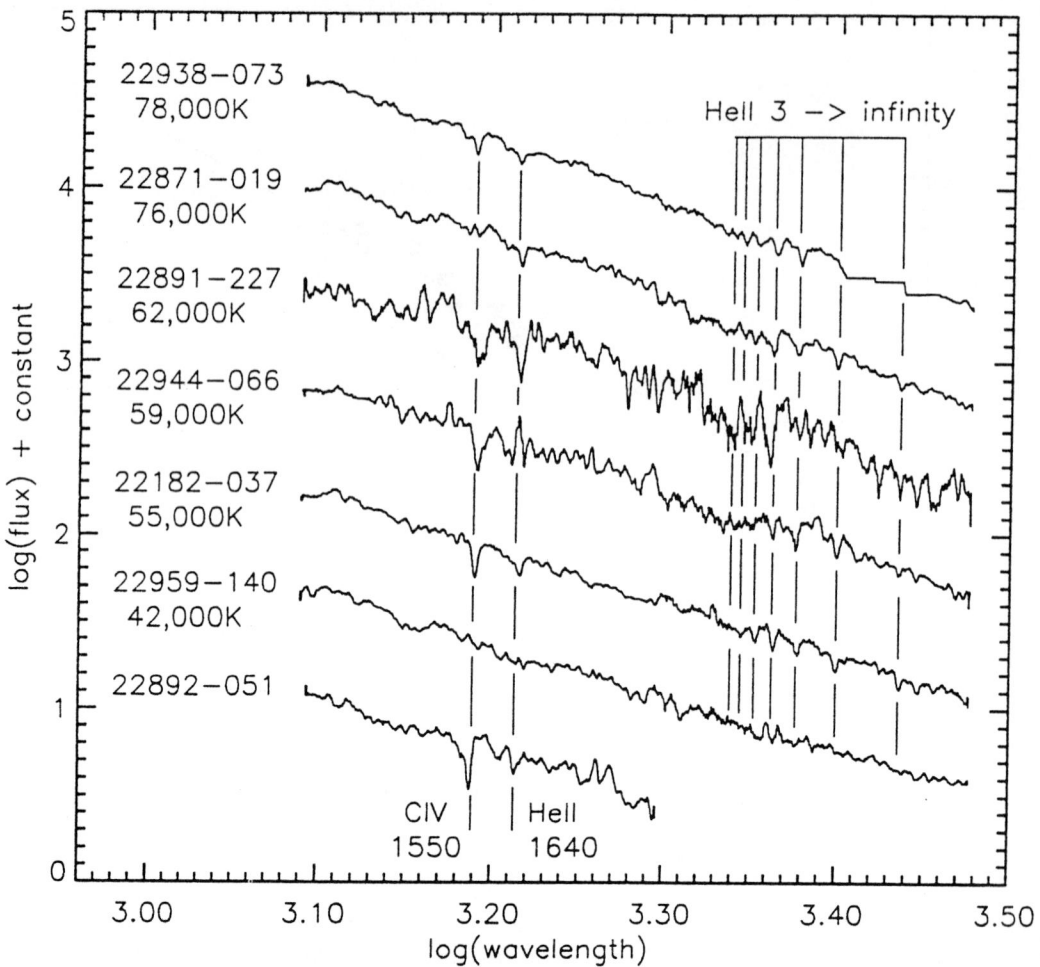

FIGURE 2. De-reddened low-resolution IUE spectra of very hot sdO stars from the Beers-Preston-Shectman survey.

We have also determined absolute magnitudes and distances for these stars from either the known color excesses (using the run of color excess with distance for stars nearby in the sky) or from the surface gravities and Bolometric corrections from the fine analyses (using masses from stellar evolution theory). The absolute magnitudes and effective temperatures show that these stars occupy a region in the HR diagram bounded by the central stars of planetary nebulae detected in surveys for planetary nebulae, other sdO stars, and the hottest white dwarfs. Indeed, careful inspection of the Palomar and ESO sky survey prints revealed faint, previously undetected planetary nebulae around three of the stars (Drilling 1983; Schönberner & Drilling 1984; Heber, Werner & Drilling 1988).

Because these stars, when added to the central stars of planetary nebulae detected in surveys for planetary nebulae, form a complete sample (with respect to apparent magnitude) of the hottest stars in the Galaxy, and because all evolutionary tracks leading from the main sequence to the white dwarfs funnel through the region of the HR diagram defined by these stars ($T_{eff} > 40,000$ K), one can use them to determine the relative birthrates of the various types of white dwarf progenitors. The results are given in Table

Central Stars of Known PN	230×10^{-14}	Phillips 1989
Other Post-AGB	46×10^{-14}	11 stars
Post-EHB	1×10^{-14}	10 stars

TABLE 1. Birthrates ($pc^{-3} yr^{-1}$) for white dwarf progenitors.

1. The first birthrate is that given by Phillips (1989) for the central stars of planetary nebulae detected in surveys. The second is for the stars in our sample which lie on the post-asymptotic giant branch tracks of Schönberner (1981) and Wood & Faulkner (1986) in the HR diagram, but do not show planetary nebulae bright enough to be detected in surveys for planetary nebulae. The third is for stars which lie on the post-extended horizontal branch tracks of Paczynski (1971) and Caloi (1989). In all cases, the space densities were determined from the known distances of the stars, and the times that it takes the stars to cross the region of the HR diagram defined by them from the evolutionary calculations. In the case of the post-EHB stars, the crossing times were estimated from the Paczynski tracks, and the birthrate is therefore actually a lower limit. We conclude that all but a negligible fraction of the white dwarfs are descended from post-AGB stars, but that 20 % of these stars did not produce planetary nebulae bright enough to be detected in surveys for planetary nebulae.

3. The Beers-Preston-Shectman Sample

We have observed 7 of the sdO stars from the Beers-Preston-Shectman sample to date at low resolution with IUE. The spectra have been dereddened using the Seaton (1979) reddening law and the color excesses given by Beers et al. (1992). The dereddened spectra are shown in Figure 2, along with the effective temperatures determined using (2.2). The UV *line* spectra are consistent with the results of Beers et al. (1992) in that those stars which Beers et al. find to be helium-rich show strong He II 1640 absorption (the apparent emission feature in the spectrum of CS 22944-66 is not real) and the $n = 3 \to \infty$ absorption series of He II. All of the helium-rich stars except for CS 22871-19 also show strong C IV 1550 absorption. CS 22959-140, the one star which Beers et al. find not to be helium-rich shows a nearly continuous spectrum in the UV. The effective temperatures derived from the *continuous* UV spectrum range from 42,000 K to 78,000 K, indicating that these are truly very hot subdwarf O stars.

This work was supported in part by grants from the National Aeronautics and Space Administration (NAG5-71) and the National Science Foundation (AST85-14574, AST90-01376, and AST92-22326).

REFERENCES

AUER, L. H., SHIPMAN, H. L., 1977, ApJ, 211, L103
BEERS, T. C., PRESTON, G. W., SHECTMAN, S. A., DOINIDIS, S. P., GRIFFIN, K. E., 1992, AJ, 103, 267
CALOI, V., 1989, A&A, 221, 27
DREIZLER, S., 1993, A&A, 273, 212
DRILLING, J. S., 1983, ApJ, 270, L13

DRILLING, J. S., 1987, in Philip A. G. D., Hayes D. S., Liebert J. W., eds, The Second Conference on Faint Blue Stars, L. Davis Press, Schenectady, p. 489

DRILLING, J. S., 1992, in Heber U., Jeffery C. S., eds, The Atmospheres of Early-Type Stars, Springer-Verlag, Berlin, p. 257

DRILLING, J. S., HEBER, U., 1986, in Hunger K., Schönberner D., Rao N. K., eds, Hydrogen Deficient Stars and Related Objects, Reidel, Dordrecht, p. 23

HAMANN, W. -R., SCHÖNBERNER, D., HEBER, U., 1982, A&A, 116, 273

HEBER, U., HUNGER, K., 1987, in Philip A. G. D., Hayes D. S., Liebert J. W., eds, The Second Conference on Faint Blue Stars, L. Davis Press, Schenectady, p. 599

HEBER, U., WERNER, K., DRILLING, J. S., 1988, A&A, 194, 223

HUNGER, K., GRUSCHINSKE, J., KUDRITZKI, R. -P., SIMON, K. P., 1981, A&A, 95, 244

HUSFELD, D., BUTLER, K., HEBER, U., DRILLING, J. S., 1989, A&A, 222, 150

JEFFERY, C. S., HEBER, U., 1992, A&A, 260, 133

KUDRITZKI, R. -P., SIMON, K. P., 1978, A&A, 70, 653

KUDRITZKI, R. -P., GRUSCHINSKE, J., HUNGER, K., SIMON, K., 1980, in Battrick B., Mort J., eds, Second European IUE Conference (ESA SP-157)

MENDEZ, R. H., KUDRITZKI, R. -P., SIMON, K. P., 1983, in Flower D. R., ed, IAU Symposium No. 103, Planetary Nebulae, Reidel, Dordrecht, p. 343

MENDEZ, R. H., KUDRITZKI, R. -P., HERRERO, A., HUSFELD, D., GROTH, H. G., 1988, A&A, 190, 113

PHILLIPS, P., 1989, in Torres-Peimbert S., ed, IAU Symposium No. 131, Planetary Nebulae, Kluwer, Dordrecht, p. 425

PACZYNSKI, B., 1971, Acta Astron., 21, 1

RAUCH, T., HEBER, U., HUNGER, K., WERNER, K., NECKEL, T., 1991, A&A, 241, 457

SCHÖNBERNER, D., 1981, A&A, 103, 119

SCHÖNBERNER, D., DRILLING, J. S., 1984, ApJ, 278, 702

SEATON, M. J., 1979, MNRAS, 187, 73P

STEPHENSON, C. B., SANDULEAK, N., 1971, Publ. Warner & Swasey Obs., 1, 1

WOOD, P. R., FAULKNER, D. J., 1986, ApJ, 307, 659

Discussion

JANES: Where are these stars? I mean to say, how far from the plane are they?

DRILLING: The radial velocities, distribution on the sky, and apparent magnitudes indicate that the sdO stars from the Case-Hamburg-LSU survey belong primarily to the disk population, and those from the Beers-Preston-Shectman survey are halo (or thick disk) objects.

Quantitative Spectroscopy of Very Hot Subluminous O-stars: K 648, PG 1159-035 and KPD 0005+5106†

By ULRICH HEBER[1]‡, STEFAN DREIZLER[1]
AND KLAUS WERNER[2]

[1]Dr. Remeis-Sternwarte Bamberg, Universität Erlangen-Nürnberg, Sternwartstraße 7, D 96049 Bamberg, GERMANY

[2]Institut für Theoretische Physik und Sternwarte der Universität Kiel, D 24098 Kiel, GERMANY

We present a progress report on quantitative analyses of high resolution UV and optical spectra of three very hot subluminous O stars obtained with the Hubble Space Telescope and the Calar Alto 3.5m telescope, respectively. The stars K 648, PG 1159-035 and KPD 0005+5106 are all post asymptotic giant branch (AGB) stars and are key objects for our understanding of the AGB – white dwarf connection. Using state-of-the art NLTE model atmospheres we determine atmospheric parameters and the abundances of H, He, C, N, and O. K 648, the central star of the Planetary Nebula Ps1 in the globular cluster M 15, is the least evolved star with T_{eff} = 37,000 K and $\log g \approx 4.0$, whereas PG 1159-035, the prototype of a new class of hydrogen-deficient pre-white dwarfs, is the hottest one (T_{eff} =140,000 K, $\log g$ = 7.0) and the unique DO white dwarf KPD 0005+5106 is the most evolved star in our sample. The chemical surface compositions of all three stars are very peculiar: K 648 is found to be slightly helium-enriched and strongly carbon-rich (by a factor of $\approx$ 300) when compared with the cluster's metallicity. The enrichment is even larger than found for the nebula. PG 1159-035 and KPD 0005+5106 are both strongly hydrogen-deficient with no hydrogen lines detectable. While PG 1159-035 has very high C and O abundances, these metals are trace elements in the atmosphere of KPD 0005+5106 only. O VIII emission lines identified in the spectra of KPD 0005+5106 are indicative of a stellar corona. Modeling of the ROSAT-PSPC-spectrum shows that the O VIII cannot be photoionized but is likely to be caused by collisional ionisation in shock fronts formed in a stellar wind.

1. Introduction

Several of the hot subluminous stars are descendants from the asymptotic giant branch (see Heber 1992 and Werner 1993 for reviews) and, therefore, are important tracers of stellar evolution from the AGB to the white dwarf grave. We have selected three stars of particular interest for detailed spectroscopic analyses using ultraviolet and optical spectroscopy and state-of-the-art NLTE model atmospheres. The programme stars are K 648, the central star of the planetary nebula Ps1 in the galactic globular cluster M 15, PG 1159-035 (= GW Vir), the prototype of a new class of hydrogen-deficient pre-white dwarfs and KPD 0005+5106, a unique DO white dwarf. The analysis of KPD 0005+5106 is published in full detail by Werner, Heber & Fleming (1994), the optical analysis of PG 1159-035 can be found in Werner, Heber & Hunger (1991, WHH), while preliminary

† Based on observations with the NASA/ESA Hubble Space Telescope, obtained at the Space Telescope Science Institute, which is operated by the Association of Universities for Research in Astronomy, Inc., under NASA contract NAS5-26666.
‡ Visiting Astronomer, German–Spanish Astronomical Center, Calar Alto, operated by the Max-Planck-Institut für Astronomie Heidelberg jointly with the Spanish National Commission for Astronomy

results from the UV spectra are reported by Werner & Heber (1993, PG 1159-035) and Heber, Dreizler & Werner (1994, K 648).

In the next section we present our observations and model atmospheres and describe the spectral analyses for the three stars in sections 3 to 5. We discuss the results in the light of stellar evolution theory in the final section.

2. Observations and NLTE model atmospheres

2.1. *Optical spectroscopy*

Optical long slit spectra of all three stars were obtained with the Twin spectrograph at the 3.5m Telescope at the Calar Alto Observatory. The observations of PG 1159-035 and KPD 0005+5106 were straightforward, whereas care had to be taken in the case of K 648 since this star lies in a very crowded field close to the center of the globular cluster M 15. Details are given by Heber, Dreizler & Werner (1994).

The spectral resolution of the optical spectra ranges from about 1 Å (K 648) to 2.5 Å (PG 1159-035).

2.2. *UV spectroscopy*

Ultraviolet spectra were obtained with the Hubble Space Telescope using the Faint Object Spectrograph (FOS) for PG 1159-035 (grating G130H) and the Goddard High Resolution Spectrograph (GHRS) for K 648 (grating G160M). The UV spectral windows were chosen to measure the N V resonance doublet and the O IV triplet at 1340Å (Mult. No. UV 17) besides additional carbon lines. The former lines are the only means to determine the N and O abundance of K 648.

The FOS observations of KPD 0005+5106 are already published by Sion & Downes (1992). We retrieved the spectra from the data archive.

Problems with the Standard pipeline reduction of the FOS spectrum of PG 1159-035 were already spotted by Werner & Heber (1993). Moreover, it became clear recently (Rosa 1993) that the pipeline procedure underestimates the background level to be subtracted. Our data for PG 1159-035 were kindly re-calibrated at the ST-ECF by M. Rosa, which included reference data valid for the epoch of the observations which were not available during initial pipeline processing. The improved background correction changed the flux level only marginally, because PG 1159-035 is an extremely blue target and the background problem is probably caused by scattered red light (Rosa 1993). Finally, the absolute flux calibration is improved by applying the correction function derived by Rosa & Benvenuti (1994) from HST observations of the white dwarf G191-B2B and Koester's model atmospheres. The recalibrated flux distribution is plotted in Figure 1. Note that the S-shaped continuum distortion found by Werner & Heber (1993) for the original calibration has been removed by the recalibration process.

2.3. *NLTE model atmospheres and atomic data*

Quantitative spectroscopic analyses of these observations have been started and grids of NLTE model atmospheres were computed with the codes written by K. Werner, S. Dreizler and T. Rauch which are based on the Accelerated Lambda Iteration technique developed by Werner & Husfeld (1985) and Werner (1986). We used the latest version of the NLTE code (Dreizler 1993, Dreizler & Werner 1993, Rauch 1993). The model atoms were tailored to the needs of the analysis of the particular star. A summary of model atoms used is given in Table 1. For O-stars usually the ionisation equilibrium of helium (He I vs. He II) is used to determine T_{eff}. This method does not work in the case of the programme stars since the He I lines are not measurable because in the

	K 648	PG 1159-035	KPD 0005+5106
No. of ions	8	17	8
	H I-II	H I-II	H I-II
	He I-III	He I-III	He I-III
	C II-V	C III-V	C III-V
		O IV-VII	
		Ne IV-VII	
No. of levels	173	240	106
No. of lines	601	249	334

TABLE 1. Model ions considered in NLTE model atmosphere calculations

cases of PG 1159-035 and KPD 0005+5106 the stars are too hot and in the case of K 648 the nebular He I emission is masking the underlying photospheric absorption completely. Hence, T_{eff} has to be determined by other means.

3. K 648 – The central star of the Planetary nebula Ps1 in the globular cluster M 15

Ps1 (Pease 1928) remained the only example of a Planetary Nebula as a member of a globular cluster until Gillet *et al.* (1989) found another example in M22. Hence, K 648 belongs to the small group of halo Planetary Nebula Nuclei (PNN), for which Clegg (1987, 1993) lists only six additional objects not associated with a globular cluster. A nineth halo PNN was found recently by Napiwotzki (1994).

About a dozen other hot post-AGB stars (besides the two PNN in M15 and M22) are known members of globular clusters but do not appear to have Planetary Nebulae (de Boer 1985, Heber & Kudritzki 1986). The cluster post-AGB stars have to be of rather low mass since they evolved from low initial main sequences masses (about 1 $M_\odot$). Hence the cluster members allow the theory of low mass post-AGB evolution to be checked (Schönberner 1987): We can investigate whether a critical luminosity for the PN ejection exists or whether a luminosity – metallicity relation exists. Schönberner (1987) found that the latter relation indeed exists for long-period cluster variables and that for three globular clusters that contain both long-period variables and a post-AGB star their respective luminosities agree. K 648, however, does not fit into this relation since its luminosity is larger than predicted. Therefore an accurate luminosity determination for K 648 is needed (Schönberner 1987). We intend to derive its luminosity and effective temperature from a spectroscopic model atmosphere analysis. Moreover, the photospheric chemical composition will shed additional light on the evolution of K 648.

3.1. Qualitative Spectral Analysis

Our optical spectra allow for the first time the line wings of H_γ and H_δ to be measured despite strong nebular line emission. While He II absorption lines are easily detectable and unaffected by any nebular emission, He I lines are completely filled in by the latter. Besides the hydrogen and helium lines several absorption lines of C III and C IV can be identified.

First qualitative results can be obtained by comparing the spectra of K 648 to those of other sdO stars that have been analysed already. The carbon line spectrum of K 648 is surprisingly similar to that of the sdO star LS IV+10°9 (see Figure 1 of Heber, Dreizler

& Werner 1994), for which a NLTE analysis (Dreizler 1993) yielded $T_{eff} = 44,500$ K, $\log g = 5.55$ and showed C and N to be four times overabundant with respect to the Sun and Mg and Si to be about solar. N and Si lines, however, are far much weaker (or absent) in K 648 than in the comparison sdO-star. This implies that their C abundances are similar but the N and Si abundances of K 648 are much lower than in LS IV+10°9.

The O IV line strengths in the UV are remarkably similar to those of the sdO star HD 128220B ($T_{eff} = 40,600$ K, $\log g = 4.5$), for which Rauch (1993) derived an O abundance of 1/10 solar (see Figure 21 of Rauch 1993).

3.2. Quantitative Spectral Analysis

In the absence of any measurable He I absorption lines, T_{eff} can be derived from the ionisation equilibrium of carbon (C III vs. C IV), which has been shown to be a very sensitive procedure when UV spectral lines are used (Rauch 1993). At the temperatures in question, the C III line strengths decrease strongly with increasing temperature (see Figure 2 of Heber, Dreizler & Werner 1994) whereas the C IV lines are almost insensitive to T_{eff} and, hence, can be used to derive the carbon abundance. The gravity and the He/H ratio is determined from the wings of H_γ and H_δ as well as from the profiles of the He II lines 4200 Å, 4542 Å and 4686 Å.

The preliminary results are as follows:

$T_{eff} = 37,000$ K; $\log g = 4.0$; $n_{He}/n_H = 0.5$ and a carbon abundance of 3 times the solar value, i.e. much larger than the cluster's metallicity.

The resulting line profile fits are not yet fully consistent and might be improved by slightly increasing T_{eff} and the carbon abundance and decreasing $\log g$. Additional model atmosphere calculations are necessary to improve the results.

The preliminary results of the model atmosphere analysis of the photospheric spectrum of K 648 can be compared to the results derived from the nebular spectrum (Adams et al. 1984). T_{eff} is in excellent agreement with the result of Adams et al. (1984): $T_{eff} = 38,000$ K (derived from Zanstra and other methods), while the abundances of He and C are considerably larger in the photosphere than in the nebula: Adams et al. found a normal He/H ratio and solar C abundance whereas the photospheric He/H is 5 times larger and the C abundance at least 3 times larger. This discrepancy might be resolved if the nebula is chemically stratified. The outermost layers of the nebula might contain H-rich, C-poor material which would dilute the mean He and C abundance determined for the nebula. Spectra of high spatial resolution are required in order to check this hypothesis.

4. PG 1159-035 – the prototype of a new class of very hot hydrogen-deficient pre-white dwarfs

PG 1159-035 is the prototype of a new class of very hot hydrogen-deficient pre-white dwarf stars. It also defines a new type of variable stars (the GW Vir stars). PG 1159-035 has been studied extensively both photometrically (Winget et al. 1991) and spectroscopically (Werner et al. 1991, WHH). The analysis of the optical spectrum, however, was rendered difficult by several problems. Several highly ionized species can be identified (He II, C IV, O VI). Due to the lack of any element observed in two stages of ionisation the T_{eff} determination was carried out by fitting the line profiles of several He II-, C IV- and O VI-lines. It turned out that the emission line cores observed in some lines are quite temperature sensitive. These emission cores are formed in the outer layers of the atmosphere under extreme NLTE conditions and great care about the details of the employed model atoms had to be taken. The gravity was determined – as usual – by matching the Stark broadened wings of spectral lines. However, the treatment of the line broadening is difficult because all of the lines in the optical arise from highly excited, almost degenerate energy levels. As a consequence the Stark broadening effect is neither linear nor quadratic, but a superposition of both. No theory is available for this complicated line broadening. Hence WHH had to rely on an approximate treatment. Despite these problems a self-consistent fit of the optical spectrum was achieved resulting in $T_{eff} = 140,000$ K ($\pm 10\%$), $\log g = 7.0 \pm 0.5$. The atmospheric chemical composition of PG 1159-035 is dominated by carbon (50 %, by mass), helium (33 %) and oxygen (17 %).

Despite the very promising results of the optical analysis, UV observations can improve the precision of the atmospheric parameters and abundances considerably, because

(i) in the UV the O V line at 1371 Å has already been detected in high resolution IUE spectra (Liebert et al. 1989). NLTE models predict it to be very temperature sensitive and, therefore, allows T_{eff} to be determined with unprecedented precision.

(ii) some spectral lines arising from low lying energy levels (e.g. the C IV resonance lines, the 3p – 4s transition of C IV at 1230 Å) can be measured in the UV. This allows the gravity determination from optical lines to be checked.

(iii) the nitrogen abundance can only be determined from the UV, since no line of this element can be found in the optical region.

Moreover we might be able to determine the blue edge of the GW Vir instability strip by improving the precision of the analysis from UV spectra. WHH found that the PG1159 star PG 1520+525 has atmospheric parameters and chemical surface composition identical with those of PG 1159-035 to within error limits. However, PG 1520+525 is non-variable. A high precision spectrum analysis of both stars is required to reveal differences between both stars and might give hints as to why one star is pulsating (PG 1159-035) and the other does not (PG 1520+525).

The observed UV fluxes are compared to the final model of WHH in Figure 1. We redetermined the hydrogen column density by fitting the interstellar Ly-α profile and derived $\log n_H = 20.4$. The model has been scaled to match the observation at 1300 Å. While the S-shaped continuum distortion is removed by recalibration, the slope of the continuum is not in accordance with the model.

Figure 2 compares one of the most important spectral ranges to the emergent spectrum calculated from the final model of WHH. The T_{eff} dependence of the theoretical spectrum is also illustrated in Figure 2. Note the strong decrease of the O V line strength with increasing T_{eff}. From Figure 2 we conclude that the HST-FOS spectrum of PG 1159-035 indeed allows T_{eff} to be determined with unprecedented precision of 3 % or better. It fully confirms the results of the optical analysis, i.e. the systematic errors of the latter are shown to be small. Some discrepancies for the C IV lines remain to be studied.

5. KPD0005+5106 – the hottest DO White Dwarf known

KPD 0005+5106 was discovered by Downes et al. (1985) as an ultraviolet (UV) excess star in the galactic plane. Optical spectroscopy revealed the absence of He I lines but the presence of a He II 4686 Å absorption line. Thus the spectrum resembles those of the "hot DO" class described by Wesemael et al. (1985). A strong central emission core in the He II 4686 Å line, which is also found in some PG 1159 stars, implied a close relationship between KPD 0005+5106 and these stars. Two other emission lines were detected, one at the position of H_γ which was assigned to circumstellar or outflowing highly ionised gas, and another one near 4945 Å, which was tentatively identified as a

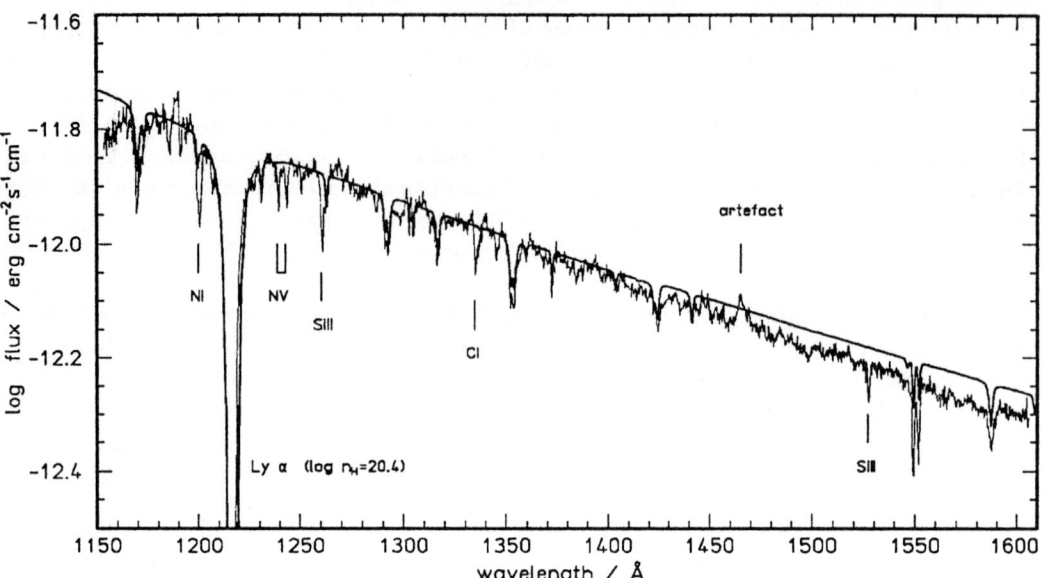

FIGURE 1. Comparison of the re-calibrated FOS spectrum (see text) of PG 1159-035 to the final model of WHH. Interstellar lines and an artefact are marked.

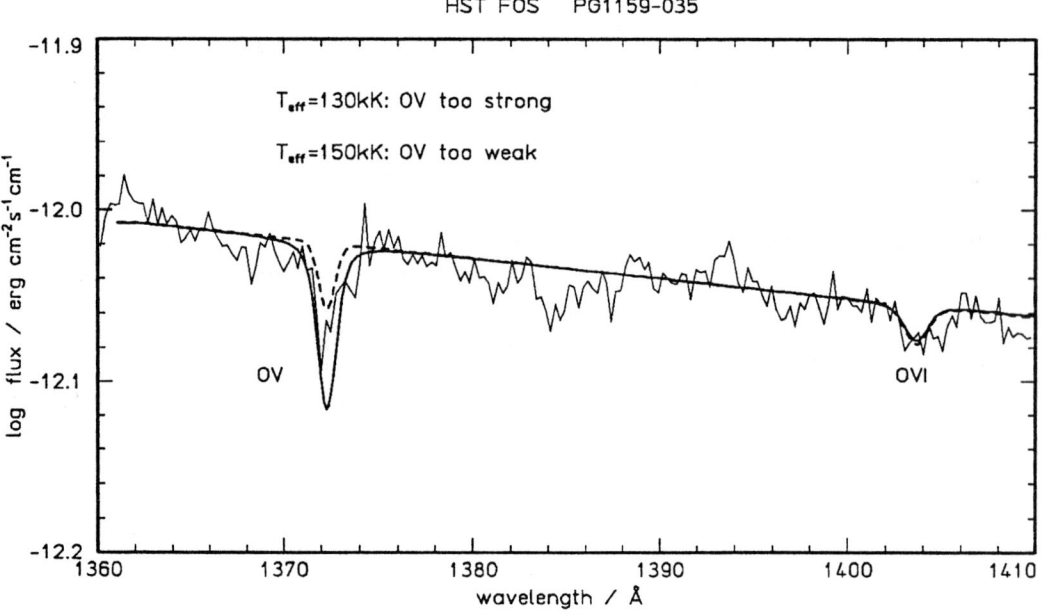

FIGURE 2. Comparison of the FOS spectrum of PG 1159-035 to model predictions near the crucial O V line.

C V or N V line. This emission line, together with another one detected near 4660 Å and identified as C IV, was suggested to be of circumstellar origin and interpreted as an indication of a recent episode of mass-loss. This scenario was also invoked to explain the emission near H_γ as due to hydrogen.

More recent optical observations of high signal-to-noise were presented by Werner &

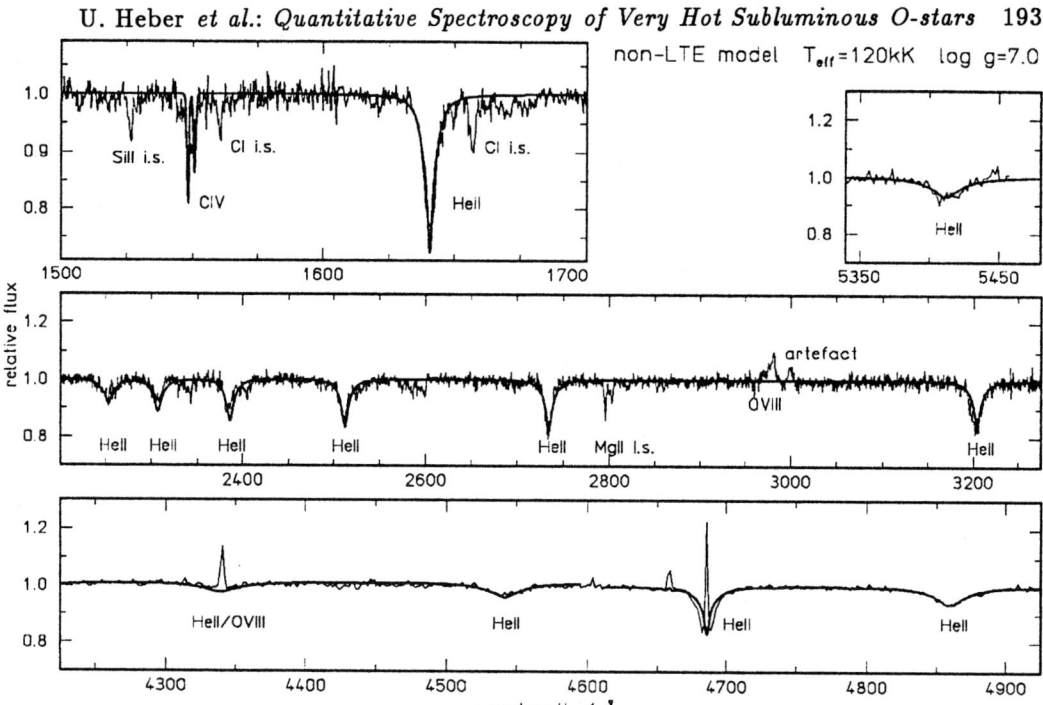

FIGURE 3. Profile fits to the He II lines of KPD 0005+5106. He II 4340 Å is dominated by the O VIII emission. The overall fit is satisfactory except for He II 4686 Å, a line that notoriously fails to match observations.

Heber (1992). Most surprising was the detection of a strong emission line at 6068 Å which could be identified as O VIII. This observation gave a natural explanation for the emission line present at the position of H_γ, namely as another O VIII line. The detection of such ultrahigh-excitation features excludes a photospheric origin. Instead it was suspected that a corona might be resonsible for this phenomenon.

Unlike for PG 1159-035 and K 648, the UV spectra of KPD 0005+5106 did not reveal any species observable with lines in two ionisation stages and, hence no ionisation equilibrium can be exploited and large errors for the values of T_{eff} and $\log g$ will have to be accepted. However, the great number of He II lines and that lower levels with three different excitation energies are involved help to confine the parameters to reasonable intervals (n=2→3 1640 Å, n=3→n' and n=4→n' series). Thanks to the high S/N-ratio of the optical spectra, the absence of He I lines (in particular 5876 Å) which were detected in some other hot DO stars (Werner 1993) already points at a temperature exceeding 100 000 K. Profile fits to the He II lines are displayed in Figure 3. We show the profiles of the best fitting model, which has T_{eff} = 120 000 K and $\log g$ = 7.0. Werner, Heber & Fleming (1993) estimated the error range to be ± 15 000 K in T_{eff} and ± 0.5 dex in $\log g$.

Abundances for C, N and O turned out to be difficult to determine due to the weakness of the lines. The mass fraction of C and N lie in the range 1/10 solar to solar whereas O is depleted by a factor of 10 or more.

KPD 0005+5106 was also detected as a X-ray source by EXOSAT and by ROSAT during the all-sky survey phase of the mission. Fleming, Werner & Barstow (1993) analysed the PSPC spectrum and found that it cannot be fit by the photospheric model derived from the spectral analysis of the optical and UV lines but can be fit by thermal bremsstrahlungs model of pure helium composition and T = 260 000 K. They concluded that the X-rays which are detected from KPD 0005+5106 are emitted by a coronal plasma around the star. This plasma is too cool for O VIII to be due to photoionisation because the ionisation potential of O VIII is 871 eV but the PSPC pulse height spectrum does not contain photons at energies larger than 400 eV. Sion & Downes (1992) presented evidence for ongoing mass outflow and Fleming, Werner & Barstow (1993) argue that the observed X-rays are emitted via thermal bremsstrahlung from a 300 000 K plasma about the star heated by a stellar wind which might be supersonic and might form shock fronts which could create O VIII by collisional ionisation.

6. Stellar Evolution

Comparing the positions of our three programme stars in the (T_{eff}, $\log g$)-plane to evolutionary tracks it is evident that all three stars are in the post-AGB phase of evolution. How can we explain the observed chemical peculiarities in terms of the stars' evolutionary history? We see the products of hydrogen- and helium-burning (the latter is probably not the case for KPD 0005+5106) at the stellar surface. These species are either dredged-up from deeper layers or stellar mass-loss had laid bare the interior layers of the progenitor stars. WHH proposed that the born-again post-AGB scenario could explain the properties of the PG1159 stars. However, discrete episodes of mass-loss that have recently been observed on the planetary nebula central star Longmore 4 (Werner et al. 1992) indicate that we do not understand post-AGB evolution sufficiently well for a conclusive answer.

Acknowledgements

We thank Michael Rosa (ST-ECF, Garching) for his assistance in preparing and reducing the HST observations and for recalibrating the FOS spectrum of PG 1159-035. S. D. and K. W. are supported by the DFG (grant He1356/16-1, We1312/6-1) and the BMFT (50 OR 9302 4)), respectively. Travel grants by the DFG (477/1872/93) are gratefully acknowledged. Calculations were performed on the CRAY X-MP of the Rechenzentrum der Universität Kiel, the CRAY Y-MP EL of the Regionales Rechenzentrum Erlangen, and the CRAY Y-MP of the Leibniz Rechenzentrum München.

REFERENCES

ADAMS, S., SEATON, M. J., HOWARTH, I. D., AURIERE, M., WALSH, J. R., 1984, MNRAS, 207, 471

CLEGG, R. E. S., 1987, ESO Conf. & Workshop Proc., 27, 543

CLEGG, R. E. S., 1993, IAU Symp. No. 155, 549

DE BOER, K. S., 1985, A&A, 142, 321

DOWNES, R. A., LIEBERT, J., MARGON, B., 1985, ApJ, 290, 321

DREIZLER, S., 1993, A&A, 273, 212

DREIZLER, S., WERNER, K., 1993, A&A, 178, 199

FLEMING, T. A., WERNER, K., BARSTOW, M. A., 1993, ApJ, 416, L78

GILLET, F. C., JACOBY, G. H., JOYCE, R. R., COHEN, J. G., NEUGEBAUER, G., SOIFER, B. T., NAKAJIMA, T., MATTHEWS, K., 1989, ApJ, 338, 862

HEBER, U., 1992, in Heber U., Jeffery C. S., eds, The Atmospheres of Early-Type Stars, Lecture Notes in Physics Vol. 401, Springer, Berlin, p. 233

HEBER, U., KUDRITZKI, R. P., 1986, A&A, 169, 244

HEBER, U., DREIZLER, S., WERNER, K., 1994, Acta Astron., Workshop on Planetary Nebula Nuclei Evolution: Models versus Observations, in press

LIEBERT, J., WESEMAEL, F., HUSFELD, D., WEHRSE, R., STARRFIELD, S. G., SION, E. M., 1989, AJ, 97, 1440

NAPIWOTZKI, R., 1994, Acta Astron., Workshop on Planetary Nebula Nuclei Evolution: Models versus Observations, in press

PEASE, F. G., 1928, PASP, 40, 342

RAUCH, TH., 1993, A&A, 276, 171

ROSA, M. R., 1993, ST-ECF Newsletter, 20, 16

ROSA, M. R., BENVENUTI, P., 1994, A&A, in press

SCHÖNBERNER, D., 1987, ESO Conf. & Workshop Proc., 27, 519

SION, E. M., DOWNES, R. A., 1992, ApJ, 396, L79

WERNER, K., 1986, A&A, 161, 177

WERNER, K., HUSFELD, D., 1985, A&A, 148, 417

WERNER, K., 1991, A&A, 251, 147

WERNER, K., HEBER, U., 1993, in Barstow M. A., ed, White Dwarfs: Advances in Observation and Theory, NATO ASI Series C Vol. 403, Kluwer, Dordrecht, p. 67

WERNER, K., HEBER, U., 1992, in Heber U., Jeffery C. S., eds, The Atmospheres of Early-Type Stars, Lecture Notes in Physics, Vol. 401, Springer, Berlin, p. 291

WERNER, K., HEBER, U. 1993, in Barstow M. A., ed, White Dwarfs: Advances in Observation and Theory, NATO ASI Series C Vol. 403, Kluwer, Dordrecht, p. 303

WERNER, K., HEBER, U., HUNGER, K., 1991, A&A, 244, 437

WERNER, K., HEBER, U., FLEMING, T., 1994, A&A, in press

WERNER, K., HAMANN, W. R., HEBER, U., NAPIWOTZKI, R., RAUCH, T., WESSOLOWSKI, U. 1992, A&A, 259, L69

WESEMAEL, F., GREEN, R. F., LIEBERT, J., 1985, ApJS, 58, 379

WINGET, D. E., NATHER, R. E., CLEMENS, J. C., et al., 1991, ApJ, 378, 326

Discussion

BOND: At the beginning of your talk you said that K648 had been believed to be too bright for the cluster's metallicity. Is this still true, and if so what are your speculations?

HEBER: We find K648 to have a luminosity comparable to the previous estimate of Adams et al. (1984, MNRAS, 207, 474): $L = 3000\ L_\odot$. This is too high for the metallicity - luminosity relation (see Schönberner, 1987, ESO Workshop Proceedings, 27, 519). It has the same luminosity as the post-AGB star ROB 162 that lacks a nebula.

SWEIGART: Can you explain your derived abundances for the nucleus of K648 by assuming that one is seeing into the core where carbon and helium were enriched during earlier helium shell flashes on the AGB?

HEBER: This would require a large amount of mixing, because hydrogen is also present in the atmosphere. Hence, one needs to mix carbon from the core to the surface. The nitrogen and oxygen abundances will give us a clue to the evolutionary history of K648, since N should be enriched in the H-burning zone, where O would be reduced by CNO-cycling. On the other hand, O could be enriched by *alpha* capture on C in the He burning zone. The analysis of the O IV and N IV lines in the ultraviolet are under way.

DORMAN: Are you planning to look at the planetary nebula (PN) is in M22? Second, what is the basis for saying that K648 violates a L-M relation? K648's parameters ($\log L/L_\odot = 3.5$) are theoretically possible, if rare, noting that most GC PAGB stars have $\log L/L_\odot < 3.3$.

HEBER: Yes, we have applied for ESO time to observe the PN in M22. The luminosity of K648 can indeed be reached by post-AGB tracks. Whether the Post Asymptotic Giant Branch (PAGB) stars lacking a PN really have $\log L/L_\odot < 3.3$ remains to be seen. We intend to determine their atmospheric parameters as precisely as we can. The luminosity-metalicity relation is found for long-period cluster variables empirically and predicted by theory if you assume that all clusters have the same age. The luminosity of K648 is somewhat too large for the metallicity of the cluster M15.

CACCIARI: You said that if the distance was larger, the mass of these BHB stars would be larger too. How much should the distance modulus be increased in order to find masses in agreement with the ZAHB?

HEBER: I do not know the number by heart. But if we shift the cluster further away, we run into another problem: The masses of the cooler stars are then too high.

Analyzing the Helium-Rich Hot sdO Stars in the Palomar Green Survey

By PETER THEJLL

Niels Bohr Institute, Blegdamsvej 17, DK-2100 Copenhagen Ø, DENMARK

The goals, methods and results of an ongoing effort to analyze all the helium-rich subdwarf O stars in the Palomar Green survey are presented. Analysis of spectra as well as radial velocities is performed. Interpretations of the results for all the very He-rich sdO (i.e., %He > 50) in the Palomar Green survey are interpreted in view of various evolutionary scenarios, and statistical arguments about the role of sdOs in late stages of stellar evolution are reviewed.

It is shown that large distances to the helium-rich sdO in the Palomar Green survey are found, a result that is supported by other published data on spectral analysis, as well as the available data on radial velocities. The number statistics of the sdO in the Palomar Green survey shows that the existing densities of the sdO in the galactic plane, and the large scale height inferred from spectral analyses, are in conflict, which might indicate a more dense population in the Old disk than what can be accounted for in a simple exponential density law.

1. Introduction

Since they were first described systematically by Greenstein & Sargent (1974) the hot helium-rich subdwarf O stars have represented a challenge for astronomers. These objects are so hot that any attempts at analyzing the radiation we see coming from these objects require advanced radiative transfer methods to build analysis-tools. Photometry is mainly useful for identifying the objects as extremely hot - photometric colors can mostly be used, above 35,000 K, to measure the objects reddening, not their temperature or gravity. The cooler hydrogen-rich sdB stars can be studied well with photometric means, see for examples Heber (1986) or Theissen *et al.* (1991).

An understanding of the characteristics of the sdO can also be had with knowledge of space motions, as well as by studying those few sdO that have been observed in binary systems where periods have been measured and masses derived. Elsewhere in these proceedings we discuss how companion stars can be used for reaching similar goals (see Jimenez *et al.* 1994).

The present paper will deal mainly with a description of our efforts to derive atmospheric parameters via spectroscopic analysis of H, He I, and He II lines in the optical range. The atmospheric parameters are needed to find the spatial distribution of the stars which is needed for our ultimate goal, an understanding of the role of the sdO in late stages of stellar evolution. In addition we present results on the analysis of the radial velocities of hot subdwarfs.

If the sdO are single-star evolution products they, like the sdB, will yield important information on the extremes of single-star mass-loss which is important for modelling the chemical evolution of the Galaxy and for understanding the mass-loss mechanism on the red giant branch. If the sdO are results of binary interactions, they may provide examples of the results of WD mergers (Iben 1990), or close evolution leading to stripping of envelopes (Mengel *et al.* 1976).

2. Models

The NLTE models we use are simple, but the nature of the observational material we have does not warrant more complicated modelling. The composition is assumed to be a mixture of helium and hydrogen - no metals are included. The radiative models are calculated assuming non local thermodynamic equilibrium (NLTE) for the bound-free and free-free transitions (so called NLTE-C) and assuming NLTE for 6 bound-bound transitions of H. Once the model structures are calculated full NLTE spectra are calculated for the optical range, i.e., 3600 Å to 6000 Å. In this range there are many He I and He II lines as well as the Balmer series mixed with He II lines.

Below about 45,000 K the temperature sensitive He I lines make it easy to determine T_{eff} very accurately and therefore also (relatively speaking) log g and the abundance of helium, although the gravity and the abundance have to some extent the same effect on line profiles and strengths. At higher temperatures only He II and H+He II lines are available and the uncertainty in T_{eff} is now large leading to additional uncertainties in log g and abundance.

A grid of models covering 35,000 K to 65,000 K in steps of 5,000 K, from log g = 4.0 to 6.5 in steps of 0.5 and the abundances 50, 60 70, 80, and 99 % He was made by Franziska Bauer and Dietmar Kunze, and interpolated for use in a least-χ^2 profile fitting program.

3. Observational data and fitting method

Medium resolution (FWHM near 5 Å) good-quality (signal-to-noise ratio near 100) spectra of about 100 of the roughly 230 sdO stars in the Palomar Green survey of blue high galactic latitude objects (Green, Schmidt & Liebert 1986: PG from now on) were obtained by Rex Saffer. An example is shown in Figure 1.

About 20 of the stars have He abundance above 50 % - the rest are hot hydrogen rich objects that will require a larger grid of models for analysis. Typical members of this group are such stars as Feige 34 and BD+28° 4211. For the really hot stars in this group (90,000 K and above) high-resolution spectroscopy of emission features can be used to advantage.

The stars that we could model with the present grid were analyzed by fitting each spectrum to the grid of interpolated model spectra and finding the best fit for a selection of the lines observed. Each selected line is normalized to continuum flux 1.0 and centered on the laboratory wavelength and some part of the profile is then fitted (see caption of Figure 2 for explanatory details).

Not all lines in a spectrum need to be chosen for fitting as there is much redundancy in for instance, the He I lines. The strongest lines are usually chosen and this is usually the lines at 4026 Å (also blended with He II), 4387 Å, 4471 Å, 4926 Å, and 4713 Å if observed and if resolvable from the He II line at 4686 Å. The He II lines used usually include 4200 Å, 4542 Å, and 4686 Å, as well as the H-blended lines at 4100 Å and 4339 Å. The depth of the cores of the 4542 Å and 4200 Å He II lines are sensitive to log g as is the width of the 4471 Å He I line.

1, 2, and 3 σ contours of χ^2 around the best fit are calculated and presented in a T_{eff} vs. log g plane for the best fitting abundance. We calculate the *joint* 3-parameter levels of confidence and thus show a crossection of the 3-dimensional '1, 2, 3-σ volume' around the best fit. We use the table of $\Delta\chi^2$ provided by Press *et al.* (1986) as definitions of the 1, 2, and 3 σ levels above the minimum χ^2. An example of the fit and the levels of confidence for a relatively cool sdO and for a relatively hot sdO are shown in Figures 2 and 3.

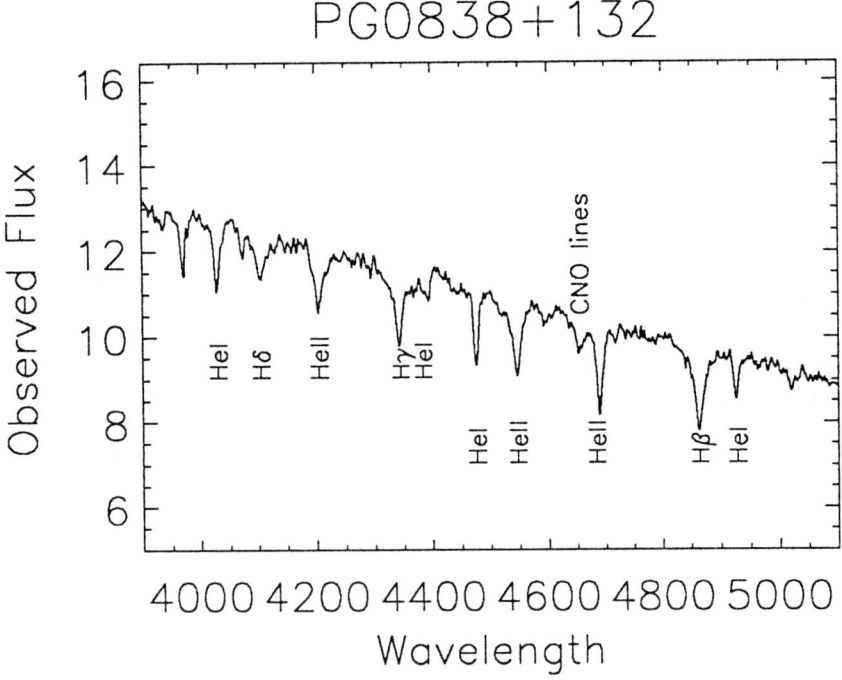

FIGURE 1. Spectrum of PG0838+132. Some of the commonly used He and H lines are indicated.

4. Fitting results

The results of fitting 21 sdOs from the PG survey using the models and techniques described here are shown in Table 1. For 5 of these stars Dreizler et al. (1990) have already published NLTE fits, based on models of similar complexity but with spectra of better resolution and the same S/N ratio as ours. Our fits for these stars do not agree, and there seems to be a systematic disagreement, see Figure 4. What is the reason for this significant difference, and what are the consequences of such a difference for the goals of our project?

The direct cause of the different fits found can be due to two things we think: Differences in the models or differences in the analysis methods. In collaborations with Dirk Husfeld and Stefan Dreizler it has not been possible yet to find if there are any important differences between our models - we have found a 1500 K temperature difference in the structure of a model near the depth where He I lines are formed, but whether this is enough to produce the differences in model spectra is unclear. Work continues on this problem.

The fitting method of Dreizler et al. (1990) is based on preliminary equivalent width fits combined with careful profile fits in the sense of requiring that the best fit from equivalent widths gives a good match between the observed line profile and the calculated one, as judged by eye. This method is therefore not quantitative, as our method is, but it does make it possible for the knowledgeable human 'fitter' to use his/her experience about line profiles to produce a good best fit. To be quantitative we chose not to use such a method, so our method suffers under the problems of having to specify to a program what parts of the line profile are to be used for the fits - we may not be able to place

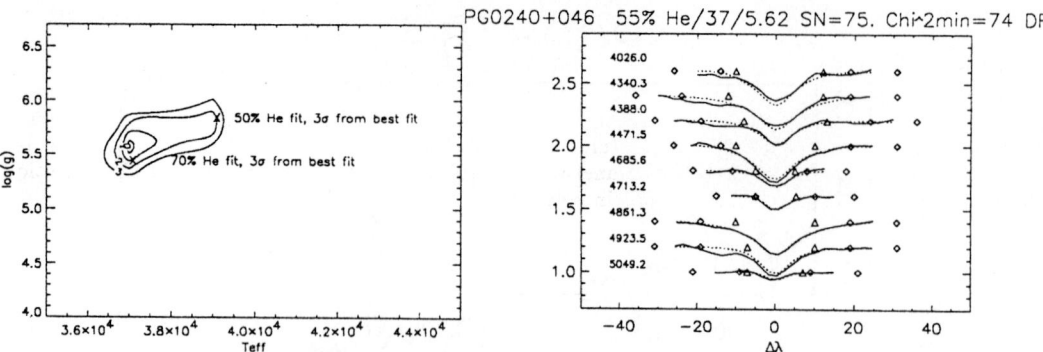

FIGURE 2. NLTE fit of PG0240+046. On the right are shown the observed lines and the model lines from the best fitting model superimposed. The diamonds on the sides of each profile indicate where the continuum was set while the triangles nearer the line cores indicate the region in which the fitting was performed. On the left is the T_{eff} vs. log g diagram showing the location of the best fit and estimates of the 1,2 and 3 σ levels of confidence. The levels represent 3-parameter joint levels of confidence, i.e. we are showing the crossection of the joint 1,2 and 3σ confidence volumes in T_{eff}, log g, and % He space. Shown, as crosses, are also the fits at other abundances, all inside the 3 σ volume but projected onto the plane.

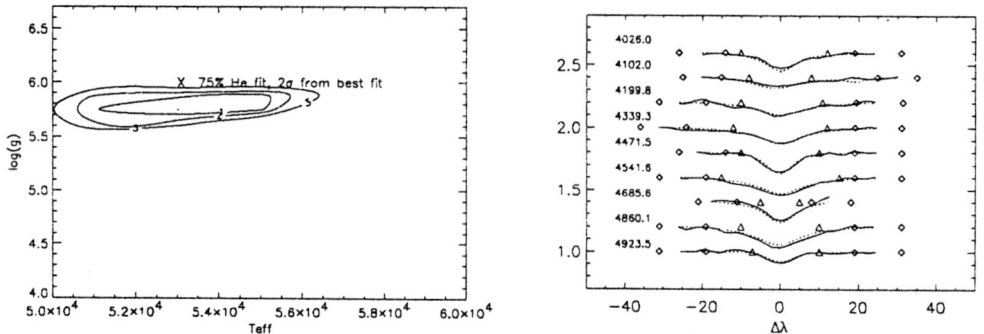

FIGURE 3. NLTE fit for PG0838+133. The best fitting abundance is 90 % He.

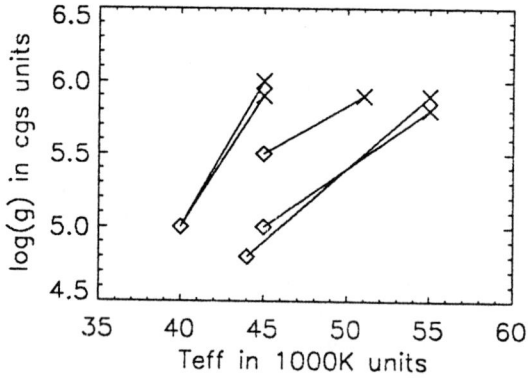

FIGURE 4. Comparing NLTE fits for 5 PG sdO stars. The diamonds represent the fits found by Dreizler et al. (1990) while the crosses represent the results presented in this paper. Lines link fits for the same star.

TABLE 1. Columns: (1) gives PG-survey name of the star, (2)—(4) gives the parameters of the best fit, (5) and (6) give the best available magnitude known for the star and its type, respectively. V refers to the Johnson V magnitude from (Green, Schmidt & Liebert 1986; GSL86), vGr and y refer to the Greenstein multichannel v and the Strömgren y also found in GSL86. Vbpg is the Johnson V magnitude calculated from the GSL86 Bpg. yM is the y value given by Moehler *et al.* (1990), while Vcmc is the Johnson V from CMC7 (1993), and yW is the Strömgren y magnitude from Wesemael *et al.* (1992). (7) is the distance above the galactic plane, in kpc, and (8) is the luminosity in solar units.

Name (1)	T_{eff} (2)	log g (3)	% He (4)	m (5)	type (6)	z (7)	$L/L_\odot$ (8)
0039+134	55000	5.8	99	13.72	yM	0.7	200
0208+016	45000	6.0	66	13.73	yM	0.6	50
0217+155	55000	5.8	90	15.15	yW	1.2	200
0240+046	37000	5.6	55	14.12	yM	1.3	130
0838+133	53000	5.8	90	13.66	yM	0.5	200
0902+057	43000	6.0	97	14.34	Vcmc	0.5	40
0921+311	41000	5.4	85	14.60	yW	0.8	70
0952+519	57000	5.9	99	12.70	V	0.4	170
0953+024	49000	5.9	97	15.00	y	0.9	90
1011+650	57000	5.9	90	14.90	V	1.0	170
1020+695	53000	5.6	99	14.63	yW	1.1	250
1030+665	47000	5.8	99	15.60	Vbpg	1.5	100
1230+068	43000	5.5	94	13.20	yW	0.8	130
1544+253	55000	5.6	99	14.40	Vcmc	1.2	290
1708+614	55000	5.5	90	15.20	Vbpg	1.5	360
2129+151	49000	5.8	95	14.23	yW	0.6	110
2201+145	45000	5.7	95	15.40	Vbpg	1.1	100
2213-006	49000	5.5	80	14.17	yW	1.1	290
2215+151	45000	5.9	99	14.53	vGr	0.5	60
2244+152	43000	5.5	95	15.72	vGr	1.9	130
2352+180	51000	5.9	85	13.40	yW	0.5	110

as much emphasis on the agreement between model and theory in, for instance, the line cores as one can do with the other method.

We hope to soon be able to know whether fitting methods are more important than model differences, but at the moment it is not possible to state which set of results are 'correct' and which are not.

It is worth noting that no comprehensive comparison of hot NLTE models has been published – as has been done with LTE codes, one example being the work on the DB white dwarf temperature scale problem (Thejll *et al.* 1991a). There exists no 'standard toolkit' for NLTE modelling of hot models, although we are aware that this is the impression by some members of the astronomical community.

As to the consequences for our final goals in this project qualitative results will not be altered by the differences discussed here as all the Dreizler *et al.* (1990) fits we can compare to provide lower log g's and therefore indicate larger distances to the sdOs. As we will show now, the distances that we find are so great that the sdO in the PG survey must be members of some thick disk or halo population, and with greater distances this conclusion is only strengthened.

5. Measuring the sdO scale height

We have been able to find distances to a subset of the PG sdO stars, but as it is a random one we hope that our conclusions based on the subsample can be extended to the whole set of sdO stars in the PG survey.

The PG survey has the shape of a cone around the North Galactic Pole going down to 30 degrees latitude and therefore we have data on stars above the plane, not in it. We can measure such things as scale heights above the plane with some confidence as the PG survey is well-designed for this. What we cannot say anything about with a great deal of confidence is what the sdO in the galactic plane are like, but we will return to this point shortly.

It is usual to assume that stars above the plane are distributed according to some exponentially decreasing density law, e.g.,

$$\rho(z) = \rho_0 exp(-z/z_0) \tag{5.1}$$

where z is the height above the plane, z_0 is the scale height and ρ_0 the density in the plane. Stars in the galactic halo have a different distribution law which has nearly constant density until you get to really great distances outside the plane. The sdO here (and those in Dreizler et al. (1990)) are all within a few kpc of the plane. One characteristic of the density law above is that when sampled with a cone as done in the PG survey, the distribution over distance peaks at 2 scale heights and has a mean distance at 3 scale heights. Since the peak of a distribution is more robust to omissions and selection effects than is the mean, we can quickly get an estimate of the scale height if we look at the position of the peak of the distribution. As it happens we do see such a peak in our distribution of objects (see Figure 5) and therefore preliminarily accept the idea that the stars are in a exponentially decreasing disk-like population and that the scale height of this distribution is in the range from 500 to 1000 parsec. Of course, any magnitude limited survey has to have a maximum distance to which it penetrates and that will automatically produce a distribution with a peak *somewhere*, but we shall proceed assuming that there is evidence for an exponentially decreasing density.

By fitting the theoretical distribution to the observed one with formal means or using the '$1/V_{max}$' method we can also estimate the scale height and we get similar results (Thejll 1989). Such a scale height is not what you find in the 'Old disk' population (250 parsec) or for the sdB stars (200 - 300 parsec).

6. Statistics of sdB and sdO

Data on sdB and sdO, including data from our own work, are summarized in Table 2. The sdB are well studied by several authors, while the available data on the sdO comes from our own work and that of Downes (1986) who detected and analyzed sdO in the galactic plane (b < 12 degrees). The distance to each star was calculated by Downes using assumptions about the absolute magnitude of sdO and by measuring the amount of interstellar reddening of each object, assuming a known intrinsic color. The space density in the plane was then calculated using the '$1/V_{max}$' method, assuming a scale height of 175 parsec.

It is appropriate to comment that the method used by Downes is not as accurate as are spectrophotometric methods because the sdO have a spread in absolute magnitudes and because the distance dependent reddening is not well known. We also note that although a very small scale height was assumed for the sdO the density derived is not very sensitive to the scale height assumed, for a sample in the plane. A similar density is found with the simple approach of counting sdOs in a spherical volume close to the

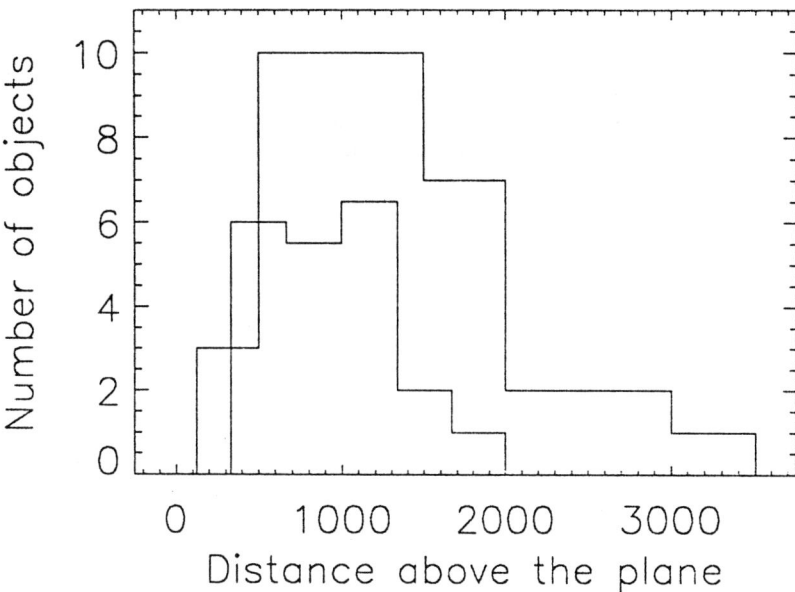

FIGURE 5. Two histograms showing the distribution of sdO stars over distance above the galactic plane. The larger histogram is from the larger sample of sdO stars (extremely helium-rich as well as other sdO stars) analyzed in Thejll et al. (1991b). The smaller, slightly overlapping histogram, represent the extremely helium-rich stars in this paper only. For an exponential density law the peak of the distribution is expected at 2 times the scale height. In both histograms the peak is near 1 kpc, indicating a scale height near 500 parsec and up.

Sun and dividing the number found by the volume. As the sample of sdOs near the Sun may not be complete this simple estimate is a lower limit to the true density of sdO near the galactic plane. A catalog of most of the known hot subdwarf O and B stars is in Kilkenny, Drilling & Heber (1988). A NLTE analysis of the spectra of the Downes sample of sdOs is in progress.

For a population well-described by the density law (equation 5.1), a relationship exists between the number of stars in a conical volume and the scale height and the density in the plane. Integrating over distance to infinity (resembling a very deep survey that covers several scale heights of the distribution), we get

$$N = z_0^3 \rho_0 tan^2(90-b) 2\pi, \tag{6.1}$$

where b is the limiting galactic latitude of the cone, and N is the total number of stars in the volume. How well do the data in Table 2 satisfy this equation when N is compared to the counts of PG survey sdBs and sdOs?

Inserting the most extreme estimates of the two quantities we get the predicted number for sdBs in the PG catalog from 125 to 6600 sdB, i.e., from one fifth to ten times as many sdB as there was found. For the sdO we see that from 600 to 20,000 sdO are predicted, i.e., from 2.5 to 90 times as many as were actually found. The available data on z_0 and ρ_0 for sdB are therefore *consistent* with the numbers in the PG survey found while this is *not* the case for the sdO.

This result can be interpreted in two ways - either there is a problem with the sdO data (say, the PG survey is incomplete, or the analysis of distances is in error) or with

TABLE 2. Summary of population data for sdO and sdB. Sources of data, given under 'src.' are: 1=Theissen et al. (1991). 2=Bixler, Bowyer & Laget (1991). 3=Thejll et al. (1991b). 4=Saffer (1991). 5=Heber (1986). 6=Downes (1986). τ is the surface birth rate, calculated by projecting the space birth-rates onto the galactic plane. For comparison, the birthrate (τ) for He-He close binary white dwarf systems, is estimated (Webbink 1984), to be 2.9×10^{-11} pc^{-2} yr^{-1}, based on Miller & Scalo (1979) IMF and a binary frequency of 0.14 systems per decade in orbital period.

	sdB	src.	sdO	src.
T_{eff}	24 - 38,000K		40 - 80,000K	
log g	5.5 - 6.2	4	4.5 - 6.5	3
Composition	H : 90 - 100%		He : 10 - 100%	3
z_o in pc	150<300	1	460<1150	3
	250<400	4		
	190	5		
	200<290	2		
ρ_o in pc^{-3}	4×10^{-6}	5	$3.3 < 6.5 \times 10^{-7}$	3
	2×10^{-6}	6	7×10^{-7}	6
	$2 < 5.5 \times 10^{-6}$	2		
τ in pc^{-2} yr^{-1}	$2 \pm 1 \times 10^{-11}$		1×10^{-11}	3

the assumptions of equation 5.1. Putting aside the first possibility let us focus on the latter: The KPD survey was well designed for finding the density of sdB and sdO in the galactic plane, and the PG survey well designed for finding the scale heights, but if there are two populations of sdO then putting the ρ_0 found in the plane together with the z_0 found in the halo will be invalid, and this interpretation of the above problem is interesting from an evolutionary viewpoint.

7. Possible evolutionary scenarios

The hot subdwarf B stars are easier to analyze and understand than the sdO stars. The sdB stars are very likely to be the results of single-star evolution with near-complete envelope loss at the first giant branch tip (RGT). To do this it is necessary to use old stars as the progenitors. Mass loss at the RGT is not well understood theoretically but fairly good empirical descriptions exist. The mass-loss 'law' of Reimers (1977) gives an accurate fit to observed mass-losses as well as containing some basically sound physical principles. With this law and realistic models of RGB stars it is possible to model near-complete envelope mass loss for old low-mass main sequence stars in the galaxy (see Thejll et al. 1992). The mass-loss is cut short by the onset of the helium core flash - the timing of the mass-loss and the onset of the helium-core flash is critical for the resulting envelope mass of the horizontal branch stars formed. Identifying the sdB with the He-core burning sequence of 0.5 $M_\odot$ is easy if you plot theoretical evolution tracks for 0.5 $M_\odot$ He-core burning objects along with derived T_{eff} and log g - the sdB coincide with the 0.5 $M_\odot$ He-core burning tracks - see Figure 6.

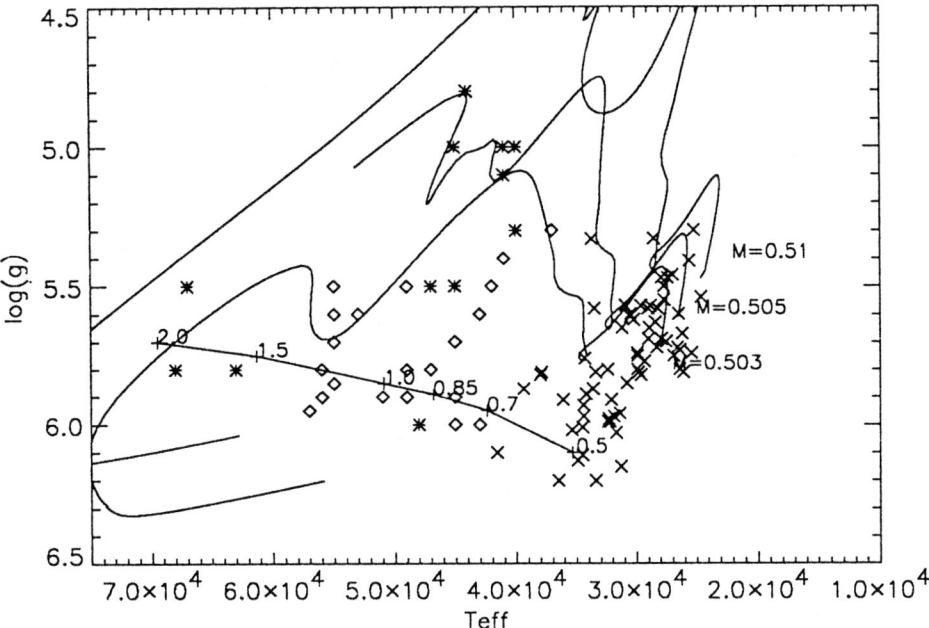

FIGURE 6. Temperature–gravity diagram showing the positions of our sdOs (diamonds) and those of Dreizler et al. (1990) (crosses), sdBs (X's) from Saffer (1991), as well as theoretical evolution tracks for 0.5 $M_\odot$ He-core objects labelled with their total masses, from Caloi (1989), and the Helium main sequence from Paczynski (1971). The numbers along the HEMS are labelled with the mass in solar units.

If envelope-loss was complete one might expect helium-rich hot stars to be formed - such as the sdOs, but they would all have masses close to 0.5 $M_\odot$ (this is the mass at which the helium-core flash occurs for low mass progenitors, i.e., those that produce a degenerate core as they evolve up the RGB), nearly independent of chemical composition. When we plot the positions of sdO in the theoretical HR diagram their positions do not coincide with any post-HB theoretical evolution tracks, so a simple identification is not possible.

We notice that most of the sdO locations we plot in Figure 6 coincide with the location of the helium main sequence - that is, the location for stars burning helium in the core for various masses, which is the evolutionary phase of longest duration after the red giant branch, before the white dwarf cooling sequence. The sdB spend about 10^8 years burning helium in the core and then about a tenth to a fifth as much time in helium-shell-burning stages, during which they may evolve in loops as shown in the figure. The sdBs are on the HEMS, but due to the effects of envelopes the HEMS is reddened so that stars of one core mass but varying envelope masses, will form a sequence such as the sdBs - stretching away from the 0.5 $M_\odot$ point of the HEMS. Lifetimes on the HEMS are shorter for larger masses, as expected, and we should expect the observed cutoff for high HEMS masses. Times spent in such post-HEMS stages are short compared to the HEMS time itself, for a given mass, so what we see is consistent with the interpretation that sdOs are HEMS objects of varying masses. However, the tracks from the sdB locations pass through the same areas so it cannot be ruled out that some of the sdO are post sdBs, or indeed high-luminosity objects cooling fast in post-AGB phases.

One problem of a scenario in which sdB evolve into sdO is to show how the H-rich envelopes of the sdB turn into He-rich envelopes of the sdO. One could consider convective mixing, but Groth et al. (1985) have shown that a H-rich envelope does not become convectively unstable and therefore cannot turn into a He-rich envelope by dredging up He from below. Another possibility is to loose the H-rich envelope. At this meeting J. MacDonald and S. Arrietta have shown how mass-loss from evolving sdBs can lead to depletion of the envelopes - a mechanism for turning sdBs into sdOs thus seems to be indicated. One necessary consequence of the scenario where 'all' sdOs are formed from post-sdBs is that the mass of the sdOs be 0.5 $M_\odot$. A handful of sdO masses are known from studies of classical binary system analysis - these vary from 0.2 to 2.0 $M_\odot$ (Ritter 1990), so there is no strong independent evidence for a narrow mass distribution in sdOs. At this meeting Jimenez et al. show how to use companion stars to sdOs (and sdBs) to measure masses.

The observed positions in the T_{eff} vs. log g plane, as well as the problems that single star evolution theories have with producing helium objects with a range of masses, favor an interpretation of the sdO as He-core burning products of binary interactions.

8. Other indicators of population membership

Measuring the spatial distribution of the sdO is not the only way to arrive at an understanding of their population characteristics – kinematics can also be used. Complete space motions require accurate proper motions as well as radial velocities, and of these two the radial velocity is decidedly easier to measure in that accurate data can be acquired in a short time. A few of the brighter subdwarfs are included in some proper motion catalogs but the majority of the objects in the PG survey do not have known proper motions. We therefore concentrated on analyzing published radial velocities.

We have found radial velocity data for sdBs and sdOs in the works of Beers et al. (1992), Drilling & Heber (1987), and Saffer (1991), and analyzed it in collaboration with Chris Flynn.

The above samples of radial velocities (RV) are not for samples of stars evenly distributed across the sky so we must take special steps to avoid selection effects. The method consists of calculating the RV velocity dispersion for hypothetical stars picked from model populations in the directions defined by the sdB or sdO stars.

We consider four populations: the Young disk with scale height 200 parsec, the Old disk with a scale height of 300 parsec, a Thick disk with a scale hight of 1200 parsec (Gilmore & Reid 1983) and a halo with an essentially constant density over the distance range of interest here. Table 3 gives the results. The lower half of the table gives the line of sight velocity dispersion (LOSVD) for stars of the four populations picked to lie in the directions defined by the sample indicated by the top label in each column. The upper half of the table gives the actual sdB or sdO sample LOSVD, its standard error and the number of stars involved.

We see, for example, that in the Rex Saffer sample of sdO (last column) the sdO LOSVD was 69 ± 7 km s^{-1} and that the Old disk, when sampled in the directions defined by that sdO sample, has a LOSVD of 30 km s^{-1}, the Thick disk 50 km s^{-1} and the halo 157 km s^{-1}, while the Young disk fell at 15 km s^{-1}. This means that the sdO LOSVD is intermediary to the Thick disk and the Halo.

As a check we analyze A0 and KV stars from the Bright Star (BS) catalog, a fainter sample of KV stars from the RV sample of Andersen & Nordström (1983), and Halo stars from the Norris, Bessel & Pickles (1985) sample, see Table 4. The A0 stars have a velocity dispersion of 11 km s^{-1} which is similar to the 17 km s^{-1} expected for members

TABLE 3. Line of sight velocity dispersion in km s^{-1} calculated from data published by Drilling & Heber (DH) (1987), by Beers et al. (TB) (1992), and by Saffer (RS) (1991). N gives the number of stars involved in each sample.

	RS sdB	TB sdB	TB sdO	DH sdO	RS sdO
sample disp	53±7	73±5	101±16	51±7	69±7
N	33	131	19	27	43
Young disk	15	15	15	16	15
Old disk	30	30	30	32	30
Thick disk	50	47	48	53	50
Halo	151	125	130	167	157

TABLE 4. Testing data. A0 and KV stars from the BS catalog have been randomly picked, as have KV stars from the Andersen & Nordström (AN) (1983) sample, and halo stars from the NBP sample. In each case the vel. disp. matches the population the stars were picked from.

	A0 BS	K0V AN	KV BS	Halo NBP
sample disp	11±1	28±3	31±3	143±13
N	33	51	54	63
Young disk	17	16	16	16
Old disk	33	31	32	31
Thick disk	54	53	52	51
Halo	166	168	162	151

of the Young disk. Both sets of KV stars have LOSVD of about 30 km s^{-1} completely consistent with their expected membership in the Old disk. The NBP stars have a LOSVD of 143 km s^{-1} which matches very well the expected 151 km s^{-1} in the halo.

In general the sdB have a smaller LOSVD than the sdOs in any magnitude limited sample. This is consistent with interpreting the sdOs as belonging to an 'older' population than the sdB, although we must in the future refine our analysis techniques as noted by Uli Heber in the questions after this paper.

Restricting the sample of RS sdOs to those stars with $|b| < 12$ degrees lowers the LOSVD to 42 ± 8 km s^{-1}. The Young disk LOSVD for that sample would be 17 km s^{-1}, and we take this as evidence that there is no Young disk component among the sdO.

We note that the DH sample was analyzed, with different methods, by DH themselves, and that they concluded that the stars were members of Population I. This is not what we find.

9. Conclusions

Our own analysis of the spectra of the helium-rich sdO stars in the PG survey indicate that the distances are large - such as are expected in a thick disk or in the halo. This interpretation of our data is supported by the spectral analyses published for a few objects, and by the available radial velocity data on sdOs in general.

The derived temperatures and gravities place the very helium-rich sdO close to the helium main sequence - i.e., near the theoretical locations of helium-core burning objects of various masses with extremely small or nonexistent hydrogen-envelopes. Some of the sdO are probably evolved sdB stars, while others are probably bright post-AGB objects. At least one mechanism exists for turning a H-rich sdB envelope into a H-depleted envelope during evolution from the core burning stage - that of mass loss through radiation driven winds, as presented by MacDonald & Arrietta (1994). We feel, however, that this cannot account for the majority of the extremely helium rich sdO that we see placed so near to the HEMS.

Our analysis cannot say anything about the density of sdO in the galactic plane, as it was a survey of objects above the plane. Simple estimates, and the published data on the density in the plane gives a relative large value for the expected numbers of objects in the PG survey of sdOs, and this can be interpreted in a number of ways. We do not think there is anything wrong with the large distances we measure - even larger distances are found by other authors for the same objects; while it remains to be shown why two separate efforts derive significantly different distances the likely outcome of settling the differences will still be that the sdO in the PG survey are distributed at large distances above the galactic plane. Another possibility for the disproportionately large density in the plane and the small density above the plane could be that the simple exponential density law does not well describe the sdO population - perhaps because there are two populations; one in the Old disk with high density and one in the halo or in a thick disk, with a lower density. This discrepancy is not seen for the sdB where the estimates of density in the plane and scale height combine to predict a number of objects in the PG survey that covers the observed value.

Our study of kinematics of the hot subdwarfs indicates that the sdO and the sdB are old - intermediary in age to the old disk and the halo. Insisting on only one old population requires that we accept only the lower estimates of the scale height for sdO – near 450 parsec – as well as only low estimates of the density - near 10^{-7} sdO pc^{-3}. These numbers are in agreement with the data of our analysis, given the uncertainties, but the low scale height required may well be a difficult obstacle to overcome in a single-population model if in the future data should be published that increases the average sdO distances above the plane.

The data favors an interpretation of the sdO as products of the binary evolution of old progenitor systems since single star evolution does not predict the formation of helium objects with masses above 0.5 M$_\odot$.

I am happy to acknowledge support from the Carlsberg Foundation. The work presented in this paper represents a long-term collaborative effort including F. Bauer and D. Kunze (FRG), R. Saffer (STScI), J. Liebert (Steward Observatory), H. L. Shipman (University of Delaware) and D. Husfeld (München, FRG).

REFERENCES

ANDERSEN, J., NORDSTRÖM, B., 1983, A&A, 122, 23-32

BIXLER, J. V., BOWYER, S., LAGET, M., 1991, A&A, 250, 370-388

BEERS, T. C., PRESTON, G. W., SHECTMAN, S. A., DOINIDIES, S. P., GRIFFIN, K. E., 1992, AJ, 103, 267-296

CALOI, V., 1989, A&A, 221, 27-35

CMC7: Carlsberg Meridian Circle, La Palma, No.7, 1993, Copenhagen University Observatory, Royal Greenwich Observatory, Real Instituto y Observatorio de la Armada, in press

DREIZLER, S., HEBER, U., WERNER, K., MOEHLER, S., DE BOER, K. S., 1990, A&A, 235, 234-241

DOWNES, R. A., 1986, ApJS, 61, 569-584

DRILLING, J. S., HEBER, U., 1987, Radial velocities of hot SdO stars, in Philip, A. G. D.,

Hayes, D. S., Liebert, J. W., eds, Second Faint Blue Stars Conference, L. Davis Press., Schenectady, p. 603-606

Gilmore, G., Reid, N., 1983, MNRAS, 202, 1025-1047

Green, R. F., Schmidt, M., Liebert, J., 1986, ApJS, 61, 305-352

Greenstein, J. L., Sargent, A. I., 1974, ApJS, 28, 157-209

Groth H. G., Kudritzki R. P., Heber U., 1985, A&A, 152, 107-116

Heber, U., 1986, A&A, 155, 33-45

Iben, I., Jr., 1990, ApJ, 353, 215

Jimenez, R., Thejll, P., Saffer, R., Jorgensen, U. G., 1994, Late type companions to hot sdO stars. in Adelman S. J., Upgren A. R., Adelman C. J., eds, Hot Stars in the Halo, Cambridge University Press, Cambridge, p. 211

MacDonald, J., Arrieta, S. S., 1994, Stellar winds and the evolution of sdB's to sdO's. in Adelman S. J., Upgren A. R., Adelman C. J., eds, Hot Stars in the Halo, Cambridge University Press, Cambridge, p. 238

Kilkenny, D., Drilling, J., Heber, U., 1988, electronic version of catalog in SAAO Circ. 12, 1

Mengel, J. G., Norris, J., Gross, P. G., 1976, ApJ, 204, 488

Miller, G. E., Scalo, J. M., 1979, ApJ, 41, 513-547

Moehler, S., Heber, U., de Boer, K. S., 1990, A&A, 239, 265-275

Norris, J. A., Bessel, M. S., Pickles, A. J., 1985, ApJS, 58, 463-492

Paczynski, B., 1971, Acta Astron., 21, 1

Press, W. H., Flannery, B. P., Teukolsky, S. A., Vetterling, W. T., 1986, Numerical Recipes, 1st ed, Cambridge University Press, Cambridge

Reimers, D., 1977, A&A, 61, 217

Ritter, H., 1990, Catalogue of cataclysmic binaries low-mass x-ray binaries and related objects

Saffer, R., 1991, PhD Thesis, University of Arizona

Theissen, A., Moehler, S., Schmidt, J. H., Heber, U., 1991, A statistically complete sample of hot subdwarfs. in Heber U., Jeffery C. S., eds, Atmospheres of Early-Type Stars, Springer-Verlag, Berlin, LNP-401, p. 264

Thejll, P. A., 1989, PhD Thesis, University of Delaware

Thejll, P. A., Vennes, S., Shipman, H. L., 1991a, ApJ, 370, 355

Thejll, P. A., 1991b, in Heber U., Jeffery C. S., eds, Atmospheres of Early-Type Stars, Springer-Verlag, Berlin, LNP-401, p. 261

Thejll, P. A. et al., 1992, in Barstow M., ed, Leicester Workshop on White Dwarfs, Kluwer, Dordrecht, NATO ASI, p. 77

Webbink, R. F., 1984, ApJ, 277, 355

Wesemael, F., Fontaine, G., Bergeron, P., Lamontagne, R., Green, R. F., 1992, AJ, 104, 203

Discussion

GARRISON: The sdO/sdB phenomenon may not be due to a single mechanism, i.e., some stars may be helium-burning and others may not. All we measure is the surface gravity, not the luminosity.

THEJLL: Yes.

BOND: An embarrassingly elementary question: what is the modern operational definition of sdO and sdB?

THEJLL: It is given in the Palomar Green catalog (Green, Schmidt & Liebert, 1986, ApJS, 61, 305). A HB star shows Balmer lines beyond H9, our sdB does not - at medium resolution. An sdO stars shows He II 4686 Å as well as He I lines.

DRILLING: That the sample of sdO stars studied by Uli Heber and me is certainly very different than your sample. Ours was a low-latitude sample complete to B = 12, where yours is a high-latitude sample complete to B = 16, which contains many more stars. I suspect that the stars in our sample are intrinsically more luminous.

THEJLL: It is interesting that your bright, low-latitude sample of sdO, analyzed with our technique, gives the same answer as higher latitude samples. The 'RS' sample of stars contains KPD stars and as explained in the text we also get an 'old disk' answer when we restrict that sample to low latitudes. This indicates to us the absence of essentially 'young' sdO stars anywhere - disk or halo.

DORMAN: What is the number ratio of sdO stars to sdB stars? The theoretical models give $\sim$ 120 Myr in EHB phase, and $\sim$ 30 Myr in the post-EHB (Agb-Manqué) phase. (The ratio being somewhere from 1:6 to 1:4 EHB: post-EHB phase). Second, are sdO stars always lower in gravity? Finally, the very lowest mass He-shell burning stars do not complete He burning (although they do complete H-burning) and so become He-rich white dwarfs.

THEJLL: 1.) The Palomar Green survey (complete to V $\sim$ 16 mag) contains $\sim$ 230 sdO and $\sim$ 680 sdB stars so the ratio is close to 1:3. The fast Agb-Manqué phase leads to somewhat high birthrates for sdOs relative to white dwarfs - and this may become an embarrassment to the planetary nebula $\rightarrow$ white dwarf evolution scheme. I think the sdO stars are not all in the AGB-Manqué phases of evolution. 2.) sdO stars tend to have gravities that are lower than sdB stars, but the range is larger than for the sdB.

HEBER: The sdB stars are ideal for space density and scale height determination, because they all have almost the same absolute visual magnitude. sdO stars show a much larger range in gravity and hence in M_v. Some of them are even as luminous as post-AGB stars. This means, one can see the more luminous sdOs to large distances than the less luminous ones. This bias has to be corrected for.

THEJLL: Yes, good point.

Late Type Companions to Hot sdO Stars

By RAUL JIMENEZ[1], PETER THEJLL[2], REX SAFFER[3], AND UFFE G. JØRGENSEN[2]

[1] Nordita, Blegdamsvej 17, DK-2100 Copenhagen Ø, DENMARK
[2] Niels Bohr Institute, Blegdamsvej 17, DK-2100 Copenhagen Ø, DENMARK
[3] STScI, Baltimore MD, USA

We show how to use late type companions to hot subdwarfs as a tool for determining accurate sdO atmospheric parameters. For the sdO star Mrk509C we find a disagreement between the distance to its late type companion and the distance to the sdO implied by the NLTE analysis if the companion is assumed to be of luminosity class V. This disagreement can be interpreted as either a problem with the NLTE models, or with the assumption of the luminosity class V for the companion - it could also be a class VI subdwarf, and if this is assumed then perfect agreement between the spectroscopically determined log g and the value for log g inferred from the distance is obtained. The distance to the system is near 1100 pc in the case that the companion is of class V and near 500 parsec if the companion is of class VI.

1. Introduction

Like most other stars, the hot subdwarf B and O stars have close companions now and then (e.g., Thejll, MacDonald & Saffer 1991). In this paper we present the observations of an sdO star with an M-type close companion, and show how the two components can be analyzed. In the analysis we show how the parameters of the sdO star can be determined on the basis of CCD photometry of the M star, and how these parameters relate to the direct NLTE spectral analysis of the sdO star.

Hot subdwarf B and O stars are generally placed near the blue end of the Horizontal Branch (see Thejll 1994). The sdBs are probably half solar mass objects burning helium in the core while the sdO class contains hotter stars and may be a collection of objects in different late evolution stages, including helium core burning, in a range of total masses.

Analysis of the sdBs can be performed successfully with LTE models and photometric or spectral analysis. These objects are cool enough ($T_{eff} < 30,000$ K) that the Balmer jump contains information on log g and that the LTE assumption for atmospheric models and model spectra is justified. Comparison of the location of field and cluster sdB stars in the theoretical HR diagram with the location of the 0.5 $M_\odot$ He-core burning sequence in envelope masses uniquely identifies the evolutionary stage of these stars.

Analysis of the sdO stars is more complicated for two reasons: First, it is necessary to use NLTE models of the spectra to measure T_{eff}, log g, and composition and, second, the derived locations of the stars in the theoretical HR diagram do not uniquely identify their evolutionary stage.

Both these problems may, in part, be overcome by using data from companion stars when they occur. The accuracy of NLTE modelling and analysis methods can be gauged with independently determined distances, and the location in the HR diagram can be interpreted better if stellar masses are available - and they can be derived if accurate distances as well as reliable sdO gravities are at hand.

	R	I
SdO	13.44 ± 0.02	13.66 ± 0.02
Comp.	18.10 ± 0.1	17.12 ± 0.1

TABLE 1. Calibrated magnitudes for Mrk509C and its companion.

2. Observations and Data reductions

During an unrelated sequence of CCD observations of globular clusters, using the Danish 1.5 m telescope at La Silla, we (RJ, PT and UGJ) were using a set of standard stars that included a blue and hot subdwarf near the galaxy Markarian A, known as Markarian 509C or "Mark A" in Landolt (1992). Jorge Melnik of ESO pointed out to us that this star had a companion visible in IR images. In our own R and I-band images the star is clearly visible (see Figure 1) and even in bluer images there is an ellipticity that reveals the presence of the companion. The red and faint companion is about 1.4"

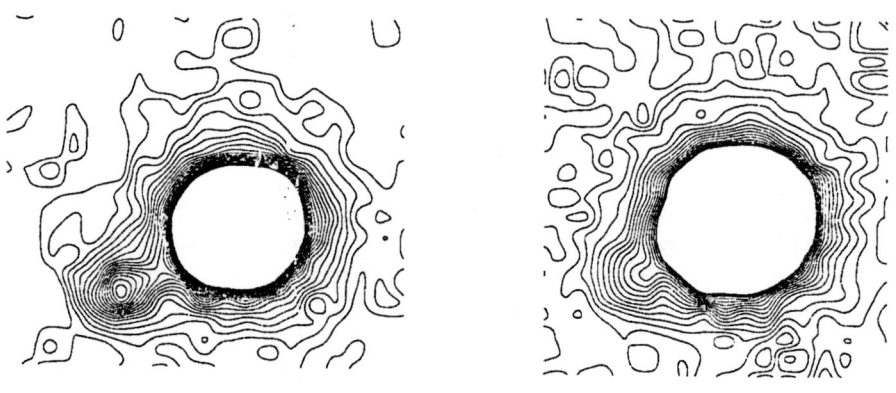

FIGURE 1. Contour plots of the R (right) and I-band (left) images for Markarian A. The companion is to the lower left. The frame is 23×20 pixels in size, or 4.6"×4". South is up and West is to the left.

NW of the blue and bright star and is thus only resolvable under good conditions.

Spectroscopy of the blue star (by RS) identifies it as a hot subdwarf O star. The strong He I lines and weak He II line at $\lambda 4686$ preliminarily set the temperature to between roughly 35,000 K and 40,000 K.

Using DAOPHOT under IRAF we measured the R and I band instrumental magnitudes in our CCD images of Mrk509C. Point spread functions were built from some of the roughly 10 stars in the field. Using the other calibration stars in our observing program we can reduce instrumental magnitudes to calibrated magnitudes - listed in Table 1.

3. Analysis

We identified the spectral type of the companion using the calibrated magnitudes, and calculated the distance to the companion using standard relations. Table 2 shows the results of this effort.

(1) R−I	(2) type	(3) M_V	(4) V−R	(5) M_R	(6) m_R-M_R	(7) D pc	(8) log g	(9) m 6.0	(10) m 5.5	(11) m 6.5
1.04	M1	9.5	1.4	8.1	9.9	950	5.3	2.3	0.7	7.4
0.94	M0	9.0	1.28	7.7	10.3	1150	5.2	3.4	1.1	11
0.84	K9	8.6	1.25	7.4	10.6	1320	5.0	4.5	1.4	14

TABLE 2. The columns contain the following: (1) R−I color observed. (2) Johnson (1966) spectral type from (1) assuming luminosity class V. (3) Absolute V magnitude from (1). (4) V−R color from (1). (5) Absolute R magnitude from (3) and (4). (6) Distance modulus from (5) and the measured R magnitude (R = 18.1±0.1) of the companion. (7) Distance in parsecs to the companion, using (6). (8) log g of the sdO assuming (7) and knowing the apparent magnitude of the sdO (V = 13.258, Landolt 1992) and the Eddington flux of the best fitting model ($H_{\nu=5500Å} = 0.0004$ erg/cm²/s/Hz/sr - rather independently of the assumed gravity) and assuming that the sdO mass is 0.5 $M_\odot$. (9), (10) and (11) are the masses of the sdO assuming (7) and the gravities log g = 6.0, 5.5 and 6.5 respectively.

In this table we also calculate the gravity of the sdO given knowledge of the best fitting model (see below) and assuming its mass is 0.5 $M_\odot$. Also shown in the table are those masses of the sdO that are required if various values for log g of the sdO are adopted. These calculations naturally assume that the two stars are true companions.

The relationship used to find the distance (D) to the sdO is based on: $4\pi H_\nu R^2 = f_\nu D^2$ where H_ν is the model Eddington flux of the best fitting model and f_ν is the observed flux of the sdO star. Inserting $g = \frac{GM}{R^2}$ in this, we get, with D in units of parsec and g in cgs units

$$D = 1.33\,10^{-5}\sqrt{\frac{M/M_\odot}{g}\frac{H_\nu}{f_\nu}}. \quad (3.1)$$

The best fit for the optical spectrum of Mrk509C is 41,000 K and log g = 6.0 ± 0.3 at 95 % He (and 5 % H). The NLTE models and analysis methods are described by Thejll (these proceedings). The error given on log g above is the 3-parameter joint 1 sigma level. The error on the abundance is a few percent, and negligible on the temperature as the He I lines change strength very quickly near 40,000 K. Figure 2 shows the 1, 2, and 3 σ joint levels of confidence for 3 fitting parameters in a T_{eff} vs. log g plane, shown at an abundance of 95 % He. For those other abundances at which the best fit falls inside the 1 σ volume in T_{eff}, log g, and % He space we have plotted the best fits as crosses only, excluding the corresponding contours.

4. Discussion

The data shown in Table 2 raises some interesting points. First of all it is clear that because the colors of late type stars change so quickly with spectral class even moderately accurate R−I measurements can lead to quite accurate classifications that in turn lead to distance estimates of interestingly high accuracy when compared to the relatively modest accuracies possible when determining atmospheric parameters of hot subdwarfs. With better data it will be possible to improve on the ± 0.15 accuracy possible for companion-determined log g of the subdwarf - it should be pointed out that ± 0.2 in log g for a hot subdwarf O stars, based on NLTE modelling, is good work indeed. If a simple spectrum

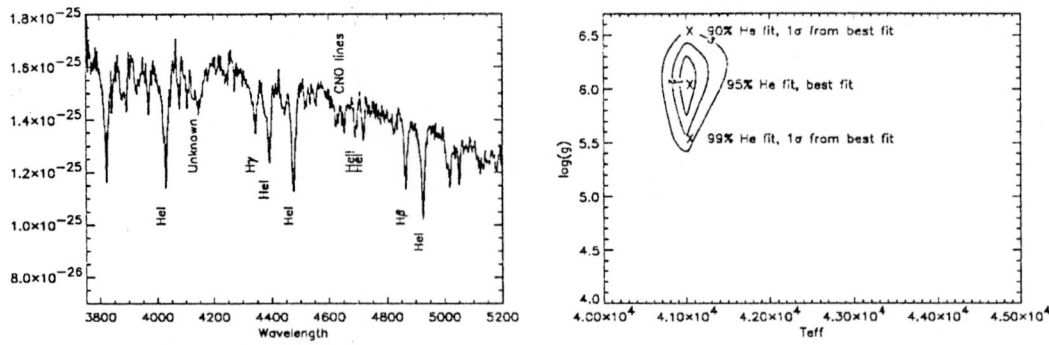

FIGURE 2. Medium resolution spectrum of Markarian A (on the left). Some photospheric lines are marked, but the spectrum is heavily contaminated by other lines, either photospheric in nature or from another source in the line of sight. The lines are not attributable to lines from the photosphere of the M0 companion. On the right are the 1, 2 and 3 σ joint 3-parameter contours around the best fit of a NLTE analysis at 95 % He. Also marked are the best fits at 90 and 99 % He, with the qualities of those fits indicated.

could be obtained of the companion it should be possible to quickly confirm or rule out membership to such late classes as K9 since strong, easily-detected, molecular bands are present in such late type stars.

Secondly, we compare the sdO gravities derived in the table to the gravity found in the NLTE analysis of the optical spectrum. If the two stars are close to each other, and if all the analysis has been done correctly then the two gravities should agree. As it is they do not, and that leads to three possibilities:

(a) The stars may not be close to each other
(b) The distance to the companion is not correct
(c) The NLTE analysis of the sdO is in error

The field we have is about 2 arc minutes by 2 arc minutes and there are only about 10 stars in this field. Two of them are Mrk509C and its companion which is 1.4 arc seconds away. It is therefore likely that the closeness of the two stars indicate a true binarity.

The distance to the companion is based on an assumption about its luminosity class. The luminosity class (V) assumed could be wrong - the star could be a giant star and very far away indeed or it could be a compact object close to us - a white dwarf for instance. However, by far most stars in late classes such as M and K are main sequence objects due to their very long lifetimes. As the stars are in the halo the companion could be a class VI (subdwarf) star, and in that case the absolute magnitude would be higher and the object closer. If the classical subdwarfs are about 2 magnitude below the main sequence the influence is to reduce the distances and increase the required gravity of the sdO (assuming its mass is 0.5 $M_\odot$) to log g = 6.0 $^{+0.1}_{-0.2}$, in perfect agreement with the spectroscopic value for log g. As you move up into the halo the classical subdwarfs do of course tend to become more abundant in space than the disk Population I objects and possibility (b) should therefore be carefully considered. Spectroscopic classification of the close companion is required to finally determine its type and luminosity class. Due to the closeness of the bright subdwarf, IR spectroscopy may be the best method.

Lastly there is the possibility that the NLTE analysis performed on the sdO optical spectrum is in error somehow. Either there is a problem in the model spectra themselves or the analysis method is in error.

NLTE models are difficult to make, and there is no such thing as a "standard tool" for NLTE model-making. No comprehensive comparison of hot models made with different

NLTE codes has been presented, and an analysis of the errors possibly introduced by the fitting methods chosen does not exist either. We use least-chi square methods where selected portions of line profiles are fitted against model profiles. Other workers (e.g., Dreizler et al. 1990) use methods based on equivalent width fitting as well as profile fits. For this and other reasons the uncertainties on the derived atmospheric parameters found by NLTE analysis of sdO stars are today not well established, and it is therefore particularly interesting to develop alternative methods for deriving these parameters independently of the NLTE-tools.

5. Conclusion

We have found that the hot sdO Mrk509C and its companion (probably of type K9 - M1) are physical binaries, that the distance to the companion is near 1100 parsecs, and that this corresponds to a gravity of 5 to 5.3 for the sdO, assuming an sdO mass of 0.5 $M_\odot$ and luminosity class V for the companion, which is in poor agreement with the best estimate of the gravity for the sdO from a NLTE analysis of the optical spectrum. However, with the assumption that the companion is of class VI, i.e., a classical subdwarf, the agreement is perfect: The distance is near 500 parsecs and the implied gravity for the sdO, assuming a mass of 0.5 $M_\odot$, is $\log g = 6.0^{+0.1}_{-0.2}$ in good agreement with the NLTE analysis of the sdO spectrum.

It is possible that the models and fitting method used for our NLTE analysis are giving gravities that are too large by about 0.8 dex, if we believe that the two stars are physically linked and if a mass for the sdO of 0.5 $M_\odot$ is assumed. If we allow larger masses for the sdO then the fit at $\log g = 6.0$ becomes acceptable at an sdO mass of 2 - 3 $M_\odot$. Data on sdO masses are sparse, but Ritter (1990) lists several estimates of sdO masses from binary systems ranging up to 2 $M_\odot$.

By determining the exact spectral type and luminosity class (i.e. absolute magnitude) of the companion to Mrk509C it will become possible to narrow down the mass interval acceptable for the sdO and this will make it easier to say whether we can trust the NLTE analysis or not.

We note finally that work on companions not only to sdO stars but also to the cooler sdB stars can be used to measure the field sdB masses with probably enough accuracy to answer the question whether they have the canonical mass of 0.5 $M_\odot$. It is also worth noting that any cluster sdBs with companions could be studied in this way in order to test whether the rather low cluster sdB masses (0.2 - 0.3 $M_\odot$) reported by Moehler et al. are real.

We acknowledge support from the EEC (grant no:920014) and from the Carlsberg Foundation. Valuable comments from Jorge Melnik (ESO, La Silla), Ana Ulla and Bertrand Plez (both NBI), and Michael Andersen (Astronomical Observatory, Copenhagen) are gratefully acknowledged.

REFERENCES

Dreizler, S., Heber, U., Werner, K., Moehler, S., de Boer, K. S., 1990, A&A, 235, 234-241

Johnson, H. L., 1966, ARAA, 4, 193-206

Landolt, A. U., 1992, AJ, 104, 340

Moehler, S., Heber, U., de Boer, K. S., 1994, in Adelman S. J., Upgren A. R., Adelman, C. J., eds, Hot Stars in the Halo, Cambridge University Press, Cambridge, p. 217

RITTER, H., 1990, Catalogue of cataclysmic binaries low-mass x-ray binaries and related objects
THEJLL, P., 1994, Analyzing the helium-rich hot sdO stars in the Palomar Green survey. in Adelman S. J., Upgren A. R., Adelman C. J., eds, Hot Stars in the Halo, Cambridge University Press, p. 197
THEJLL, P. A., MACDONALD, J., SAFFER, R. A., 1991, A&A, 248, 448-452

Discussion

CARNEY: I found this paper on infrared detection to be very clever and potentially very useful for distance estimates, as they noted. Has anyone done systematic work on precision radial velocities for sdO/sdB stars looking for binaries? Morse has shown that using synthetic spectra and modest - S/N echelle spectra that 2 km s^{-1} precision is quite feasible.

GARRISON: Some people are starting to look at a couple of the apparently F supergiants which are binary, and the masses seem to be high. These would be the progenitors of the sdO stars.

Hot Stars in Globular Clusters †

By SABINE MOEHLER[1]‡, U. HEBER[2]
AND K. S. de BOER[3]

[1]Landessternwarte, Königstuhl, D 69117 Heidelberg, FRG

[2]Dr. Remeis-Sternwarte, Universität Erlangen-Nürnberg, Sternwartstraße 7, D 96049 Bamberg, FRG

[3]Sternwarte der Universität Bonn, Auf dem Hügel 71, D 53121 Bonn, FRG

In many globular clusters one can find gaps in the CMD along the blue horizontal branch. There have been many explanations proposed for this long known phenomenon one of which also severely affects work done on hot subdwarfs in the field: The stars below the gaps may not be BHB but sdB stars produced by white dwarf mergers (Iben 1990). Such an origin for sdB stars could also apply to field sdB/sdOB stars but predicts a very different mass distribution than the classical EHB model (Heber 1986). To find out which of the two models holds true we started observations of the BHB stars in the clusters M 15 and NGC 6752. From spectroscopic and spectrophotometric data we determine their physical parameters and thus can see, whether the stars below the gap are sdB stars or still BHB stars. We find that in M 15 the gap separates BHB from BHB stars while in NGC 6752 it separates HB and sdB stars.

Knowing the distances of the stars we then derive their masses. The masses we determine this way for the HB stars are significantly lower than predicted by HB theory. We compare our results to data from the literature and discuss the implications.

1. Introduction

The most commonly known hot stars in globular clusters are stars belonging to the blue horizontal branch (BHB). The phenomenon of gaps in this region of the CMD has been detected quite a long time ago (Racine 1971). Nevertheless no real answer to the questions, where these gaps come from and why they are at different positions along the HB, has been found up to now. Many possible explanations for the gaps have been suggested during the last years:

(*a*) The gap is created by diverging evolutionary paths from the ZAHB, turning a uniform distribution on the ZAHB into a bimodal one (Newell 1973, Lee *et al.* 1988).

(*b*) The HB is a mixture of two different groups of stars (differing for example in core rotation, abundances, ..., Rood & Crocker 1985). In this case one would expect gaps in other regions of the CMD as well, mainly in the RGB region.

(*c*) A more extreme variety of (b) is the idea that the stars separated by the gap are truly different, i.e. the stars on the blue side of the gap are not BHB stars at all, but were produced by other mechanisms than those, e.g., merging of two helium-rich white dwarfs (Iben 1990).

The latter scenario was much discussed during the recent years as it leads to stars with physical parameters similar to those of the sdB and sdO stars in the field of the Milky Way and, therefore, has consequences beyond the question of the gaps in globular cluster CMDs. The main difference between the merger and the classical EHB (Extended Horizontal Branch, Heber 1986) model is the mass distribution predicted by the two

† Based on observations obtained at the European Southern Observatory, La Silla, Chile
‡ Visiting Astronomer, German–Spanish Astronomical Center, Calar Alto, operated jointly by the Max-Planck-Institut für Astronomie, Heidelberg jointly with the Spanish National Commission for Astronomy

models. The EHB hypothesis predicts a sdB mass of (0.5 ± 0.02) M$_\odot$, while the merging model predicts a mass varying between 0.3 and 0.7 M$_\odot$, where the stars with the lower mass live considerably longer than the heavier ones (Iben 1990).

2. Observations and Reduction

For both clusters spectra of intermediate and low resolution were obtained to derive the surface gravities from Balmer line profile fits and the effective temperatures from the Balmer jump and the continuum slope. The observations and their reduction will be described in detail elsewhere; below we give a brief summary.

2.1. M 15

We took intermediate resolution spectra of blue stars in M 15 during 1989 – 1992 with the 3.5m telescope of the Calar Alto Observatory using the Cassegrain Twin Spectrograph. We used a dispersion of 72 Å mm^{-1} and a slit width of 2". The setups covered about 3850 – 5000 Å in the blue channel with a resolution of about 3.7 Å. The red spectra are not used as there were several problems with the red channel during those years. In 1992 we also recorded low resolution spectra with a dispersion of 144 Å mm^{-1} covering 3400 – 5300 Å. Here we used a slit width of 3.6" and achieved a resolution of about 6 Å. For these spectra we took special care to orient the slit along the atmospheric dispersion direction to minimize its effects and loose as little flux as possible in the UV.

2.2. NGC 6752

In 1992 we took intermediate and low resolution spectra of the blue stars in NGC 6752 with the 3.6m telescope at La Silla using EFOSC1. For the intermediate resolution spectra we used a dispersion of 120 Å mm^{-1} and a slit width of 1" resulting in a resolution of 4.7 Å. These spectra cover 3800 – 5600 Å. The low resolution spectra were taken with a dispersion of 220 Å mm^{-1} and a slit width of 5" resulting in a resolution of 24 Å. The spectral coverage is 3200 – 6500 Å.

2.3. Reduction

All frames were corrected for bias and pixel-to-pixel variations. In some cases we used RCA chips and, therefore, had to correct the column structure. If the dark current was below 5 counts/pix/hr we omitted dark correction. The wavelength calibration was done two-dimensionally to allow a correct sky-subtraction. The sky was fitted two-dimensionally using interactively defined sky regions. The spectra were extracted using Horne's algorithm (Horne 1986). All spectra were corrected for atmospheric extinction and relatively flux calibrated using flux standard stars.

3. Analysis

To determine effective temperatures and surface gravities of our stars we used fully line-blanketed LTE model atmospheres, namely the ATLAS6 models of Kurucz (1979) for $T_{eff} < 20,000$ K and Kiel models for the hotter stars. The Kiel models consist of a grid of model atmospheres and line profiles that was computed for sdB field stars with codes described by Heber et al. (1984) and references therein. The codes have been modified to allow for full metal line blanketing using the opacity distribution functions of Kurucz (1979). For the Kiel as well as the ATLAS6 models, solar metal abundances have been adopted, while the helium abundance is solar in the ATLAS6 and 1/10 solar in the Kiel models. The latter value appears more appropriate for the hot blue stars in

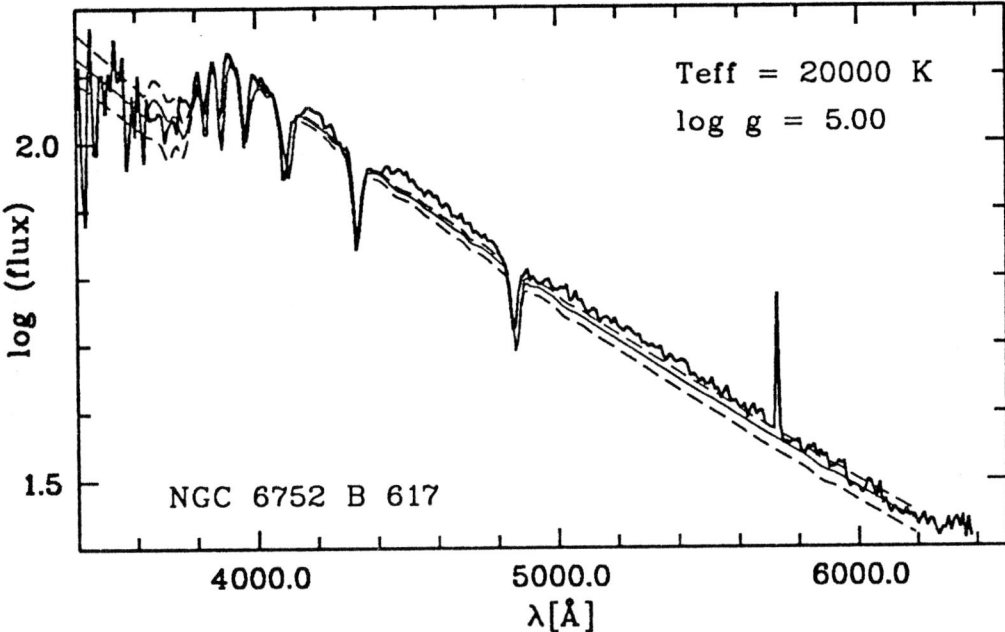

FIGURE 1. The low resolution spectrum of B 617 (V = 17^{m}84) in NGC 6752. The solid line represents a Kiel model with T_{eff} = 20,000 K and log g = 5.00; the dashed lines are models with 18,000 K and 22,000 K, respectively

globular clusters. The influences of the metal lines as well as the new ODF tables of Kurucz (1992) on the flux distribution and the Balmer line profiles remain to be checked.

The intermediate resolution spectra were corrected for velocity shifts, then coadded and normalized by fitting a third-order spline to the continuum. To allow a comparison between model atmospheres and observed data we convolved the model profiles with the resolution of the spectra. Then we computed for each of the blue Balmer lines (H_β, H_γ, H_δ) the squared difference between the observed spectrum and the theoretical line profile within ± 40 Å around the line center. We used the sum of these differences as estimator for the quality of the fit. Taking the five smallest values gives a range of effective temperature and surface gravity.

Taking the (T_{eff}, log g) combinations found by the method described above we tried to fit the Balmer jump and the continuum of the low resolution spectra. As can be seen from Figure 1 we had some problems with the low resolution data because the red continuum seems to be somewhat too high for the adopted temperature. As we see this effect also in the spectra of field stars taken with the same setup at the same run we think that the possibility of residual red light due to the background stars of the cluster can be ruled out. Any cool companion should also show up in the intermediate resolution spectra leading to a flattening of the continuum towards redder wavelengths which we do not observe.

Lowering the effective temperature in order to fit the red continuum would not only lead to contradictions with the observed Balmer jump but also with the fits obtained from the intermediate resolution spectra: As it turned out, the fits of the intermediate resolution spectra on the average predicted a range extending to higher temperatures than the fits of the low resolution spectra to the Balmer jump. As a compromise T_{eff}

number	T_{eff} from Balmer jump T_{eff}	log g	M_V	M [$M_\odot$]
1	18000	4.25	2.45	0.267
20	17000	4.25	2.35	0.322
208	16000	4.25	2.65	0.269
276	18000	4.25	3.12	0.144
421	18500	4.50	3.12	0.246
574	18000	4.25	2.44	0.270
686	17000	4.25	3.02	0.174

number	T_{eff} from Q index T_{eff}	log g	M_V	M [$M_\odot$]
27	12000	3.50	1.22	0.288
258	11000	3.25	1.10	0.213
325	16000	4.00	1.38	0.486
348	12000	4.25	1.27	1.540

TABLE 1. Results in M 15. The numbers of the stars refer to Buonanno et al. (1983) and Battistini et al. (1985)

was identified with the temperature of the model that fitted the Balmer jump best and was within the lowest error models of the intermediate resolution fits. The surface gravity was simultaneously obtained from the fits to the intermediate resolution spectra.

For some stars no low resolution spectra are available but only UBV photometry by Buonanno et al. (1983). In these cases we have chosen the reddening-free Q index as temperature indicator by comparing it to the theoretical values of Kurucz (1979), which define one curve in the (T_{eff}, log g)-plane. The Balmer line profile fits provide another curve when we use the fits with minimal errors. For three stars (B 18, B 440, and B 484) this method does not work as these stars are too cool. Below about 9500 K the Balmer line profiles are insensitive to log g. Therefore, we omitted these stars from further analysis for now.

We assume that using these methods we get an error in log T_{eff} of about 0.035 (corresponding to ± 1500 K at T_{eff} = 18,000 K). For log g we take half the spacing between two points in our model grid as formal error, which is 0.13. The results for M 15 and NGC 6752 are given in Table 1 and 2, respectively. In Table 2 we also list the results from Heber et al. (1986), but with numbers for the stars taken from Buonanno et al. (1986).

4. Results

4.1. Physical parameters

The results of the analysis are plotted in Figure 2. Also shown are the results from Heber et al. (1986) for NGC 6752. Up to about 20,000 K all stars tend to lie above the ZAHB, a phenomenon already known from field sdB stars (Schmidt et al. 1992; Moehler et al. 1990).

One can see that all stars in M 15 are below about 20,000 K and, therefore, are

	T_{eff} from Balmer jump			
number	T_{eff}	log g	M_V	M [M$_\odot$]
210	22500	5.25	3.66	0.629
617	20000	5.00	4.46	0.204
1288	22500	5.00	4.52	0.158
1509	16000	4.25	2.14	0.430
2126	25000	5.25	3.90	0.415
3655	17000	4.25	3.02	0.174
4380	22500	4.00	2.55	0.092

	T_{eff} from IUE spectra			
number	T_{eff}	log g	M_V	M [M$_\odot$]
4719	15000	4.00	2.27	0.238
1754	40000	5.00	2.61	0.315
331	25000	5.60	3.74	1.099
763	26000	5.50	3.75	0.797
5151	23500	5.20	3.84	0.437

	T_{eff} from Q index			
number	T_{eff}	log g	M_V	M [M$_\odot$]
577	12700	3.90	1.47	0.519
2454	9700	3.50	0.89	0.597
4104	16000	4.00	1.82	0.324

TABLE 2. Results in NGC 6752. The numbers of the stars refer to Buonanno et al. (1986)

BHB stars according to their physical parameters independently of their position in the color-magnitude diagram. The gap is not convincingly reproduced in the (T_{eff}, log g)–diagram but the stars B 18, B 440, and B 484 lie on the cool side of the gap as was expected from their position in CMD. The apparent gap in Figure 2 at log $T_{eff} \approx 4.15$ is due to observational selection effects.

The gap of the HB in NGC 6752 is well reproduced in the diagram and separates HBB from sdB stars. While the HBB stars are displaced from the ZAHB in the same way as the blue stars in M 15 the sdB stars lie rather close to the ZAHB (in contrast to what we found in the field (cf. Moehler et al. 1990)).

4.2. Masses

Knowing the surface gravities and the effective temperatures of the stars we can determine the visual brightness at the surface of the star using model atmospheres (Kurucz 1979) and, combining it with the measured apparent visual brightness, the angular diameter of the star. (At first we had used the Bolometric correction to obtain the absolute visual magnitudes of our stars but as this is a somewhat tricky parameter we changed to the method described above.) As we also know the distances of the stars we then can derive the physical diameter of the star and from that deduce its mass. The apparent visual magnitudes of the stars are taken from the photographic photometry of Buonanno et al.

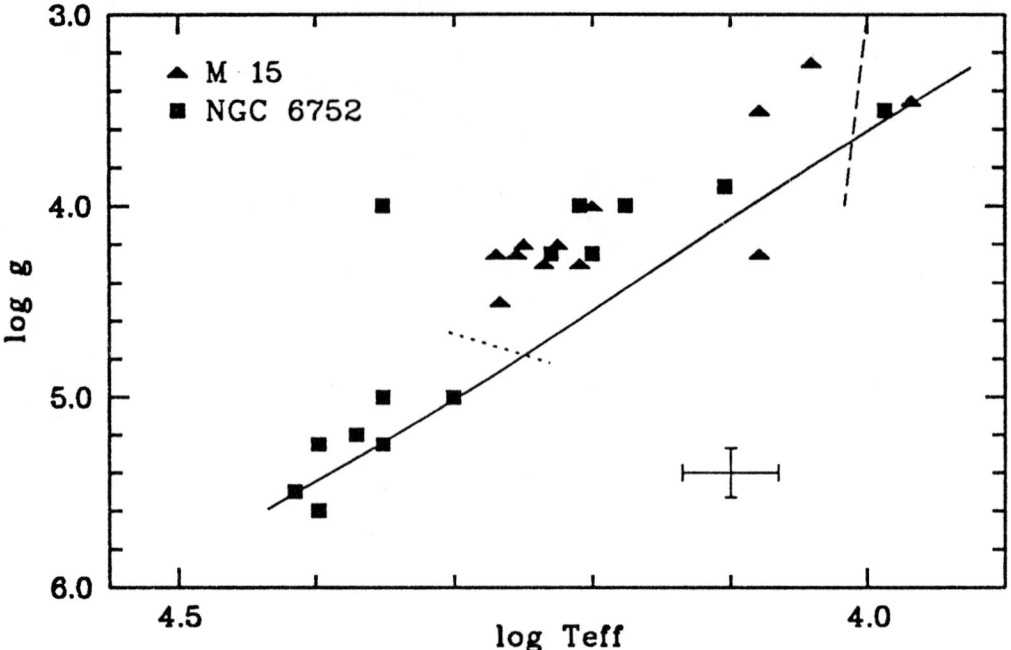

FIGURE 2. (T_{eff}, $\log g$)-plot of our results for M 15 and NGC 6752. The solid line represents a ZAHB model taken from Sweigart (1987) for $Y_{MS} = 0.25$ and $Z_{MS} = 10^{-4}$. The dashed line represents the gap in the CMD of M 15, the dotted line the one of NGC 6752.

(1983, 1986) and Battistini et al. (1985). Taking into account the errors of our results and assuming errors of $0^{m}\!.1$ for distance moduli and visual brightnesses and $0^{m}\!.02$ for $E_{(B-V)}$ we get an internal error of about 50 % for the individual masses.

The values taken for the distance moduli of M 15 and NGC 6752 can be found in Table 3, the results are listed in Tables 1 and 2 and plotted in Figure 3. As can be seen from the table the masses are rather low. Comparing them in Figure 3 to theoretical ZAHBs taken from Sweigart (1987) dramatically shows that the derived masses are sytematically below the predicted values for almost all stars below 20,000 K. A decrease in mass with increasing temperature is predicted by the standard HB models but we find a much steeper slope of this relation than would be expected. Again, the sdBs in NGC 6752 differ from the rest, showing a very large scatter around the ZAHB instead of any systematic trend. It should, however, be noted that the sdB stars in NGC 6752 as well as the hotter stars in M 15 are the faintest ones and part of the scatter may be due to observational errors in the photographic photometry used.

4.3. Data from other groups

In view of these results it seemed advisable to compare them to results obtained by other groups for this problem. Unfortunately, (to our knowledge) there are only two sources for such data: The paper by Crocker et al. (1988) and the work by de Boer et al. (1994). Crocker et al. deal with five globular clusters, namely M 3, M 5, M 15, M 92, and NGC 288 (of which M 5 does not show any gap along the BHB). Of these clusters we cannot use their data of M 15 as they did not identify their targets. Their data of M 3 are of no use to us since the photometry they used for this cluster is still unpublished. de Boer et al. did a mass determination for the BHB stars in NGC 6397 (which does not show any gap or EHB) using intermediate resolution spectra and Strömgren photometry.

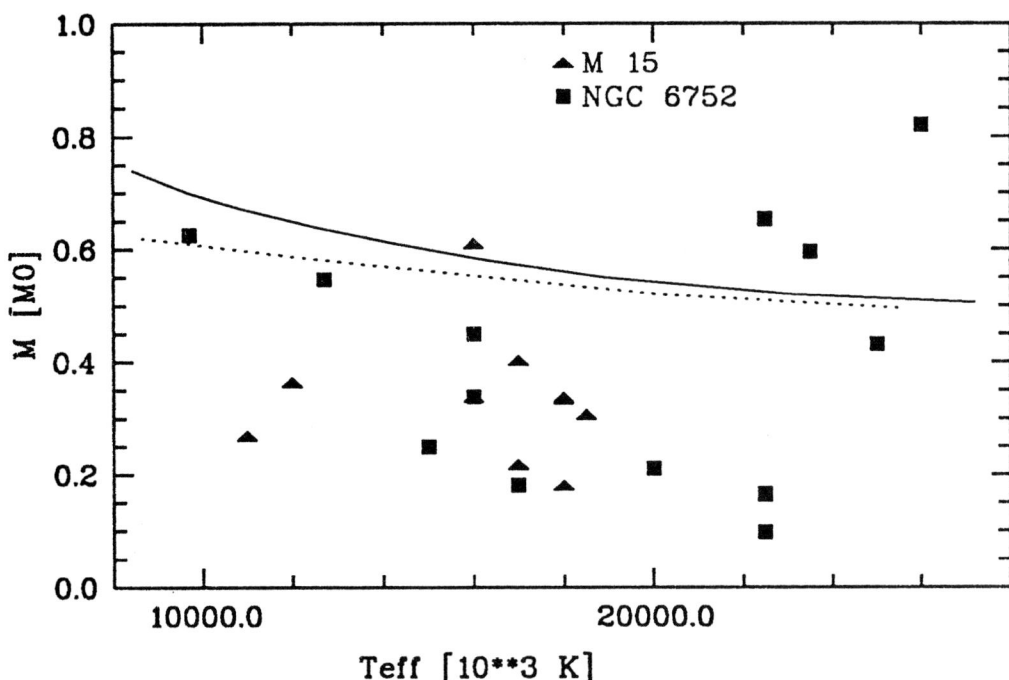

FIGURE 3. The resulting masses for the stars in M 15 and NGC 6752, plotted against T_{eff}. The solid line represents a ZAHB model taken from Sweigart (1987) for $Y_{MS} = 0.25$ and $Z_{MS} = 10^{-4}$, the dashed one represents a ZAHB model with $Y_{MS} = 0.25$ and $Z_{MS} = 10^{-3}$.

Cluster	(m-M)$_{V,0}$ [mag]	E_{B-V} [mag]	[Fe/H]	V_r [kms^{-1}]	σ_0 [kms^{-1}]	log (ρ_0) [M$_\odot$pc^{-3}]	m$_{gap}$ [mag]
NGC 6752	13.25	0.05	-1.54	-32	4.5	5.2	+3.1
M 15	15.09	0.10	-2.17	-107	12.0	6.2	+1.0
NGC 288	14.67	0.01	-1.40	-46	2.9	2.1	+1.4
M 92	14.47	0.01	-2.24		5.9	4.4	+1.0
M 5	14.42	0.03	-1.40	+54	5.7	4.0	-
NGC 6397	11.71	0.18	-1.91	+18	4.5	5.3	-

TABLE 3. Cluster parameters used for the mass determinations

The physical parameters of the blue stars in M 5, M 15, M 92, NGC 288, NGC 6397, and NGC 6752 are plotted in Figure 4. In this figure one can see an indication of a low-temperature gap at log $T_{eff} \approx 4.08$. Much more obvious is the fact that NGC 6752 is still the only cluster where sdBs have been analysed so far. In all other clusters with a gap this gap is within the BHB. There seems to be a slight indication that the stars above the gaps as well as the stars in clusters without gaps lie closer to the ZAHB than the stars below the gaps (again except NGC 6752). The stars in M 5 (which does not show any gap in the CMD) that lie below the gap at log $T_{eff} \approx 4.08$ are also displaced from the ZAHB to lower gravities.

Using the physical parameters plotted in Figure 4, the photometric papers cited by Crocker et al. (1988) for the apparent visual magnitudes and the distance moduli and

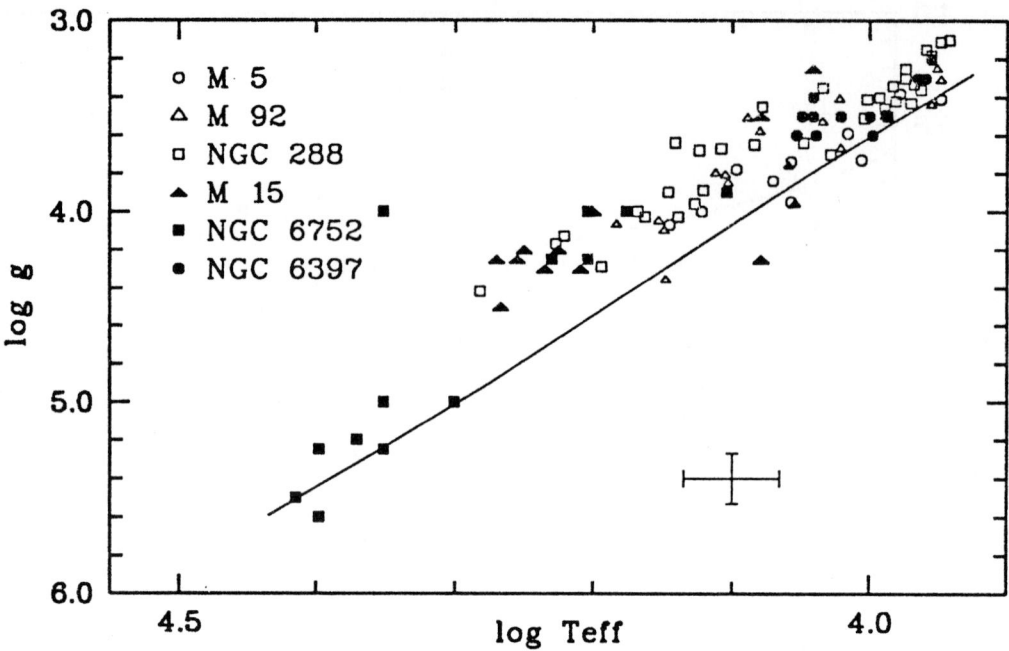

FIGURE 4. (T_{eff}, $\log g$) - plot of the results for M 15, NGC 6752 (this paper), M 5, M 92, NGC 288 (Crocker et al. 1988), and NGC 6397 (de Boer et al. 1994). The solid line represents a ZAHB model taken from Sweigart (1987) for $Y_{MS} = 0.25$ and $Z_{MS} = 10^{-4}$.

reddenings listed in Table 3, we derived the masses of the blue stars in M 5, M 92, and NGC 288. For the stars in NGC 6397 we used the results provided by de Boer et al. (1994). The results are plotted in Figure 5 and obviously confirm the trend we saw already in the data of M 15 and NGC 6752.

Masses have already been quoted before in the literature, especially by Crocker et al. (1988). However, these authors proceeded in a different way: First they determined the position of a star along the theoretical HB using V and (B-V) data and then used the mass predicted by theory for the position of this star. For M 92 their mass distribution extends down to about 0.60 $M_\odot$ (Crocker et al. 1988), while our method, based on their data, yields a mass distribution for the blue part of the HB that starts at about 0.55 $M_\odot$ and extends down to 0.2 $M_\odot$, which is a clear contradiction to the result found by the "projection" method. In our opinion this is due to the fact that using this "projection" method does not produce independent results but only results consistent with theory: you put theory in and you get the theory out again!

5. Discussion

The masses we found are in most cases too low for normal horizontal branch stars, because they are smaller than the canonical mass for the core helium flash. On the other hand these stars seem to be too cool to be explained by Iben's merger model. Here, we like to briefly discuss possible reasons for the mass discrepancy. To compensate for the offset in mass the photometric error of the visual brightnesses would have to be about $0^m\!.5$. An error in distance of 25 % would be necessary to shift the masses back to the ZAHB. Both errors, however, could not explain why the slope of the mass-temperature

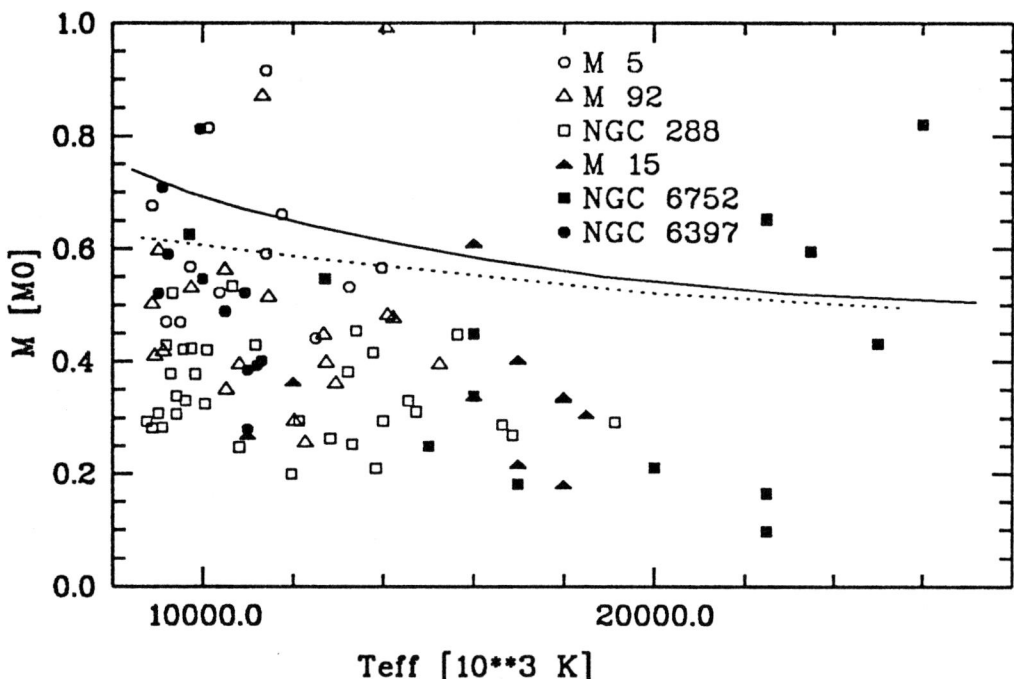

FIGURE 5. The resulting masses for the stars in M 15, NGC 6752 (this paper), M 5, M 92, NGC 288 (Crocker et al. 1988), and NGC 6397 (de Boer et al. 1994) plotted against T_{eff}. The solid line represents a ZAHB model taken from Sweigart (1987) for $Y_{MS} = 0.25$ and $Z_{MS} = 10^{-4}$, the dashed one represents a ZAHB model with $Y_{MS} = 0.25$ and $Z_{MS} = 10^{-3}$.

relation is so much steeper than expected. Anyway it seems highly unlikely that five different (and mostly well studied) globular clusters would have such large errors in their photometric data.

To shift the masses back to the ZAHB an increase in gravity by about 0.2 dex or a decrease in temperature by about 4500 K (for $T_{eff} = 18,000$ K) would be necessary. An error in T_{eff} or $\log g$ would almost be compensated by the corresponding change in $\log g$ and T_{eff}, respectively. In addition it would be necessary that four different determinations of T_{eff} and $\log g$ all have the same systematic error. Since the methods employed by the four groups differ from each other it is rather improbable that the error should result from the way in which the physical paramters were determined. As all groups used solar metallicity models for the stars with $T_{eff} > 10,000$ K the remaining possibility could be that the effect of the metallicity on the determination of the physical parameters is much stronger than thought up to now. In this case the effect would have to be strongest around 15,000 K to 20,000 K and then almost vanish for the sdB/sdOB stars. We will investigate this question using the Kurucz (1992) model atmospheres that have become available recently.

That the sdBs in NGC 6752 fit the theoretical ZAEHB quite well although the BHB stars in this clusters show the same deviations as the BHB stars in the other clusters is very confusing. It would be very interesting to see how sdB stars in other clusters behave in the (T_{eff}, $\log g$)-diagram. So far the masses of the sdB stars in NGC 6752 just show a rather large scatter but no trend. The mean of their masses is (0.52 ± 0.35) $M_\odot$ which allows no final decision between the model of Iben (1990) and the EHB model.

6. Summary

We have analysed stars in M 15 and NGC 6752 above and below the gaps along the BHB. We have found that in M 15 the gap lies within the BHB and in NGC 6752 it separates BHB and sdB stars. In the (T_{eff}, $\log g$)–diagram the BHB stars in both clusters are displaced from the ZAHB in contrast to the sdB stars in NGC 6752. The masses for the BHB stars in both clusters are significantly lower than predicted by canonical HB theory and the slope of the mass-temperature relation is much steeper than predicted by theory. This holds true also for the clusters M 5, M 92, NGC 288 (Crocker et al. 1988) and NGC 6397 (de Boer et al. 1994).

The sdB stars in NGC 6752 do not fit in this picture as they stick rather close to the ZAEHB and their masses exhibit a large scatter instead of any trend with temperature.

We want to thank the staff from the Calar Alto and the La Silla Observatories for their support during our observations. S. M. thanks the Deutsche Forschungsgemeinschaft for financial support through grants Bo 779/5 and Mo 602/2.

REFERENCES

BATTISTINI, P., BREGOLI, G., FUSI PECCIO, F., LOLLI, M., 1985, A&AS, 61, 487

BUONANNO, R., BUSCEMA, G., CORSI, C. E., IANNICOLA, G., FUSI PECCI, F., 1983, A&AS, 51, 83

BUONANNO, R., CALOI, V., CASTELLANI, V., CORSI, C. E., FUSI PECCI, F., GRATTON, R., 1986, A&AS, 66, 79

CROCKER, D. A., ROOD, R. T., 1985, in Philip A. G. D., ed, Horizontal Branch and UV Bright Stars, L. Davis Press, Schenectady, NY, p. 107

CROCKER, D. A., ROOD, R. T., O'CONNELL, R. W., 1988, ApJ, 332, 236

DE BOER, K. S., SCHMIDT J. H., HEBER U., 1994, in Adelman S. J., Upgren A. R., Adelman C. J., eds, Hot Stars in the Halo, Cambridge University Press, Cambridge, p. 277

HEBER, U., 1986, A&A, 155, 33

HEBER, U., HUNGER, K., JONAS, G., KUDRITZKI, R. P., 1984, A&A, 130, 119

HEBER, U., KUDRITZKI, R. P., CALOI, V., CASTELLANI, V., DANZIGER, J., GILMOZZI, R., 1986, A&A, 162, 171

HORNE, K., 1986, PASP, 98, 609

IBEN, I., JR., 1990, ApJ, 353, 215

KURUCZ, R. L., 1979, ApJS, 40, 1

KURUCZ, R. L., 1992, IAU Symp. 149, p. 225

LEE, Y.-W., DEMARQUE P., ZINN R., 1988, IAU Symp. 126, p. 505

MOEHLER, S., HEBER, U., DE BOER, K. S., 1990, A&A, 239, 265

NEWELL, E. B., 1973, ApJS, 26, 37

RACINE, R., 1971, AJ, 76, 331

ROOD, R. T., CROCKER, D. A., 1985, in Philip A. G. D., ed, Horizontal Branch and UV Bright Stars, L. Davis Press, Schenectady, NY, p. 99

SCHMIDT, J. H., MOEHLER, S., THEISSEN, A., DE BOER, K. S., 1992, in Heber U., Jeffery C. S., eds, The Atmospheres of Early-Type Stars, Lecture Notes in Physics, 401, Springer, Berlin, p. 254

SWEIGART, A. V., 1987, ApJS, 65, 95

Discussion

CARNEY: Have you tried observing cluster RR Lyrae stars? There you can utilize double-mode pulsators to estimate masses "directly".

MOEHLER: Sorry, no.

LANDSMAN: The sdB stars in NGC 6752 show a tight correlation with the Zero Age Horizontal Branch (ZAHB), yet your derived masses show a large scatter. Why is this?

MOEHLER: Their absolute visual magnitudes cover a range of about a magnitude and their log g values cover a range of 0.2.

BOND: What is the phenomenology of these gaps? Do they occur at the same absolute magnitude and are they statistically significant?

MOEHLER: They occur mostly at about $M_v = +1$ mag, but the one in NGC 6752 occurs at $M_v = +3.1$ mag. At least the gap in NGC 6752 has been found independently by two groups.

CACCIARI: We have performed a similar analysis using UV IUE data on a few BHB stars in M3 and NGC 6752 (see my paper on p. 282), and the log g-log T_{eff} diagram from the most recent Dorman et al. (1993, ApJ, 419, 596) models. In this diagram the evolutionary tracks cover a region where stars can be found depending also on parameters such as core mass, envelope mass, evolutionary rate etc. It may be difficult to derive a unique answer on the evolutionary status of a star from the log g - log T_{eff} diagram, since there are a few other parameters that complicate the situation.

MOEHLER: I agree with you, but this has no effect on the masses we determine, as we do not use any HB theory there.

PETERSON: D. Crocker has noted the difficulty of calibrating the fluxes/photometry of the very blue stars in globular clusters, since there are very few standards of similar blue type that are not themselves horizontal branch stars. The calibration thus may depend on which stars are used, and how well their own ultraviolet fluxes are known or modeled. Comments?

MOEHLER: As flux standards we used Feige 110, BD +28° 4211, and HZ 44. These stars are sdO's with temperatures well above 35,000 K. Due to the high effective temperatures the flux blueward of the Balmer jump is still high and should therefore be sufficient to well determine the response curve down to 3400 Å. Also for the sdB stars in NGC 6752 our temperatures and surface gravities and the ones derived by Heber et al. (1986, A&A, 162, 171) from IUE spectra give the same range of values.

THEJLL: Do you think your Bolometric correction is correct? On what is it based?

MOEHLER: We used solar-metallicity ATLAS6 (Kurucz, 1979, ApJS, 40, 1) models. To explain the offset in log M_v, the Bolometric correction would have to be wrong by 0.4 mag.

Faint Blue Stars from the Hamburg Schmidt Survey

By STEFAN DREIZLER[1]†, U. HEBER[1]†, S. JORDAN[2]† AND D. ENGELS[3]†

[1]Dr. Remeis-Sternwarte Bamberg, Universität Erlangen-Nürnberg, Sternwartstraße 7, D 96049 Bamberg, GERMANY

[2]Institut für Theoretische Physik und Sternwarte der Universität Kiel, D 24098 Kiel, GERMANY

[3]Hamburger Sternwarte, Gojenbergsweg 112, D 21029 Hamburg, GERMANY

We report on the follow-up spectroscopy of stellar candidates in the Hamburg Schmidt survey, an objective prism survey primarily aiming at quasars. More than 250 stellar candidates have been observed with medium spectral resolution of 3-5 Å. We demonstrate that this survey is a very rich source of faint blue stars in general and for very interesting individual objects in particular. The mix of spectral subclasses we find is similar to that of previous surveys, except that we find a larger fraction of helium rich sdO stars and very few DB white dwarfs.

The results of quantitative analyses of six stars are highlighted: Three binaries are of special interest. HS0824+2854 = PG0824+289 is a unique spectroscopic binary composed of a DA white dwarf and a dwarf carbon companion and may be the Rosetta stone for the nature of the dwarf carbon stars. HS0507+0435 and HS2240+1234 seem to be the first visual binaries composed of two DA white dwarfs that have been found by a faint blue star survey. The DAB star HS0209+0832 may give a clue to the white dwarf spectral sequences because it is the only white dwarf with a detection of He I from optical spectroscopy in the DB-gap. Two newly discovered PG1159 stars could be key objects for understanding the post-AGB – white dwarf connection, because HS0704+6153 turned out to be the coolest and, hence, most evolved PG1159 star known, while HS2324+3944 is an unusually hydrogen-rich PG1159 star of rather low gravity and may just start to enter the PG1159 phase of evolution.

1. Introduction

The success of the Palomar Green Survey (PGS, Green et al. 1986) evoked a "boom" of comparable surveys on several observatories. One of them is the Hamburg Schmidt Survey (HSS) which is carried out with the 80cm Calar Alto Schmidt telescope (Engels et al. 1988). Although it was primarily initiated as a quasar survey, it is no surprise that it is also a very rich source for faint blue stars. Unlike the PGS and most other surveys, objective prism spectra (1390 Å mm^{-1} at H$_\gamma$) are obtained in a similar way as in the Case Blue Survey (CBS) (Pesch & Sanduleak 1983). The major difference between the two surveys is the procedure of candidate selection, which is done by visual inspection in the CBS, whereas the HSS plates are digitized in Hamburg with a PDS 1010G microdensitometer. A search software selects from the 30-50 000 spectra per plate in the magnitude range $13\overset{m}{.}5$ to $18\overset{m}{.}5$ quasar candidates as well as faint blue star candidates on the base of blue continua and/or emission lines (Hagen 1987; Hagen et al. 1987; Hagen et al. 1994). The discrimination between quasar and stellar candidates is done by classifying the density spectra on a vector graphics terminal in Hamburg, the further follow-up observations and analyses are shared in a collaboration between the

† Visiting Astronomer, German–Spanish Astronomical Center, Calar Alto, operated by the Max-Planck-Institut für Astronomie Heidelberg jointly with the Spanish National Commission for Astronomy

institutes in Hamburg, Kiel, and Bamberg, where the colleagues in Hamburg work on the extragalactic objects and the stellar component is analysed in Bamberg and Kiel. The HSS allows objects redder than the colour cut-off of the PGS to be found and many DA white dwarf with Balmer lines close to maximum line strength (i.e., temperatures in the range 10,000 K to 15,000 K) can easily be identified already from the Schmidt spectra.

The sky coverage of the HSS is now comparable to the PGS, it is, however, up to two magnitudes deeper. The semi-automatic candidate selection from objective prism spectra makes the HSS very efficient and can provide a substantially deeper and statistically complete sample of hot subluminous stars and white dwarfs.

The motivation for the follow-up spectroscopic analysis of the stellar component of the HSS are manifold and cover a wide range of astrophysical questions.

Many details of the late phases of stellar evolution are still poorly understood. The most striking problem is the existence of two spectroscopic sequences among the white dwarfs and the subluminous B and O stars. In both cases hydrogen-rich (DA; sdB/sdOB) and helium-rich (DO, DB; sdO) subclasses exist. It is not yet clear, if these two spectroscopic sequences are, indeed, due to a disjunct evolution or if transitions occur during the evolution as discussed by several authors (see e.g., Sion 1986) for the white dwarfs and by Caloi (1989) for the subdwarfs. Detailed analyses for many objects are therefore required to answer this question.

Until recently, subdwarf B and O stars have been regarded as "rare" objects, the results of the PGS indicated, however, that they actually dominate the population of blue stars at high galactic latitudes down to at least $18^{m}.0$. Their origin is still under debate. Three alternative scenarios have been proposed. First, these objects could be formed by the merging of two low mass helium degenerates (Iben & Tutukov 1984). The other alternatives propose an enhanced mass loss on the RGB either in single star evolution (Caloi 1989) or by mass transfer to a companion (Mengel et al. 1976, see also Moehler et al. 1994, for a discussion)

Rare subtypes of white dwarfs like the hot helium rich DOs or the magnetic white dwarfs are still poorly understood since only a few candidates are known. The HSS provides several objects of these subclasses. The PGS lead to the discovery of two previously unknown classes of stars, the sdOD and the PG1159 stars, where the latter ones turned out to be key objects in the understanding of the late stages of stellar evolution. The HSS is substantially deeper and more "exotic" types of stars could be expected.

Since the HSS has the potential to provide complete samples of blue stars for $B < 17^{m}.5$, scale heights, space densities, and birth rates can be determined for several classes of stars. The scale height of white dwarfs is very uncertain. Only one determination from a deep survey is available (Boyle 1989). The deepness of the HSS should also allow to determine the scale height at least for hot white dwarfs which was not possible with the PGS. The spatial distribution of sdB stars are investigated in five extensive studies (Heber 1986, Green & Liebert 1988, Moehler et al. 1990, Saffer et al. 1991, and Bixler 1991). However, the determined scale heights differ substantially. Birth rates and space densities for sdO stars are determined by Schönberner & Drilling (1985) using a very small sample. For other types of faint blue stars no statistical data exist.

The spatial distribution of the sdB/sdO stars could also be used to test a very interesting prediction of Kaluzny & Udalski (1992) and Liebert et al. (1994) derived from the analysis of the sdB stars in the metal-rich galactic Cluster NGC 6791, that only the very metal rich RGB stars form sdB stars. The distribution of sdB stars (and also sdOs if evolutionarily connected with the sdBs) should then be comparable to the young disk

type	number	type	number	type	number
DA	61	sdB	42	HBA	18
DB	2	sdOB	9	HBB	5
DO	5	sdO	19	sdF/G	18
DAB/DBA	4	He sdO	20		
DA binaries	6			QSO	2
magn. wd	2	PG1159	3	CV	2

TABLE 1. Summary of spectral types of the Hamburg Schmidt survey.

population. No such objects should therefore be found in the halo except from members of globular clusters, where a different formation process is possible due to environmental effects.

Faint blue stars have an intensive ionizing radiation field which can influence the energy balance of the interstellar matter (Deupree & Raymond 1983) and the galactic halo (de Boer 1985). It is also under discussion if the integrated light of these stars can explain the observed UV excess of many elliptical galaxies (Greggio & Renzini 1990).

2. Observations

Several observing runs have been performed to classify the stellar candidates and to provide medium resolution spectra with a good signal-to-noise ratio for a spectral analysis. Follow-up spectroscopy was started in 1989 and is continued up to now. All observations have been carried out at the German–Spanish Astronomical Center on Calar Alto, Spain, mainly using the 3.5m telescope. Different instrumental set-ups were used. Most observations were done using the Cassegrain Twin Spectrograph (see Dreizler et al. 1994), which turned out to be a powerful instrument for this purpose, or with the focal reducer (see Jordan et al. 1993). A complete wavelength coverage from the atmospheric transmission limit to ~ 8000 Å depending on the spectral resolution and the size of the CCD is recorded with one observation. This includes all important hydrogen and helium lines and also enables us to detect cool companions by a red excess or typical spectral features in the red spectral range.

Data reduction was carried out either in Kiel using the program package IDAS written by G. Jonas or in Bamberg using MIDAS. We use standard procedures to ensure a homogeneous data set. For flat-fielding we use dome-flats, wavelength calibration is performed with a He-Ar lamp. Flux calibration is done using observations of Feige 34, BD+28° 4221, or BD+33° 2642.

Up to now more than 250 objects have been observed. Due to the careful candidate selection possible with the information from the objective prism spectra this sample is only slightly contaminated ($\sim$ 15 %) with objects not relevant to this programme (sdF/G, HBA stars). Together with the cool DAs these objects can easily be identified by their strong Balmer absorption detectable in the objective prism spectra. In the beginning two quasars were found among the stellar candidates which escaped early detection because no prominent emission feature fell into the region of the Schmidt spectra (Groote et al. 1991). One of these quasars, HS0624+6907, turned out to be the brightest quasar discovered by optical selection. A fine-tuning of the selection procedure now also accounts for those quasars.

In general we find the same types of stars as in the PGS, however, their relative numbers

star	type	B	α (1950)	δ (1950)	comment
HS0111+0012	DO	14.5	01 11 12.8	+00 12 35	
HS0727+6003	DO	15.8	07 27 00.4	+60 03 57	
HS0742+6520	DO	16.0	07 42 41.0	+65 20 23	
HS1830+7209	DO	17.0	18 30 56.6	+72 09 22	
HS0704+6153	PG1159-A	16.8	07 04 59.2	+61 53 09	
HS1517+7403	PG1159-A	15.9	15 17 08.7	+74 03 00	
HS2324+3944	PG1159-hybrid	14.6	23 24 50.0	+39 44 52	
HS0209+0832	DAB	13.9	02 09 25.5	+08 32 48	
HS0824+2854	DA+dC	14.4	08 24 01.8	+28 53 55	PG0824+289
HS0949+4508	DA+dM	–	09 49 12.4	+45 07 36	
HS1558+6140	DA+dM	–	15 58 08.3	+61 40 30	
HS2237+8154	DA+dMe	–	22 37 30.2	+81 54 50	
HS0507+0435	DA+DA	14.3/15.6	05 07 34.5	+04 35 15	18″
HS2240+1234	DA+DA	V=16.2/16.5	22 40 02.5	+12 34 23	10″
HS1254+3440	magn. wd	V≈ 17	12 54 53.0	+34 30 52	B=9.5 MG
HS1412+6115	magn. wd	14.4	14 12 53.9	+61 15 04	B≈ 8 MG

TABLE 2. Stars of special interest from the Hamburg Schmidt survey.

are different (see Table 2). While the subdwarfs dominated the PGS, white dwarfs and subdwarfs are more or less equally abundant, even if we restrict the sample to a similar colour cut-off as that of the PGS. This is just an effect of the lower limiting magnitude of the HSS and was therefore expected. However, the relative numbers of helium-rich and helium-poor objects in these classes are not as expected. Compared to the PGS we detected many more extremely helium-rich sdO (He sdOs) stars and fewer helium-rich white dwarfs of spectral type DB but several DO or DOZ stars (see below). Note, however, that a final evaluation is only possible with a statistically complete sample.

Besides the more common subclasses, the HSS has also significantly contributed to the rarely populated subclasses (see Table 2). Two magnetic white dwarfs have been detected, one has been published by Hagen et al. (1987). One, HS1412+6115 = PG1412+614 was already observed by Green et al. (1986), but misclassified as a sdOC star. A highlight was the discovery of HS0824+2854 = PG0824+289 as an unique binary system composed of a DA and a dwarf carbon star (see also next section). Also four transition objects between the DAs and DBs could be found, e.g., the DAB star HS0209+0832, which lies in the so called DB gap (see also next section). Among the DA white dwarfs several very hot DAs, comparable to G191-B2B, were found. Like in other very hot DAs these objects should still contain metals in their atmospheres. The number of known hot helium rich white dwarfs (DOs) is doubled by the HSS. Wesemael et al. (1985) separate the DOs in cool and hot DO depending on the detection of He I. Four DOs (see Table 2) are found by our follow-up spectroscopy until now, which all seem to belong to the hot subgroup. Together with the two "cool" PG1159 stars found by the HSS, this enables us to study the transition from the PG1159 stars into DO white dwarfs. Additionally, one hydrogen-rich PG1159 star was found (see below).

3. Quantitative Spectral Analyses

In this section we want to summarize some of our quantitative results and highlight the analysis of the two PG1159-type stars HS0704+6153 and HS2324+3944.

3.1. HS 0209+0832: A DAB white dwarf in the DB gap

White dwarfs are basically divided into two separate classes: spectral type DA with almost pure hydrogen atmospheres and type non-DA (DO, DB, ...) with outer layers mainly consisting of helium. Several new hybrid objects were found in the HSS being key objects for the understanding of this dichotomy. Of special interest is HS0209+0832, a DAB star, being the only white dwarf in the so called "DB gap" between 30 000 K and 45 000 K showing helium in the optical spectrum (Jordan et al. 1993).

Additional to the optical observations from the focal reducer at the prime focus of the 3.5m telescope of the Calar Alto observatory we obtained IUE spectra in the low-dispersion mode which revealed the presence of C IV resonance absorption line. Homogeneously mixed and chemically stratified LTE model atmospheres with an ultrathin hydrogen layer ($M_H \approx 7 \cdot 10^{-14}\ M_\odot$) on top of the He envelope (Koester et al. 1979, Jordan & Koester 1986) were used to analyse the spectra. Parameters derived from *homogeneous* atmospheres, $T_{eff} = 36\,000 \pm 2000$ K, $\log g = 8.0$ $n_H/n_{He} = 0.02$ $n_C/n_H = 1 \cdot 10^{-4}$ by number, can reproduce the slope of the IUE spectra consistently with the Balmer and He I lines. *Stratified* atmospheres were rejected, because the result from the fit of Balmer lines is inconsistent with the continuum slope in the IUE range. In addition, the wings of the He I 4471 Å line are too broad in the stratified model.

It is not yet clear how to place HS0209+0832 in the present white dwarf evolution scenario. Two hypotheses are possible, the formation of a stratified atmosphere may have been disturbed by counteracting processes like meridional circulation or recent mass accretion from an undetected companion or the interstellar medium filled up the atmosphere of HS0209+0832 with helium and carbon.

3.2. HS 0824+289 = PG 0824+289: A unique dC+DA binary

Only carbon *giants* were known until Dahn et al. (1977) discovered G77-61, a carbon star with main sequence luminosity. It remained the only carbon *dwarf* until four other examples where discovered by Green et al. (1991). PG 0824+289 is unique because it is the only carbon dwarf having a spectroscopically detectable white dwarf companion (Heber et al. 1993). Therefore this object might provide the key to understand the dwarf carbon stars.

Optical observations were performed with the Cassegrain-Twin spectrograph at the 3.5m and with the Cassegrain spectrograph at the 2.2m telescope of Calar Alto observatory. While the blue spectral range is dominated by the Balmer lines of the DA white dwarf with no signs of a companion star, the red spectral range is dominated by C_2 and CN bands of the dC star, demonstrating the great value of the large wavelength range of the Twin-spectrograph. Homogeneously mixed LTE model atmospheres (Koester et al. 1979) give an effective temperature of $T_{eff} = 40\,000$ K, $\log g = 8.0$ for the DA. The dwarf carbon star contributes 25 % to the flux at 5500 Å.

PG0824+289 could be explained by close binary evolution similar to G77-61 (Dearborn et al. 1986) where the original secondary has accreted carbon rich material when the original primary has reached its largest radius avoiding the common envelope phase. Light curves are required to see whether PG 0824+289 is a *close* binary system or not. It should also be pointed out that PG0824+289 probably is a Population I object due to small radial velocity and proper motion, unlike all other dC stars known.

3.3. Binaries

A non-negligible fraction of the stars from the Hamburg survey are found to be binaries as already pointed out by Jordan & Heber (1993). The case of the DA+dC system PG0824+289 was described above. An example for a spectroscopic binary, consisting of a sdB and a cool companion was reported by Heber et al. (1991). Several other sdB stars are probably also spectroscopic binaries. Two visual DA+DA binary systems and three spectroscopic binaries (see Table 2) consisting of a DA white dwarf with the same proper motion and a M or Me companion are currently analysed by Jordan et al. (1994). To our knowledge it is for the first time that visual white dwarf binaries have been discovered by a faint blue star survey. Other examples are from proper motion surveys. At least, the PGS did not turn up one such case.

A preliminary analysis of the double DA HS2240+1234 is found in Jordan & Heber (1993). However, the effective temperatures (13 000 K and 15 000 K) were determined under the assumption that $\log g = 8$. Since both components are at the same distance the magnitude difference together with theoretical mass-radius relations can be used to check the consistency of the spectroscopically determined atmospheric parameters. One component of the second DA+DA pair HS0507+0435 also has a temperature of about 13 000 K, i.e. close to the blue edge of the ZZ Ceti instability strip. So it would be interesting to test these stars for photometrical variations.

3.4. HS2324+3944 and HS0704+6153: Two unusual PG1159 stars

PG1159 stars are key objects to understand the late phases of stellar evolution. They populate the transition region between the central stars of planetary nebulae (CSPN) and the white dwarfs. The spectra of these stars are characterized by the absence of any H and He I lines as well as a broad absorption trough due to He II 4686 Å and neighbouring C IV lines. Latest results about these stars are reviewed by Werner (1993). Spectral analyses with line blanketed NLTE models for eight PG1159 stars (Werner et al. 1991, Werner & Heber 1991, Werner 1991, Werner et al. 1992, and Motch et al. 1993) have shown that these stars display the helium-, carbon-, and oxygen-rich helium buffer layer of the former double shell burning and thermally pulsing post-AGB star at their surface. Therefore these stars have lost their entire hydrogen-rich envelope and part of the helium-rich layers. The most extreme case, H1504+65, has even lost the complete helium-rich layer and exhibits a pure C/O core. Effective temperatures and surface gravities cover a wide range. The coolest member (HS0704+6153) has 65 000 K, the hottest (H1504+65) 170 000 K. Surface gravities range from $\log g = 5.7$ in the subtype of CSPN-PG1159 stars to $\log g = 8.0$.

Two of the PG1159 stars discovered by the HSS are unusual objects and shall therefore be discussed in more detail.

3.4.1. HS2324+3944

While in "normal" PG1159 stars hydrogen was not detected, in HS2324+3944 the Balmer series is clearly visible. Only three other stars of this hybrid type are known up to now (Napiwotzki 1992) which are, unlike HS2324+3944, CSPNs. For one of it, NGC 7094, Werner (1992) gave a coarse estimate of the atmospheric parameters: $T_{eff} = 100\,000$ K, $\log g = 5.0$, He/H ~ 1, and C/H ~ 0.2.

A spectral analysis of those objects is very difficult since the effective temperature, the surface gravity, and the hydrogen, helium and carbon abundances have to be determined simultaneously. A very detailed model grid is therefore required to analyse these stars. First results for HS2324+3944 can be given here, it should, however, be kept in mind that they are preliminary since no optimum fit can yet be presented. We estimate an effective temperature of $T_{eff} = 130\,000 \pm 15\,000$ K and a surface gravity of $\log g = 5.7 \pm 0.5$, where at low T_{eff} the lower gravity is required to fit the emission in the C IV 5801/5812 Å doublet and vice versa. The helium abundance lies in the range of He/H $= 0.3\ldots1.0$, the carbon abundance is of the order of C/H ~ 0.2. In Figure 1 we compare two models

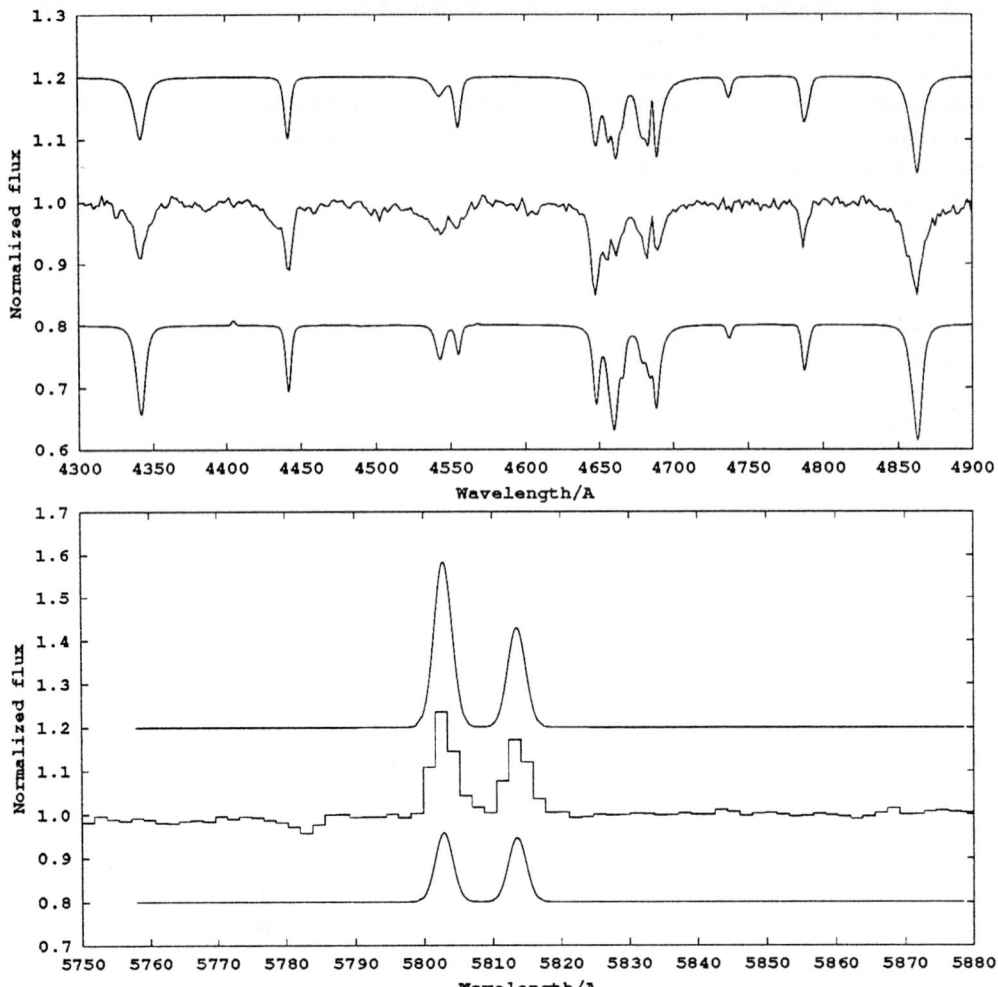

FIGURE 1. Comparison of HS2324+3944 with two theoretical models:
upper spectrum: $T_{eff} = 140\,000$ K, $\log g = 6.0$, He/H = 0.3, C/H = 0.2,
lower spectrum: $T_{eff} = 120\,000$ K, $\log g = 5.5$, He/H = 0.3, C/H = 0.2.

($T_{eff} = 140\,000$ K, $\log g = 6.0$, He/H = 0.3, C/H = 0.2 and $T_{eff} = 120\,000$ K $\log g = 5.5$, He/H = 0.3, C/H = 0.2) with the observations of HS2324+3944. The fit shows that the above given parameters are quite correct, fine tuning is, however, still required. The deep absorption component and the relative strength of the He II and Balmer lines can not yet be reproduced sufficiently well. Future analyses of the other members of the hybrid type will help to decide if a continuous transition from H-rich post-AGB stars to the H-deficient and C-rich PG1159 stars exist.

3.4.2. HS0704+6153

HS0704+6153 is a very interesting object since this star turned out to be the coolest known PG1159 star with atmospheric parameters very similar to those of DOs (Dreizler et al. 1994).

In the analysis of HS0704+6153 the He I/II ionisation equilibrium can be evaluated to derive the effective temperature since the star is cool enough to show a weak He I 5876 Å line. Together with the temperature sensitive C IV 5801/5812 Å doublet an effective temperature of $65\,000 \pm 5\,000$ K is determined. For the surface gravity determination we mainly rely on the trough region and He II 4860 Å and a gravity of $\log g = 7.0 \pm 0.5$ is found. It should be noted that the fit is not independent from the determination of the carbon abundance which is also based on the analysis of the trough region, He II 4860 Å, and C IV 4786 Å. An abundance of C/He = 30 % by number is the upper limit since the He II lines become too weak at the higher carbon abundance. C/He = 10 % is the lower limit, set by the strengths of the carbon lines. The temperature derived above is found to be consistent with the carbon abundance. The absence of C III lines gives an additional lower temperature limit of 60 000 K. They would become clearly visible at substantially lower effective temperatures.

An upper limit for the hydrogen abundance is determined from the H_β/He II 4860 Å blend. Due to the high effective temperature of HS0704+6153 an abundance of H/He < 1.0 is undetectable at this resolution. An increase of the hydrogen abundance to H/He = 1.0 has still no effect on other line profiles.

The effective temperature of 65 000 K and a surface gravity of $\log g = 7.0$ together with the extreme helium abundance (H/He< 1 by number) place HS0704+6153 in the region of hot DO white dwarfs. The high carbon and oxygen abundances (C/He = 0.2, O/He ~0.05 by number), however, are typical for PG1159 stars. Its effective temperature is by far lower than that of any other PG1159 stars. HS0704+6153 therefore is in the transition stage between the PG1159 stars and the DO white dwarfs, hence it is an ideal object to study the gravitational settling of metals in helium-rich atmospheres.

Comparing its position in the (g, T_{eff}) diagram to evolutionary predictions shows that its mass must be lower than that of any other known PG1159 star (see Dreizler et al. 1994). It is possible that HS0704+6153 evolved from the EHB rather than from the AGB, from which the other PG1159 stars are likely to have evolved.

4. Conclusion

We presented an overview of our follow-up spectroscopy of faint blue stars selected from the Hamburg Schmidt survey. The quantitative analysis of six individual stars has been reviewed. All these are key objects to understand the late phases of stellar evolution in more detail: The DAB star HS0209+0832 is the only white dwarf in the DB-gap in which helium can be detected from optical spectra, PG0824+286 is the only dwarf carbon star with a visible DA white dwarf companion. Two other visual binary systems composed of each two DAs, HS0507+0435 and HS2240+1234, have been analysed. HS0704+6153 turned out to be the coolest PG1159 star analysed up to now, it therefore is a key object to understand the transition from the PG1159 stars to the DO white dwarfs. HS2324+3944 is one out of four known hydrogen-rich PG1159 stars, which might be in the transition from a H-rich post-AGB into a hydrogen-deficient, carbon-rich PG1159 star.

Acknowledgements

We thank our colleagues in Hamburg, especially N. Bade and H. Hagen, for the object selection and the staff from the Calar Alto Observatory for their support during our observations. The Hamburg Schmidt survey has been supported by the Deutsche

Forschungsgemeinschaft (DFG) through grants Re 353/11-1,11-2,11-3 and 22-1,22-2,22-3. S. D. acknowledges the DFG for financial support through grant He 1356/16-1 and for a travel grant 477/1872/93. Calculations were performed on the CRAY X-MP of the Rechenzentrum der Universität Kiel, the CRAY Y-MP EL of the Regionales Rechenzentrum Erlangen, and the CRAY Y-MP of the Leibniz Rechenzentrum München.

REFERENCES

BIXLER, J. V., BOWYER, S., LAGET, M., 1991, A&A 250, 370
BOYLE, B. J., 1989, MNRAS, 240, 533
DE BOER, K. S., 1985, A&A, 142, 321
CALOI, V., 1989, A&A, 221, 27
DAHN, C. C., LIEBERT J., KRON, R. G., SPINRAD, H., HINTZEN, P. M., 1977, ApJ, 216, 757
DEARBORN, D. S. P., LIEBERT, J., AARONSON, M., DAHN, C. C., HARRINGTON, R., MOULD, J., GREENSTEIN, J. L., 1986, ApJ, 300, 314
DEUPREE, A. K., RAYMOND, J., 1983, AJ, 275, 171
DREIZLER, S., WERNER, K., JORDAN, S., HAGEN, H. J, 1994, A&A in press
ENGELS, D., GROOTE, D., HAGEN, H. J., REIMERS, D., 1988, ASP Conf. Ser., 2, 143
GREEN, E. M, LIEBERT, J., 1988, in Philip A. G. D., Hayes D. S., Liebert J. W., eds, The Second Conference on Faint Blue Stars, IAU Coll. No. 95, L. Davis Press, Schenectady, p. 261
GREEN, R. F., SCHMIDT, M., LIEBERT, J., 1986, ApJS, 61, 305
GREEN, P., MARGON, B., MACCONNELL, D. J., 1991, ApJ, 380, L31
GREGGIO, L., RENZINI, A., 1990, ApJ, 364, 35
GROOTE, D., HEBER, U., JORDAN, S., 1991, A&A, 223, L1
HAGEN, H. J., 1987, PhD thesis, Hamburg University
HAGEN, H. J., GROOTE, D., ENGELS, D., HAUG, U., REIMERS, D., 1987, Mitt. Astron. Ges., 67, 184
HAGEN, H. J., GROOTE D., ENGELS, D, REIMERS D., 1994, A&A, submitted
HEBER, U., 1986, A&A, 155, 33
HEBER, U., JORDAN, S., WEIDEMANN, V., 1991, in Vauclair G., Sion E., eds, NATO ASI Series C Vol. 336, Kluwer, Dordrecht, p. 109
HEBER, U., BADE, N., JORDAN, S., VOGES, W., 1993, A&A, 267, L31
IBEN, I., JR., TUTUKOV, A. V., 1984, ApJ, 284, 719
JORDAN, S., KOESTER, D., 1986, A&AS, 65, 367
JORDAN, S., HEBER, U., 1993, in Barstow M. A., ed, Proc. of the 8th European Workshop on White Dwarfs, White Dwarfs: Advances in Observation and Theory, NATO ASI Series C Vol. 403, Kluwer, Dordrecht, p. 47
JORDAN, S., HEBER, U., ENGELS, D., KOESTER, D., 1993, A&A, 273, L27
JORDAN, S., DREIZLER, S., HEBER, U., HAGEN, H., KOESTER D., 1994, to be submitted
KALUZNY, J., UDALSKI, A., 1992, Acta Astron., 42, 29
KOESTER, D., SCHULZ, H., WEIDEMANN, V., 1979, A&A, 76, 262
LIEBERT, J., SAFFER, R. A, GREEN, E. M, 1994, preprint
MENGEL, J. G., NORRIS, J., GROSS, P. G., 1976, ApJ, 204, 488
MOEHLER, S., HEBER, U., DE BOER, K. S., 1990, A&A, 239, 265
MOEHLER, S., HEBER, U., DE BOER, K. S., 1994, in Adelman S. J., Upgren A. R., Adelman C. J., eds, Hot Stars in the Halo, Cambridge University Press, Cambridge, p. 217
MOTCH, C., WERNER, K., PAKULL, M. W., 1993, A&A, 268, 561

NAPIWOTZKI, R. 1992, in Heber U., Jeffery C. S., eds, Lecture Notes in Physics Vol. 401, The Atmospheres of Early-Type Stars, Springer, Berlin, p. 310
PESCH, P., SANDULEAK, N., 1983, AJS, 51, 171
SAFFER, R. A, BERGERON, P., LIEBERT, J., KOESTER, D., 1991, in Vauclair G., Sion E., eds, Proc. of the 7th European Workshop on White Dwarfs, White Dwarfs, NATO ASI Series C, Vol. 336, Kluwer, Dordrecht, p. 53
SCHÖNBERNER, D., DRILLING, J. S., 1985, ApJ, 290, L49
SION, E. M., 1986, PASP, 98, 821
WERNER, K., 1991, A&A, 251, 147
WERNER, K., 1992, in Heber U., Jeffery C. S., eds, Lecture Notes in Physics Vol. 401, The Atmospheres of Early-Type Stars, Springer, Berlin, p. 273
WERNER K., 1993, in Barstow M. A., ed, Proc. of the 8th European Workshop on White Dwarfs, White Dwarfs: Advances in Observation and Theory, NATO ASI Series C, Vol. 403, Kluwer, Dordrecht, p. 67
WERNER, K., HEBER, U., 1991, A&A, 247, 476
WERNER, K., HEBER, U., HUNGER, K., 1991, A&A, 244, 437
WERNER, K., HAMANN, W. R., HEBER, U., NAPIWOTZKI, R., RAUCH, T., WESSOLOWSKI, U., 1992, A&A, 259, L69
WESEMAEL, F., GREEN, R. F., LIEBERT, J., 1985, ApJS, 58, 379

Discussion

DORMAN: Can you remind us where on the log g - log T_{eff} diagram you showed is occupied by the sdO's? Is there any real separation in location between He-rich sdO's and the others?

DREIZLER: Subdwarf O (sdO) stars cover a large range in the log g - log T_{eff} diagram: 40,000 K $<$ T_{eff} $<$ 100,000 K and 4 $<$ log g $<$ 6.5. This covers the area between the post Asymptotic Giant Branch (AGB) tracks and the Extended Horizontal Branch (EHB) (see Dreizler, 1993, A&A, 273, 212). Compared to PG 1159 stars, the sdO stars are cooler and have a lower surface gravity. There is no clear separation between sdO stars and the He-rich sdOs. The He-rich sdO stars tend, however, to be slightly cooler.

Stellar Winds and the Evolution of sdB's to sdO's

By JAMES MACDONALD AND STEVEN S. ARRIETA

Department of Physics and Astronomy, University of Delaware, Newark, Delaware 19716, USA

The proximity of the sdB stars to the sdO stars in the Hertzsprung-Russell Diagram, coupled with the differences in surface He/H ratio, suggests that sdB's evolve into sdO's. Although post-Horizontal Branch evolutionary tracks without mass loss do pass through the relevant regions of the HRD, they have not been able to explain the differences in surface abundances, primarily because convective dredge-up of helium is not found to occur. Here we present first results of a study of an alternative scenario in which stellar wind mass loss strips away much of the hydrogen-rich envelope to reveal helium-rich layers. Since Beers et al. (1992) have found that 60 % of subdwarfs have velocities < 60 km s^{-1}, indicative of an old disk population, we have adopted a metalicity $Z = 0.02$.

Evolutionary tracks have been calculated for 0.5 and 0.6 $M_\odot$ Population I stars, with and without inclusion of stellar wind mass loss. The starting point for each calculation is a Zero-Age Horizontal Branch star consisting of a helium core (Y = 0.98, Z = 0.02) surrounded by a hydrogen-rich layer (X = 0.7, Y = 0.28, Z = 0.02) of specified mass, M_H. The mass rate loss, which is based on the theoretical formula of Abbott (1982) for radiatively driven hot star winds, is related to the stellar luminosity, L, by

$$\dot{M} = 1.4 \, 10^{-15} \frac{(L/L_\odot)^2}{M_{eff}/M_\odot} M_\odot yr^{-1}$$

where the effective mass, $M_{eff} = M(1 - L/L_{ed})$, and L_{ed} is the Eddington luminosity. The evolution of the stars is followed to the white dwarf cooling track, except for those that evolve to the asymptotic giant branch or experience a late helium flash.

The evolutionary tracks in the Log T_{eff}-Log g diagram are shown in Figures 1 and 2 together with the positions of sdO's from observations by Hunger et al. (1981), Méndez et al. (1981), Husfeld et al. (1987), Dreizler et al. (1990), Thejll et al. (1991) and sdB's from the data of Saffer, Bergeron & Liebert (1991). The tracks with and without mass loss are qualitatively the same.

To show the pronounced differences in evolution of the surface composition, we compare in Table 1 the initial and final envelope parameters for each evolutionary sequence. We conclude that *for Population I stars, mass loss at rates determined by hot star wind theory is capable of removing a sufficient fraction of the sdB hydrogen envelope to reveal helium-rich layers, transforming the star into a sdO*. The majority of the mass loss is found to occur during the stable thick shell helium burning phase and it is during this phase that the transition to a hydrogen-deficient atmosphere takes place. Thus a rough estimate of the expected ratio of space density of sdO's to that of sdB's is the relative amount of time spent in shell helium burning to that spent in core helium burning. This ratio for our calculations is 0.18–0.20, which is in fair agreement with the space density ratio, 0.11– 0.21, estimated from the data of Beers et al. (1992), and within the observational errors of the value, 0.35, found by Downes (1986) for a complete sample.

Note: WDCT = White Dwarf Cooling Track, LHeF = Late Helium Flash, AGB = Asymptotic Giant Branch

$M/M_\odot$	$M_H/M_\odot$	Wind?	Total Mass Lost ($M_\odot$)	Final Mass of H ($M_\odot$)	Final Surface X	Final Surface Y	Final State
0.5	$1.39\ 10^{-3}$	No	0.0	$1.47\ 10^{-4}$	0.700	0.280	WDCT
0.5	$1.39\ 10^{-3}$	Yes	$1.3\ 10^{-3}$	$2.50\ 10^{-5}$	0.164	0.816	WDCT
0.5	$2.67\ 10^{-3}$	Yes	$1.5\ 10^{-3}$	$4.99\ 10^{-5}$	0.340	0.640	WDCT
0.5	$1.00\ 10^{-2}$	No	0.0	$1.15\ 10^{-4}$	0.700	0.280	WDCT
0.5	$1.69\ 10^{-2}$	Yes	$2.4\ 10^{-3}$	$7.98\ 10^{-4}$	0.677	0.302	WDCT
0.6	$1.57\ 10^{-3}$	No	0.0	$3.93\ 10^{-5}$	0.700	0.280	WDCT
0.6	$1.57\ 10^{-3}$	Yes	$5.2\ 10^{-3}$	$1.57\ 10^{-7}$	0.001	0.979	LHeF
0.6	$1.24\ 10^{-2}$	No	0.0	$3.70\ 10^{-3}$	0.699	0.281	AGB
0.6	$1.25\ 10^{-2}$	Yes	$6.2\ 10^{-3}$	$2.62\ 10^{-5}$	0.234	0.746	LHeF

TABLE 1. Comparison of Initial and Final Envelope Parameters

REFERENCES

ABBOTT, D. C., 1982, ApJ, 259, 282

BEERS, T. C., PRESTON, G. W., SHECTMAN, S. A., DOINIDIS, S. P., GRIFFIN, K. E., 1992, AJ, 103, 267

DOWNES, R. A., 1986, ApJS, 61, 569

DREIZLER, S., HEBER, U., WERNER, K., MOEHLER, S., DE BOER, K. S., 1990, A&A, 235, 234

HUNGER, K., GRUSCHINSKE, J., KUDRITZKI, R. P., SIMON, K. P., 1981, A&A, 95, 244

HUSFELD, D., 1987, in Philip A. G. D., Hayes D. S., Liebert J. W., eds, IAU Colloquium No. 95, The Second Conference on Faint Blue Stars, L. Davis Press, Schenedtady, NY, p. 237

MÉNDEZ, R. H., KUDRITZKI, R. P., GRUSCHINSKE, J., SIMON, K. P., 1981, A&A, 101, 323

SAFFER, R. A., BERGERON, P., LIEBERT, J., 1991, in Vauclair G., Sion E. M., eds, White Dwarfs, Kluwer, Dordrecht

THEJLL, P., BAUER, F., SAFFER, R., KUNZE, D., SHIPMAN, H., LIEBERT, J., 1991, in Heber U., Jeffery C. S., eds, The Atmospheres of Early Type Stars, Springer-Verlag, Berlin, p. 261

Discussion

DRILLING: I would be very interested in knowing the times that it takes the stars to traverse these evolutionary tracks.

MACDONALD: The ratio of time spent in core He burning to shell He burning is in rough agreement with the ratio of space densities of sdB's to sdO's. The transition to sdO's by mass loss occurs during the shell burning phase.

BEERS: To reiterate a point made earlier – the reported differences in the kinematics of sdO and sdB stars cannot be used to uniquely assign them to different populations - say disk and halo. The reason is that the much larger range in absolute magnitude covered by the sdO stars means that a magnitude limited sample of these stars may be comprised of a very different population mix than the sdB stars.

CARNEY: I would like to reiterate the possibility that two metal-poor populations may exist in the solar neighborhood, one with a large (and constant) scale height, and one with a potential scale height-metallicity dependence. Use of a single scale height for metal-poor stars is potentially very misleading.

0.5 M$_{sun}$ — Wind

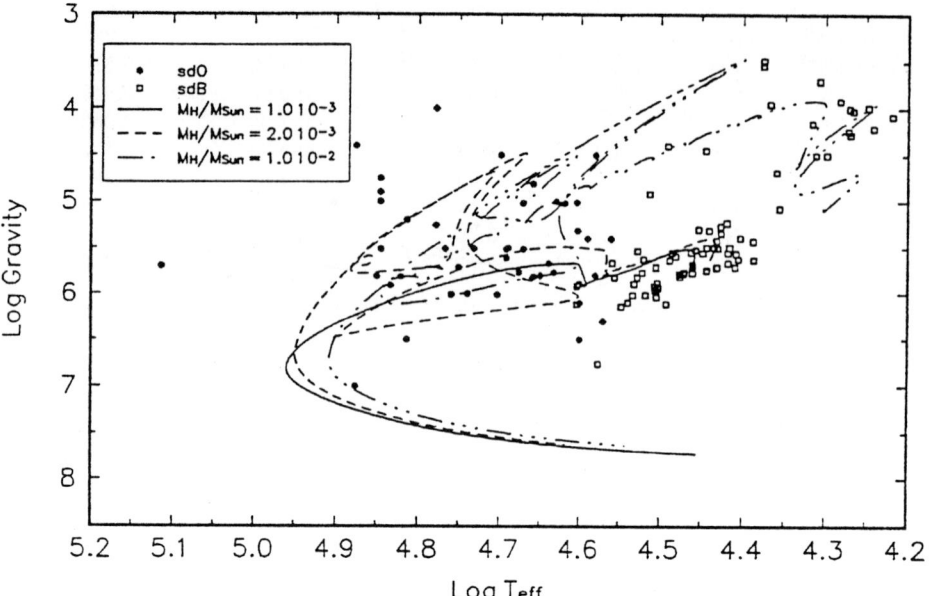

0.5 M$_{sun}$ — No Wind

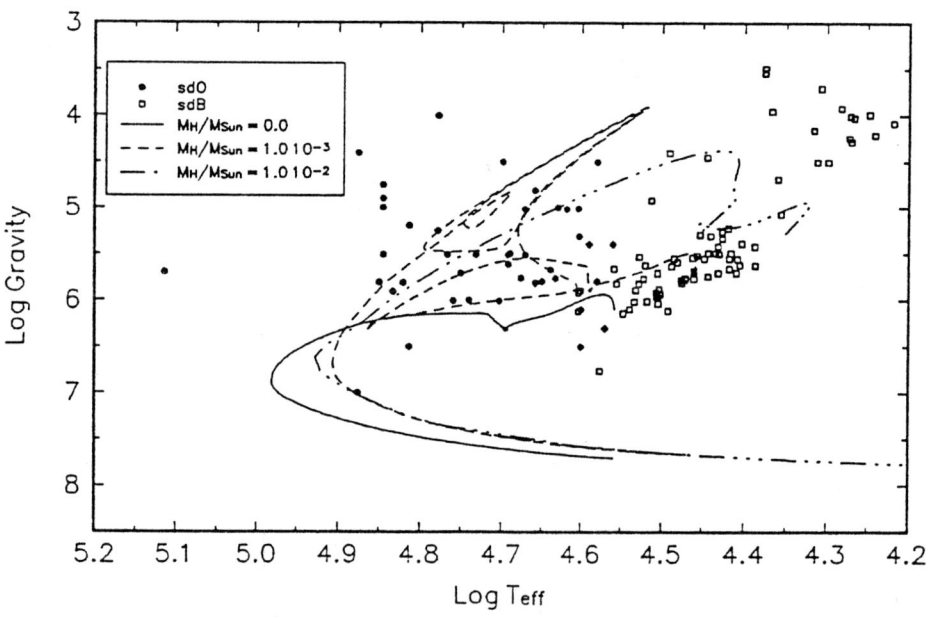

FIGURE 1. Evolutionary tracks in the Log T_{eff}-Log g diagram for 0.5 M$_\odot$ are shown together with observed values.

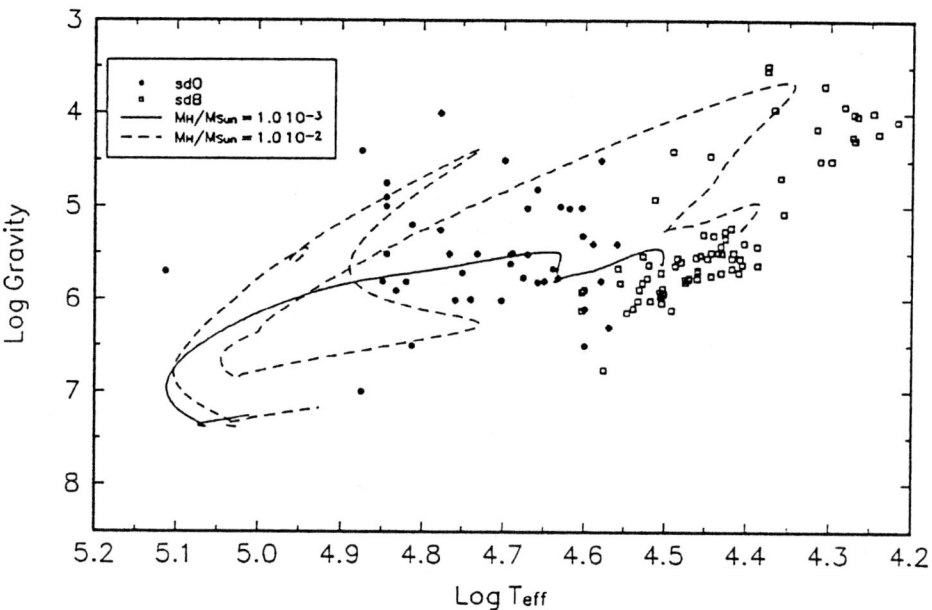

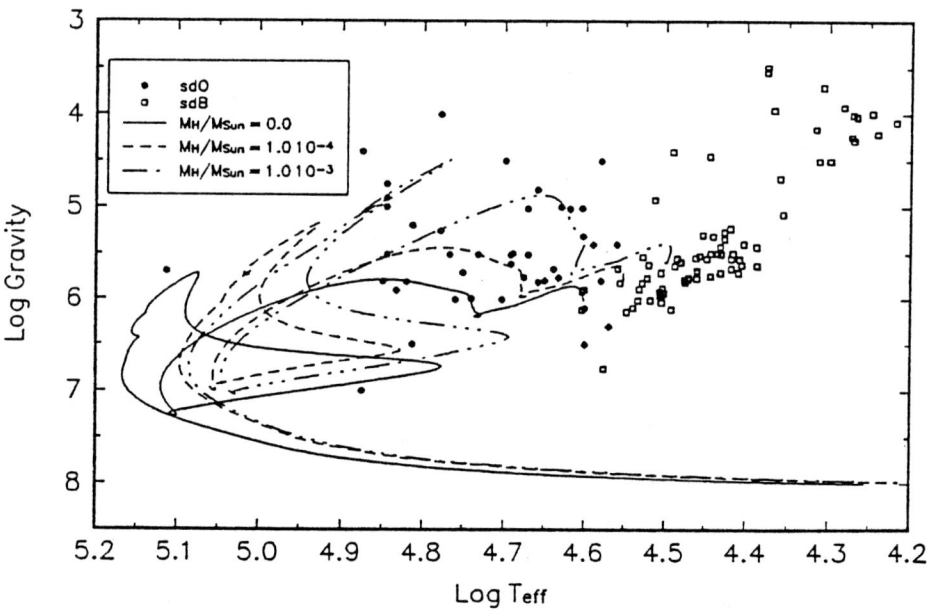

FIGURE 2. Evolutionary tracks in the Log T_{eff}-Log g diagram for 0.6 M$_\odot$ are shown together with observed values.

Halo Stars in the Vilnius Photometric System

By VYTAS STRAIŽYS

Institute of Theoretical Physics and Astronomy, Gostauto 12, Vilnius 2600, LITHUANIA

Possibilities for the Vilnius seven-color photometric system to identify halo or Population II stars, which are generally metal-deficient objects, are discussed.

---

1. The Vilnius photometric system

The Vilnius seven-color photometric system was established for the two-dimensional classification of stars affected by interstellar reddening. Later on, the system was shown to be capable of determining the metallicity of F, G, and K stars, as well as of identifying photometrically the stars of different peculiar types, including the Be, Bp, Ap, and Am stars, F-G-K subdwarfs, G and K metal-deficient giants and subgiants, carbon-rich stars, white dwarfs, and many types of the unresolved binaries. The identification and classification methods in the Vilnius system are described in detail in my monograph (Straižys 1992). Here we shall discuss only the possibilities for this system to identify halo or Population II stars which generally are metal-deficient objects.

The Vilnius photometric system consists of seven bandpasses U, P, X, Y, Z, V, and S with mean wavelengths of 345, 374, 405, 466, 516, 544, and 656 nm and halfwidths of 20-40 nm. The system can be realized both by glass and interference filters and with photoelectric and CCD detectors. The usual accuracy necessary for stellar classification is about ±0.01 mag (in some spectral ranges, a lower accuracy is acceptable). The limiting magnitude for a 1 meter-class telescope and a CCD detector is about 17 mag in V.

For the identification of peculiar stars, the interstellar reddening-free Q,Q diagrams are used. The definition of the Q-parameter is:

$$Q(1,2,3,4) = (m_1-m_2) - E(1,2)/E(3,4)\, (m_3-m_4)$$

where m_1-m_2 and m_3-m_4 are color indices and $E(1,2)$ and $E(3,4)$ are the corresponding color excesses. The ratios $E(1,2)/E(3,4)$ are usually either constant or slightly dependent on the spectral type and luminosity of a star.

2. Halo stars

For the identification of the halo main sequence stars (subdwarfs), we use the Q(UXY), Q(UYV) diagram. This diagram may be used for the determination of temperatures and metallicities of F-G-K subdwarfs, with the appropriate corresponding calibration.

For the identification of the metal-deficient halo giants and subgiants, we have no single Q,Q diagram. Instead, a three-dimensional space with the Q(UPYV), Q(PXYV), and Q(XYV) parameters must be used.

The identification of the reddened horizontal-branch stars is even more complicated. Since at the higher galactic latitudes the interstellar reddening is small, for the identification of the halo stars we may use two-color or color-difference vs. color diagrams.

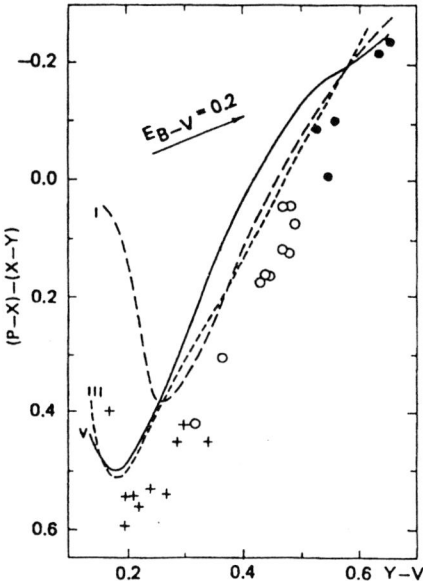

FIGURE 1. Blue horizontal branch stars of spectral class A (crosses), RR Lyrae-type stars near the minimum light (circles), and red horizontal branch stars (dots) in the (P-X)-(X-Y) vs. Y-V diagram. All stars are dereddened.

One such diagram for the identification of the blue horizontal branch stars is the (P-X)-(X-Y) vs. Y-V (Figure 1). In addition to the BHB stars of spectral class A, some RR Lyrae type stars near their minimum light and some red horizontal branch stars are also plotted. All of them form a sequence displaced with respect to the sequences of different luminosity classes of the disk population (Sperauskas 1987a). The BHB stars of spectral class A have smaller values of X-Y due to metal-deficiency and larger values of P-X due to higher Balmer jump.

Since the RR Lyrae variables seem to follow the same sequence as the other stars of the horizontal branch, Sperauskas (1987b) suggested a method for determination of their color excesses by using the interstellar reddening-free parameter Q(PXY) and the Y-V color index:

$$E(Y\text{-}V) = (Y\text{-}V) + 0.70\, Q(PXY) - 0.71.$$

It has been supposed that this relation in the range of A-F stars does not depend on the temperature, the luminosity, and the metallicity. The relation was verified for RR Lyrae in different phases by Straižys et al. (1989). It was shown that in the Q(PXY),Y-V diagram the star moves counterclockwise along an open elongated loop. At every value of Y-V = const, there are two values of Q(PXY) and, consequently, this equation cannot be applied to the determination of E(Y-V) without knowledge of the phase. The falling part of light curve, including the minimum phase, gives the most realistic values of the color excess. Observations of more RR Lyrae-type stars of different periods and metallicities are necessary for more exact calibration of this equation.

Other diagrams useful for identification and classification of the BHB stars are analysed by Bartkevičius & Straižys (1970) and Sperauskas (1987a).

REFERENCES

BARTKEVIČIUS, A., STRAIŽYS, V., 1970, Bull. Vilnius Obs., No. 30, 3
SPERAUSKAS, J., 1987a, Bull. Vilnius Obs., No. 77, 3
SPERAUSKAS, J., 1987b, Bull. Vilnius Obs., No. 79, 36
STRAIŽYS, V., 1992, Multicolor Stellar Photometry, Pachart Publishing House, Tucson, Arizona
STRAIŽYS, V., ČERNIS, K., VANSEVIČIUS, V., MEIŠTAS, E., 1989, Bull. Vilnius Obs., No. 83, 43

Horizontal-Branch Stars in the Geneva Photometric System

By B. HAUCK

Institut d'Astronomie de l'Université de Lausanne 1290 Chavannes-des-Bois, SWITZERLAND

The location of field horizontal-branch stars in some diagrams of the Geneva photometric system is studied. Application to a list of proposed FHB stars is made.

1. Introduction

Some years ago, Hauck & Phillip (1981) showed that the field horizontal-branch stars (hereafter FHB stars) are located in a specific area of some photometric diagrams of the Geneva system. Consequently, as in the case of the Strömgren photometric system, we can use photometric diagrams to segregate the horizontal-branch stars from other stars. The main properties of FHB stars are a big Balmer jump, producing a large Δd or Δc_1 value, and a metal deficiency that is responsible for small m_2 or m_1 values. In addition, these stars occupied a particular location in a d vs Δ diagram.

To check these previous conclusions only one source (Philip 1984) has been considered in this paper. The calibration of Kobi & North (1990) was used to derive T_{eff} and $\log g$ for some FHB stars. Finally, a new list of suspected FHB stars (Philip & Adelman 1994) was also considered.

2. Location of FHB stars in three photometric diagrams

Table 1 gives data for the FHB stars from Philip (1984) in the Geneva system, the source of the data being Rufener (1988). We have plotted d vs $B2 - V1$ and m_2 vs $B2 - V1$ (open or full dots) in Figures 1 and 2 respectively. Three stars deserve a short discussion, all having a too small d value in d vs $B2 - V1$.

BD 32° 2188 (1) was excluded from the FHB stars by Hauck & Phillip (1981). Its location in a d vs Δ diagram is that of an A supergiant, perhaps weakly deficient in metals.

BD 33° 2171 (2) and HD 83041 (3) lie very near the mean location of giant stars, far below HD 12293 or XZ Cet (4). HD 83041 is classified by Houk (1982) as A2II/III(w) and Hauck (1986) has proposed it as a λ Boo-type candidate. BD 33° 2171 shows the same photometric properties as HD 83041. We propose that these two stars should not be considered as FHB stars.

In a d vs Δ diagram (Figure 3), the FHB stars are also situated in a special location, mainly due to their great d value, a parameter well correlated to the Balmer jump, and Δ, a parameter sensitive to the Balmer jump and the metal deficiency. The location of the three stars discussed above confirms that these stars are not - photometrically at least - FHB stars. The location of HD 12993 in the area of the BV stars is due to its metal deficiency. So as to check if we can use the calibration of Kobi & North (1990) to

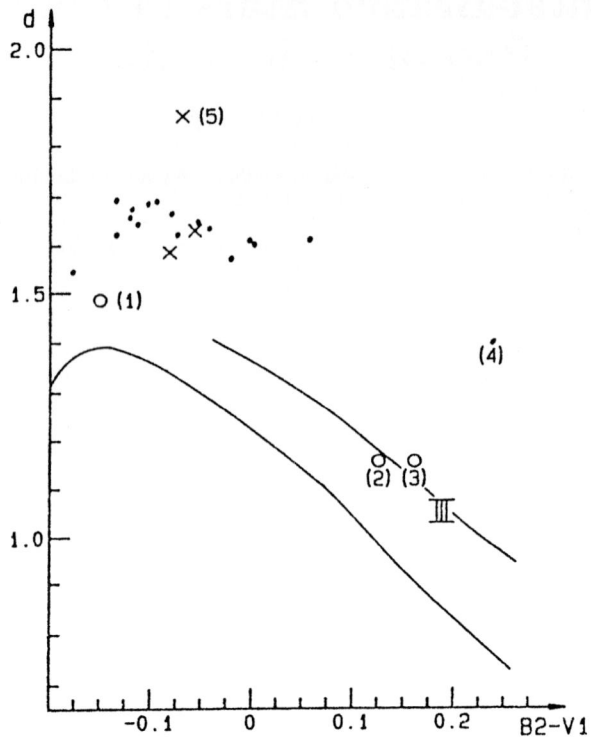

FIGURE 1. d vs $B2 - V1$ for FHB stars : - (1) BD $32°$ 2188, - (2) BD $33°$ 2171, - (3) HD 83041, - (4) HD 12293, - (5) HD 94509. Full lines are the sequence of reference (or lower part of class V sequence) and the mean location of class III .

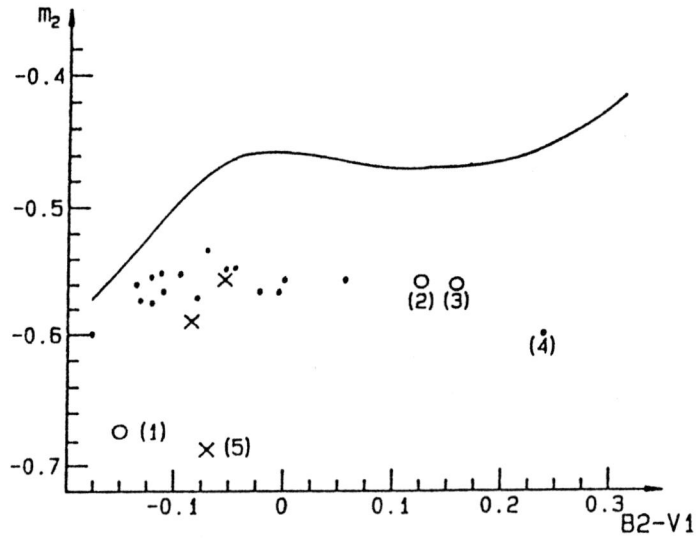

FIGURE 2. m_2 vs $B2 - V1$ for FHB stars. The full line is the Hyades sequence.

BD/HD	$B2-V1$	d	m_2	Δ	T_{eff}	$\log g$
32° 2188	-0.151	1.485	-0.674	0.221	-	-
33° 2171	0.125	1.157	-0.559	0.153	7070	3.77
42° 2309	-0.109	1.681	-0.565	0.674	8500	3.27
2857	0.054	1.609	-0.558	0.630	7270	2.68
12293	0.236	1.398	-0.601	0.291	6339	2.19
14829	-0.121	1.671	-0.575	0.645	8650	3.39
64488	-0.114	1.644	-0.551	0.072	8615	3.43
74721	-0.121	1.696	-0.554	0.681	8680	3.43
83041	0.157	1.157	-0.558	0.175	6800	3.51
86986	-0.055	1.642	-0.548	0.673	7260	3.10
109995	-0.098	1.689	-0.552	0.716	8310	3.22
130095	-0.079	1.664	-0.572	0.647	8140	3.14
161817	-0.004	1.605	-0.566	0.589	7410	2.82
SS 191 II	-0.134	1.692	-0.560	0.736	8775	3.41
SS 193 II	-0.131	1.620	-0.573	0.583	8120	3.61
SS 197 II	-0.178	1.540	-0.600	0.472	-	-
SS 199 II	-0.073	1.618	-0.532	0.685	8120	3.30
SS 202 II	-0.023	1.571	-0.565	0.525	7650	3.11
SS 209 II	-0.047	1.636	-0.546	0.665	7780	3.06
HLF 5-10	0.000	1.599	-0.557	0.636	7390	2.82

TABLE 1. Data in the Geneva system for FHB stars in Philip (1984)

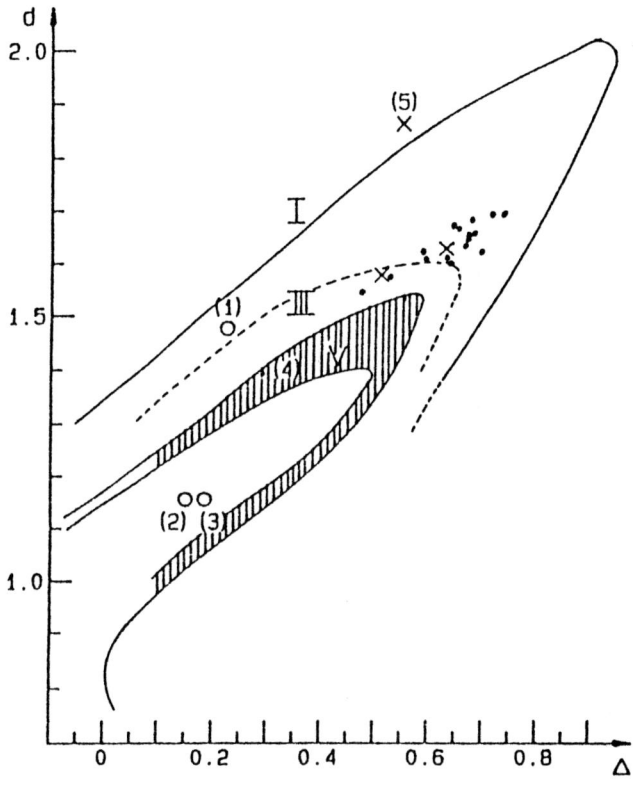

FIGURE 3. d vs Δ for FHB stars

HD	$B2-V1$	d	m_2	Δ	T_{eff}	$\log g$
15042	-0.083	1.579	-0.591	+0.009	7670	3.24
94509	-0.069	1.860	-0.689	-0.095	7790	2.38
181119	-0.056	1.621	-0.558	+0.042	7900	3.17

TABLE 2. Data in the Geneva system for FHB stars in Philip & Adelman (1994)

derive T_{eff} and $\log g$ for FHB stars, we have compared our values with those given by Hayes & Philip (1988). Our values are given in Table 1. The agreement is excellent.

The [M/H] values obtained from Kobi & North (1990)'s calibration cannot be used. Kobi & North have used the calibration of Δm_2 in terms of [Fe/H] derived by Berthet (1990). This calibration is established with stars of spectral types between F5 and G5.

Hauck (1978) has shown that the $[Fe/H]$ vs Δm_2 relation for the FHB is not the same as for the F5-G5 type stars.

3. Examination of proposed FHB stars

Philip & Adelman (1994) proposed 18 candidate FHB stars, on the basis of their $uvby\beta$ colours. We have searched for these stars in the data for the Geneva photometric system, but unfortunately only three have been measured in this system. They are given in Table 2. Their locations are indicated by a cross in the d vs $B2-V1$ and m_2 vs $B2-V1$ and d vs Δ diagrams.

HD 15042 and HD 181119 are clearly in the area of FHB stars in each diagram, but not HD 94509 (5). The latter is classified as Bp shell by Houk (1975). In the d vs Δ diagram this star is located near the locus of the supergiant stars. That corresponds to the MK type found in SIMBAD. It is also far from the location of the shell stars in such a diagram Hauck (1987). This star should be studied in more detail in order to clarify its status.

REFERENCES

BERTHET, S., 1990, A&A, 236, 440

HAUCK, B., 1978, A&A, 63, 273

HAUCK, B., 1986, A&A, 154, 349

HAUCK, B., 1987, A&A, 177, 193

HAUCK, B., PHILIP, A. G. D., 1981, in Philip A. G. D., Hayes D. S., eds, Astrophysical parameters for globular clusters, IAU Coll. No. 68, p. 571

HAYES, D. S., PHILIP A. G. D., 1988, PASP, 100, 801

HOUK, N., 1975, Michigan Catalogue of Spectral Types, Vol. 1, University of Michigan, Ann Arbor, Michigan

HOUK, N., 1982, Michigan Catalogue of Spectral Types, Vol. 3, University of Michigan, Ann Arbor, Michigan

KOBI, D., NORTH, P., 1990, A&AS, 85, 999

PHILIP, A. G. D., 1984, Commun. Van Vleck Obs., No 2, p. 1

PHILIP, A. G. D., ADELMAN, S. J., 1994, in Philip A. G. D., Hauck B., Upgren A., eds, Databases for Galactic Structure, L. Davis Press, Schenectady

RUFENER, F., 1998, Catalogue of stars measured in the Geneva Observatory photometric system, 4th ed., Geneva Obs.

Zeeman Observations of FHB and Hot Subdwarf Stars

By V. G. ELKIN

Special Astrophysical Observatory of the Russian Academy of Sciences, Nizhnij Arkhys 357147, RUSSIA

Observations have been made of seven FHB and eight hot subdwarf stars to search for magnetic fields. But there is no convincing evidence for such fields in any of the stars surveyed.

---

1. Introduction

Magnetic activity in the main sequence band stars has been well studied. The magnetic Ap/Bp stars were the first stars in which large magnetic fields were discovered. Some F, G, and K stars have solar-like magnetic activity. Near the end of their stellar evolution are the magnetic white dwarfs and the other degenerate objects with magnetic fields. Despite considerable progress in the observations of magnetic fields, the questions of the origin and evolution of the stellar magnetic fields still remain.

A star settles on the horizontal branch (HB hereafter) after the red giant stages and the helium flash in its core. The lifetimes of star on the horizontal-branch are about 10^8 years. (Sweigart & Gross 1976). As discussed by Michaud et al. (1983), this time could be enough for the production of abundance anomalies by diffusion processes.

The elemental abundance analyses of their spectra showed that the some HB stars have abundances which are very similar to the chemical peculiar stars of the main sequence. Sargent & Searle (1967) noted the similarity of the abundance anomalies observed in the HB star Feige 86 and in the Bp star 3 CenA. The blue HB star HD 97859 resembles to the Bp silicon type stars (Stalio 1974) and the possible HB star 38 Dra is similar to the peculiar manganese stars (Adelman & Sargent 1972). Klochkova & Panchuk (1987) found that the FHB star HD 161817 have anomalous abundances of chromium and strontium relative to that of iron. The majority of blue HB stars and sdB stars have helium underabundances. The sdO stars which are the very hot extension of the horizontal branch have normal or overabundant helium (Greenstein & Sargent 1974).

Magnetic fields in Ap/Bp stars play an essential role in the origin and development of the chemical peculiarities. Borra et al. (1983) searched for a magnetic field in the HB star Feige 86 and in hot subdwarf HD 149382. They did not find any magnetic fields which were larger than the standard deviations of the measurements.

This paper describes the results of my magnetic field investigations of the FHB stars and the hot subdwarfs.

2. Observations

The spectra of seven FHB stars and seven hot subdwarfs were obtained with the main stellar spectrograph of the 6-m telescope in the blue region with a reciprocal dispersion of $D = 9$ Å mm^{-1}, except for the star HD 4539, which was observed with $D = 28$ Å mm^{-1} also in blue region. A circular polarization analyzer was placed in front of the spectrograph slit. For the observations of the star Feige 87 the Balmer line magnetometer was used.

star	JD(2448000+)	B_e(G)	error	v_r(km s^{-1})	error
HD 60778	701.224	+150	115	+74.2	0.6
HD 74721	585.528	+240	150	+39.9	1.1
HD 86986	992.527	-430	580	+2.6	0.8
HD 109995	052.358	-820	470		
	992.607	+700	520	-139.2	1.3
HD 117880	994.598			+141.9	
HD 161817	344.522	+30	110		
	382.402	-550	145	-367.7	0.6
	423.355	-90	80		
	700.516	+100	160	-366.3	0.7

TABLE 1. Magnetic fields and radial velocities of FHB stars

3. Magnetic measurements of the FHB stars

As suggested by Michaud et al. (1983) chemical separation may play an important role in the blue horizontal branch stars with $T_{eff} > 15,000$ K, but not in the cooler HB stars. Kodaira (1973), Kodaira & Philip (1984), Klochkova & Panchuk (1987), Adelman & Hill (1987), and Adelman & Philip (1990) investigated the elemental abundances of the FHB stars and confirmed that these stars have abundance deficiencies of the observed chemical elements.

The FHB stars have sufficiently strong spectral lines inspite of the chemical underabundances. The slow rotation of these stars helps in measuring the longitudinal magnetic fields with the photographic method. My measurements of the magnetic fields and the radial velocities are presented in Table 1.

Klochkova & Panchuk (1987) found the chemical peculiarities of the star HD 161817 compared with other FHB stars, as was noted above. Thus I studied HD 161817 more carefully and obtained four Zeeman spectra. As seen from Table 1, only one spectra of HD 161817 showed a magnetic field which exceeded three times the standard deviation. The other three spectra do not show any magnetic fields. Therefore, this one result cannot be considered as the evidence of a magnetic field in HD 161817.

The two spectra of HD 109995 show magnetic fields, which were slightly in excess of one standard deviation. However, I not have sufficient observational data of this star to provide convincing evidence of a magnetic field. All of the other FHB stars did not show any magnetic fields. I also obtained one spectrogram of HD 64488, but this star rotates too rapidly for magnetic measurements. The spectrogram of HD 117880 is not very good. For this star I estimated only its radial velocity.

The radial velocities (v_r) of HD 109995 and HD 161817 do not differ from previous measurements. However, the v_r of other FHB stars exhibit some differences from those in the Bibliography of Stellar Radial Velocities (Abt & Biggs 1972).

4. A search for magnetic fields in hot subdwarfs

The OB subdwarfs are the hot extension of the HB stars (Greenstein & Sargent 1974). These stars lie several magnitudes below the main sequence. There is some similarity between the O subdwarfs and He-rich stars of the main sequence (Osmer & Peterson 1974).

Observations of the magnetic field in the hot subdwarfs are more difficult. Their spectra have absorption lines of hydrogen and neutral and ionized helium. The lines

Star	Sp. Type	B_e(G)	error	v_r(km s^{-1})	error	
HD 4539	sd B	-1290	2120	+3.3	3.7	28 Å mm^{-1}
Feige 87	sd B	-1790	3495			Balmer line
HD 76431	sd B	-50	130	+44.5	1.1	
BD+75° 325	sd O	+1260	870	-52.2	1.6	
BD+25° 2534	sd Op					
HD 128220	sd O+G	-520	950	+13.4	2.7	
HD 133001	sd O+G					
HD 149382	sd OB	-10	890	+23.2	2.0	

TABLE 2. Magnetic fields and radial velocities of the hot subdwarfs

are too weak for photoelectric observations, and a number of lines are too few for the photographic method. Nevertheless, I attempted to measure the longitudinal magnetic fields in the brightest hot subdwarfs, because Borra et al. (1983) proposed that magnetic fields in the magnetic subdwarfs might be expected to be larger than in the magnetic Ap/Bp stars.

The results of the magnetic field and the radial velocity measurements of the hot subdwarfs are presented in Table 2. I obtained Zeeman spectra of seven hot subdwarfs and one star Feige 87 with the photoelectric method. For two stars BD+25° 2534 and HD 133001 I was not successful in measuring the magnetic field. Magnetic observations of the other six subdwarfs reveal no evidence of magnetic fields.

5. Conclusions

Longitudinal magnetic fields have not been detected in any of the observed stars. However, I used a method which permits one to discover only large-scale fields. These stars can still have weaker magnetic fields (< 500 Gauss) or fields with complicated structures.

I wish to thank the International Science Foundation for awarding me a travel grant to attend Hot Stars in the Halo, and Dr. Saul Adelman and the organizing committee for this very useful conference.

REFERENCES

ADELMAN, S. J., HILL, G., 1987, MNRAS, 226, 581
ADELMAN, S. J., PHILIP, A. G. D., 1990, MNRAS, 247, 132
ADELMAN, S. J., SARGENT, W. L. W. 1972, ApJ, 176, 671
BORRA, E. F., LANDSTREET, J. D., THOMPSON, J., 1983, ApJS, 53, 151
GREENSTEIN, J. L., SARGENT, W. L. W., 1974, ApJS, 28, 157
KLOCHKOVA, V., PANCHUK, V., 1987, Astrofyz. Issled. (Izv. SAO), 26, 14
KODAIRA, K. 1973, A&A, 22, 273
KODAIRA, K., PHILIP, A. G. D., 1983, ApJ, 278, 208
MICHAUD, G., VAUCLAIR, S., VAUCLAIR, G., 1983, ApJ, 267, 256
OSMER, P. S., PETERSON, D. M., 1974, ApJ, 187, 117
SARGENT, W. L. W., SEARLE, L., 1967, ApJ, 150, L33
STALIO, R., 1974, A&A, 31, 89

SWEIGART, A. V., GROSS, P. G., 1976, ApJS, 32, 367

Discussion

SAFFER: The frequency of strongly magnetic white dwarfs is very small, 1-2 %. In my subdwarf survey of 80 stars, one star, PG 2218 + 051, could not be well fitted (in spectroscopy) - could it be magnetic? If so, the frequency of magnetism in subdwarfs would be similar to that of the WDs.

ELKIN: The number of possible magnetic subdwarfs must not differ significantly from number of magnetic white dwarfs if they have an evolutionary relationship. The problem is how to separate the magnetic from the nonmagnetic subdwarfs. Perhaps it is necessary to investigate those subdwarfs with chemical peculiarities and photometric variability, which could be connected with magnetic fields. This conjecture is by analogy with the CP stars.

BIDELMAN: Babcock and Preston have given some evidence of a magnetic field in RR Lyrae. Does anyone know if there has been any recent confirmation of this?

ELKIN: Babcock (1958, ApJS, 3, 141) found a magnetic field in RR Lyrae. Preston (1967, in Cameron R. S., ed, The Magnetic and Related Stars, Mono Book Corp., Baltimore, p. 3) could not confirm this result. But Romanov et al. (1988, in Glagolevskij Yu., Koplov, I., eds, Magnetic Stars, Leningrad, p. 51) found in RR Lyrae a magnetic field which varied for every set of observations.

What Does a FHB Star's Spectrum Look Like?

By C. J. CORBALLY[1] AND R. O. GRAY[2]

[1] Vatican Observatory Group, University of Arizona, Tucson, AZ 85721, USA

[2] Physics and Astronomy Department, Appalachian State University, Boone, NC 28608, USA

We are puzzled by a set of 67 early A-type stars in the Galactic Halo whose Strömgren photometric indices imply that they should be good Field Horizontal Branch candidates. However, their MK spectral classsfications show only two to have hydrogen lines with the relatively narrow wings (and weak metal lines) characteristic of conventional FHB stars. We have some evidence from theory that in the rest of this set we may indeed be looking at FHB stars, but ones with helium-rich atmospheres.

1. Introduction and Data

As reported in Corbally & Gray (1993) we have obtained blue-region CCD spectra at 2.8 Å resolution for 67 early A-type stars located at high galactic latitudes. These stars were listed by Philip (1984) as candidate field horizontal branch (FHB) stars on the basis of their small m_1 and large c_1 Strömgren indices. We classified their spectra on the MK System and were expecting most of them to have the relatively narrow hydrogen-line wings and slightly weak metallic lines that would characterize classical blue horizontal branch (BHB) stars (Slettebak et al. 1961). Instead, we found that about one third had spectra like those of normal dwarf or subgiant stars and only two would normally be picked out in the 3800-4900 Å region as BHB stars. Table 1 summarizes our classifications.

2. Previous clues

Some discomfort with the classical description of BHB (and so FHB) stars as having sharp and deep hydrogen lines is found in papers such as those by Searle & Rogers (1966), Sargent (1967), and MacConnell et al. (1971). In the preceding paper (Philip 1994, and also Philip & Lee 1985) we are shown spectra from the 6-meter telescope of three prototype FHB stars. A glance at these spectra clearly shows them to have broad hydrogen wings, characteristic of A-stars near the main sequence. Our own medium resolution spectra confirm this for all four prototype FHB stars of Oke, Greenstein & Gunn (1966), stars which have Balmer jumps characteristic of low-gravity atmospheres.

Dr. E. Böhm-Vitense (1993, private communication) has provided us with a likely explanation of the conflict between the Balmer jumps and the hydrogen-line wings. Her previous work on the atmospheres of helium stars (Böhm–Vitense 1967) showed that, as the hydrogen abundance decreased in an atmosphere, so the hydrogen-line wings became wider. If the FHB stars we observed have helium-rich atmospheres, then broader hydrogen lines and a large Balmer jump would be reasonable consequences of the reduced continuous opacity that is coupled with an increased pressure and electron density.

3. An initial model

Dr. R. L. Kurucz has kindly computed a model atmosphere for an initial helium-rich composition: He 0.9, M/total 0.1, $\log g = 2.5$, $T_{eff} = 9500$ K. A synthetic spectrum

Number	Description
20	normal dwarfs and subgiants
1	normal giant
27	weak-lined dwarfs and subgiants
10	λ Boötis type
3	Ap (mostly Si)
4	protoshell and peculiar
2	classical FHBs

TABLE 1. Summary of classifications for the FHB stars

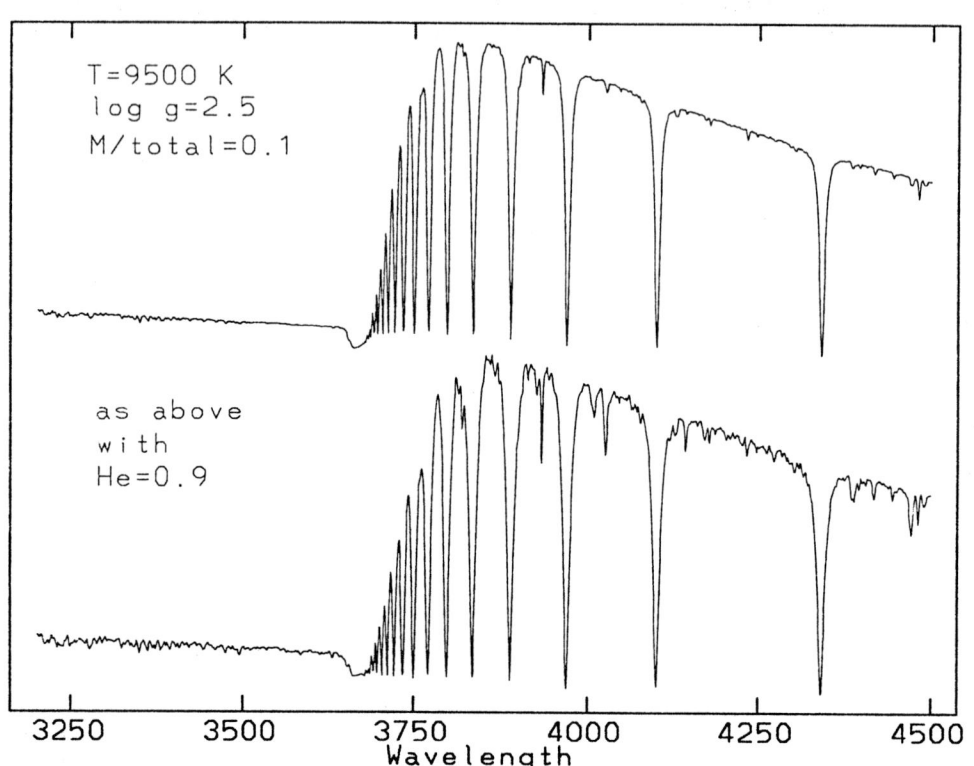

FIGURE 1.
Normal (top) and hydrogen-poor (bottom) synthesized spectra for the temperature, $\log g$, and metallicity indicated.

for this model was computed using SPECTRUM (Gray & Corbally 1994), and this is shown against the spectrum for a similar model, save with a normal hydrogen content, in Figure 1. It can be readily seen how the decrease in hydrogen abundance broadens the hydrogen-line wings, in fact to the breadth of about a B9 III star. It also enhances the metallic-line strengths for this metal-weak model. Of significance to the large photometric c_1 indices of the FHB stars is that Figure 1 shows the Balmer jump for a hydrogen-deficient star at this temperature is, if anything, increased.

Hence, we find the initial results encouraging. A hydrogen-deficient atmosphere might well be the explanation for our finding apparently genuine FHB stars that show broad hydrogen-line wings. However, the initial model gives overly strong helium lines (see Figure 1, bottom spectrum), so further modelling attempts will be made with lower helium abundances and somewhat cooler temperatures. Before we add a new class of hydrogen-poor BHB stars to the classical, normal-hydrogen ones, we need to have better evidence that the models reproduce what we see in the spectra.

We thank A. G. D. Philip for his list of FHB stars and some unpublished photometric data. We are indebted to R. L. Kurucz for his helium-rich model atmosphere.

REFERENCES

BÖHM-VITENSE, E., 1967, ApJ, 150, 483

CORBALLY, C. J., GRAY, R. O., 1993, in Dworestsky M. M., Castelli F., Faraggiana R., eds, Peculiar Versus Normal Phenomena in A-Type and Related Stars, ASP Conf. Ser., Vol. 44, p. 432

GRAY, R. O., CORBALLY, C. J., 1994, AJ, 107, 742

MACCONNELL, D. J., FRYE, R. L., BIDELMAN, W. P., BOND, H. E., 1971, PASP, 83, 98

OKE, J. B., GREENSTEIN, J. L., GUNN, J., 1966, in Stein R. F., Cameron A. G. W., eds, Stellar Evolution, Plenum, New York, p. 399

PHILIP, A. G. D., 1984, VVO Contributions, No. 2, p. 1

PHILIP, A. G. D., 1994, in Adelman S. J., Upgren A. R., Adelman C. J., eds, Cambridge University Press, Cambridge, p. 41

PHILIP, A. G. D., LEE, J., 1985, in Philip A. G. D., ed, Horizontal-Branch and UV-Bright Stars, L. Davis Press, Schenectady, p. 57

SARGENT, W. L. W., 1967, ApJ, 148, L147

SEARLE, L., ROGERS, A. W., 1966, ApJ, 143, 809

SLETTEBAK, A., BAHNER, K., STOCK, J., 1961 ApJ, 134, 195

Discussion

BIDELMAN: I have been interested for years in the possible existence of stars that are rather less hydrogen deficient than such "classic" objects as Upsilon Sagittarii. It is apparent that the hydrogen lines themselves would not be greatly affected but I did not realize that their wings would behave as you have described. Why do such stars have stronger H wings than usual?

CORBALLY: Böhm-Vitense has explained to us that due to the lower absorption coefficient in hydrogen-poor stars and due to the higher atomic weight for the remaining material, the layer down into which we see in hydrogen depleted giants has a higher pressure and electron density than in solar abundance giants. The hydrogen wings, therefore, become wider, looking like those in higher gravity stars. Further, since these atmospheres have a low continuous opacity, the line opacity to continuous opacity ratio is large. So, the lines, whether of hydrogen or of metals, will be stronger than expected for the abundances of these stars.

PHILIP: If Kodaira and Wallerstein were to redo their abundances for HD 161817 and HD 109995 with the idea that the H abundance was lower do you have an idea of what sort of differences they would find?

CORBALLY: We should ask them this; but I should expect that while their log g would remain about the same, as the H abundance was lowered, so would the metal abundances have to come down.

BEERS: It appears to me that the apparent increase in the metallic lines is a necessary consequence of your suggested mechanism for the increased breadth of the Balmer lines. The question is - if one were to identify broad Balmer lines in the presence of extremely weak metal lines, would this be an entirely different beast altogether?

CORBALLY: On the basis of the hydrogen poor models that we have considered, yes, this would be a different beast.

?: I want to congratulate you on an extraordinary paper that illustrates the power of the morphological approach of spectral classification. Can Sweigart and Demarque explain how the helium gets up to the surface (or is it a purely superficial effect of diffusive separation)?

PETERSON: Have you compared your calculated spectra with observations of the entire Balmer Series, including the Balmer jump?

CORBALLY: Not yet; however, we shall compare c_1 indices derived from the calculated spectra with those observed by Philip. This is certainly a good example of where the complementary aspects of photometry and spectroscopy lead to more complete spectroscopy becoming necessary.

A Technique for Distinguishing FHB Stars From A-Type Stars

By RONALD WILHELM[1], TIMOTHY C. BEERS[1]
AND RICHARD O. GRAY[2]

[1] Department of Physics and Astronomy, Michigan State University, East Lansing, MI 48824, USA

[2] Department of Physics and Astronomy, Appalachian State University, Boone, NC 28608, USA

A technique has been developed to calculate the stellar parameters, T_{eff}, log g, and [Fe/H] for field horizontal branch stars and main sequence A-type stars.

The effective temperatures and surface gravities for the FHB/A stars are calculated by comparing the broadband UBV colors to the color calibrations from Kurucz, and by comparing the measured width for Hδ and Hγ, from medium resolution spectroscopy, to a grid of 833 synthetic spectra encompassing the run of stellar parameters. The equivalent width of the Ca II K line is then compared to the grid at the appropriate T_{eff} and log g and a metal abundance is estimated.

Comparison to standard stars show a scatter in the stellar parameters of $\Delta T_{eff} = \pm 257$ K, $\Delta log\, g = \pm 0.24$, and $\Delta[Fe/H] = \pm 0.09$. The derived values for surface gravity from standard stars show a clear separation of the FHB and A stars. Derived [Fe/H] values for eight globular cluster BHB stars also show reasonably good agreement with published values. Comparison to the previously used photometric method indicates that the new technique will help resolve many ambiguous identifications. Potential problems, such as peculiar and metallic lined A stars and low S/N spectra, will also be discussed.

1. Motivation for the Technique

The ongoing HK objective prism survey of Beers *et al.* (1985, 1988), has been very successful in identifying "hot" stars in the halo and thick disk of the Galaxy. The vast majority of these stars, which are found by virtue of their weak or non-existent Ca II K line and strong Hϵ line, are either field horizontal branch or A-type main sequence stars. To date the survey has identified approximately 6700 FHB/A stars. Follow-up broadband UBV photometry and medium resolution (1-2 Å), low-signal-to-noise ($10 \leq S/N \leq 20$) spectroscopy has been completed on nearly 1000 of these stars.

Our ultimate goal is to use the FHB and A stars as kinematic tracers of both the halo and thick disk. To this end it is important to separate the two populations independent of either their kinematics or metallicities. An obvious technique is to utilize the inherent differences in surface gravity between the FHB and main sequence stars over most of the interesting temperature range. A procedure has been developed to compute the three physical parameters, T_{eff}, *log g*, and [Fe/H], by comparing the observed UBV colors and spectral line widths to a grid of synthetic values.

2. The Procedure

2.1. *Constructing the Grid*

A total of 833 synthetic spectra have been computed using the ATLAS9 Kurucz model atmospheres Kurucz (1992) and the spectral synthesis routine SPECTRUM by Richard Gray. SPECTRUM uses as input the pressure and temperature at each optical depth from the Kurucz models, and computes the normalized emergent flux over the wavelength region of 3900 Å $\leq \lambda \leq$ 4500 Å. The spectrum is then smoothed using a Gaussian filter (SMOOTH by Gray) to match the resolution of the observed spectra. Figure 1a shows a typical synthetic spectrum, smoothed to 1.9 Å resolution.

The parameter space coverage of the synthetic spectra is 6000 $\leq T_{eff} \leq$ 10,000 K, in increments of 250 K; 2.0 $\leq log\ g \leq$ 5.0 dex, in increments of 0.5 dex; and -3.0 $\leq [Fe/H] \leq$ 0.0, in increments of 0.5 dex.

After generating a synthetic spectrum for all combinations of the parameter space, a Voigt profile is fit to each Hδ, Hγ, and Ca II K feature. The width at 80 % of the local continuum level ($D_{0.2}$) is then calculated for the two Balmer lines and an equivalent width (W_k) is calculated for Ca II K. Figures 1b and 1c show the Voigt fit to the Hδ and Ca II K lines from figure 1a, as well as the derived line width for each. The Kurucz color calibration was then used to find synthetic colors for all possible combination of parameter values. A three dimensional grid containing the 5 synthetic observables; Hδ ($D_{0.2}$), Hγ ($D_{0.2}$), W_k, (B-V) and (U-B) at each point in T_{eff}, log g, and [Fe/H] was then constructed.

Finally, the observed spectra is fit with a Voigt profile in an identical manner as the synthetic spectra.

2.2. The Grid Comparison

As noted earlier, the program star spectra are of rather low S/N. Also the photometry suffers from small internal uncertainties and uncertainties in the adopted reddening values that amount to a scatter of about ± 0.03 in each of the colors. Therefore, it is advantageous to take into account the uncertainties in the observed values before matching to the synthetic grid. To accomplish this a grid point is considered a match to the observed values if the corresponding grid values are within a tolerance of $D_{0.2} = \pm 0.5$ Å, $W_k = \pm 0.3$ Å, and Colors = ±0.04. Figure 2 is an example of all possible matches in T_{eff} and log g, within the given tolerance values, for the prototypical FHB star HD 161817 and the main sequence A star HD 201601. Since the Ca II K line is a strong function of the overall metal abundance, it is not used in the determination of the temperature or surface gravity. Although the inclusion of the tolerance value has allowed many possible matches for an individual observable there is a clearly recognizable region in which most or all of the observables overlap. It is immediately apparent that the overlap regions indicate HD 161817 is a low surface gravity star while HD 201601 has a surface gravity indicative of a main sequence star.

To quantify the true overlap region, each observable/grid point match in T_{eff} and log g is given a weight between 1 and 4. The weight is in direct proportion to the dependency of that observable on each parameter at that point on the T_{eff}, log g plane. This weighting assures that the observables which best constrain a given region of parameter space are emphasized. Figures 3a-c are plots of the synthetic observables as a function of T_{eff}. It is clear from Figure 3a that (B-V) is a strong function of T_{eff} over most of the range in T_{eff}. It is therefore given a large dependency value. Figure 3b shows (U-B) to be a strong function of log g for $T_{eff} <$ 8000 K. In the case of 3c we see that $D_{0.2}$ is a strong function of T_{eff} over a range of 6000 - 8000 K. Above 8000 K, $D_{0.2}$ becomes a strong function of log g, which is, fortuitously, the temperature where (U-B) loses its log g dependency. The combination of all three observables result in fairly good T_{eff} and log g indicators over the entire range of parameter space.

A subspace interval is defined to be a grid point and its eight nearest neighbors. The value of each subspace interval is the sum of weights from each of the observables. The

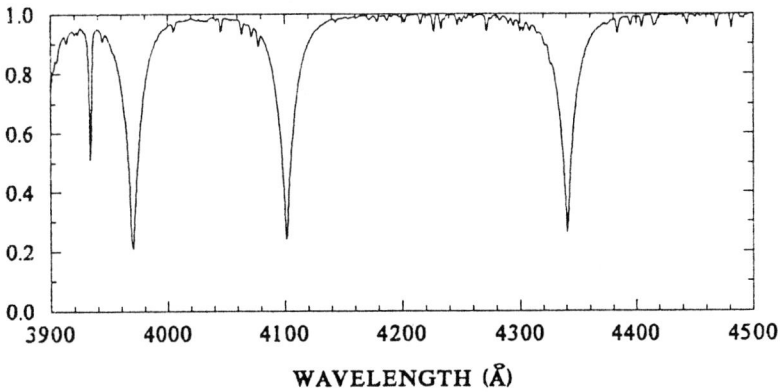
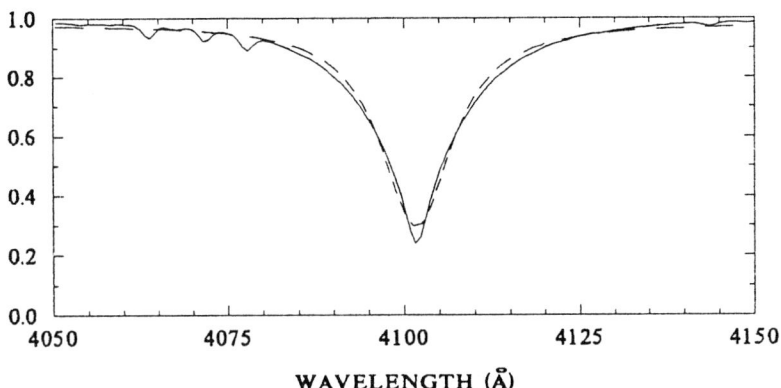
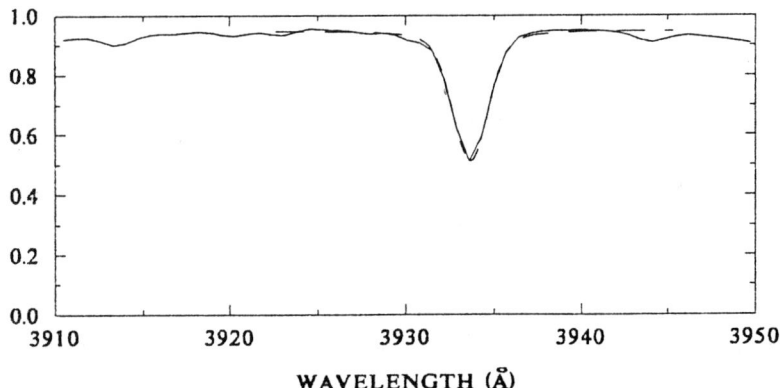

FIGURE 1. Top: Synthetic Spectrum with $T_{eff} = 7500$ K, log g $= 2.50$, and [Fe/H] $= -2.00$. Middle: Voigt fit to Hδ with $D_{0.2} = 19.5$ Å. Bottom: Voigt fit to Ca II K with $W_k = 1.52$ Å.

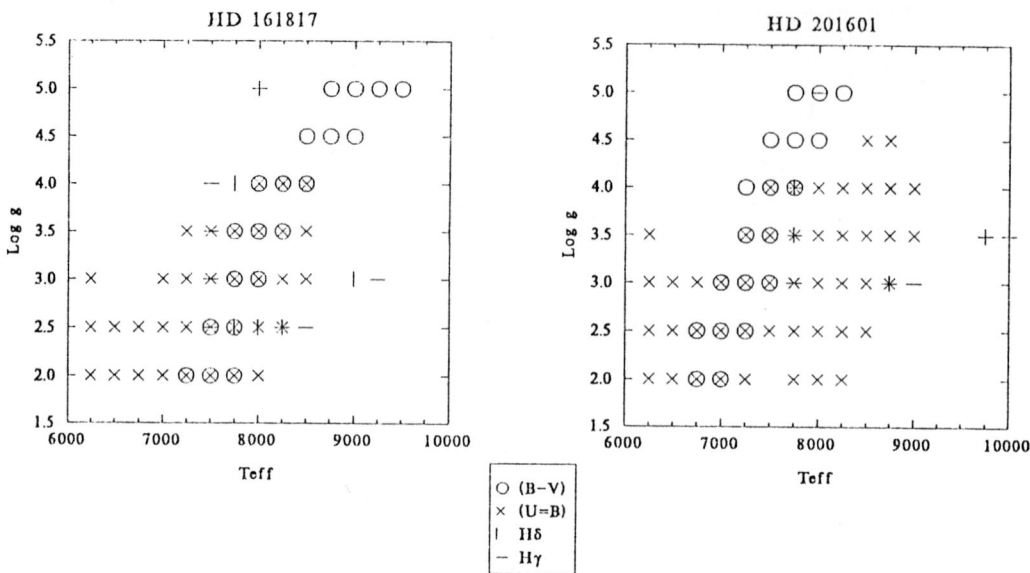

FIGURE 2. All possible observable matches for FHB star HD 161817 and main sequence A star HD 201601.

interval with a maximum value is then the true position of the overlap region from which an initial value for T_{eff} and log g is found.

The observed parameter, W_k, serves as our metal abundance indicator. We have presumed that a monotonic relation between [Ca/Fe] and [Fe/H] exists for the stars under consideration, and that the strength of Ca II K is indicative. The procedure finds the best grid match for W_k at the initial values of T_{eff} and log. This match gives an estimate of the metal abundance, parameterized by [Fe/H].

The color matches are then constrained by the initial value of [Fe/H], and a new overlap region in the T_{eff}, log g plane is found. To insure the overlap region is consistent with the actual observed parameters the simultaneous fit to all the observables at a given T_{eff} and log g is calculated and divided by the value of the subspace interval at that point. This division prevents a severely deviant observable from "dragging" the solution too far from the identified overlap region. The minimum of this calculation gives the new T_{eff} and log g values. The process is iterated one more time and the final values of T_{eff}, log g and [Fe/H] are found.

3. Results

To test both the ability to predict the correct stellar parameters and separate the FHB and A star populations, stars with parameters determined from high dispersion spectroscopy were observed at KPNO using the Goldcam spectrograph on the 2.1 meter telescope. A total of 20 FHB/A stars were observed at high S/N (> 50) and at a resolution of 1.9 Å.

Eight of these stars have parameter values from more than 4 sources. These were used as the primary standards to test the errors in our calculated values. The published values for the A stars are from Cayrel de Strobel et al. (1992), as well as the T_{eff} and log g values for the FHB stars. The metal abundance for the FHB stars were adopted

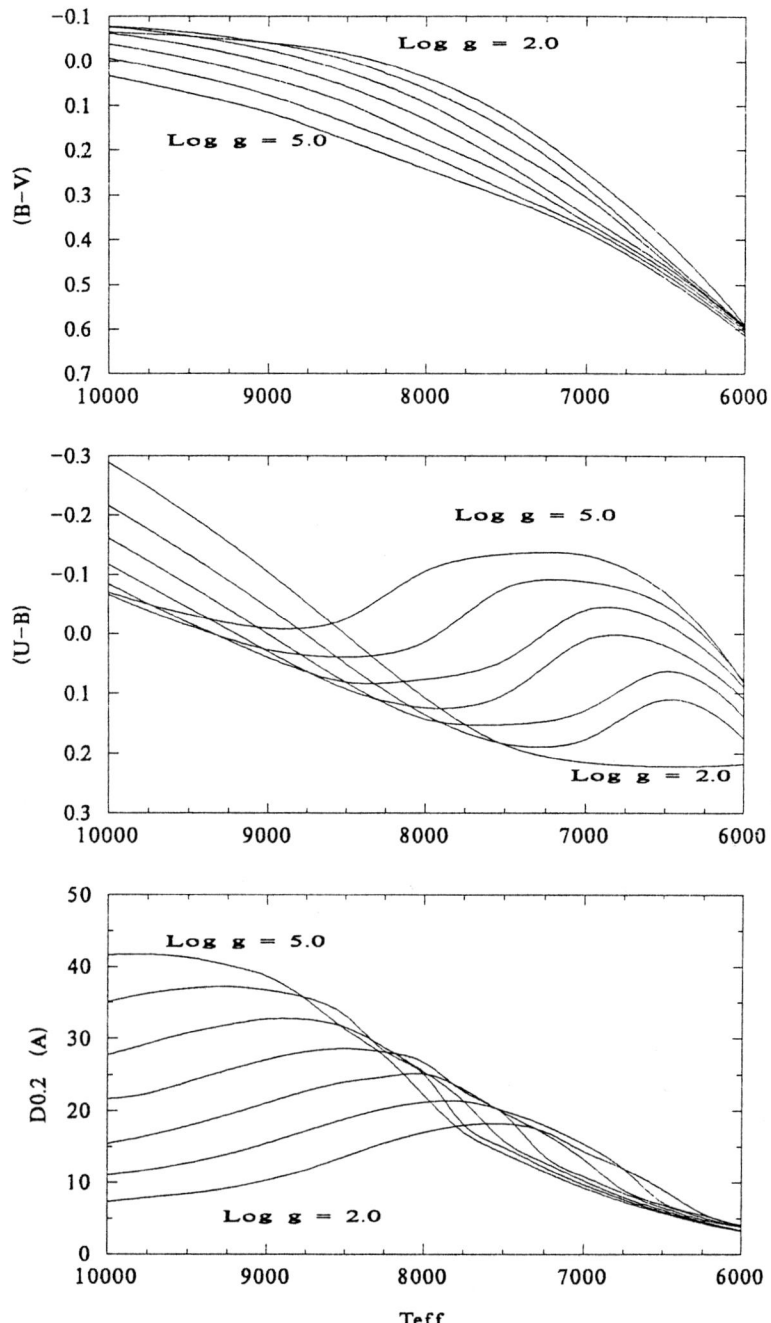

FIGURE 3. The three temperature and surface gravity indicators as a function of T_{eff}.

	Publ.			Calc.			Diff.		
HD Star	T_{eff}	log g	[Fe/H]	T_{eff}	log g	[Fe/H]	ΔT_{eff}	ΔLog g	Δ[Fe/H]
2857	7636	2.90	-1.50	7672	3.13	-1.54	36	0.23	-0.04
86986	7800	3.10	-1.82	7939	3.22	-1.97	139	0.12	-0.15
109995	7950	3.10	-1.87	8218	3.07	-2.00	268	-0.03	-0.13
161817	7750	2.90	-1.64	7635	2.49	-1.59	-115	-0.41	+0.05
189849	7675	3.50	0.00	8590	3.26	0.00	915	-0.24	0.00
193432	10285	3.80	0.00	9761	3.36	-0.05	-524	-0.44	-0.05
201601	7750	3.75	0.00	7721	3.87	0.00	-29	0.12	0.00
214994	9500	3.60	0.20	9294	3.39	0.00	-206	-0.21	-0.20

TABLE 1. Published and calculated parameters for the standard FHB and main sequence A - type stars

from Adelman & Philip (1994) with the exception of HD 2857 which was from Cayrel de Strobel. The scatter in the published values for the eight standards is $\sigma(T_{eff}) = \pm$ 100 K, $\sigma(\text{Log } g) = \pm 0.1$ dex, and $\sigma([Fe/H]) = \pm 0.13$ dex.

Table 1 lists the star names, the average of the published values, the calculated values from our technique and the differences between the two. The calculated values show good agreement with the exception of the T_{eff} value for HD 189849. This star has a somewhat uncertain reddening correction, which is probably the main cause of the discrepent value. The mean and standard deviation for the difference values of all 8 standards is, $\Delta T_{eff} = 60 \pm 419$ K, $\Delta \log g = -0.14 \pm 0.24$, and $\Delta[Fe/H] = -0.07 \pm 0.09$. Excluding HD 189849 from the T_{eff} calculation drops the scatter to $\pm$ 257 K.

Figure 4a is a plot of the differences in log g, as a function of log g, for the 20 observed standards. Many of the stars in this plot have only one source or a large scatter between sources. Still the range in differences fall between $\pm$ 0.5 dex for both the FHB and "normal" A stars. This is quite good, considering the grid spacing in log g is 0.5 dex. The metallic A stars, however, give values considerably too low. This is due to excess line blanketing in the U band pass. Since (U-B) is used as one of the primary surface gravity indicators, the result is a low value in log g (see Figure 3b).

To further test the procedure's ability to predict metal abundance, spectra of eight blue horizontal branch stars (hereafter, BHB) in the globular clusters, M5, M10, M13, and M92, were obtained during two separate runs at KPNO. Because these stars are intrinsically faint the range of signal-to-noise is, $8 < S/N < 22$. This range is also typical for most program FHB/A spectra from the Beers et al. survey. The mean and standard deviation for the difference values in metallicity is $\Delta[Fe/H] = -0.07 \pm 0.43$ dex. At present we have no U magnitudes for 5 of these stars. Therefore, (U-B) colors were estimated from fiducial fits to the horizontal branch from published two color diagrams. With the inclusion of the actual (U-B) values we expect the scatter in [Fe/H] to be reduced and the above value should be considered an upper limit for the technique. Figure 4b shows the difference values for [Fe/H] as a function of [Fe/H] for the eight BHB stars, as well as, the 20 standards. The scatter about the zero line is quite small for the standards. The larger scatter in the BHB values is due to uncertainties in the U magnitude and the low S/N spectra.

With the stellar parameters now calculated it is possible to test the procedures ability to separate the FHB and main sequence A stars. Figure 5 is a log g, T_{eff} plot using the calculated values of all the standard stars and the eight globular cluster stars. The upper dashed line, at log g = 3.5, is the nominal cut-off for a main sequence star in this temperature range. The lower dashed line at log g = 3.26 reflects the uncertainty

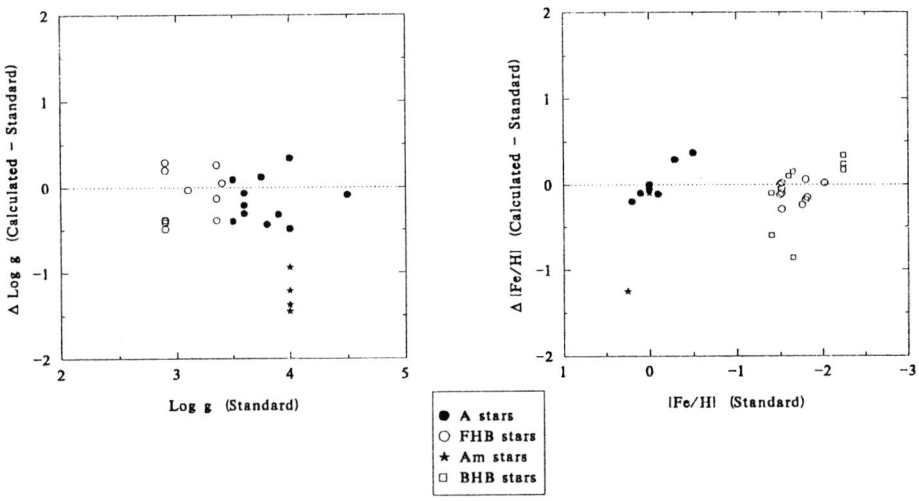

FIGURE 4. Calculated minus Standard values for log g and [Fe/H]

in the calculated surface gravity. The lower solid line is the theoretical ZAHB line from Sweigart (1987), for stars with a helium abundance of Y = 0.25. Since the general trend of stars on the horizontal branch is to move to lower surface gravities as they evolve, this line is the upper limit for log g at a given T_{eff}. The upper solid line is again the uncertainty of 0.24 dex in log g. From these lines we expect unambiguous separation of the two populations for T_{eff} < 7500 K. Temperatures below 9000 K should separate reasonably well since the majority of FHB stars will not be on the ZAHB. Above 9000 K the two populations are no longer separable in log g.

The open circles and squares in Figure 5 are the FHB and BHB stars respectively. The filled circles are normal main-sequence A/F stars and the filled stars are metallic-lined A stars. The two populations separate quite well for T_{eff} < 9000 K with the exception of the Am stars. These stars will be identified by a different technique. As expected both populations overlap at T_{eff} > 9000 K and can not be separated using this plot.

4. Conclusion and Other Considerations

From the preceeding results it appears this technique will be quite successful in separating FHB stars from main sequence A stars for temperatures less than 9000 K. Owing to the ability to get reasonably good metal abundances, $\sigma([Fe/H]) < \pm 0.5$ dex, this technique is a great improvement over the two color diagram separation technique of Beers et al. (1992), where overlaps in metallicity between the two populations prevented an unambiguous classification for a significant number of stars. Once full analysis of the kinematic and chemical characteristics of the separable populations are completed it is hoped these characteristics can be used as a critria for separation of those stars with $T_{eff} > 9000$ K.

The matter of greatest concern is the ability to predict good stellar parameters from low S/N spectra. To quantify the effect low S/N has on the derived parameters, Gaussian noise was added to the synthetic spectra and the line widths calculated. Results indicate

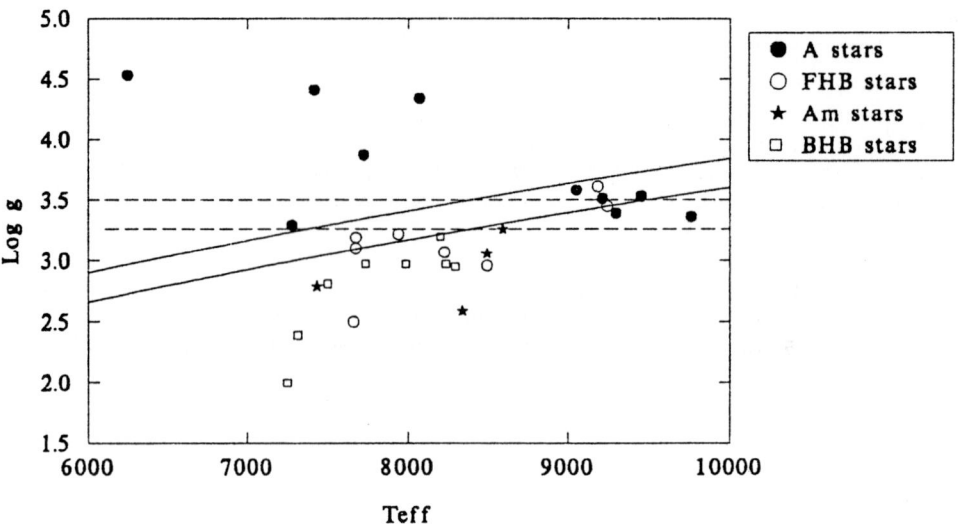

FIGURE 5. Separation of FHB and A stars by virtue of differing surface gravities

that that for S/N < 20 an appreciable scatter ($\sigma(log\ g)$ and $\sigma([Fe/H])$) of about 0.3 dex) in the derived parameters is introduced. The main contributor to this scatter is the uncertainty in the local continuum level. Several techniques for smoothing the continuum are currently being tested and should help to reduce this scatter significantly. A further shortcoming of this technique is the inability to separate out the metallic-lined A stars. These are known to comprise about 15 % of the A star population. To identify these stars we hope to use a metallic-line averaging technique similiar to that of Pier (1983), which sums 10 metallic lines and flags those stars which show resultant signals much greater than the continuum noise.

REFERENCES

ADELMAN, S. J., PHILIP, A. G. D., 1994, Elemental abundances of the field horizontal-branch stars III. MNRAS, in press

BEERS, T. C., PRESTON, G. W., SHECTMAN, S. A., 1985, A search for stars of very low metal abundance. AJ, 90, 2089–2102

BEERS, T. C., PRESTON, G. W., SHECTMAN, S. A., 1988, A catalog of candidate field horizontal-branch stars. ApJS, 67, 461–501

BEERS, T. C., PRESTON, G. W., SHECTMAN, S. A., DOINIDIS, S. P., GRIFFIN, K. E., 1992, Spectroscopy of hot stars in the halo. AJ, 103, 267–296

CAYREL DE STROBEL, G., HAUCK, B., FRANCOIS, P., THEVENIN, F., FRIEL, E., MERMILLIOD, M., BORDE, S., 1992, A catalogue of [Fe/H] determinations: 1991 edition. A&AS, 95, 273–336

KURUCZ, R. L., 1992, private communication

PIER, J. R., 1983, AB stars in the southern galactic halo II. Spectroscopy and radial velocities. ApJS, 53, 791–813

SWEIGART, A. V., 1987, Evolutionary sequences for horizontal-branch stars. ApJS, 65, 95–135

Discussion

BOND: Two questions: 1.) Did you consider using Strömgren photometry, which would be much more sensitive to log g than UBV? 2.) Do you have any preliminary reactions to the Corbally paper, which might indicate that getting log g from the Balmer wings could give the wrong answer?

WILHELM: 1.) I know the Strömgren system is much better for log g determinations but all the photometry to date has been UBV so we are working within that system. 2.) Do you mean besides not wanting to come up here? I do not think its a real problem. We did get the correct values for the FHB stars. By using tolerance values and depending on all 5 observables it really carries us over small variations in the Balmer line width.

CORBALLY: I noticed that the prototype FHB star HD 161817, which was classified by Gray and me as A8V weak-metals, was used by you (Wilhelm) as a test object and came out with a log g = 2.5. So, its broader Balmer line wings (possibly produced by a hydrogen-poor atmosphere) were presumably countered by its colors in your fitting procedure and so it was assigned a low rather than high gravity.

WILHELM: We do get a low surface gravity since at this temperature (7750 K) the log g values are chosen mainly from the photometry. The log g dependence of the Balmer lines really kick in at $T_{eff} > 8000$ K.

CARNEY: Since you used Kurucz's synthetic colors and Gray's synthetic spectra, have you checked to see that Kurucz's synthetic spectra agree with those calculated by Gray?

WILHELM: Yes, I have compared the Balmer lines produced by SPECTRUM and the Balmer lines of Kurucz and I find virtually no difference. SPECTRUM uses the temperatures and pressures at each optical depth from a Kurucz model atmosphere as input and gives the same Balmer line profiles as calculated by Kurucz. Richard has really done a wonderful job with SPECTRUM.

KINMAN: We must be aware that there may be Population I stars that can mimic Field Horizontal Branch stars. What we are currently trying to do, either empirically or with the help of models is to find stars in the field that match as best as we can those that are found on the horizontal branches of globular clusters. There may be small differences between the spectra of those local stars and those found in globular clusters and this requires further investigation.

Elemental Abundances of Halo A and Interloper stars

By SAUL J. ADELMAN[1] and A. G. DAVIS PHILIP[2,3]

[1]Department of Physics, The Citadel, Charleston, SC 29409, USA

[2]Institute for Space Observations, 1125 Oxford Place, Schenectady, NY 12308, USA

[3]Union College, Schenectady, NY 12308, USA

We discuss the elemental abundances of several halo A type stars derived using high dispersion spectra with signal-to-noise ratios of 50 or more. These include field horizontal-branch A stars and the A0 V star 7 Sex. We also analyzed two stars HD 60825 and HD 64488 with the uvby colors of FHB stars, but upon examination of their high dispersion spectra proved not to be such stars. The FHB A type stars have somewhat different chemical compositions than other Population II stars. Both 7 Sex and HD 60825 have nearly solar derived elemental abundances while HD 64488 is slightly metal-poor.

1. Introduction

We summarize our studies of the elemental abundances of halo A type stars, especially field horizontal-branch (FHB) A stars and the A0 V star 7 Sex. By using high dispersion spectra with S/N ratios of 50 or more we have utilized much weaker lines than previous workers. This has lead to more accurate and precise abundances. We also discuss interlopers in our sample with the uvby colors of FHB stars.

Photometric, spectrophotometric, and spectroscopic studies have shown that field horizontal-branch stars are the analogs of globular cluster stars on blue horizontal-branches (Philip 1985, Adelman & Hill 1987). Thus studies of the brighter field HB stars can tell us much about the fainter globular cluster HB stars. Further comparative studies of field and globular cluster HB stars might indicate if they were formed under similar circumstances.

Astronomers are interested in identifying horizontal-branch (HB) stars and determining their elemental abundances as:

1. In the HR diagram the morphology of theoretical HB sequences and, most importantly, the ages derived for globular clusters are both sensitive to the assumed chemical composition. The ages, for example, obtained by matching theoretical isochrones, or from the luminosity difference between the horizontal-branch and main-sequence turnoff, depend on the assumed O/Fe ratio.

2. Elemental abundance differences between HB stars and Population II dwarfs and giants may indicate the results of the first dredge-up, the He-flash and/or mixing beyond that predicted by current evolutionary models. Thus, accurately determined HB-star abundances can provide an important test of canonical stellar evolution theory.

3. Elements whose abundances remain unaffected by processing on the giant branch presumably reflect the state of nucleosynthesis of galactic material at a very early epoch, and thus are an important probe of Galactic chemical evolution.

4. Hydrodynamical process such as diffusion in the stellar envelope are expected to become progressively more important towards the blue end of the HB. Thus abundance determinations may provide sensitive probes of the hydrodynamical state of subphotospheric layers (Michaud et al. 1983).

2. Identifying FHB A stars

Low dispersion spectroscopic methods are an important way to identify FHB stars. For example, MacConnell et al. (1971) presented a list of possible FHB stars identified from Schmidt objective-prism spectra. This list also contains some stars with higher luminosities than normal HB stars.

Four-color uvby measures are another good way to identify FHB stars. The higher luminosity of FHB stars, relative to the Population I main-sequence A stars, means these stars have a lower surface gravity. The c_1 index thus is larger than that for comparable Population I stars by about 0.2 mag or larger. For stars later than spectral type A3, the lower metallicity results in a smaller m_1 index, but only by a few hundredths of a magnitude, but increases for the cooler A-type stars. Kilkenny & Hill (1975) located the various types of early-type stars in the m_1, (b-y) and c_1, (b-y) diagrams. As white dwarfs and FHB stars fall furthest from the main-sequence stars, the four-color system is such a good detector of these two stellar types. Philip (see Philip 1978 for references) published six finding lists of FHB stars that have been identified by means of four-color photometry.

The combination of spectroscopic and photometric observations is a better way to select candidates. Even with four-color measures it is still possible for stars of other types to fall in the photometric area occupied by FHB stars. Notable exceptions (stars which have the four-color characteristics of FHB stars, but on later analysis turn out to be something else) are HD 57336 (UV spectra show strong absorption lines, characteristic of Population I), BD +18° 4873 (found to be a Population I A star in the halo by Philip & Adelman 1990), HD 64488 (Adelman & Hill 1987), and HD 60825 (found to be a Population I A star by Adelman & Philip 1994).

The ultraviolet energy distributions of horizontal-branch A stars show significant differences compared with those of Population I stars (Huenemoerder, de Boer & Code 1984). The horizontal-branch A stars have weak spectral features. Further the flux from these stars at wavelengths smaller than 1800 Å is larger than the flux from Population I stars of the same temperature. This difference decreases at larger effective temperatures. Usually the ultraviolet and optical results agree. But some stars show discrepant results such as HD 57336.

Philip (1978) presented Hβ measures of FHB stars and showed that the Hβ index is smaller for a FHB star than it is for a main-sequence star. As the Hβ index is unaffected by interstellar reddening this index is an important way, in conjunction with four-color measures, to determine the color excesses of A-type stars. Unfortunately the narrow Hβ filter does not let much light through. For stars fainter than 11 magnitude the probable errors of the Hβ index (made with a 1 meter telescope) are too large to get a precise enough index to accomplish a reliable reddening determination.

Stetson (1991) used uvbyβ photometry to selected candidate FHB stars. Adelman & Philip (1994) obtained high dispersion spectra at KPNO which showed that three of his stars are true FHB stars. Their IUE low dispersion exposures of two of these stars and seven others confirmed that they are class members. This strongly indicates the value of the β index.

Inspection of high dispersion spectra is an important way to confirm that a star is a member of Population II. It has to be metal weak and relatively slow rotator. Peterson (1985) found that the maximum v sin i for horizontal-branch stars is about 30 km s^{-1}.

Elemental abundance analyses of good quality high dispersion spectra are the ultimate way to confirm membership in Population II. Unfortunately many of the older published studies have serious problems. As the stellar metal lines are usually quite weak, errors in continuum placement will result in systematic errors in the equivalent widths. Spectra with at least modest signal-to-noise ratios (50+) and coudé type dispersions are needed. They can be obtained with solid state detectors such as CCDs and Reticons or by the coaddition of photographic plates.

3. A New generation of elemental abundance analyses

The use of higher signal-to-noise data at high resolution has been a primary factor in the improvement of elemental abundance analyses over the past decade. If one wants to obtain coverage of the photographic region, one can either coadd digitized photographic region spectrograms with a computer or obtain many spectral regions with electronic detectors such as CCDs. Previous to our current studies only one FHB star, HD 86986, was analyzed with spectrograms have a signal-to-noise ratio greater than 25 (Kodaira 1973). These are weak lined stars so that at least moderate signal-to-noise spectrograms are need for their analyses.

Several years ago Adelman & Hill (1987) performed fine analyses of two prototype field horizontal-branch (FHB) A-Type stars, HD 109995 and HD 161817 and the interloper A star HD 64488 using of order 10 coadded 6.8 Å mm^{-1} IIaO spectrograms obtained at the Dominion Astrophysical Observatory. The analyses of the FHB stars used lines with equivalent widths as weak as 7 mÅ. Adelman, Fisher & Hill (1987) produced a spectral atlas which shows the resulting spectra. These analyses have been updated for this paper, using ATLAS9 model atmospheres (Kurucz 1993) and making minor changes in the gf values mainly to incorporate newer more accurate values. We refit the spectrophotometry and the Hγ profiles with the predictions of the new models and found T_{eff} = 8150 K, log g = 3.25 for HD 109995 with [M/H] = -2.0 and T_{eff} = 7225 K, log g = 2.80 for HD 161817 with [M/H] = -1.5. These values are slightly cooler and of slightly lower gravity than found by Adelman & Hill (1987).

Adelman & Philip (1990a, 1992a, 1994) obtained several spectra of the FHB stars HD 74721, HD 86986, HD 93329, HD 128801, HD 130095, and HD 167105 as well as a one each of HD 117880 and HD 202759 during three observing runs with the coudé feed telescope of Kitt Peak National Observatory. They used a TI CCD, camera 5, and grating A which in the blue gave a reciprocal dispersion of 4.7 Å mm^{-1}, a resolution of 0.14 Å, and a spectral coverage of about 55 Å. Observations of HD 109995 and HD 161817 and of several superficially normal B and A stars established that there are small systematic wavelength differences between the equivalent width scales of this camera-grating combination and that of the spectrographs of the Dominion Astrophysical Observatory (Adelman & Philip 1990b, 1992b, 1994). For this paper we give results based on the DAO equivalent width scale.

The spectroscopic material is analyzed with Kurucz's (1993) ATLAS9 model atmospheres. These models have more accurate line opacities than previous generations of model atmospheres, especially in the ultraviolet. The gf values for all analyses are based on the most accurate sources. A program by Hubeny (private communication) was used to calculate He I line profiles for the λ4472 line in HD 128801. We found He/H = 0.05 $\pm$ 0.015. This is slightly smaller than Buzzoni et al. 1983 value of 0.075 for Population II stars. The other FHB stars were too cool to produce He I lines in the photographic region. Table 1 summarizes the results. The solar values are from Anders & Grevesse (1989) except for the more recent iron value from Biemont et al. (1991) and Holweger et al. (1991). To convert the derived abundances which we found relative to the total number of atoms per unit volume to the number of hydrogen atoms we assume N_{He}/N_H = 0.075, the Buzzoni et al. (1983) value for Population II stars except for HD 128801.

| | | | | HD Number | | | | | |
Species	74721	86986	93329	109995	128801	130095	161817	167105	Sun
He I	...	...	...	...	-1.30	...	...	...	-1.00
O I	...	...	...	...	-4.02	...	...	...	-3.09
Mg I	-5.53	-5.78	-5.51	-5.91	...	...	-5.69	...	-4.42
Mg II	-5.26	-5.44	...	-5.49	-5.21	-5.97	-5.13	...	-4.42
Al I	-7.43	-7.92	-7.56	-7.93	-6.91	-8.07	-7.76	-8.07	-5.54
Si II	-5.55	-5.90	-5.76	-5.85	-5.77	-5.98	-5.86	-5.87	-4.35
Ca I	-7.52	-7.29	-7.18	-7.42	...	-8.81	-7.13	-7.81	-5.64
Ca II	-6.95	-7.12	-6.82	-7.42	-7.94	-8.39	-7.00	-7.45	-5.64
Sc II	-10.63	-10.74	-10.35	-10.83	...	...	-10.54	-11.06	-8.90
Ti I	...	...	...	...	...	...	-8.51	...	-6.98
Ti II	-8.10	-8.36	-8.05	-8.37	-8.31	-9.07	-8.32	-8.40	-6.98
V II	-9.19	-9.34	-9.32	-9.23	...	...	-9.44	...	-8.00
Cr I	...	-8.33	-8.01	-8.35	...	...	-8.36	-8.06	-6.33
Cr II	-7.75	-7.94	-7.63	-7.92	-7.86	-8.19	-7.90	-7.32	-6.33
Mn I	...	...	...	...	...	...	-8.79	...	-6.55
Fe I	-6.09	-6.40	-6.03	-6.30	-5.69	-6.44	-6.21	-6.41	-4.52
Fe II	-5.92	-6.26	-5.89	-6.48	-5.83	-6.63	-6.11	-6.19	-4.52
Co I	...	...	...	...	...	...	-8.56	...	-7.08
Ni I	...	...	...	...	...	...	-7.39	...	-5.76
Sr II	-10.94	-11.42	-11.33	-11.65	...	...	-10.40	...	-9.10
Y II	...	-11.63	-10.88	-11.55	...	...	-11.18	...	-9.76
Zr II	-10.43	-10.83	-10.44	...	...	...	-10.51	...	-9.44
Ba II	-11.68	-12.10	...	...	...	...	-11.88	...	-9.87
Eu II	...	...	...	...	...	...	-12.68	...	-11.49

Note: Other FHB stars with derived abundances follow:
HD 117880   Si II -5.35
HD 202759   Ti II -8.83, Fe I -6.75, and Fe II -6.97

TABLE 1. FHB star abundances

As noted by Adelman & Philip (1992) the values for HD 86986 and HD 10999 agree quite well. Hence their other abundances might be quite similar. That for oxygen needs to be checked as HD 109995 has one of the largest values of [O/Fe] known for Population II stars (Adelman, Hayes & Philip 1986), outside the range of values found for most other Population II stars. Figure 1 compares their abundances anomalies and those for HD 167105. HD 167105 is somewhat similar to HD 86986 but is somewhat metal-poorer than both HD 86986 and HD 109995.

HD 74721 is somewhat more metal rich than HD 86986 and HD 109995 which means it is more like HD 161817. The greatest difference is in the Sr abundances. HD 93329 is similar to both of these stars, but is somewhat Sr poorer. Figure 2 compares the abundance anomalies for HD 74721, HD 93329, and HD 161817.

Figure 3 compares HD 128801 with HD 130095 and HD 109995, which is intermediate in metallicity. HD 130095 is more metal poor than the other better studied FHB stars although HD 202759 for which we have just the Ti and Fe abundance may be even more metal-poor. HD 128801, which is somewhat hotter than the other FHB stars, shows a different abundance pattern than the other FHB A stars studied. It is the least Fe deficient; yet is one of the most Ca deficient.

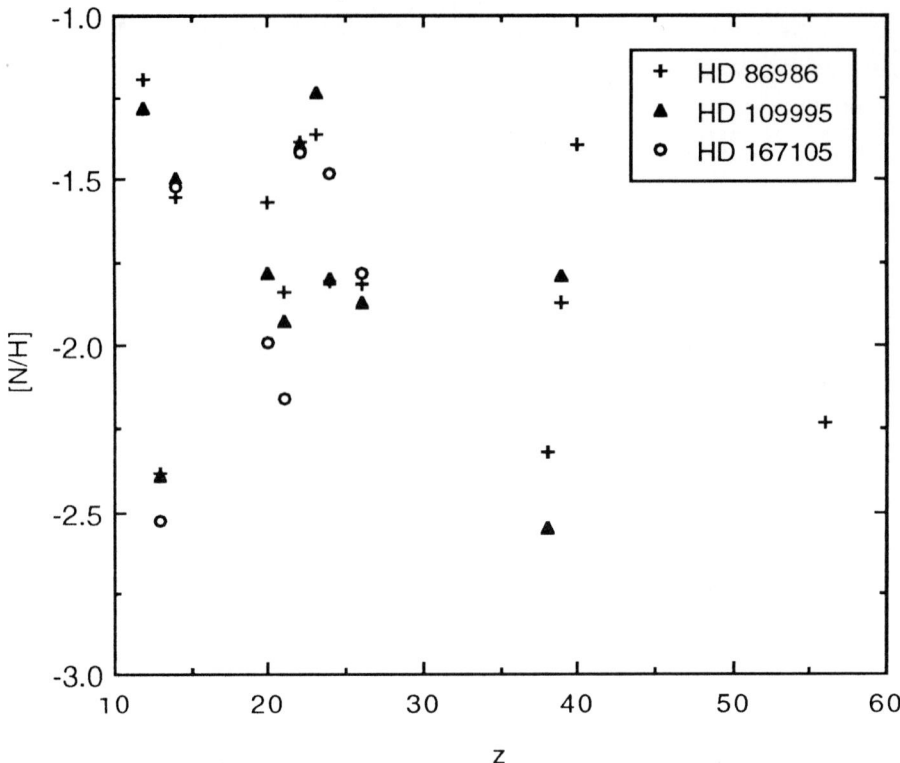

FIGURE 1. The elemental abundance anomalies [N/H] of the FHB A stars, HD 86986, HD 109995, and HD 167105.

Unfortunately, we have only obtained six abundances for all of the 8 FHB stars: Al, Si, Ca, Ti, Cr, and Fe. For these elements, we performed a correlation analysis of their abundance anomalies with one another and with the values of effective temperature and surface gravity. For eight items, there is a one in twenty probability (P = 0.050) of the correlation being produced by chance if the linear correlation coefficient r = 0.707 (Bevington & Robinson 1992). We found positive results for Al vs. Fe (r = 0.934), Ca vs. Ti (r = 0.832), Ti vs. Cr (r = 0.719), and Ti vs. Fe (r = 0.712). Most of which are of close equilibrium group elements whose abundance should be correlated. That for Al and Fe is not of this class. That the Ca and Fe abundance anomalies are uncorrelated (r = 0.301) means that one cannot use either abundance to infer the other.

Table 2 compares the abundance anomalies ([N/H]) of our FHB A stars with those of giants in five globular clusters (Gratton 1982, Pilachowski, Sneden & Wallerstein 1983). Although all stars are metal poor, there are clearly differences. In particular the FHB A stars are for a given [Fe/H] value more silicon and calcium deficient. Further none of the FHB A stars match the abundance patterns of the globular cluster giants in detail.

Luck & Bond (1985) analyzed metal-deficient field red giants and showed the trends for [O/Fe], [Ca/Fe], [Ti/Fe], [Cr/Fe], [Ni/Fe] as a function of [Fe/H]. For our 8 FHB A stars these values also given at the bottom of Table 2. Most fall within or close to the trends found by Luck & Bond (1985). However, the [Ca/Fe] values for HD 74721, HD

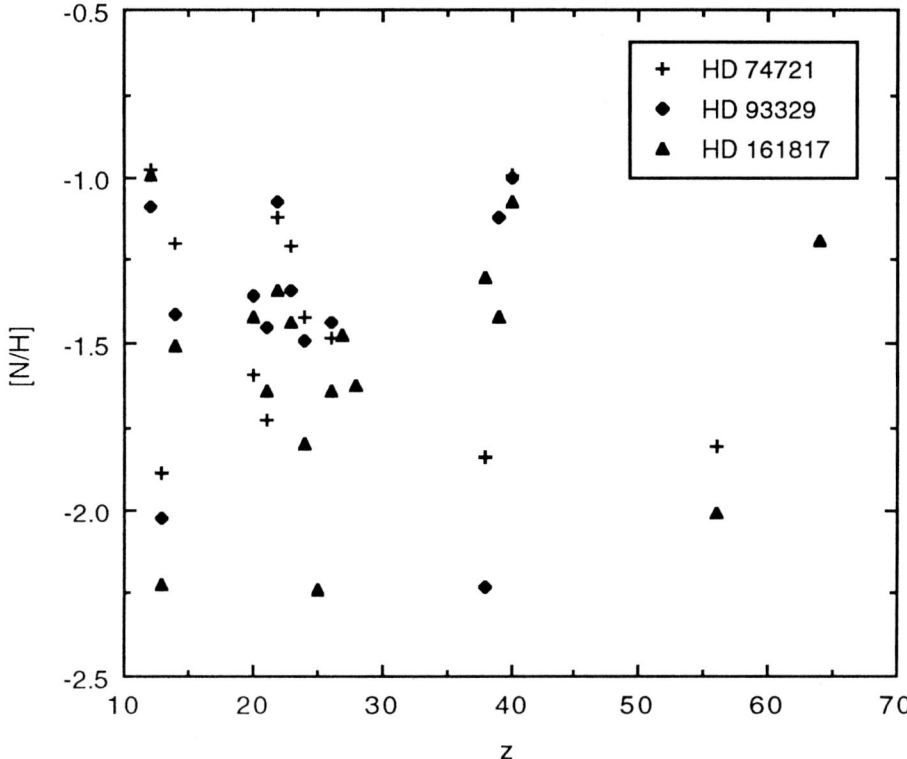

FIGURE 2. The elemental abundance anomalies [N/H] of the FHB A stars, HD 74721, HD 93329, and HD 161817. These stars are less metal-poor than the 3 stars of Figure 1.

109995, HD 128801, HD 130095, and HD 167105 are clearly below the trend of values with those for HD 122801 and HD 130095 being substantially different, of order 1.0 dex. This suggests that some of the FHB stars have different abundance anomalies from metal-deficient field red giants with similar values of [Fe/H].

These comparisons use only a few FHB A stars with accurate abundances while there are many more stars of other types. Thus we probably have not seen the full range of FHB A star abundances although most probably have [Fe/H] values between -1.4 and -2.0 dex. We are also limited as not all of these stars have spectra covering most of the photographic region.

4. 7 Sextantis

Rodgers & Wood (1970) suggested that 7 Sex (HR 3906 = HD 85504), which has a relatively high proper motion combined with a radial velocity of $+97$ km s^{-1}, is an old disk horizontal-branch star. Sargent, Searle & Wallerstein (1964) found that its abundances were near solar and Rodgers & Wood (1970) derived normal helium and iron abundances. Cacciari (1985) noted its UV spectrum at the blue end is steeper than in normal A0 V Population I stars, indicating an earlier spectral type or the presence of a hot companion. But, Carney (1983) found no evidence for same. Adelman & Philip (1992a) using a DAO Reticon observation (2.4 Å mm^{-1}) with 67 Å of spectral coverage

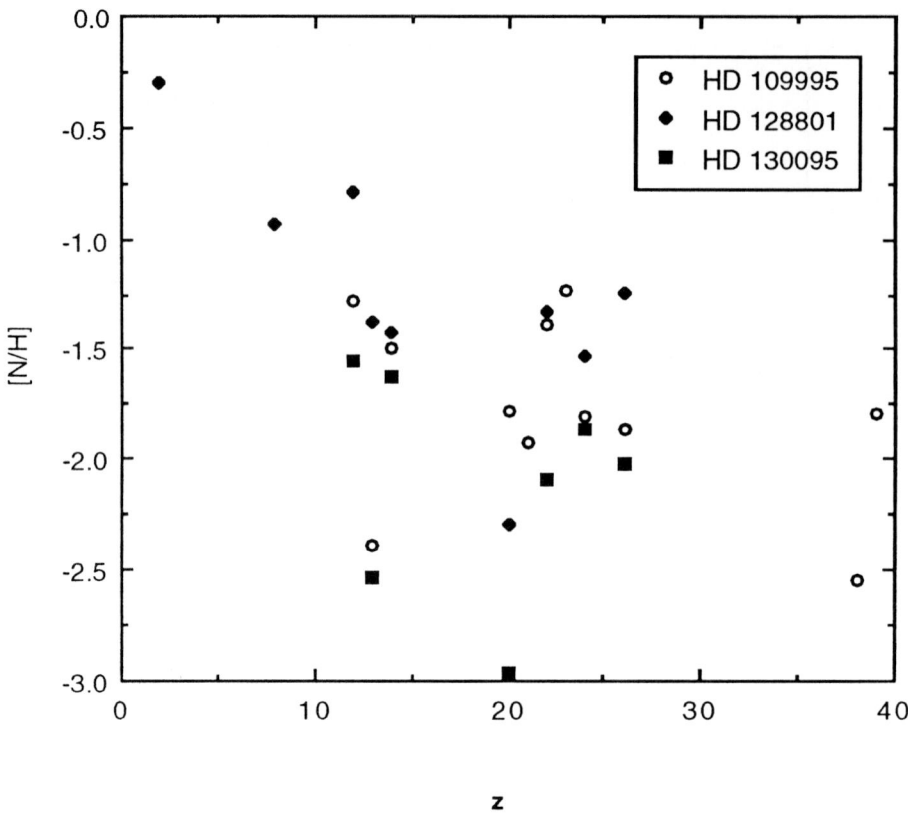

FIGURE 3. The elemental abundance anomalies [N/H] of the FHB A stars, HD 109995, HD 128801, and HD 130095. HD 128801 is for the most part the least metal-deficient star of our sample while HD 130095 is the most metal-deficient star of our sample with a number of derived abundances. The values of HD 109995 are shown for comparison.

and several KPNO CCD observations similar to those for the FHB A stars found that its derived abundances were generally close to solar with the Mg and S abundances tending to be greater than solar. Adelman & Philip (1994) have obtained and analyzed additional KPNO spectra. The results on the DAO equivalent width scale are in Table 3 along with several other superficially normal B and A stars studied by Adelman (1991) and the Sun.

Most of the derived abundances of 7 Sex are close to solar. Excluding S which is somewhat uncertain the mean [N/H] = 0.18 ± 0.21 which suggests 7 Sex is marginally metal rich. The S as the Zr abundance is based only on one line. Of greater certainty both C and Cr have abundances +0.5 dex greater than solar. The overall pattern suggests a marginal metal richness. At temperatures near 10,000 K, one can find among the superficially normal Population I stars, those which are metal-rich, the HgMn and hot Am stars, those which have near solar metals like 7 Sex, and those which are metal poor, for example α Dra and Vega. For the metal rich stars some sort of hydrodynamic processes appear to be operating in the stellar envelope. Some of, but not all of, the metal poor stars can be explained as rapid rotators seen pole-on (e.g., Vega) (Gulliver, Hill & Adelman 1994). Almost all of the stars with near solar abundances have at least one abundance definitely different from solar. Our understanding what is happening is still fragmentary. But when one finds a star with near solar abundances and a large

Species	FHB Star HD Number								Gratton		Pilachowski et al.		
	74721	86986	93329	109995	128801	130095	161817	167105	NGC 3201	M22	NGC 4833	NGC 6254	NGC 6752
He	...	...	...	...	-0.30	...	...	...	...	...	...	...	...
O	...	...	...	...	-0.93	...	...	...	-1.0	-1.8	-1.51	-1.23	-1.50
Mg	-0.98	-1.19	-1.09	-1.28	-0.79	-1.55	-0.99	...	-0.9	-0.8	...	...	...
Al	-1.89	-2.38	-2.02	-2.39	-1.37	-2.53	-2.22	-2.53	...	...	...	...	...
Si	-1.20	-1.55	-1.41	-1.50	-1.42	-1.63	-1.51	-1.52	-0.8	-1.1	...	...	...
Ca	-1.60	-1.56	-1.36	-1.78	-2.30	-2.96	-1.42	-1.99	-1.2	-1.5	-1.04	-0.90	-0.90
Sc	-1.73	-1.84	-1.45	-1.93	...	...	-1.64	-2.16	-1.2	-1.9	-1.37	-1.40	-1.27
Ti	-1.12	-1.38	-1.07	-1.39	-1.33	-2.09	-1.34	-1.42	-1.0	-1.5	-0.98	-0.97	-1.01
V	-1.21	-1.36	-1.34	-1.23	...	...	-1.44	...	-1.2	-1.7	-1.25	-1.06	-1.33
Cr	-1.42	-1.80	-1.49	-1.80	-1.53	-1.86	-1.80	-1.48	-1.0	-2.0	-1.12	-1.44	-1.38
Mn	...	...	...	...	...	...	-2.24	...	-1.3	-1.8	...	...	...
Fe	-1.48	-1.82	-1.44	-1.87	-1.24	-2.02	-1.64	-1.78	-1.2	-1.9	-1.34	-1.28	-1.26
Co	...	...	...	...	...	...	-1.48	...	-1.0	-1.5	-0.73	...	-1.37
Ni	...	...	...	...	...	...	-1.63	...	-1.2	-1.7	-1.49	-1.50	-1.35
Sr	-1.84	-2.32	-2.23	-2.55	...	...	-1.30	...	...	...	...	...	...
Y	...	-1.87	-1.12	-1.79	...	...	-1.42	...	-1.6	-1.8	...	...	...
Zr	-0.99	-1.39	-1.00	...	...	...	-1.07	...	-1.6	...	...	...	-1.56
Ba	-1.81	-2.23	...	...	...	...	-2.01	...	-1.6	-0.8	-1.57	...	-1.47
Eu	...	...	...	...	...	...	-1.19	...	...	...	-1.53	...	-1.74
[O/Fe]	...	...	...	...	+0.31	...	...	...					
[Ca/Fe]	-0.12	+0.26	+0.08	+0.00	-1.06	-0.94	+0.22	-0.21					
[Ti/Fe]	+0.36	+0.44	+0.37	+0.48	-0.09	-0.07	+0.30	+0.36					
[Cr/Fe]	+0.06	+0.02	-0.05	+0.07	...	...	-0.16	+0.30					
[Ni/Fe]	...	...	...	...	...	...	+0.01	...					

TABLE 2. Comparison of abundance anomalies [N/H] of FHB A stars with other population II objects

radial velocity explaining what is happening is difficult. That Adelman & Philip (1992a) found it had a v sin i of 22 km s^{-1} might be a clue to its history. Perhaps it is related to the blue straggler phenomenon.

5. Interlopers

Fine analyses have been performed for two interloper A type stars. Adelman & Philip (1994) identified HD 60825 as a Population I interloper A star using KPNO CCD exposures similar to those used for the FHB A stars. Adelman & Hill (1987) found HD 64488, whose four-color photometry, optical spectrophotometry, and ultraviolet fluxes were similar to those of horizontal-branch stars, to be a relatively fast rotating star (v sin i = 147 km s^{-1}) whose abundances were, on the whole, slightly less than solar.

We used the homogeneous uvbyβ colors of Hauck & Mermilliod (1980) and uvby photometry from Philip (private communication) with the procedure of Moon & Dworetsky (1985) as revised by Napiwotzki et al. (1993) to obtain new estimates of the effective temperature and surface gravity for HD 64488. We found T_{eff} = 8775 K, log g = 3.53,

Spec.	π Cet	21 Aql	134 Tau	ν Cap	7 Sex	α Dra	o Peg	Vega	γ Gem	θ Leo	HD 60825	HD 64488	Sun
He I	-1.07	-1.05	-1.00	-1.19	-1.00	-1.40	-1.26	-1.52	...	-1.22	...	...	-1.00
C II	-3.77	-3.92	-3.45	-3.39	-2.84	-3.78	-4.40	...	...	...	...	...	-3.31
O I	-3.30	-3.24	...	-3.33	...	-3.60	-3.36	...	...	...	...	...	-4.01
Mg I	-4.84	-4.89	-5.19	-4.71	-3.95	-4.75	-4.49	-5.07	-4.30	-4.53	...	-5.16	-4.42
Mg II	-4.52	-4.59	-4.53	-4.61	-4.55	-4.82	-4.54	-5.11	-4.71	-4.53	-4.64	-4.78	-4.42
Al I	-5.81	-6.14	-5.85	-6.03	-5.81	-6.11	-5.58	...	-5.98	-6.03	...	-6.81	-5.53
Si II	-4.52	-4.40	-4.51	-4.69	-4.37	-4.94	-4.43	...	-4.41	-4.46	...	-4.24	-4.45
S II	-4.82	-5.04	-4.53	-4.85	-3.81	-5.03	-4.00	...	-4.60	-4.32	-3.41	...	-4.79
Ca I	...	...	-5.92	-5.98	-5.22	-6.16	-5.61	-6.21	-5.73	-5.76	-5.73	-5.48	-5.64
Ca II	-5.72	-5.66	-5.33	-5.55	-5.29	-5.64	-5.43	...	...	-5.57	...	-5.43	-5.64
Sc II	...	...	-9.25	-9.34	-9.00	-9.81	-9.30	-9.62	-9.13	-9.27	-8.87	-9.59	-8.90
Ti I	...	...	...	...	...	...	...	...	-6.62	...	...	...	-6.98
Ti II	-7.17	-7.46	-7.06	-7.05	-6.75	-7.27	-6.86	-7.47	-6.74	-6.95	-6.90	-7.18	-6.98
V II	...	...	-8.11	-7.64	-7.90	-7.92	-7.31	...	-7.79	-7.45	-7.74	...	-8.00
Cr I	...	...	-5.87	-6.20	-5.79	-6.36	-6.16	...	-5.97	-6.31	-6.18	...	-6.33
Cr II	-6.54	-6.64	-6.41	-6.13	-5.95	-6.61	-6.17	-6.77	-5.99	-6.32	-6.09	-6.28	-6.33
Mn I	...	...	...	...	...	-6.73	-6.42	-7.16	...	...	-6.72	...	-6.55
Mn II	...	...	...	-6.69	-6.36	...	-6.22	-7.20	-6.35	-6.31	-6.28	...	-6.55
Fe I	...	-4.71	-4.58	-4.50	-4.21	-4.91	-4.32	-5.05	-4.41	-4.52	-4.48	-5.01	-4.52
Fe II	-4.62	-4.80	-4.63	-4.47	-4.35	-4.93	-4.35	-5.12	-4.44	-4.43	-4.52	-5.13	-4.52
Ni II	-5.98	-6.04	-5.85	-5.67	-5.61	-5.92	-5.00	-6.29	-5.57	-5.34	-5.39	...	-5.75
Sr II	...	...	-9.23	-8.77	-8.99	-9.58	-8.01	...	-9.07	-8.31	-9.22	-9.74	-9.10
Y II	...	...	...	...	...	...	-9.13	...	-10.05	-9.48	...	...	-9.76
Zr II	...	...	...	...	-8.94	...	...	...	-9.11	...	-8.92	...	-9.44
Ba II	...	...	...	-9.29	-9.66	...	...	-10.58	-9.40	...	-9.89	...	-9.87
Teff	13150	12900	10825	10250	10135	10075	9600	9400	9260	9250	9000	8700	
log g	3.85	3.35	3.83	3.90	3.69	3.30	3.60	4.03	3.60	3.55	3.50	3.70	
ξ(km/s)	0.0	0.0	0.0	0.0	0.7	0.4	1.8	0.6	1.2	1.7	2.0	3.0	

TABLE 3. Comparison of derived and solar abundances (log N/H)

values which are similar to those derived by Adelman & Hill (1987) who found T_{eff} = 8700 K, log g = 3.70. The predicted energy distribution for a [M/H] = -0.5 ATLAS9 model agrees well with the observed fluxes (Philip & Hayes 1983), but the observed Hγ profile has to be shifted slightly in residual intensity to agree with the predictions. The amount was only slightly more than needed for HD 109995 and for HD 161817. Thus we adopted the new model parameters. A rederivation of the microturbulence results in 3.0 km s^{-1} which is similar to previous results. The abundances are listed also in Table 3. 0.04 dex is added to convert log N/N_T values to log N/H values. It appears to be star with slightly weaker metals on the whole than the Population I stars of this table with its abundances tending to fall within the range of the other stars. Silicon, calcium, titanium, and chromium are nearly solar while the other abundances especially that of aluminum are less than solar. The errors in the equivalent widths are larger than those of the other stars due to its higher rotational velocity. Thus it is probably a slightly metal poor Population I star.

As the derived abundances for HD 60825 are close to solar, it is appropriate to assume a solar He/H ratio for them and add +0.04 dex to the log N/N_T values to obtain the log N/H values. These abundances on the DAO scale are compared in Table 3 with those of other superficially normal B and A stars from this paper and from Adelman (1991) and with solar values. Of the derived abundances only sulfur and zirconium, results based on weak lines, appear to be than 0.5 dex greater than solar for HD 60825. This star appears to have slightly more solar abundances than 7 Sex.

Among the cooler stars in Table 3, γ Gem and HD 60825 contrast with α Dra and Vega which are metal poor and with o Peg and θ Leo, the hot Am stars which are metal rich. Only a few stars among the sharp-lined late B and early A stars have been found with nearly solar abundances while the majority have exhibited abundance anomalies. Thus we need to understand why γ Gem and HD 60825 are so different. Improving their abundance analyses to see whether with more spectral coverage and higher signal-to-noise spectra the degree of agreement with the solar values increases or decreases is important.

We thank Kitt Peak National and Dominion Astrophysical Observatories for their assistance. This research was supported in part by a grant to SJA from The Citadel Development Foundation.

REFERENCES

ADELMAN, S. J., 1991, MNRAS, 252, 116
ADELMAN, S. J., FISHER, W. A., HILL, G., 1987, Publ. Dom. Astrophys. Obs. Victoria, 16, 203
ADELMAN, S. J., HILL, G., 1987, MNRAS, 226, 581
ADELMAN, S. J., PHILIP, A. G. D., 1990a, MNRAS, 247, 132
ADELMAN, S. J., PHILIP, A. G. D., 1990b, PASP, 102, 842
ADELMAN, S. J., PHILIP, A. G. D., 1992a, MNRAS, 254, 539
ADELMAN, S. J., PHILIP, A. G. D., 1992b, PASP, 104, 316
ADELMAN, S. J., PHILIP, A. G. D., 1994, MNRAS, in press
ANDERS, E., GREVESSE, N., 1989, Geochim. Cosmochim. Acta, 53, 197
BEVINGTON, P. R., ROBINSON, D. K., 1992, Data Reduction and Error Analysis for the Physical Sciences, 2nd edition, McGraw-Hill, New York
BIEMONT, E., BAUDOUX, M., KURUCZ, R. L., ANSBACHER, W., PINNINGTON, E. H., 1991, A&A, 249, 539
CACCIARI, C., 1985, A&AS, 61, 407
CARNEY, B. W., 1983, AJ, 88, 623
GRATTON, R. G., 1982, A&A, 115, 171
GULLIVER, A. F., HILL, G., ADELMAN, S. J., 1994, ApJ, in press
HOLWEGER, H., BARD, A., KOCK, A., KOCK, M., 1991, A&A, 249, 545
HUENEMOERDER, D. P., DE BOER, K. S., CODE, A. D., 1984, AJ, 89, 851
KODAIRA, K., 1973, A&A, 22, 273
KILKENNY, D., HILL, P. W., 1975, MNRAS, 173, 625
KURUCZ, R. L., 1993, ASP Conf. Series, 44, 87
LUCK, R. E., BOND, H. E., 1985, ApJ, 292, 559
MACCONNELL, D. J., FRYE, R. L., BIDELMAN, W. P., BOND, H. E., 1971, PASP, 83, 98
MICHAUD, G., VAUCLAIR, G., VAUCLAIR, S., 1983, ApJ, 267, 256
MOON, T. T., DWORETSKY, M. M., 1985, MNRAS, 217, 305
NAPIWOTZKI, R., SCHÖNBERNER, D., WENSKE, V., 1993, A&A, 268, 653
PETERSON, R. C., 1985, in Philip A. G. D., ed, Horizontal-Branch and UV-Bright Stars, L. Davis Press, Schenectady, p. 85

PHILIP, A. G. D., 1978, in Philip A. G. D., Hayes D. S., eds, The HR Diagram: The 100th Anniversary of Henry Norris Russell, Reidel, Dordrecht, p. 209
PHILIP, A. G. D., 1985, in Philip A. G. D., ed, Horizontal-Branch and UV-Bright Stars, L. Davis Press, Schenectady, p. 41
PHILIP, A. G. D., 1983, ApJS 53, 751
PHILIP, A. G. D., HAYES, D. S., 1983, ApJS 53, 751
PILACHOWSKI, C. A., SNEDEN, C., WALLERSTEIN, G., 1983, ApJS, 52, 241
RODGERS, A. W., WOOD, P. R., 1970, ApJ, 161, L145
SARGENT, W. L. W., SEARLE, L., WALLERSTEIN, G., 1964, ApJ, 139, 1015
STETSON, P. B., 1991, AJ 102, 589

Discussion

CORBALLY: Richard Gray and I noticed for HD 130095 at classification resolution that besides being slightly metal weak, especially for the Ca, it showed a slightly enhanced λ 4233 line. Did you notice any other proto-shell characteristics in this star's spectrum?

ADELMAN: I did not. But we did not observe λ 4233 in all stars.

BEERS: When you point out peculiar Ca characteristics in FHB stars, naturally it catches our attention! We do find that for the FHB stars in the HK survey that have available UBV photometry and 1 Å resolution spectroscopy, there appears to be good agreement between [Fe/H] and the expected Ca II K strength.

ADELMAN: HD 128801 of our FHB stars is the most discrepant.

CARNEY: How sensitive is your [Ca/Fe] result to your adopted temperature, compared to the other element-to-iron ratios?

ADELMAN: My [Ca/Fe] is relatively insensitive to 100 K changes in effective temperature or 0.2 dex changes in log g parameter. Some of the element to iron ratios vary by 0.05 dex for similar changes.

PETERSON: Uncertainties in T_{eff} and in transitions probabilities need to be carefully evaluated. Especially near 10,000 K, the effective temperatures derived from the photometric indices become uncertain. Laboratory transition probabilities frequently contain ±0.2 dex errors which may depend on strength and excitation.

ADELMAN: Most of our FHB stars have effective temperatures near 8000 K. HD 128801 with a value of 10,000 K is certainly affected as you say. But one can require ionization equilibrium for say Fe I and Fe II to be achieved.

The Mass of Blue Horizontal-Branch Stars in the Globular Cluster NGC 6397

By KLAAS S. DE BOER[1], JELANA H. SCHMIDT[1], AND ULI HEBER[2]

[1]Sternwarte, University of Bonn, D-53121 Bonn, GERMANY

[2]Remeis Sternwarte, Univ. of Erlangen, D-96049 Bamberg, GERMANY

From new spectra and existing photometry of horizontal-branch stars in the globular cluster NGC 6397 we have derived temperature and gravity. Using the distance derived from colour-magnitude diagrams the luminosity of the stars and thus their mass can be calculated.

Our preliminary results show that the HB stars have masses of $0.2\ M_\odot < M_{HB*} < 0.6\ M_\odot$, within the very small temperature range of $9000 < T_{eff} < 12,000$ K. They have a systematic trend from low temperature and large mass to high temperature and small mass. This mass range does not fit classical models of HB-type stars, nor does it fit more exotic suggestions for the origin of the HB stars.

1. Introduction

Stars on the blue end of the zero-age horizontal branch (ZAHB) in colour-magnitude diagrams (CMDs) are thought to have a mass in a small range above the minimum He-core mass of $0.45\ M_\odot$. These stars are burning He and the thinness of the outer hydrogen atmosphere determines their blueness. They are thought to have evolved from the red giant stage. However, it has become evident that stars hotter than those on the classical HB, stars of sdB, sdOB, or sdO type, which are said to form the extended HB (EHB) in the CMD, do not fit to models of HB stars. The past history of these stars is not so clear anymore and several scenarios for their origin have been proposed. These include: the classical evolution from the giant branch to the HB but now with enhanced mass loss (possibly in a binary system; Iben & Tutukov 1985), and the formation of a new object from two merging stars (e.g. white dwarfs; Iben 1990). The latter scenarios lead to a range of possible masses different from the range expected from classical evolution. Related with the history is also the problem of the 'gaps' in observed HBs (see Moehler, Heber & de Boer 1994).

To discriminate between these theories one has to determine the mass of actual HB and EHB stars. This is, in principle, straightforward. Using good photometric (*ubvy*) and spectroscopic (Balmer line profiles) data one can determine T_{eff} and log g of the stars. The luminosity of the star follows if one knows its distance and the mass can be calculated.

For field stars such studies have used the reversed reasoning, one then finds the distance of the star by assuming a mass (usually $0.5\ M_\odot$). In directionally selected statistically complete samples scale heights then can be derived (e.g., for sdB stars: Heber 1986; Moehler, Heber & de Boer 1990; Theissen et al. 1993) and, including proper motions, also space motions (Colin et al. 1994).

Using globular clusters with a known distance one can derive the mass of the stars. Such studies are available for NGC 6752 (Heber et al. 1986) and M 15 (Moehler, de Boer & Heber 1992), and are reviewed by Moehler, Heber & de Boer (1994).

Here we present preliminary results for a study of the HB stars in NGC 6397, stars which are rather cool and probably similar to the classical field HB stars (see e.g., Philip

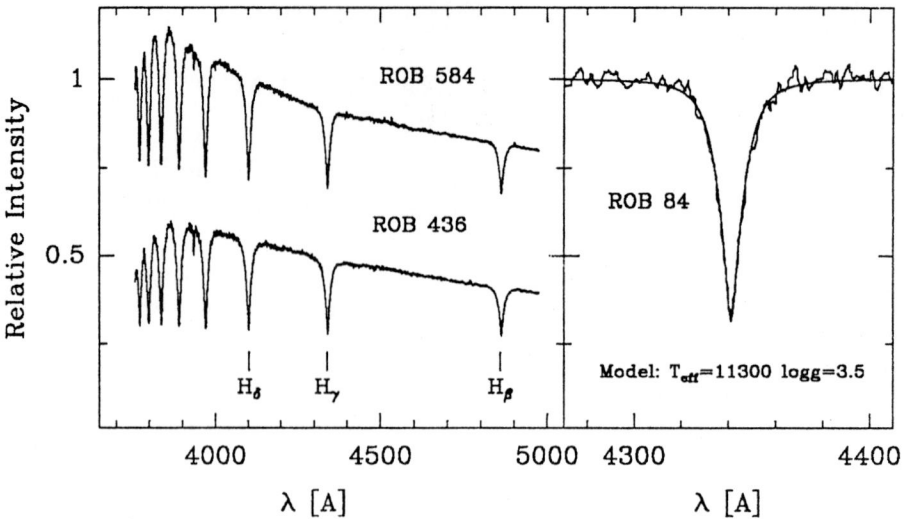

FIGURE 1. Sample spectra of HB stars in NGC 6397 are shown; full spectra for two stars (at left), and the fitted Hγ line with derived parameters for one star (at right)

1980; Kodaira & Philip 1984; Huenemoerder, de Boer & Code 1984). Spectra were obtained in June 1991 at the 1.5m B&C equipped telescope of ESO on La Silla. Strömgren photometry was available from Graham & Doremus (1968). A full account will be given by us elsewhere.

2. Spectra and stellar parameters

The spectra have been reduced in the standard way. Comparing the observed Balmer-line profiles with theoretical ones calculated with the ATLAS8 code (Kurucz 1979) best fits are made. Our preliminary results are obtained using solar metallicity models. Normally, a range of pair-values (T_{eff}, log g) gives a good fit. The Strömgren photometry was used in comparison with theoretical Strömgren indices (Lester et al. 1986) to also find a range of pairs (T_{eff}, log g) permitted by the photometry. From the combined data the best solution was found. Figure 1 shows a representative fit. The derived parameters are given in Table 1.

The stellar masses have been determined in the following way: The angular diameters of the stars were calculated comparing theoretical visual fluxes and apparent visual brightnesses. From the angular diameters and the distance of the stars the stellar radii were obtained. The mass of each star follows from the stellar radius and the surface gravity. Additionally, for several stars IUE spectra had been obtained (de Boer 1981), and these can be used to derive the integrated flux from the actual spectral photometry. In this case also a correction for reddening has to be applied. Both methods agreed in a satisfactory way.

3. HB-star masses

The masses derived for the HB stars in NGC 6397 lie between 0.2 and 0.6 $M_\odot$ (see Figure 2), while theoretical models give at least 0.45 $M_\odot$ for the masses of HB stars.

Star[a]	V	T_{eff}	log g	$M_{HB*}[M_\odot]$[b]	$+\Delta M_{HB*}[M_\odot]$	$-\Delta M_{HB*}[M_\odot]$
48	13.09	9050	3.3	0.54	+0.22	−0.19
56	13.95	11300	3.6	0.27	+0.11	−0.09
84	13.74	11300	3.5	0.29	+0.12	−0.10
102	13.90	11050	3.4	0.19	+0.08	−0.07
127	13.80	11100	3.5	0.24	+0.10	−0.08
147	13.25	9250	3.3	0.44	+0.18	−0.15
210	13.63	10000	3.5	0.36	+0.15	−0.12
436	13.20	9000	3.2	0.38	+0.15	−0.13
486	13.46	9950	3.6	0.56	+0.23	−0.19
495	13.73	10950	3.6	0.38	+0.15	−0.13
604	13.64	10500	3.5	0.37	+0.14	−0.12

[a] Star name as given by Woolley et al. (1961)
[b] A distance modulus of $(m-M)_0 = 11.71$ (2.2 kpc) (Drukier et al. 1993) has been used.
[c] The estimated errors of T_{eff} and log g are $\Delta T_{eff} \simeq 300\,K$ and $\Delta\,logg = 0.1\,dex$.

TABLE 1. HB-stars of NGC 6397 and their properties

The first fact we note is that the stars lie in a well defined range in the M_{HB*} versus T_{eff} diagram. The hotter the star the less mass it has. This trend agrees with the theory for ZAHB stars. Secondly, the stars cover the mass range of 0.6 to 0.2 $M_\odot$ within a range of just 9,000 to 12,000 K (!). On the other hand, the stars selected for the programme cover the entire range in B−V of the blue part of the HB of NGC 6397 (as seen in the CMD of Alcaino & Liller 1980).

The masses derived depend, of course, on all the assumptions as sketched above. One uncertain parameter is the distance modulus. Anthony-Twarog et al. (1992) have an extensive discussion of the uncertainties involved. To see what effect an error in $(m-M)_0$ may have we have also determined the masses in case $(m-M)_0 = 11.92$ (Alcaino & Liller 1980) instead of $(m-M)_0 = 11.71$ (see Table 1).

To make the effect of a different distance modulus clear we have plotted the mass ranges side by side in Figure 2. An increase of $(m-M)_0$ by 0.2 mag shifts the smallest mass from 0.19 to 0.23 $M_\odot$, the largest from 0.56 to 0.68 $M_\odot$.

4. Discussion

We are faced with the result that the blue HB stars in NGC 6397 have a range in mass reaching definitely below that predicted by ZAHB models.

Since 0.45 $M_\odot$ is required for He to ignite, and since the blue HB stars seem to look normal in their spectra and in the CMD, we can only speculate about explanations. Is there extra mass loss after He ignition? Have the bluest HB stars suffered from stripping encounters?

We like to thank the staff on La Silla for their help with the observations. This research is in part supported by the DFG under grant Bo 779/11.

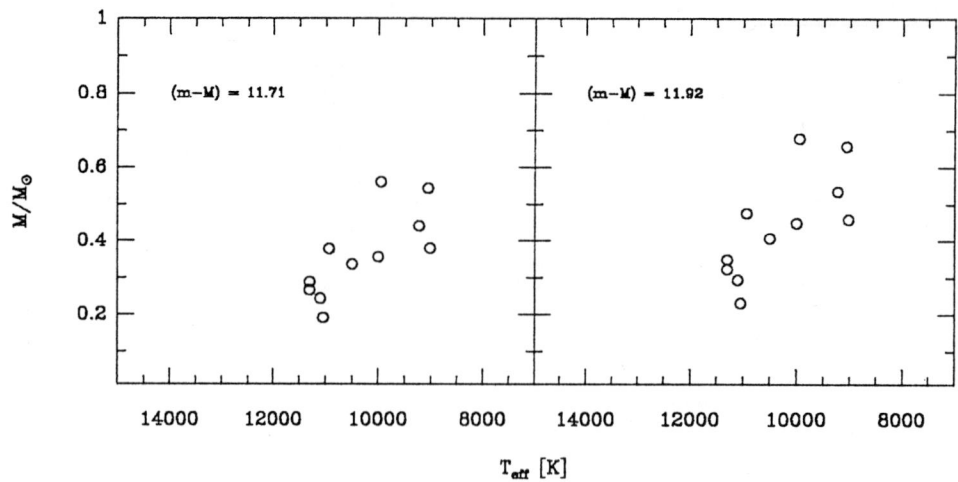

FIGURE 2. The masses derived for the HB stars in NGC 6397 are plotted against T_{eff}. The data show a remarkably tight M_{HB*} vs. T_{eff} range, reaching well below the theoretical lowest HB-star mass of 0.45 $M_\odot$. The left diagram shows the masses for the nominal $(m-M)_0 = 11.71$, the right diagram for a distance modulus 0.2 higher.

REFERENCES

ALCAIONO, G., LILLER, W., 1980, AJ, 85, 680

ANTHONY-TWAROG, B. J., TWAROG, B. A., SUNTZEFF, N. B., 1992, AJ, 103, 1264

COLIN, J., DE BOER, K. S., DAUPHOLE, B., DUCOURANT, C., DULOU, M. R., GEFFERT, M., LECAMPION, J. -F., MOEHLER, S., ODENKIRCHEN, M., SCHMIDT, J. H., THEISSEN, A., 1994, A&A, in press

DE BOER, K. S., 1981, in Philip A. G. D., Hayes D. S., eds, IAU Coll. 68, Astrophysical parameters for globular clusters, L. Davis Press, Schenectady, p. 25

DRUKIER, G. A., FAHLMAN, G. G., RICHER, H. B., SEARLE, L., THOMPSON, I., 1993, AJ, 106, 2335

GRAHAM, J. A., DOREMUS, C., 1968, AJ, 73, 226

HEBER, U., 1986, A&A, 155, 33

HEBER, U., KUDRITZKI, R. P., CALOI, V., CASTELLANI, V., DANZIGER, J., GILMOZZI, R. 1986, A&A, 162, 171

HUENEMOERDER, D. P., DE BOER, K. S., CODE, A. D., 1984, AJ, 89, 851

IBEN, I., 1990, ApJ, 353, 215

IBEN, I., TUTUKOV, A. V., 1985, ApJS, 58, 661

KODAIRA, K., PHILIP, A. G. D., 1984, ApJ, 278, 208

KURUCZ, R., 1979, ApJS, 40, 1

LESTER, J. B., GRAY, R. O., KURUCZ, R. L., 1986, ApJS, 61, 509

MOEHLER, S., DE BOER, K. S., HEBER, U., 1992, in Heber U., Jeffery C. S., eds, The Atmospheres of Early-Type Stars, Springer, Berlin, p. 251

MOEHLER, S., HEBER, U., DE BOER, K. S., 1990, A&A, 239, 265

MOEHLER, S., HEBER, U., DE BOER, K. S., 1994, in Adelman S. J., Upgren A. R., Adelman C. J., eds, Hot Stars in the Halo, Cambridge University Press, Cambridge, p. 217

PETERSON, C. J., 1993, in Djorgovski S. G., Meylan G., eds, Structure and Dynamics of Globular Clusters, ESO, in press

PHILIP, A. G. D., 1985, in Philip A. G. D., ed, Horizontal-Branch and UV-Bright Stars, L. Davis Press, Schenectady, p. 41
THEISSEN, A., MOEHLER, S., HEBER, U., DE BOER, K. S., 1993, A&A, 273, 524
WOOLLEY, R., ALEXANDER, J. B., MATHER, L., EPPS, E., 1961, R. Obs. Bull., No. 43

Discussion

SAFFER: In the Iben and Tutukov merger scenarios, core He ignition is possible at 0.3 $M_\odot$, but those lower mass stars have lower temperatures, not higher ones as indicated in the cluster data.

THEJLL: What happens to a He core that did not ignite on the RGB and is now evolving across the HR diagram after losing its envelope? What happens when the track crosses the Helium Main Sequence? For example, take an M = 0.3 $M_\odot$ core.

SWEIGART: This type of evolution has been recently discussed by Castellani & Castellani (1993, ApJ, 407, 649). Essentially, a post-RGB star that does not ignite helium evolves across the HR diagram at approximately constant luminosity on a time scale determined by the time required for the hydrogen shell to burn the remaining hydrogen in the envelope. Eventually such a star will settle onto the cooling curve appropriate for helium white dwarfs.

MOEHLER: To achieve *canonical* masses the distance modulus would have to be larger by about 0.4 to 0.5 mag.

IUE Observations of Blue HB Stars in the Globular Clusters M3 and NGC6752

By C. CACCIARI

Osservatorio Astronomico, Bologna, ITALY

Within the ongoing program of studying Horizontal Branch stars in detail to understand the HB stellar population(s) in globular clusters, we have observed with IUE 3 stars in M3 and 4 stars in NGC6752. In addition, also unpublished archival data for 5 more stars in NGC6752 have been used, and the published data for 3 stars in M3 and 6 stars in NGC6752 have been reanalysed. By comparison with Kurucz latest model atmospheres, temperatures and gravities have been estimated, and then compared to the most recent O-enhanced HB evolutionary tracks by Dorman et al. (1993). All the stars, including those that appear to be evolved off the ZAHB, are consistent with the theoretical predictions; there is no clear indication of multiple populations on the HB. However the simultaneous effect of various parameters may make this identification rather difficult in the log$g$-log T_{eff} plane.

1. Introduction

The observed morphology of the Horizontal Branch (HB) in the color-magnitude diagram (CMD) of galactic globular clusters presents features that are still poorly understood, namely:

• the non-monotonic relation between HB morphological type and metal abundance, which results in the presence of widely different HB stellar distributions in clusters with essentially the same chemical abundance (e.g., the classical case of M5, NGC362, NGC288, and NGC2808, see Rood et al. 1993). This is known as the "second parameter problem", but the most recent analyses of this issue would suggest the need of a "third" parameter (or a combination of various parameters) to better account for the observed HBs. For recent reviews we refer to Lee (1993), Buonanno (1993), Fusi Pecci et al. (1993), Van den Bergh & Morris (1993), and references therein.

• the discontinuous distribution of stars along the HBs of some clusters, which produces "gaps" in the observed HB population.

• In addition, some non-genuine HB stars may fall in the HB area of the C-M diagram as a result of their evolution, e.g., progeny of Blue Straggler stars (Fusi Pecci et al. 1992), binary mergers (Iben 1990, Bailyn 1992) or because of color saturation and large Bolometric correction (e.g., post-AGB and post-HB stars, Buzzoni et al. 1992).

It is very important to understand the nature of the stellar population(s) on the HB, because of the impact on issues such as: (i) the parameters (e.g., helium abundance) that can be determined using the HB stellar population ratios (Buzzoni et al. 1983), (ii) the relation of the HB morphology with the structural and dynamical conditions of the clusters, and possibly with other types of stars such as the Blue Stragglers (Fusi Pecci et al. 1992), (iii) the contribution of the hottest HB stars to the integrated UV light of the cluster, which is relevant in population synthesis studies of unresolved stellar systems (Burstein et al. 1988, Greggio & Renzini 1990, Ferguson et al. 1991).

For these purposes we are studying blue HB (BHB) stars in the globular clusters M3 and NGC6752, with special attention to the stars just above and below the gap that appears on the blue part of the HB in both clusters. We intend to use visual (blue) and UV (IUE) spectra to estimate effective temperatures and gravities of these stars, by comparison with the most recent Kurucz model atmospheres. The reduction and analysis of the visual spectra are still not completed, here we present a preliminary report on IUE data only.

2. The Observations

We have observed with IUE (6 Å resolution in the wavelength range 1200 - 1900 Å) the following BHB stars:

• In M3: stars 621, 352, and 843 (see Buzzoni et al. 1992 and references therein for the identification). In addition, we have reanalysed the published IUE data for 3 more stars: 156, 524, and 7561.

• In NGC6752: we have observed the stars 4009, 4548, 4951, and 5151, and used the unpublished data from the IUE archive for 5 more stars: 491, 722, 916, 534, and 2932. All the identifications are from Buonanno et al. (1986), except for stars 534 and 2932 which have been identified only by Caloi et al. (1986) and Heber et al. (1986). From these studies we have also taken the results obtained for 6 more stars, namely 2128, 3507, 3781, 3675, 3118, and 2167, and compared them with the results of our own reanalysis using the new Kurucz (1992) model atmospheres.

For similar studies in other globular clusters, see also de Boer (1985) (IUE data of UV-bright, mostly post-AGB stars) and Crocker et al. (1988) (visual data of BHB stars).

3. Analysis and Preliminary Results

For each star we have estimated the colors (13-18) and (18-V) as:

$$(13 - 18) = -2.5(\log F_{1300} - \log F_{1800})$$

and

$$(18 - V) = -2.5(\log F_{1800} - \log F_V)$$

where F_{1300} and F_{1800} are the fluxes at 1300 and 1800 Å, averaged over a bandwidth of 100 Å and 150 Å, respectively. The flux at V has been derived as:

$$\log F_V = -0.4V - 8.41$$

No reddening correction has been applied to M3, while the NGC6752 data have been corrected for E(B-V) = 0.04 using Seaton (1979) reddening law for the Galaxy.

For hot stars the color (18-V) is a much better temperature indicator than (B-V), while (13-18) is an indicator of gravity. Gravity can also be estimated from H line profiles and the Balmer jump, when visual data are available.

The colors derived for the program stars have been compared with the most recent Kurucz (1992) model atmospheres for [m/H] = -1.50, and preliminary estimates of temperature and gravity have been obtained. Luminosities have also been obtained, based on the assumptions that: (i) $M_{bol}(\odot) = 4.72$, (ii) $(m - M)_0(M3) = 15.00$, (iii) $(m - M)_0(NGC6752) = 13.07$.

4. The log L-log T_{eff} and log$g$-log T_{eff} diagrams

We have compared the program stars and the previously analysed HB stars with the O-enhanced HB models from Dorman et al. (1993) in the log L-log T_{eff} plane.

• M3: the three program stars 621, 352 and 843 appear as normal Zero-Age HB (ZAHB) stars, or at most slightly evolved, while the three UV-bright stars previously analysed (Buzzoni et al. 1992) are confirmed as Post-AGB or AGB-manqué stars.

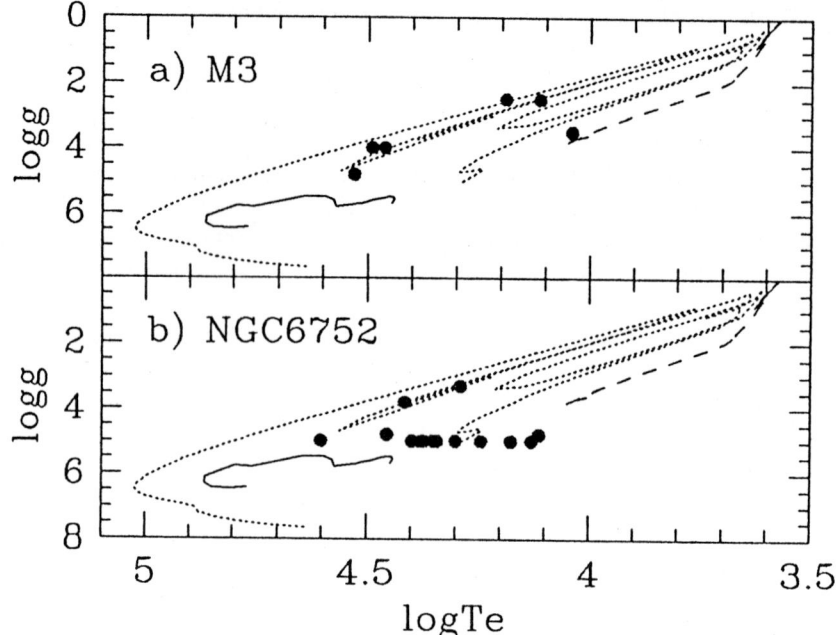

FIGURE 1. Fig.1a-b: O-enhanced evolutionary tracks of HB stars in the log$g$-log T_{eff} plane (Dorman et al. 1993) and the observed stars in a) M3, and b) NGC6752. The tracks are for [Fe/H] = -1.48, [O/Fe] = 0.63, Y(HB) = 0.25, M_c = 0.485$M_\odot$ and three values of the envelope mass, i.e. 0.003 $M_\odot$ (full line), 0.025 $M_\odot$ (dotted line) and 0.105 $M_\odot$ (dashed line).

- NGC6752: only the previously analysed star 2128 (Heber et al. 1986) shows AGB-manqué characteristics. All the other stars appear as normal ZAHB stars, or possibly slightly evolved.

In Fig.1a-b we show the results we have obtained in the log$g$-log T_{eff} plane.

- M3: all stars have consistent gravities with the theoretical models. As in the log L-log T_{eff} plane, star 621 appears to be very near the ZAHB, while the other stars are at significantly lower gravity than the theoretical ZAHB. This result was already found by Crocker et al. (1988) who suggested the possibility of multiple populations on the HB. This may not be the only explanation, however, as the comparison with the theoretical models suggests that evolution off the ZAHB may play an important role, and parameters such as the core mass and the envelope mass may further complicate the issue.
- NGC6752: most stars indicate larger values of gravity (5.0 or more) than expected from theoretical models of ZAHB stars. Note, however, that: a) for stars hotter than 20,000 K (log T_{eff} = 4.3) the gravity could be larger than 5.0 (see the results from visual data by Heber et al. 1986) but this cannot be found from Kurucz' models, since they stop at log$g$ = 5.0; and b) for stars cooler than 15,000 K (log T_{eff} ~ 4.15) the gravity derived from UV colors could be overestimated, and be better determined from visual data (see star 2167). Therefore the apparent flat distribution at log g = 5.0 may be spurious, and these stars could actually be located on the ZAHB.

5. Summary and Conclusions

- We have observed 3 BHB stars in M3 and 4 BHB stars in NGC6752 with IUE in the wavelength range 1200 - 1900 Å, the low resolution mode. In addition, we have used the unpublished archival data for 5 more BHB stars in NGC6752 and the published data for 3 stars in M3 and 6 stars in NGC6752. Using the colors (13-18) and (18-V) and the most recent Kurucz models we have estimated gravities, temperatures and luminosities and compared them with O-enhanced ZAHB evolutionary tracks.

• All these stars, including those with clear Post-AGB or AGB-manqué characteristics, seem to be consistent with the theoretical predictions, allowing for evolution and/or different envelope mass. Although the present data are still insufficient to reach firm conclusions, they do not seem to imply the existence of multiple populations on the HB (Crocker et al. 1988). However, the identification of different populations in the $\log g$-$\log T_{eff}$ plane is made difficult by the simultaneous effect of various parameters, i.e. mass of the core, mass of the envelope, and evolutionary rate along the tracks.

REFERENCES

BAILYN, C. D., 1992, ApJ, 392, 519

BUONANNO, R., 1993, in Smith G. H., Brodie J. P., eds, The Globular Cluster - Galaxy Connection, ASP Conf. Ser., 48, 131

BUONANNO, R., CALOI, V., CASTELLANI, V., CORSI, C. E., FUSI PECCI, F., GRATTON, R., 1986, A&AS, 66, 79

BURSTEIN, D., BERTOLA, F., BUSON, L., FABER, S. M., LAUER, T. R., 1988, ApJ, 328, 440

BUZZONI, A., FUSI PECCI, F., BUONANNO, R., CORSI, C. E., 1983, A&A, 123, 94

BUZZONI, A., CACCIARI, C., FUSI PECCI, F., BUONANNO, R., CORSI, C.E., 1992, A&A, 254, 110

CALOI, V., CASTELLANI, V., DANZIGER, J., GILMOZZI, R., CANNON, R. D., HILL, P. W., BOKSENBERG, A., 1986, MNRAS, 222, 55

CROCKER, D. A., ROOD, R. T., O'CONNELL, R. W., 1988, ApJ, 332, 236

DE BOER, K. S., 1985, A&A, 142, 321

DORMAN, B., ROOD, R. T., O'CONNELL, R. W., 1993, ApJ, 419, 596

FERGUSON, H. C., DAVIDSEN, A. F., KRISS, G. A., et al., 1991, ApJ, 382, L69

FUSI PECCI, F., FERRARO, F. R., BELLAZZINI, M., DJORGOVSKI, S. G., PIOTTO, G., BUONANNO, R., 1993, AJ, 105, 1145

FUSI PECCI, F., FERRARO, F. R., CORSI, C. E., CACCIARI, C., BUONANNO, R., 1992, AJ, 104, 1831

GREGGIO, L., RENZINI, A., 1990, ApJ, 364, 35

HEBER, U., KUDRITZKI, R. P., CALOI, V., CASTELLANI, V., DANZIGER, J., GILMOZZI, R., 1986, A&A, 162, 171

IBEN, I., 1990, ApJ, 353, 215

KURUCZ, R. L., 1992, in Barbuy B., Renzini A., eds., The stellar populations of galaxies, Kluwer, Dordrecht, p. 225

LEE, Y. W., 1993, in Smith G. H., Brodie J. P., eds, The Globular Cluster - Galaxy Connection, ASP Conf. Ser., 48, 142

ROOD, R. T., CROCKER, D. A., FUSI PECCI, F., FERRARO, F. R., CLEMENTINI, G., BUONANNO, R., 1993, in Smith G. H., Brodie J. P., eds, The Globular Cluster - Galaxy Connection, ASP Conf. Ser., 48, 218

SEATON, M., 1979, MNRAS, 187, 73P

VAN DEN BERGH, S., MORRIS, S., 1993, AJ, 106, 1853

Discussion

LANDSMAN: What is the quality of the IUE spectra of the BHB stars in M3? These stars would seem to push the sensitivity limit of IUE.

CACCIARI: The signal to noise ratio is not great, but the spectra are clearly visible, well above the background; and we are interested in the colors averages (over 100 to 150 Å wide bands), so the spectra are quite adequate.

SWEIGART: In NGC 6752, two stars are below the Zero Age Horizontal Branch (ZAHB), do you believe this?

CACCIARI: These are still preliminary results; we have to check them again and estimate the error bars. At the moment I cannot swear on this result, but if it is true then it would be similar to what is found by Whitney et al. (this proccedings) in Omega Centauri, namely underluminous stars on the bluest part of the HB.

WHITNEY: You have IUE spectra of one star in NGC 6752 which falls below the ZAHB. In our Far Ultraviolet Color Magnitude Diagram of Omega Centauri we also find a significant sample of very hot stars below the ZAHB.

Metallicities and Kinematics of the Local RR Lyrae Stars: Lukewarm Stars in the Halo

By ANDREW C. LAYDEN †

Cerro Tololo Inter-American Observatory, Casilla 603, La Serena, CHILE

New spectra have been used to compute accurate metal abundances and radial velocities for over 300 RR Lyrae stars within 2.5 kpc of the Sun. Analysis shows a very sharp change from halo to thick disk kinematics at [Fe/H] = −1.0. There are very few "metal-weak thick disk" stars among the RR Lyrae stars. The kinematics are uncorrelated with abundance in the halo, favoring accretion pictures of halo formation. Evidence for a retrograde net rotation of the halo is discussed. The space density of the local RR Lyrae stars is confirmed to be about twice that found at high latitude. This and the kinematics are discussed in terms of Zinn's two-component Old/Younger Halo model.

1. Introduction

I refer to RR Lyrae variables as "lukewarm" stars because, as mid-A to mid-F spectral type objects, they are significantly cooler than the "hot" blue horizontal branch (HB) stars, sdO stars, and halo A stars that make up the main subject of this workshop. Nevertheless, I feel there is a place for RR Lyrae stars our discussion, especially with regard to the study of blue HB stars. As "yellow" HB stars, RR Lyrae stars compliment studies of blue HB stars in terms of determining the kinematics, the metal abundance distribution, and the HB morphology (blue or red-ness of the HB) of the halo. In particular, the two types of stars provide a mutual check, since the methods used to discover them are so different. While careful photometry is required to separate field blue HB stars from the foreground thin disk dwarf stars (see Kinman; Wilhelm, Beers & Gray; Corbally & Gray; and other papers in this volume), RR Lyrae stars are are easily picked up in variable stars surveys by their characteristic, large-amplitude light curves. Typically, ΔB is 0.6 to 1.8 mags, and periods range from 0.35 to 0.80 days for the ab-type RR Lyrae stars.

One advantage of using horizontal branch stars, both RR Lyrae and blue HB stars, to examine the halo is that they are *not* selected by their velocities. The process by which they are selected imparts no (overt) bias to the kinematic results obtained from the sample, in contrast to samples selected from proper motion catalogs. This is not to say that HB star selection is bias-free. The location of a star on the HB (i.e., whether it appears as a blue HB star, an RR Lyrae, or a red HB star) is determined by a number of factors, including metal abundance, age, relative abundances of CNO to Fe, core rotation, mass-loss history, lifetime spent on the HB, etc. Thus the abundance distribution of a sample of RR Lyrae stars may not represent the abundance distribution of the larger population from which it is drawn. However, what is important to this paper is that, if we observe a group of RR Lyrae stars at a given metal abundance, their kinematics will be representative of the kinematics of the larger population from which they are drawn, at that abundance.

With this in mind, five years ago I began a dissertation, with Bob Zinn as my advisor, to use the RR Lyrae stars as kinematic probes of the Galactic halo. It was clear at the time that the existing database of RR Lyrae metallicities and velocities was of rather

† Visiting Astronomer, Kitt Peak National Observatory, which is operated by the Association of Universities for Research in Astronomy, Inc., under contract with the National Science Foundation.

heterogeneous quality. It was equally clear that using modern detectors and digital analysis techniques, it would be possible not only to reobserve all the previously-observed RR Lyrae stars using a single, self-consistent technique, but also to enlarge the sample significantly. The result is a final sample of 302 ab-type RR Lyrae stars, most lying within 2.5 kpc of the Sun. The metallicities, arrived at using a variation on Preston's (1959) ΔS technique, are accurate to 0.15–0.20 dex in [Fe/H]. The velocities, which I supplemented with data taken from the literature, vary in quality between 2 and 30 km s^{-1} accuracy. I took the photometry from the General Catalog of Variable Stars (Kholopov 1985).

2. Kinematics

I have computed the kinematic properties of the RR Lyrae star sample using two different techniques. The first uses the "constant rotational velocity solution" of Frenk & White (1980). This is a statistical method, using only the stars' radial velocities, their positions on the sky and their distances. All the stars in a given group (metal abundance bin, in this case) are used in a least squares fit of radial velocity (relative to a stationary observer at the Sun's position) versus a geometric factor which relates a star's galactic rotation vector with the line-of-sight vector to the star (for details, see Zinn 1985; Armandroff 1989; Morrison, Flynn & Freeman 1990). The slope of the fit is the net rotation of the group about the galactic center (V_{rot}), and the RMS scatter about the fit is a measure of the "line-of-sight" velocity dispersion of the group (σ_{los})

Figure 1 shows the kinematics of each group as a function of its metal abundance, [Fe/H]. Clearly there is an abrupt change in the kinematics at [Fe/H] = −1.0. The stars more metal-rich than this value have a net rotation of 195 ± 14 km s^{-1}, and a line-of-sight velocity dispersion of 48 ± 6 km s^{-1}. These values are in good agreement with the kinematics of the disk globular cluster system (Armandroff 1989, 1993) and with recent estimates of the field thick disk population (e.g., Norris 1986; Morrison et al. 1990; Carney, Latham & Laird 1989; and Majewski 1992). The RR Lyrae stars with [Fe/H] more metal-poor than −1.0 have kinematics in good agreement with other tracers of the galactic halo: V_{rot} = 18 ± 12 km s^{-1} and σ_{los} = 109 ± 5 km s^{-1}.

The second method I use to compute the RR Lyrae star kinematics is the more traditional "space velocity" solution, where proper motions are combined with the radial velocity and distance data to produce a three dimensional velocity for each star (I use cylindrical coordinates; ρ, ϕ, Z). Figure 2 shows the velocity components for each star as a function of [Fe/H]. Again we see the abrupt change in kinematics at [Fe/H] = −1.0, with typical thick disk properties for the metal-rich group, and halo properties for the metal poor group.

The data clearly do not show the roughly linear transition from halo to disk seen by Sandage & Fouts (1987) in their sample of high proper motion dwarfs. Those authors concluded from their observational results that the halo and disk formed in a continuous, coherent process of collapse and spin-up. The halo-to-disk transition among the RR Lyrae stars in Figures 1 and 2 is even more abrupt than seen in the field star samples of, for example, Norris (1986), Carney et al. (1990), and Morrison et al. (1990). All of these authors concluded from their kinematics-abundance data that the halo and disk systems are kinematically distinct, and that the small slopes in the halo-to-disk transitions seen in plots of mean kinematics vs abundance are due to the overlapping abundance distributions of these two populations (recall Bruce Carney's exhortation in this workshop to look at the individual data points, rather than just the mean points).

That the kinematic change from halo to disk is so abrupt among the RR Lyrae stars is

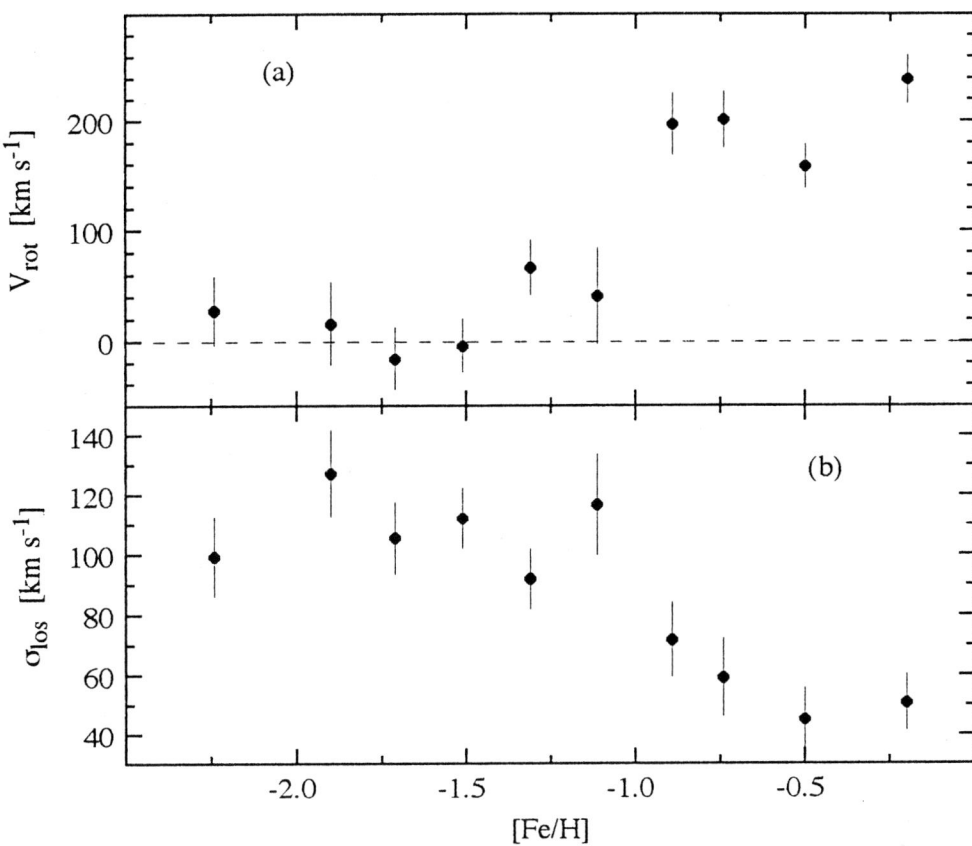

FIGURE 1. RR Lyrae star kinematics as a function of abundance, from the constant rotational velocity solutions. (a) net rotational velocity about the galactic center; (b) line-of-sight velocity dispersion.

probably due to the selection effects to which the RR Lyrae stars are subject. Taking a lesson from the well-defined halo population presented by the globular clusters, we know that the production efficiency of RR Lyrae stars decreases rapidly above [Fe/H] = -1.0, as the bulk of the HB stars move out of the instability strip and appear as stable red HB stars (e.g., Suntzeff et al. 1991). This and the declining abundance distribution of halo stars in general tend to "guillotine" the abundance distribution of the halo RR Lyrae stars as [Fe/H] increases past −1. More metal-rich RR Lyrae stars exist among the field disk population no doubt because the number ratio of disk-to-halo stars is large (e.g., Norris & Ryan 1991 and references therein); the low production efficiency of metal-rich RR Lyrae stars is more than offset by the large number of "parent stars" from which to make them.

On the other hand, it is rather surprising that there is not a large population of disk RR Lyrae stars with [Fe/H] < −1.0, blurring the halo-to-disk transition. Morrison et al. (1990) discovered a large number of "metal-weak thick disk" red giants (MWTD; stars having thick disk kinematics but with [Fe/H] between −1.0 and −1.6, and perhaps lower, as discussed in this workshop). They found that, near the plane, the number density of

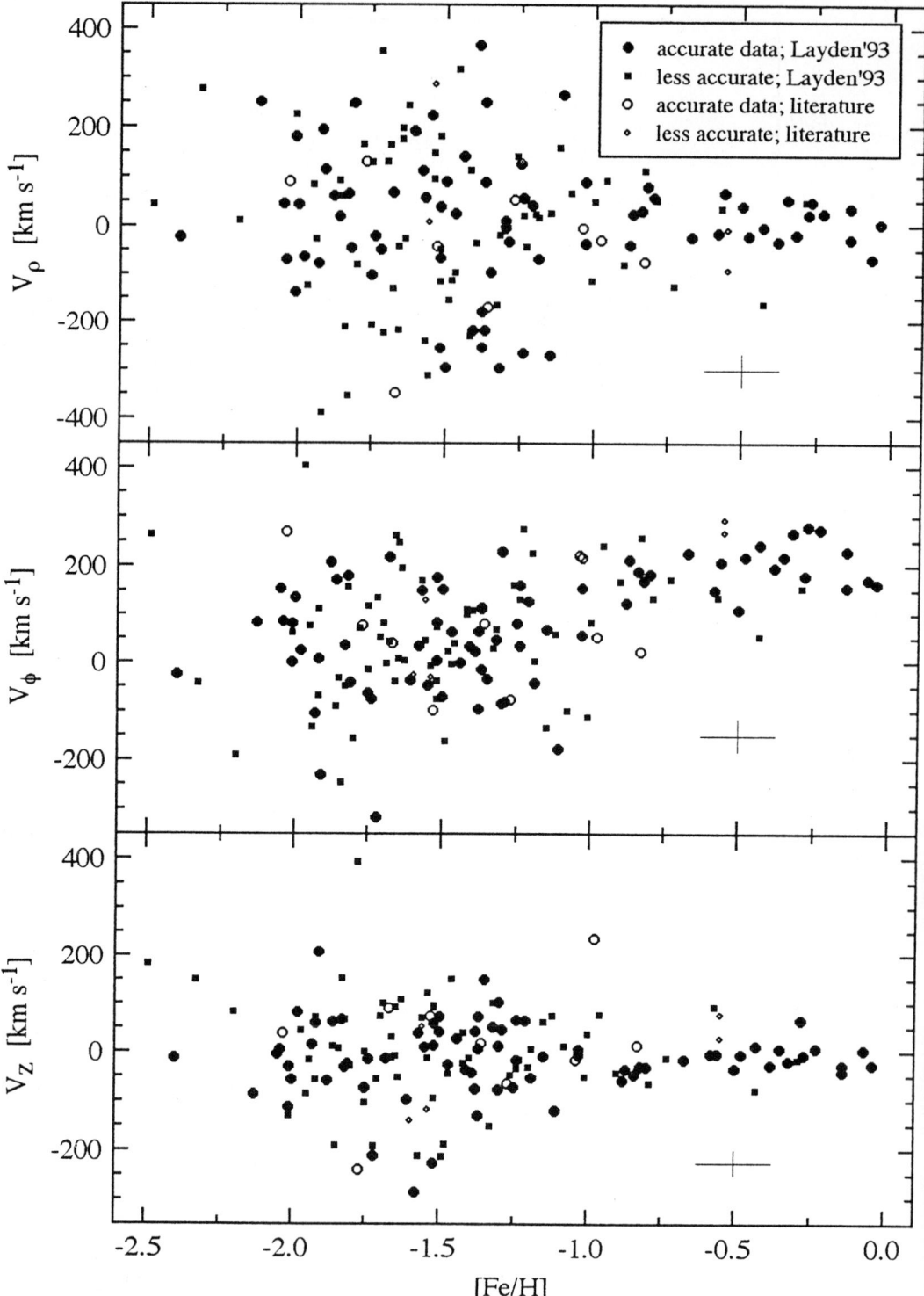

FIGURE 2. RR Lyrae star kinematics as a function of abundance, from the space velocity solutions. (a) radial, (b) rotational, and (c) vertical components.

MWTD giants was roughly equal to the number density of halo giants. Assuming the RR Lyrae star production efficiency (e.g., the ratio of RR Lyrae stars to red giants) is the same between the MWTD and the halo, one would then expect equal numbers of RR Lyrae stars in the MWTD and halo. However, using a rather heterogeneous set of data taken from the literature, Morrison et al. (1990) found a small MWTD-to-halo ratio among the RR Lyrae stars. The present improved set of metal abundance and velocity information allows a more quantitative evalution of the situation. Careful analysis of the RR Lyrae stars with space motions shows that *at most* 12 of the 41 stars with $-1.6 <$ [Fe/H] < -1.0 have "disk-like" motions (see Figure 3). Given the broad velocity dispersions of the halo, a few true halo stars are liable to fall in the "disk-like" box. Also, some of the metal-rich disk stars may have scattered into the MWTD abundance range, due to errors in abundance measurement. Thus, a more probable estimate is that 5 of 39 stars belong to the MWTD. Furthermore, it is not inconsistent with the data that *no* MWTD RR Lyrae stars exist. Clearly, the proportion of MWTD to halo RR Lyrae stars near the plane is much smaller than the near-unity proportion found by Morrison et al. for red giants in this abundance range. That is, the production efficiency of RR Lyrae stars in the MWTD is lower than that in the halo at these abundances. MFF noted that if the MWTD is several billion years younger than the halo, its redder HB morphology would lower the production efficiency of MWTD RR Lyrae stars. However, it would be correspondingly *more* difficult to form metal-rich disk RR Lyrae stars, since the bulk of the metal-rich thick disk stars would lie *redward* of the bulk of the MWTD HB stars. Clearly, accurate estimates of the RR Lyrae star production efficiency in the halo, MWTD, and metal-rich thick disk are required before more progress can be made in this field. But, this is complicated by our sketchy knowledge of the abundance distributions and density normalizations of these populations (and if fact, whether such populations as the thin and thick disks are themselves discrete entities!). So, for now, the mystery of the missing MWTD RR Lyrae stars must go unsolved.

In summary, the very slow net rotation of the halo, the lack of a significant correlation between kinematics and abundance among the halo stars, and the abrupt transition from the disk to the halo at [Fe/H] $= -1.0$ all suggest that the disk and the halo are distinct populations, which formed more or less independently of each other. In particular, the lack of a kinematics-metallicity correlation in the halo is usually taken as a sign that the halo formed in a "chaotic" manner. That is, either, (i) the star formation sites were intermixed in the forming galaxy by highly random orbits, (ii) star formation proceeded at vastly different rates in different places in the forming halo, independent of galactocentric distance or distance from the plane, or (iii) a combination of both randomizing effects. The Searle & Zinn (1978) fragment accretion picture is the classical example of this type of formation scenario.

In Table 1, I show the best estimates for the kinematics of the halo and thick disk RR Lyrae stars. The exact procedure used to derive these values is given in Layden (1993). I exclude from the halo solution stars with $-1.3 <$ [Fe/H] < -1.0, since a small fraction of the stars in this abundance range may be members of the metal-weak thick disk.

3. Two Halo Components?

In 1992, Steven Majewski presented the results of a complete, *in situ* proper motion study of F through early-K dwarfs in a cone extending toward the North Galactic Pole. He found that the net galactic rotation of the stars more than 5 kpc above the galactic plane, stars which he associated with the halo, was retrograde: $V_{rot} =$ -55 $\pm$ 16 km s^{-1}. This came as a surprise, since all local samples, like the present RR Lyrae star sample,

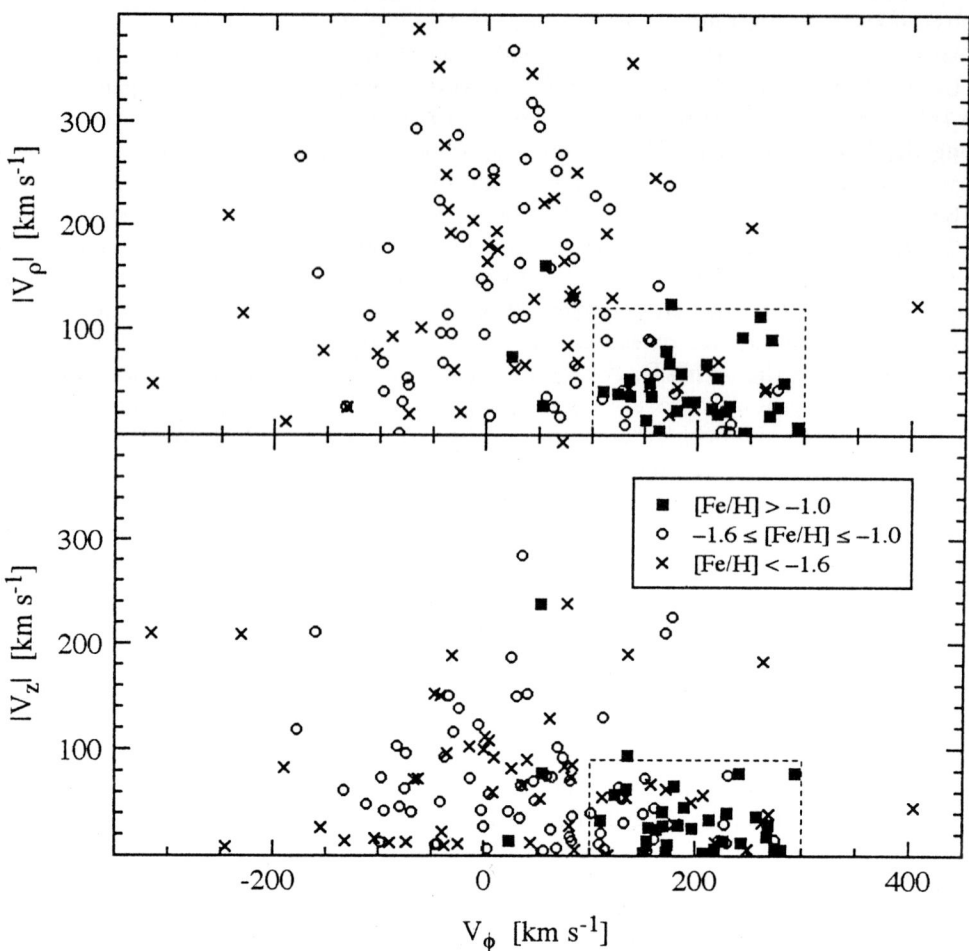

FIGURE 3. Stars within the box are defined as having "disk-like" kinematics. The box is centered at 200 km s^{-1}, and the half-width of each side is $(\rho, \phi, Z) = (120, 100, 90$ km s$^{-1})$, roughly twice the disk velocity dispersions listed in Table 1 for the respective velocity components.

Quantity	Halo	Disk
[Fe/H] range	< -1.3	> -1.0
$V_{rot} \pm$ err	17 ± 14	193 ± 9
$\sigma_{los} \pm$ err	115 ± 7	48 ± 6
$\sigma_\rho \pm$ err	160 ± 15	48 ± 8
$\sigma_\phi \pm$ err	112 ± 10	42 ± 7
$\sigma_Z \pm$ err	99 ± 10	33 ± 6

TABLE 1. RR Lyrae star net galactic rotation and velocity dispersions, in km s^{-1}.

show prograde net rotations of 20–40 km s^{-1}. He suggested that, because most of these studies use a metal abundance criterion to define the halo, their halo kinematics could easily be biased by the presence of a few MWTD stars. That is, the retrograde "true" halo and the strongly prograde MWTD cancel out in the mean to mimic a slow prograde rotation in these local samples.

I find this interpretation unlikely for several reasons. First, even the most metal-poor stars still show prograde net rotation. For [Fe/H] < -1.9, the net rotation of the RR Lyrae stars is 26 ± 27 km s^{-1}. For the non-kinematically selected sample of Carney & Latham (1986), $V_{rot} = 32 \pm 32$ km s^{-1}, and for Morrison et al. (1990), $V_{rot} = 40 \pm 16$ km s^{-1} over the same abundance range. Thus a significant number of *very* metal poor MWTD stars would be needed to balance a retrograde halo. It seems remarkable that these stars have not been found, if they exist at these abundances. However, Carney and Beers in this workshop have presented evidence for a fast-rotating, low abundance population, but at a much lower net rotation than the metal-rich thick disk described above. It is conceivable that the relatively small sample sizes of non-kinematically selected samples, and the kinematic bias of local proper-motion selected samples, to date have hindered the discovery of these stars.

Second, the net galactic rotation as a function of abundance remains effectively flat over $-2.3 <$ [Fe/H] < -1.5 (or even -1.0 in the RR Lyrae stars; Figure 1), means that the number ratio of MWTD to halo stars in this scenario must remain constant over this entire abundance range. In other words, the two populations must have a similar abundance distribution over this interval. This is inconsistent with the abundance distributions currently presented in the literature (e.g., Norris & Ryan 1991 and references therein). Furthermore, if the mean kinematics of the MWTD population become more halo-like with decreasing abundance, as suggested by Carney (1994) and Majewski (1993), the number of MWTD stars relative to true halo stars must *increase* to preserve the observed flat kinematics–abundance relation.

As a final test, it is possible to remove *all* the stars in the "disk-like" box of velocity-space described above, and recompute the kinematics. If MWTD stars are biasing the "true" halo kinematics, then the net rotation should appear significantly retrograde. For the RR Lyrae star sample, however, the recomputed V_{rot} is 10 ± 10 km s^{-1}. This in fact is an over-compensation, since the broad velocity dispersions of the halo will naturally place some true halo stars within the "disk-like" box (see Figure 3). If, as Carney and Majewski suggest, the mean rotation and velocity dispersions of the MWTD at [Fe/H] < -1.5 are more halo-like (as one might expect if this population is part of a dissipational structure marking the earliest phase of disk formation; Majewski 1993), then these stars may lie outside of the "disk-like" box defined by the more metal-rich disk stars. However, it is difficult to apply this test using larger boxes centered at lower net rotations, because one then removes relatively larger numbers of true halo stars which have prograde rotations. In the extreme example, if one were to remove all the stars with $V_{rot} > 0$ km s^{-1} from a halo having a Gaussian distribution of velocities centered on $V_{rot} = +20$ km s^{-1}, one would clearly find a spurious retrograde net rotation. Thus, this test cannot be used to rule out the possiblity of a slower-rotating, hotter MWTD population.

So it seems that a simple, metal-poor extension of the classical, fast-rotating thick disk, having an exponential abundance tail, cannot be invoked to balance Majewski's retrograde halo. However, a substantial population of less kinematically extreme disk stars, having an abundance distribution similar to that of the classical halo, may suffice. Such a population would be difficult to separate from the classical halo, since its kinematics, as well as its abundance distribution, would significantly overlap and mimic those of the classical halo. Might there be other means to separate these two populations?

Zinn (1993) may have supplied that means. He divided the globular clusters into "old" and "younger" groups based on their HB morphology (redder HB-types corresponding to younger clusters, on the assumption that age is the dominant "second parameter" determining HB morphology). He found that the old halo clusters have a faster net rotation and smaller line-of-sight velocity dispersion than the younger clusters ($V_{rot} = 70 \pm 22$ kms^{-1} and $\sigma_{los} = 89 \pm 9$ kms^{-1}, versus $V_{rot} = -64 \pm 74$ kms^{-1} and $\sigma_{los} = 149 \pm 24$ kms^{-1}, respectively). Moreover, the old clusters tend to be concentrated to the plane in a somewhat flattened system, while the younger cluster system is spherical, but has few members within about 6 kpc of the galactic center. Zinn suggested that the old halo system collapsed and spun up via a process not unlike that described by Eggen, Lynden-Bell & Sandage (1962) and Sandage (1990), eventually creating the disk system we see today. The younger halo, in Zinn's view, was then accreted about this nucleus over a period of several billion years, much like the Searle & Zinn (1978) picture.

How does this fit in with Majewski's retrograde halo scenario? If the old/younger dichotomy extends to the field stars, and if the old halo (and/or transitional halo-disk population) is sufficiently concentrated to the galactic plane, then Majewski's sample of dwarfs at $Z > 5$ kpc above the plane sampled predominantly the younger halo. The kinematics of Zinn's younger halo clusters hint that the net rotation of this population may be retrograde, as one might expect if some fraction of the prograde accreted material was dragged first into the disk plane, and evetually circularized, by dynamical friction (Quinn, Hernquist & Fullagar 1993). Meanwhile, the local kinematic studies sample members of both the younger and old halo populations, and thus find mean kinematics intermediate between the retrograde younger halo and the strongly prograde old halo. Majewski (1993) has painted a similar picture, though he identifies the fast-rotating population more with an intermediate old halo – thick disk population than with the old halo directly.

Other evidence fits this picture as well. The space density distribution of RR Lyrae and blue HB stars far from the galactic plane, as studied by Preston, Schectman & Beers (1991), predicts the space densities of these two types of stars in the solar neighborhood to be 5 ± 1 and 31 ± 20 stars kpc^{-3}, respectively. From blue HB stars in similarly distant fields, Kinman (1992) predicts a local space density of 4 blue HB stars kpc^{-3}. True local samples of these stellar types show an overabundance relative to these predictions: 12 ± 3 and 51 ± 17 stars kpc^{-3} (Layden 1993, Green & Morrison 1993, for RR Lyrae and blue HB stars, respectively). One way to explain these local overdensities is to say that the distant samples (Preston et al. 1991 and Kinman 1992) are composed primarily of the spherically-distributed younger halo stars, while the local samples contain stars from both the younger and the old halos.

Clearly however, much of the evidence presented in this section is sketchy and circumstantial. Much needs to be done before this picture of the structure and formation of the Galaxy can be accepted. In particular, it is not clear whether the halo field population participates in the younger/old halo dichotomy suggested by the clusters. Because age information is very difficult to obtain for field stars, it seems unlikely that local samples will be clearly divisible into these groups. One way to clarify the situation is to choose field star samples in the regions of the Galaxy where each of the purported halos dominates. If the globular clusters are a reasonable guide, stars with galactocentric distances between 3 and 6 kpc should mainly represent the older halo, and thus have a net rotation substantially faster than the local samples. Similarly, stars more than 5 kpc from the galactic plane should belong mainly to the younger halo, and have slow or retrograde net rotation.

RR Lyrae stars would make good halo tracers for both of these tests. Their distances can be determined more accurately than red giant or dwarf stars, thus minimizing potential errors in determining their galactic locations. Also, accurate abundances and radial velocities can be determined, so that these properties can be compared directly with the local RR Lyrae sample described above. I am currently involved in projects to observe RR Lyrae stars in both these galactic regions. Though the data aquisition is well underway, it is still to early to draw any conclusions from the data.

4. Conclusions

RR Lyrae variables, easily detectable highly-evolved stars, provide an efficient means of determining the kinematics and metal abundance of the halo. Results based on recent new data show that the RR Lyrae stars separate rather cleanly into (at least) two populations at $[Fe/H] = -1$. The net rotation (193 km s^{-1}) and small velocity dispersions of the more metal-rich RR Lyrae stars indicate their membership in the thick disk. They do *not* represent an intermediate kinematic population ($V_{rot} \sim 100$ km s^{-1}), as suggested by previous studies which employed less homogeneous data sets. The more metal-poor RR Lyrae stars have kinematics in good agreement with other local tracers of the halo ($V_{rot} = 17$ km s^{-1}).

The abrupt kinematic transition from halo to disk with increasing metal abundance suggests that the halo and disk are kinematically, and presumably formationally, distinct populations. The small number of "metal-weak thick disk" RR Lyrae stars, relative to the number of halo RR Lyraes, is surprising in light of their relative abundance among red giants (Morrison et al. 1990). More detailed study of the horizontal branch morphology of these populuations is required before firm conclusions can be reached. The lack of correlation between kinematics and abundance in the halo is consistent with the idea that the halo formed in a "chaotic" manner, perhaps from the accretion of protogalactic fragments (Searle & Zinn 1978), rather than a coherent collapse and spin-up of a single protogalactic gas cloud (Sandage & Fouts 1987).

Evidence for a retrograde halo, as suggested by Majewski (1992), was considered. I showed that metal-weak thick disk stars having kinematics similar to that of the metal-rich thick disk are *not* biasing the kinematic results of local halo samples. However, it is possible that a slower-rotating, hotter "disk" component, which has an abundance distribution similar to that of the classical halo at $[Fe/H] < -1.5$, is concealed within the classical halo. I tentatively associated this hotter population with the old halo globular clusters identified by Zinn (1993). Majewski's (1992) retrograde halo at large distances from the galactic plane may be related to Zinn's "younger" halo (see also Majewski 1993). Finally, I discussed some kinematic tests that are being performed to verify the old/younger halo model.

The results presented in this paper are largely taken from my Ph.D. dissertation, for which Bob Zinn served as advisor. I would like to thank Bob for his direction and encouragement. This research was partially supported by NSF grant AST-8914519.

REFERENCES

ARMANDROFF, T. E., 1989, The properties of the disk system of globular clusters. AJ, 97, 375

ARMANDROFF, T. E., 1993, The disk population of globular clusters. in Smith G. H., Brodie J. P., eds, The Globular Cluster – Galaxy Connection, ASP Conf. Series, 48, p. 48

CARNEY, B. W., 1994, What is the galaxy's halo population? in Adelman S. J., Upgren A. R., Adelman C. J., eds, Hot Stars in the Halo, Cambridge University Press, Cambridge, p. 3

CARNEY, B. W., LATHAM, D. W., 1986, The kinematics of halo red giants. AJ, 92, 60

CARNEY, B. W., LATHAM, D. W., LAIRD, J. B., 1989, A survey of proper-motion stars. VIII. On the galaxy's third population. AJ, 97, 423

CARNEY, B. W., LATHAM, D. W., LAIRD, J. B., 1990, A survey of proper-motion stars. X. The early evolution of the galaxy's halo. AJ, 99, 572

EGGEN, O. J., LYNDEN-BELL, D., SANDAGE, A. R., 1962, Evidence from the motions of old stars that the galaxy collapsed. ApJ, 136, 748

FRENK, C. S., WHITE, S. D. M., 1980, The kinematics and dynamics of the galactic globular cluster system. MNRAS, 193, 295

GREEN, E. M., MORRISON, H. L., 1993, The local BHB stars: BHB density, HB morphology and halo flattening. in Smith G. H., Brodie J. P., The Globular Cluster – Galaxy Connection, ASP Conf. Series, 48, p. 318

KINMAN, T. D., 1992, in Warner B., ed, The stars of spectral types A and F as probes of galactic structure. ASP Conf. Series, 30, p. 19.

LAYDEN, A. C., 1993, The metallicities and kinematics of the local RR Lyrae variables. PhD thesis, Yale University

MAJEWSKI, S. R., 1992, A complete multicolor survey of absolute proper motions to $B \sim 22.5$: galactic structure and kinematics at the north galactic pole. ApJS, 78, 87

MAJEWSKI, S. R., 1993, Galactic structure surveys and the evolution of the Milky Way. ARAA, 31, 575

MORRISON, H. L., FLYNN, C., FREEMAN, K. C., 1990, Where does the disk stop and the halo begin? Kinematics in a rotation field. AJ, 100, 1191

NORRIS, J. E., 1986, Populations studies. II. Kinematics as a function of abundance and galactocentric position for [Fe/H] < −0.6. ApJS, 61, 667

NORRIS, J. E., RYAN, S. G., 1986, Populations studies. XI. The extended disk, halo configuration. ApJ, 380, 403

PRESTON, G. W., 1959, A spectroscopic study of the RR Lyrae stars. ApJ, 130, 507

PRESTON, G. W., SCHECTMAN, S. A., BEERS, T. C., 1991, Detection of a galactic color gradient for blue horizontal-branch stars of the halo field and its implications for the halo age and density distributions. ApJ, 375, 121

QUINN, P. J., HERNQUIST, L., FULLAGAR, D. P., 1993, Heating of galactic disks by mergers. ApJ, 403, 74

SANDAGE, A., 1990, On the formation and age of the galaxy. JRASC, 84, 70

SANDAGE, A., FOUTS, G., 1987, New subdwarfs. VI. Kinematics of 1125 high-proper-motion stars and the collapse of the galaxy. AJ, 92, 74

SEARLE, L., ZINN, R., 1978, Compositions of halo clusters and the formation of the galactic halo. ApJ, 225, 357

SUNTZEFF, N. B., KINMAN, T. D., KRAFT, R. P., 1991, Metal abundances of RR Lyrae variables in selected galactic star fields. V. The Lick astrographic fields at intermediate galactic latitudes. ApJ, 367, 528

ZINN, R. J., 1985, The globular cluster system of the galaxy. IV. The halo and disk subsystems. ApJ, 293, 424

ZINN, R. J., 1993, The galactic halo cluster systems: evidence for accretion. in Smith G. H., Brodie J. P., eds, The Globular Cluster – Galaxy Connection, ASP Conf. Series, 48, p. 38

Discussion

CARNEY: I want to defend Bob Zinn's somewhat bold assertion that the "young" clusters are in retrograde motion. As I noted, Majewski has found a statistically more compelling result for stars over 5 kpc from the plane. While his result has been questioned due to his choice of distance scale, our "low halo" and "high halo" results, which share the same distance scale, supports the "high/young halo" retrograde motion. One comment relevant to your RR Lyrae stars. Our data suggest the "high halo" mean metallicity may be shifting to [Fe/H] = -2.0, complicating the use of RR Lyrae stars unless their possible younger ages compensate for the lower metallicities.

LAYDEN: Yes. At [Fe/H] = -2.0, few RR Lyrae stars are formed (most are BHB) stars. Thus the number of RR Lyrae stars available is small.

BOND: What is the current thinking on how metal-rich populations are able to populate the RR Lyrae star instability strip?

LAYDEN: Y. W. Lee (1992, AJ, 104, 1780) has reiterated the findings of Taam, Kraft & Suntzeff (1976, ApJ, 207, 201) that these stars could be produced by the extreme tail of the mass-loss distribution. The small number of thick disk RR Lyrae stars observed, relative to the very large number of thick disk stars near the Sun, indicates that the production efficiency of metal-rich stars is, in fact, very low.

DORMAN: The mass loss processes are not well understood and different mass distributions for HB stars have not been explored. For metal rich compositions very few evolutionary tracks spend significant time in the instability strip. It is difficult to make firm theoretical predictions to be tested empirically because the mass distribution is not understood theoretically and the observational constraints are weak.

Baade-Wesselink Analyses of Field vs. Cluster RR Lyrae Variables

By JESPER STORM[1], BRUCE W. CARNEY[2], BIRGITTA NORDSTRÖM[3], JOHANNES ANDERSEN[3], AND DAVID W. LATHAM[4]

[1]European Southern Observatory, Casilla 19001, Santiago, CHILE

[2]Department of Physics & Astronomy, University of North Carolina, Chapel Hill, NC 27599, USA

[3]Niels Bohr Institute for Astronomy, Physics and Geophysics, Copenhagen University, Brorfelde, DK-4340 Tølløse, DENMARK

[4]Center for Astrophysics, 60 Garden Street, Cambridge, MA 02138, USA

We compare the results of Baade-Wesselink analyses for field and cluster RR Lyrae variables: M_V vs. [Fe/H], M_K vs. log P, and in $M/M_\odot$. The cluster variables appear to be marginally brighter, but the effect is not significant except for the highly evolved V9 in 47 Tuc.

1. Introduction

RR Lyrae stars are potentially very powerful standard candles within the Local Group, and potentially beyond, due to the correlation between M_V and [Fe/H] and between M_K and log P (see Sandage 1982, Longmore et al. 1986, 1990, and Carney et al. 1992). The slopes and zero points of the correlations are matters of much debate, however, and various groups using a variety of methods find differences in the slopes of up to a factor of two and differences in the zero point of 0.3 mag.

Our main concern here is to investigate the possible differences between field and cluster RR Lyrae stars. If there are differences, they can have a significant impact on the conclusions that one draws because the correlations in many cases are determined from field stars but are applied to cluster stars assuming that they are identical. Also, if such differences exist, we might gain some insight on the various problems of HB morphology and the second parameter effect and the influence of the environment on the HB morphology.

2. The Data

The cluster RR Lyrae stars are significantly fainter than the field RR Lyrae stars previously studied making it necessary to use large telescopes. The cluster fields are also much more crowded than those of the field stars making it crucial in many cases to use panoramic detectors to properly take into account neighbouring objects.

For all the stars we have aquired optical light curves using CCDs and point spread function fitting techniques to extract the photometric information. The optical photometry for the stars can be found in Storm et al. (1992) and Carney et al. (1992b, 1993). The infrared (K-band) lightcurves were obtained using a classical photometer for the first run and IR-imagers for the rest of the data. These data were processed with proper corrections for non-linearities of the devices and again using PSF-fitting techniques for the photometric analysis. These data, obtained at the 4m at KPNO, the IRTF and at the CTIO 1.5m telescope, are presented in Storm et al. (1992) and Storm et al. (1994b).

The final piece of observational data necessary for the analysis is the radial velocity curves. These were obtained from radial velocity measures from high resolution ($R \approx 30,000$) echelle spectroscopy using the MMT for the northern stars and the Danish 1.54m at ESO for the southern star V9 in 47 Tuc. The data and reductions for the northern stars can be found in Storm, Carney & Latham (1992) and for V9 it can be found in Storm et al. (1994b).

3. The Analysis

The surface brightness variant of the Baade-Wesselink method that we have used for the present work was conceived by Wesselink (1969) and further developed by Manduca & Bell (1981). The basic principle is to estimate the relative expansion of a pulsating star from photometry and the simultaneous absolute expansion by integrating the radial velocity curve. Combining these two numbers then gives the radius of the star which in turn can be used in the Boltzman equation together with a temperature estimate from photometry to derive the absolute luminosity of the pulsating star. This value combined with the observed luminosity then gives the distance.

The main concerns regarding systematic errors with the method are the temperature scale and the conversion of radial velocities into pulsational velocities.

It as been shown that optical colors do not produce an angular diameter curve that reproduces well that derived from radial velocities (Carney & Latham 1984; Jones 1988). The $(V-K)$ color, on the other hand, has proven to be an excellent temperature indicator for these stars (e.g., Fernley 1993).

The pulsational velocities have been determined using conversion factors from radial velocity to pulsational velocity following the discussions of Jones et al. (1992) and Burki et al. (1982). Although the factors might vary slightly as a function of phase Jones et al. (1992) estimates the resulting uncertainty on the derived distances to be about 0.1 mag.

The actual matching of the photometric and spectroscopic data is done by determining angular diameter curves as a function of phase for a variety of assumed distances for the photometric angular diameter curve and comparing this curve with the curve derived from spectrocopic data and find the best match using χ^2 methods.

4. The Results

For all the stars we have determined distances from χ^2 fits. For the clusters M5 and M92 with two stars observed in each we have derived mean distances from these individual estimates and used this distance estimate when deriving the physical parameters of the individual stars. The detailed analysis of the data can be found in Storm et al. (1994a, b).

When performing the fit one has to carefully select the phase interval over which the fit is performed to make sure that phase regions where the atmosphere model cannot properly take into account the dynamical aspects of the pulsation (e.g. shockwaves) are not included in the fit. Furthermore, we have rejected certain phase intervals for some of the stars where the χ^2 fit showed a significantly worse fit than for the rest of the phase interval, which we attribute to bad data and/or other problems with the proper modelling of the atmosphere.

In Table 1 we show the resulting physical parameters for the five cluster stars analysed. The derived equilibrium temperature, T_{eq}, is determined from the Boltzman equation from the Baade-Wesslink radius and the luminosity as described by Carney et al. (1992a).

	47 Tuc	M5		M92		
	V9	V8	V28	V1	V3	
$\log P$	-0.1326	-0.2626	-0.2645	-0.1532	-0.1956	P in days
[Fe/H]	-0.71	-1.40	-1.40	-2.24	-2.24	dex
A_B	1.35	1.32	1.30	1.06	1.34	mag
$<V>-<K>$	0.89	1.07	1.06	1.21	1.13	mag
v_{rad}	-15.70	$+48.14$	$+53.51$	-125.69	-122.72	km s^{-1}
$(m-M)_0$	13.23	14.37	14.37	14.60	14.60	mag
$<M_V>$	0.32	0.65	0.67	0.43	0.47	mag
$<M_{bol}>$	0.32	0.61	0.63	0.34	0.38	mag
$<M_K>$	-0.57	-0.41	-0.38	-0.79	-0.66	mag
T_{eq}	6828	6549	6576	6355	6482	K
M_{vAB}	0.46	0.64	0.61	0.70	0.70	
$\log(g_{eq})$	2.63	2.81	2.81	2.69	2.74	g in cm s^{-2}
$T_{eq}(P,A,m)$	6227	6573	6564	6362	6517	K

TABLE 1. New Results for Cluster Variables

The mass estimate, M_{vAB}, is derived from the pulsation equation of van Albada & Baker (1971) using the physical parameters presented in Table 1. In Figure 1 we show the resultant M_V vs. [Fe/H] and M_K vs. log P relations for the field and cluster variables. For reference, the slope of the latter, derived from studies of variables within clusters, should be -2.3 (see Longmore et al. 1990).

We estimate the uncertainty in the distance modulus to the individual cluster variables to be of the order 0.2 mag (internal), depending on the star, to which one should add quadratically a contribution from systematic errors of 0.13 mag. The errors are larger for the cluster stars than for the field stars primarily because of the difficulties in obtaining high-precision infrared magnitudes.

The mass estimates are rather uncertain because the estimates depend critically on the adopted distance. We estimate the uncertainty to be of the order 0.15 to 0.20 $\mathcal{M}_\odot$ for the cluster variables, and somewhat less for the field variables, from the uncertainty in the distance estimates. The M92 stars are at the high end of this range, mainly due to the fact that the K-photometry could not be performed relative to comparison stars within the field of view of the IR arrays.

5. Discussion

From various indicators, the color-magnitude diagram and $\log P$-A_B diagram in particular, we conclude that V9 in 47 Tuc is a very evolved object and thus is unsuited for constraining the magnitude correlations. Still, the M_K-log P relation does predict very well the absolute magnitude of V9, emphasizing the strength of using infrared magnitudes for distance determinations. We suspect the good agreement for the highly evolved V9 is the result of the conflicting high luminosity and low mass (leading to a lower density and a longer period).

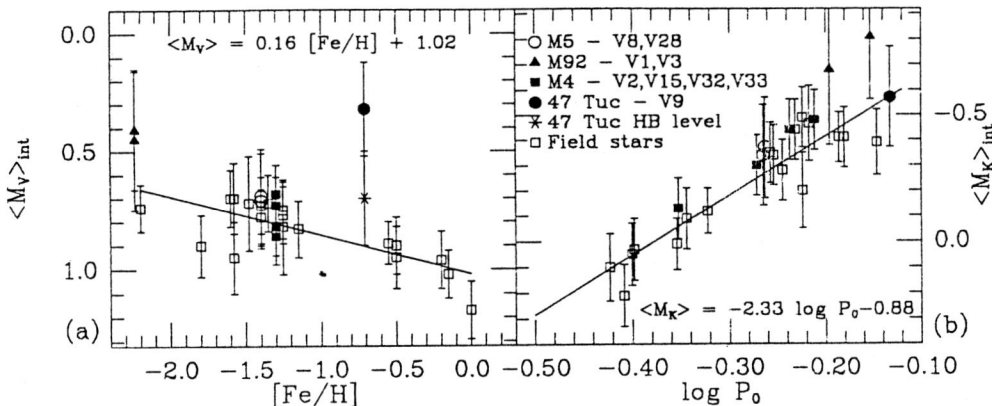

FIGURE 1. The M_V-[Fe/H] relation and the M_K-log P relation. log P_0 refers to the fundamental period. The legend in panel (b) is valid for both panels and the formulae for the drawn lines are also shown.

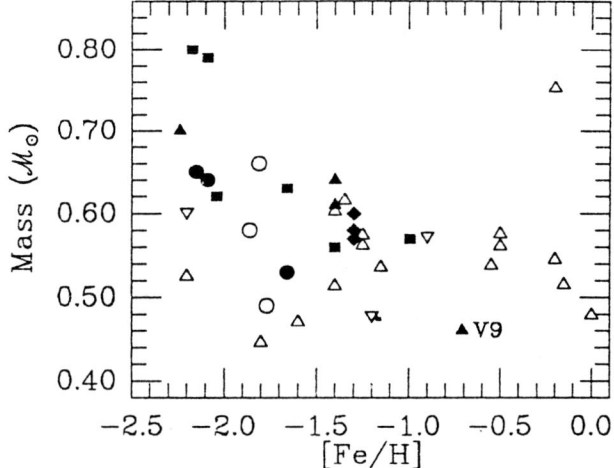

▲ M92, M5, & 47 Tuc RRab (this work)
◆ M4 RRab (Liu & Janes 1990; B-W)
△ Field RRab (CSJ; B-W)
▽ Field RRc (CSJ; B-W)
■ RRc ϕ_{31} (Simon & Clement)
○ RRd field (old opacities)
● RRd clusters (old opacities)

FIGURE 2. The mass estimates from various investigations plotted against metallicity

V9 is an extreme if not peculiar object in other respects as well. The period is exceedingly long for the color and metallicity of the star and the temperatures estimated from the $(B-V)$ and $(V-K)$ indices differ by more than 500 degrees, but still the Baade-Wesselink analysis gives a distance estimate in good agreement with that from ZAMS fitting. We estimate the cluster's true distance modulus is 13.23 ± 0.17 mag, whereas Hesser et al. (1987) found a "best subdwarf-inferred distance" to be about 13.30 mag.

Our primary goal here is to compare the cluster stars with the field stars to see if there are systematic differences present between the two samples and if a calibration based on field stars is indeed applicable to cluster stars. The field star sample used here is the one presented in Jones et al. (1992) with data from various groups transformed to a common system. The cluster data presented here has been expanded by including the results from Liu & Janes (1990) on M4 and these data has been transformed to the same system as used by Jones et al. (1992).

The cluster data appear to be slightly brighter (0.1 mag) than the field stars but this is not a very significant result, considering the uncertainties in the zero points. The slopes do not appear to disagree significantly although the baseline for the M_V-[Fe/H] relation is rather limited because of the evolved nature of 47 Tuc-V9.

The results for the masses are not well understood at present. In Figure 2 we have plotted the masses estimated from the Baade-Wesselink results, plus those from analyses of the light curves of cluster RRc variables by Simon & Clement (1993). Agreement with our cluster results is good in spite of quite different luminosity scales. V9 remains an anomaly. We have also plotted results of analyses of double-mode (RRd) variables in M15 (Cox et al. 1983), M3 (Nemec & Clement 1989), M68 (Clement 1990), and for three field RR Lyrae stars (Clement et al. 1991). Agreement with our results is again good, although it must be noted that these RRd masses were computed using the "old" Livermore opacities. Use of the "new" opacities apparently raises the mass estimates by up to 0.1 $\mathcal{M}_\odot$ (Cox 1991; Kovacs et al. 1991), although use of non-solar element-to-iron mixtures may lower these mass estimates by a comparable amount (Kovacs et al. 1992).

REFERENCES

BURKI, G., MAYOR, M., BENZ, W., 1982, A&A, 109, 258
CARNEY, B. W., LATHAM, D. W., 1984, ApJ, 278, 241
CARNEY, B. W., STORM, J., JONES, R. V., 1992a, ApJ, 386, 663
CARNEY, B. W., STORM, J., TRAMELL, S. R., JONES, R. V., 1992b, PASP, 104, 44
CARNEY, B. W., STORM, J., WILLIAMS, C., 1993, PASP, 105, 294
CLEMENT, C. M., 1990, AJ, 99, 240
CLEMENT, C. M., KINMAN, T. D., SUNTZEFF, N. B., 1991, ApJ, 372, 273
COX, A. N., 1991, ApJ, 381, L71
COX, A.N., HODSON, S. W., CLANCY, S. P., ApJ, 266, 94.
FERNLEY, J., 1993, A&A, 268, 591
HESSER, J. E., HARRIS, W. E., VANDENBERG, D. A., ALLWRIGHT, J. W. B., SHOTT, P., STETSON, P. B., 1987, PASP, 99, 739
JONES, R. V., 1988, ApJ, 326, 305
JONES, R. V., CARNEY, B. W., STORM, J., LATHAM, D. W., 1992, ApJ, 386, 646
KOVACS, G., BUCHLER, J. R., MAROM, A., 1991, A&A, 252, L27
KOVACS, G., BUCHLER, J. R., MAROM, A., IGLESIAS, C. A., ROGERS, F. J., 1992, A&A, 259, L46
LIU, T., JANES, K. A., 1990, ApJ, 360, 561

LONGMORE, A. J., FERNLEY, J. A., JAMESON, R. F., 1986, MNRAS, 220, 279
LONGMORE, A. J., DIXON, R., SKILLEN, I., JAMESON, R. F., FERNLEY, J. A., 1990, MNRAS, 247, 684
MANDUCA, A., BELL, R. A., 1981, ApJ, 250, 306
SANDAGE, A., 1982, ApJ, 252, 553
SIMON, N. R., CLEMENT, C. M., 1993, ApJ, 410, 526
STORM, J., CARNEY, B. W., LATHAM, D. W., 1992, PASP, 104, 159
STORM, J., CARNEY, B. W., BECK, J. A., 1991, PASP, 103, 1264
STORM, J., CARNEY, B. W., LATHAM, D. W., 1994a, A&A, submitted.
STORM, J., NORDSTRÖM, B., CARNEY, B. W., ANDERSEN, J., 1994b, in preparation
VAN ALBADA, T. S., BAKER, N., 1971, ApJ, 169, 311
WESSELINK, A. J., 1969, MNRAS, 144, 297

Discussion

JANES: What is the problem with the star in 47 Tuc? Why does not it fit with the normal metallicity-luminosity relation?

CARNEY: It is highly evolved, I presume. The mean HB level is, however, at the "right" luminosity. However, its mass is unusually low.

SARAJEDINI: Where does the RR Lyrae star in 47 Tuc (V9) lie among the tracks of Dorman, VandenBerg & Laskarides (1989, ApJ 343, 750)?

LU: A simple question. Why do some globular clusters have large numbers of blue stragglers, RR-Lyrae and horizontal branch stars and the other globular clusters have relatively few?

CARNEY: In the HB case, all clusters have about the same ratio relative to red giants. Why red HB, RR Lyrae, or blue HB stars dominate due to the first parameter (metallicity) and "second parameter" (age? rotation?). A metal-rich variable like V9 is a bit surprising, though, even allowing for age or rotation, I believe.

SARAJEDINI: In the case of blue stragglers, their numbers in globular clusters are related to the cluster's mass. In general, more massive globular clusters have more blue stragglers. Other important cluster parameters are the dynamical characteristics such as central concentration.

LAYDEN: The period of V9 is substantially longer than any of the periods of the metal-rich field RR Lyrae stars. One then wonders whether V9's peculiarity is due to the denser cluster environment in which it lives. Perhaps a close encounter during its red giant phase stripped off some of its atmosphere. Later, as a HB star, its lower mass would leave it blueward of the other RHB stars, in the instability strip.

SARAJEDINI: How precise is the M_k-log P relation for the derivation of globular cluster distances?

CARNEY: The field star and clusters work suggest it is quite reliable, modulo the still-uncertain halo distance scale zero point. V9 is especially interesting in that its high luminosity is offset by its lower mass and hence lower density, longer pulsation period. It falls right on the middle of the M_k-log P relation.

The Rotation of Population II A Stars

By WILLIAM P. BIDELMAN

Warner and Swasey Observatory, Case Western Reserve University, Cleveland, OH 44106-7215, USA

The rotational velocities of Population II A stars are briefly discussed from a historical perspective.

---

1. Introduction

It has been apparent since the first abundance analyses of the F- and G- type Population II stars were done in the 1950's that the rotational velocities of these were small, but no one was surprised at this since most Population I stars of the same temperature shared this characteristic. However, the realization that rotation is also small in the Population II stars of earlier type has come considerably more recently. Peterson (1993), for example, has illustrated this in a paper presented at the recent Luminous High-Latitude Stars conference, where she discussed the rotational velocities of several horizontal-branch stars in M3 and M13 as well as the classical HB A stars HD 74721 and HD 109995.

However, this concept does have somewhat older roots, which I thought it might be of interest to present here. If we limit ourselves to the A-type Population II stars, we have two classes: (1) the RR Lyrae stars, and (2) the somewhat bluer HB A stars near A0.

2. RR Lyrae stars

The first high-dispersion plates of RR Lyrae itself were taken by R. F. Sanford, who incidentally also took the coudé plates used by Chamberlain & Aller (1951) in their classic paper on HD 19445 and HD 140283. Sanford's (1949) paper reproduces tracings from which it is clear that the rotational velocity of this object is low, though he does not mention this fact. This was emphasized some years later by Babcock's (1956) paper in which he presented evidence of a variable magnetic field in that object. He would never have even attempted such an investigation unless the stars's lines were very sharp. Of course, RR Lyrae is only one RR Lyrae star, but no one has ever reported a broad-lined one.

3. Horizontal-branch A stars

As far as the A-type horizontal-branch stars are concerned, we need to look only at Merrill's (1947) paper where he reports, on the basis of a 10 Å mm^{-1} plate, that the "spectrum ... (of HD 161817) resembles that of a sharp-line main-sequence A4 star (M = +1) more closely than it does that of the intermediate white dwarfs ..." During the early and mid-1950's, Greenstein made an extensive study of the spectra of the hotter Population II stars, and in a paper with Cuffey (1954) on a faint B5 star near the cluster NGC 1647 it is stated that, "at this dispersion (38 Å mm^{-1}), the most obvious characteristic is the lack of rotation; the lines are strikingly sharp, compared to the bright standard stars. (Small rotation is also a characteristic of all other Population II stars so far observed at Palomar)."

This comment is reiterated in Greenstein's (1956) Berkeley Symposium paper and elsewhere. Eventually, I even got into the act myself; in the spring of 1962 I obtained Lick Observatory 16 Å mm^{-1} coudé spectrograms of seven supposed Population II A stars, mainly taken from Roman's (1955) list: the stars observed were HD 74721, HD 86986, BD +32° 2188, HD 109995, HD 117880, HD 136161, and HD 161817. It is noted in Whitford (1962) that "six out of seven A-type high-velocity and horizontal-branch stars had sharp lines: the exception was HD 136161." It happens that the exceptional star had by far the lowest space velocity of the group, and I have since convinced myself that it is not really a Population II star. I guess a few people must have read my remark, for HD 136161 has not surfaced in the last 30 years.

4. Conclusions

Thus, I am quite happy to go along with Ruth Peterson's suggestion that a low rotational velocity can be considered a Population II discriminant for A Stars. Most of us tend to slow down a bit with advancing age!

REFERENCES

BABCOCK, H. W., 1956, PASP, 68, 70
CHAMBERLAIN, J. W., ALLER, L. H., 1951, ApJ, 114, 52
GREENSTEIN, J. L., 1956, in Neyman J., ed, Proc. Third Berkeley Symposium on Mathematical Statistics and Probability, University of California, Berkeley, p. 11
GREENSTEIN, J. L., CUFFEY, J., 1954, PASP, 66, 187
MERRILL, P. W., 1947, PASP, 59, 256
PETERSON, R. C., 1993, in Sasselov D. D., ed, ASP Conf. Ser., 45, p. 195
ROMAN, N. G., 1955, ApJS, 2, 195
SANFORD, R. F., 1949, ApJ, 109, 208
WHITFORD, A. E., 1962, AJ, 67, 645

Horizontal-Branch Stars and Possibly Related Objects

By WILLIAM P. BIDELMAN

Warner and Swasey Observatory, Case Western Reserve University, Cleveland, OH 44106-7215, USA

In Tables 1 and 2 I have listed 53 stars brighter than 11th magnitude, excluding RR Lyrae variables, that appear to be either rather certain field horizontal-branch A stars or, if not, possibly related to them. While the distinction between the two lists is rather arbitrary, and they are certainly not exhaustive, further observations of some of the objects would be clearly worthwhile.

Name	α (1950) δ	gal. lat. (deg)	mag.	RV	μ ("/yr)	C_1	Fe/H	Remarks
HD 2857	0 29.3 -5 32	-68	9.9	-145	.10	1.21	-1.3	
HD 4850	0 47.7 -47 34	-70	9.6	-29	.08	1.28		
BD +0°145	0 53.8 +1 27	-61	10.6	-259		1.03		
HD 8376	1 20.7+31 32	-31	9.3:		.07	1.27		
HD 13780	2 11.0 -49 17	-63	9.8		.07	1.28		
HD 14829	2 20.7 -10 54	-63	10.3	-178		1.24	-0.7	
BD +30°623	4 6.1+30 39	-15	10:	+48	.00			ApJS 28, 157, in NGC 1514
HD 31943	4 56.1 -43 6	-38	7.4:		.11	1.12		
BD +26°1120	6 8.5+26 28	+4	9.1		.07	1.20		HD 252940
HD 60778	7 33.6 -0 1	+10	9.1	+43	.10	1.28	-0.5	
HD 74721	8 43.2+13 27	+31	8.7	+36	.12	1.25	-1.1	
HD 78913	9 6.3 -68 17	-14	9.3	+320	.05	1.28		
HD 86986	9 59.8+14 48	+49	8.0	+9	.26	1.27	-1.0	
HD 87047	10 0.3+31 18	+53	10.0:		.09	1.27		
F 41	11 23.3 +6 47	+61	11.0	-25	.03	0.94		
BD +32°2188	11 44.5+32 17	+75	10.8	+100		0.92		
BD +33°2171	11 53.7+33 11	+77	10.6	+57	.22	1.22		
HD 106304	12 11.3 -40 36	+22	9.1	+96	.15	1.16		
BD +42°2309	12 25.8+41 56	+75	10.8	-137		1.26	-1.2	
HD 109995	12 36.4+39 35	+78	7.6	-135	.17	1.29	-0.5	
HD 117880	13 30.8 -18 15	+43	9.1	+141	.15	1.20	-1.7	
HD 130095	14 43.9 -27 2	+29	8.1	+50v	.22	1.27	-0.6	
HD 130201	14 45.0 -45 28	+12	10.3:		.06	1.24		
HD 139961	15 39.4 -44 47	+8	8.8	+145	.20	1.31		lines diff., NSV 7204
HD 161817	17 44.7+25 46	+25	7.0	-361	.06	1.21	-1.3	NSV 9679
HD 180903	19 16.2 -24 29	-17	9.6:		.09	1.26		
HD 213468	22 29.4 -42 51	-58	10.9	-180	.05	1.26	-0.8	

TABLE 1. Early A horizontal-branch type stars excluding RR Lyrae stars, brighter than 11th magnitude

Name	α (1950) δ	gal. lat. (deg)	mag.	RV	μ ("/yr)	C_1	Remarks
CoD -39°356	1 15.6 -38 52	-77	10.9:	+41		0.56	AJ 78, 295
BD +58°559	3 4.1 +59 7	+1	10.6	+28			perh. sdF (WPB)
HD 52057	6 56.4 -36 12	-14	9.3	+38	.19	0.67	
HD 58764	7 24.1 -20 48	-2	7.1		.03		class II or shell (WPB)
HD 63463	7 45.9 -30 1	-2	8.7		.01		HB? (WPB)
	8 55.3 +29 24	+39	10.3:				HBB (PG)
HD 83041	9 32.7 -28 39	+17	8.5			0.79	PASP 83, 98, sdF?
7 Sex	9 49.6 +2 41	+40	6.0	+97	.21	1.04	Fe/H normal, NSV 4656
HR 4049	10 15.8 -28 44	+23	5.5	-38v	.02	1.60	binary, AG Ant
HD 105262	12 4.6 +13 16	+72	7.1	+41	.04	1.36	ApJ 258, L71
	12 5.4 +22 48	+79	9.8:				HBB (PG)
BD +36°2242	12 6.7 +35 59	+78	9.9	-3	.02	0.77	ApJS 28, 157
HD 106373	12 11.6 -27 58	+34	8.9	+96	.10	0.72	PASP 83, 98, shell?
HD 107369	12 18.1 -32 17	+30	9.2	-45	.02	1.61	PASP 83, 98, [Fe/H] normal
	12 46.2 +2 52	+65	10.3:				HBB (PG)
HD 123884	14 7.8 -17 45	+41	9.3	+7	.02	0.88	ApJ 111, 498
HD 128336	14 33.9 -12 17	+43	9.2		.02	0.63	PASP 83, 98, called HD 130156
	15 11.4 +36 39	+59	10.6:				HBB (PG)
	15 30.6 +21 14	+53	10.0:				HBB (PG)
BD -15°4515	17 16.3 -15 52	+12	9.1		.04	0.85	PASP 83, 98, sdF?
HD 180587	19 14.4 +10 53	-1	8.3	-2v	.02		shell
HD 184779	19 34.4 -44 2	-26	9.2	-26	.04	0.86	
HD 188960	19 55.4 -16 6	-22	7.6		.04		HB? (WPB)
HD 202056	21 11.3 -19 4	-40	9.2:		.02		HB? (WPB)
BD +18°4873	21 48.6 +19 1	-26	9.5			1.17	PASP 103, 63
HD 214539	22 37.3 -67 57	-45	7.2	+333	.04	1.08	post-HB

TABLE 2. A-type stars possibly related to horizontal-branch stars, brighter than 11th magnitude

A New Group of Post-AGB Objects - The Hot Carbon-Poor Stars

By E. S. CONLON

APS Division, School of Physics and Mathematics, The Queen's University of Belfast, Belfast, BT7 1NN, N. Ireland, U. K.

Recent analyses with LTE model atmospheres of high resolution spectra have shown that for an increasing number of high latitude hot stars, the derived compositions are incompatible with their previous classification as young massive stars. These objects are thought to be evolved low mass stars on the post-asymptotic giant branch (post-AGB). However, their anomalously large carbon deficiencies (of typically 0.2 dex below solar) are not typical of previously observed post-AGB stars. The similar carbon underabundance found in the well-known post-AGB object Barnard 29 establishes these stars as a group of carbon-poor objects. In this paper the properties of this new and increasingly numerous group are discussed in terms of current evolutionary theories and observations. It would appear that they are lower mass objects that come off the AGB before the onset of the thermal pulsing stage. The implications of these finding for our understanding of post-AGB evolution are briefly discussed.

---

1. Introduction

The majority of low and intermediate mass stars (M $\approx$ 1 - 8 M$_\odot$) are believed to pass through the post-asymptotic giant branch (post-AGB) on their way to becoming planetary nebulae. This brief transition phase of typically 10^4 years (Schönberner 1983) is one of the least well understood stages of stellar evolution. As well as improving our understanding of the final stages of a star's lifetime, the nucleosynthesis, dredge-up, and mass loss processes that occur during this evolutionary stage are principally responsible for the enrichment of the interstellar medium, in particular of the lighter elements carbon and oxygen (Wheeler *et al.* 1989).

During a spectroscopic study into *young* hot B-type stars at high galactic latitudes (Conlon *et al.* 1992), we at the Queen's University Belfast have observed a small number of blue stars with chemical compositions more compatible with an it evolved nature (Conlon *et al.* 1991, McCausland *et al.* 1992, Conlon, Moehler & Theissen 1992b). They appear to be hotter analogues to the high latitude A, F, and G supergiants, now believed to be early post-AGB stars (for a recent review, see Luck 1993). However, although both groups are metal-weak and have atmospheric parameters compatible with a post AGB-nature, there are a number of observational differences. Most striking is the severe underabundance of carbon observed in the blue stars, typically by -2.0 dex compared with the young B-type objects. In contrast, the cool post-AGB stars are normally carbon-rich.

To investigate the discrepancies between the hot and cool post-AGB candidates, a comprehensive observational program is underway. The selection criteria used are: (i) they are high latitude blue stars with atmospheric parameters that coincide with the post-AGB tracts as show in Figure 1, i.e., log T$_{eff}$ = 4.0 - 4.5 and corresponding log g in the range 2.0 - 4.0 dex or (ii) they are B-type objects that have been classified as post-AGB stars on the basis of their infrared flux distributions, IRAS colours, and CO emission (Manchado *et al.* 1989, Loup *et al.* 1990). In addition a *known* hot post-AGB star, Barnard 29, in the globular cluster M13 was analysed using high resolution échelle data (Conlon *et al.* 1993b). Model atmosphere analysis results for a number of these blue objects has firmly established the existence of a carbon-poor group of hot post-AGB

Star	T_{eff}	log g	Source
LSIV -4° 0.1	11000	2.0	McCCDK92
LB 3193	13000	2.2	McCCDK92
PHL 174	18000	2.7	CDKMcC91
PG 1704+222	19000	3.0	CMT92
Barnard 29	20000	3.0	CDK93
CPD -68° 91	20000	3.0	C93
LB 3219	21250	2.8	McCCDK92
LSIV -12° 111	24000	2.7	CDKK93
HD 119608	24000	3.0	C93
PHL 1580	24000	3.6	CDKMcC91
HD 177566	30000	3.8	CDKK93

TABLE 1. Atmospheric paramters of the program stars

objects. In this paper the properties of this newly-identified group are discussed in terms of current theories for this important evolutionary stage.

2. Observational Data

Currently a total of eleven blue carbon-poor objects have been analysed. These stars are listed in Table 1 together with the derived atmospheric parameters taken from the sources as follows: CDKMcC91, Conlon et al. (1991); McCCDK92, McCausland et al. (1992); CMT92, Conlon, Moehler & Theissen (1992b); CDK93, Conlon et al. (1993a); CDKK93, Conlon et al. (1993b); C93, Conlon (1993c).

In addition, the locations of these stars are plotted in the $log\ g - log\ T_{eff}$ plane in Figure 1. This Figure shows the post-AGB evolutionary tracks of Schönberner (1983), Schönberner (1987), and Wood & Faulker (1986) for central core masses of 0.54, 0.565, 0.598, 0.644, 0.79, and 0.89 $M_\odot$, the location of the ZAMS for B-type stars, and the horizontal branch for a star of core mass 0.48 $M_\odot$. Known carbon-rich objects are plotted as stars and known or possible carbon-poor objects are shown as filled or empty squares (B- type) and triangles (non B-type), respectively. Further details of these objects may be found in Conlon (1993c).

3. Results and Discussion

In Figure 1 most of the carbon-poor post-AGB objects are clustered near the lower mass tracks. Note that the post-AGB evolutionary tracks are not optimal as (i) they are calculated for Population I compositions and most of these objects are metal-poor and (ii) they are computed through the thermally pulsating stages on the AGB. However, it is likely that the carbon-poor objects leave the AGB before this stage begins. The tracks are used solely to illustrate that the majority of the carbon-poor post-AGB stars appear to be of low mass.

3.1. Evolutionary Scenario

The severe carbon deficiencies found for these post-AGB objects implies that they do not remain on the AGB long enough for the third dredge-up to occur. Theoretical calculations have indicated that the core mass of an AGB star must exceed some critical minimum value before the third dredge-up takes place (de Jong 1986, Iben 1991). The

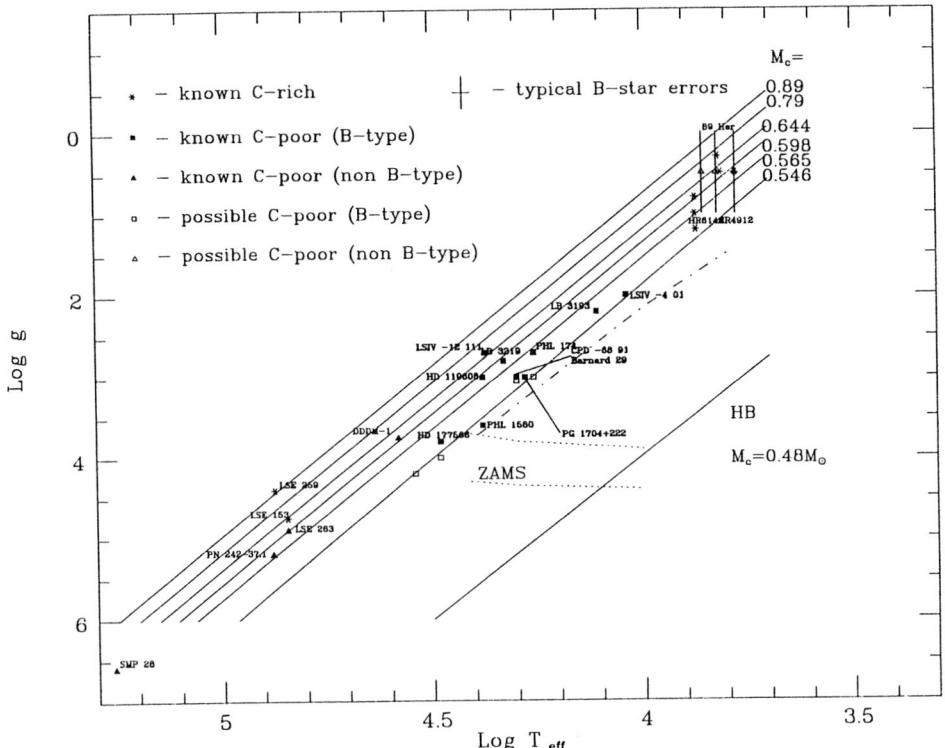

FIGURE 1. Locations of B-type post-AGB stars (filled stars) in the log g - log T_{eff} diagram.

predominantly low masses of these carbon-poor objects tend to support this idea. For example, Boothroyd & Sackmann (1988) find the lowest core mass model to experience the third dredge-up to be 0.81 $M_\odot$. However, the one exception is a low mass model of Gingold (1976), which terminates the AGB before the thermally pulsating stage is reached and is shown in Figure 1 as a dot-dashed line. Similar calculations (with more realistic AGB mass loss rates) are now required to test the carbon-poor evolutionary scenario.

The idea of low mass stars leaving the AGB before the thermally pulsating stage has recently been successful in explaining the characteric ultraviolet excess of elliptical galaxies (Brocato et al. 1990). In addition, the pronounced absence of C IV lines in their ultraviolet spectra (Burstein et al. 1988; Ferguson et al. 1991) provides further support for the proposed evolution of the carbon-poor post AGB-stars.

3.2. Progenitors

Possible immediate progenitors of this group are the RV Tauri stars. These are low luminosity objects of spectral type A, F, and G. Their typical high mass loss rates (10^{-5} $M_\odot$ yr^{-1}) imply that they have just left the AGB (within 200 to 500 years). Compared with the more massive A, F, and G supergiant post-AGB stars, their much weaker CO emission (Alcolea & Bujarrabal 1991) could be the result of low carbon abundances. The progenitors of *these* stars could be a subset of sufficiently low mass M-stars. Lazaro et al. (1991) found a group of 70 M-stars to be carbon deficient with a small number revealing underabundances up to -1.5 dex. It is thought that these objects may have descended from the weak G-band giants observed by Sneden & Pilachowski (1984).

4. The Future

Current observations are difficult to reconcile with theoretical predictions. According to Schönberner's (1983) lifetimes, the ratio of hot to cool post-AGB stars should be approximately 10:1. However, only a small number of hot B-type post-AGB candidates have been found and none of these are carbon-rich! Selection effects have meant that optically bright blue stars with evidence of circumstellar material were preferentially observed. These are more likely to be lower mass (hence carbon-poor) and slowly evolving post-AGB objects. The continuation of this extensive observational program, in particular using the second selection criterion, is essential to resolve these discrepancies.

I would like to thank F. P. Keenan and P. L. Dufton for their continued intersted in this work and acknowledge financial support from the SERC.

REFERENCES

ALCOLEA, J., BUJARRABAL, V., 1991, A&A, 245, 499

BOOTHROYD, A. I., SACKMANN, I. -J, 1988, ApJ, 328, 671

BROCATO, E., MATTEUCCI, F., MAZZITELLI, I., TORNAMBÈ, A., 1990, ApJ, 349, 458

BURSTEIN, D., BERTOLA, F., BUSON, L. M., FABER, S. M., LAUER, T. M., 1988, ApJ, 328, 440

CONLON, E. S., DUFTON, P. L., KEENAN, F. P., MCCAUSLAND, R. J. H., 1991, MNRAS, 248, 823 (CDKMcC91)

CONLON, E. S., DUFTON, P. L., KEENAN, F. P., MCCAUSLAND, R. J. H., HOLMGREN, D., 1992a, ApJ, 400, 273

CONLON, E. S., MOEHLER, S., THEISSEN, A., 1992b, A&A, 269, L1 (CMT92)

CONLON, E., DUFTON, P. I., KEENAN, F. P., 1993a, A&A, submitted (CDK93)

CONLON, E., DUFTON, P. I., KEENAN, F. P., KENDALL, T., 1993b, A&A, submitted (CDKK93)

CONLON, E., 1993c, ApJ, submitted (C93)

DE JONG, T., 1987, in Mennessier M. O., Omont A., eds, From Miras to Planetary Nebulae: Which path for stellar evolution?, Editions Frontieres, p. 289

FERGUSON, H. C., DAVIDSEN, A. F., KRISS, G. A., et al., 1991, ApJ, 382, L69

GINGOLD, R. A., 1976, ApJ, 204, 116

IBEN, I., 1991, ApJS, 76, 55

LAZARO, C., LYNAS-GRAY, A. E., CLEGG, R. E. S., MOUNTAIN, C. M., ZDROZNY, A., 1991, MNRAS, 249, 62

LOUP, C., FORVEILLE, T., NYMAN, L. A., OMONT, A., 1990, A&A, 227, L29

LUCK, R. E., 1993, in Sasselov D. D., ed, ASP Conf. Ser., 45, p. 87

MANCHADO, A., GARCIA LARIO, P., POTTASCH, S. R., 1989, ApSS, 156, 57

MCCAUSLAND, R. J. H., CONLON, E. S., DUFTON, P. L., KEENAN, F. P., 1992, ApJ, 394, 298 (McCCDK92)

SCHÖNBERNER, D., 1983, ApJ, 394, 298

SCHÖNBERNER, D., 1987, in Kwok, S., Pottasch S. R., eds, Late Stages of Stellar Evolution, Reidel, Dordrecht, p. 341

SNEDEN, C., PILACHOWSKI, C. A., 1984, PASP, 96, 44

WHEELER, J. C., SNEDEN, C., TRURAN, J. W., 1989, ARAA, 27, 279

Wood, P. R., Faulkner, D. J., 1986, ApJ, 307, 659

MK Classification of Hot Stars in the Halo

By R. F. GARRISON

David Dunlap Observatory, University of Toronto, Box 30, Richmond Hill, ON L4C4Y6,
CANADA

At high galactic latitude, there are about a dozen yellow supergiants whose spectra are very similar to those of Population I stars in the disk (e.g., 89 Her, HD 161796). The rest of the few hundred candidates suggested by various people have turned out to be obvious post-AGB stars, with evidence of mixing, s-process elements, shells, etcetera (e.g., HR 7671 and most of the so-called UU Her stars). If the former are true, massive supergiants, they must have originated in the Halo. If they originated in the Halo, there must be a proportionate number of O, B, or A supergiants as well. The numbers at each type should be in approximate agreement according to time steps in stellar evolution models.

Discussion

BIDELMAN: HD 161796 does show a very unusual feature not found in ordinary F supergiants: namely, very large and peculiar infrared emission. This alone sets it apart from most stars of its spectral type.

GARRISON: It does show a large excess from IRAS observations, but that does not prove it is a post AGB star. It does indicate that it has gone through a mass loss phase, but other high mass stars ($M > 2\ M_\odot$) lose mass before the red giant stage and they are also IRAS sources. However, Parthaserathy has also looked at the IUE spectra and sees no evidence for present mass loss in HD 161796, so again it is ambiguous.

DORMAN: If the numbers of F type post-AGB stars do agree well with tracks of a single mass it would be strong evidence that they come from a single mass population origin. However, the evolutionary lifetime is extremely mass sensitive with $0.02\ M_\odot$ making a difference of a large factor (100) in this rate at least at the low mass end of the range.

GARRISON: That is very interesting, but I hope that consideration of some extremes will still provide a good test of some sort.

BOND: It is worth nothing that Earle Luck has done detailed abundance analyses of several of the high latitude F supergiants. He finds, for example, [Fe/H] = -0.4 in HD 161796 and -1.1 in HR 7671, as well as overabundances of C, B, N, and O. These are all distinctly not typical of Population I supergiants.

GARRISON: I have no problem with HR 7671 being a post AGB star. Its peculiarities are consistent with that interpretation and I can see the metal weakness. However, the abundances of HD 161796 are close enough to solar that small adjustments in $\log g$, T_{eff}, or microturbulence will make it consistent with Population I. Most of the other stars we looked at have the characteristics of post AGB stars that we would expect, so they are not the problem. However, there are a few like HD 161796 and 89 Her that mimic Population I too successfully. We would like to find a way to recognize them at MK resolution. (For example, if HD 161796 were not located in the halo, if it were a low velocity interloper in the disk, would you recognize it as post AGB star or does its location predispose you to interpret it that way?)

Photometry of XX Virginis and V716 Ophiuchi and the Period Luminosity Relations of Type II Cepheids

By D. H. McNAMARA
AND M. D. PYNE

Department of Physics and Astronomy, Brigham Young University, Provo, Utah 84602, USA

The surface gravity of V716 Oph, the shorter-period variable, is smaller than the surface gravity of XX Vir, the longer-period variable. This fact combined with the $P\sqrt{\rho} = Q$ relation indicates that there must be a considerable cosmic scatter in the P-L relation of Type II Cepheid stars.

The two stars V716 Ophiuchi ($16^h28^m09^s$,-5°23.'17, 1950, 11.28-12.60v, $1.^d116$) and XX Virginis ($14^h14^m13^s$,-06°03', 1950, 11.55-12.78v, $1.^d348$) are classified as BL Her variables or Population II W Vir stars (CWB) with periods less than eight days in the *General Catalog of Variable Stars*. Reviews of Population II variables include Wallerstein & Cox (1984), Harris (1985), and Gingold (1985). According to current ideas the Population II variables with periods in the range of one day to eight days are in an advanced stage of evolution. They are stars that have reached the instability strip by evolving off the horizontal branch. Apparently they are low-mass stars with an inner He "burning" shell and an outer shell where H fusion occurs.

The variable stars with periods between one and three days exhibit a bewildering variety of light curve shapes and metallicities. Diethelm (1983) has proposed a scheme of classification based purely on morphological grounds. The latest version of this scheme is the recognition of three distinct classes. He proposes calling all the stars AHB variables (above-horizontal-branch stars), subdividing into them AHB1 (stars similar to XX Vir and V716 Oph), AHB2 (CW stars), and AHB3 (BL Her stars). Although his original classification is based on light-curve shapes, additional studies (Diethelm 1986, 1990) indicate the distinction between the different types carries over into the physical parameters of the star as well. The reader is referred to the three Diethelm references given above for the detailed characteristics of the three classes.

Photometry ($uvby\beta$) of V716 Oph and XX Vir has been secured over several years. An analysis of the photometry indicates they are both metal-poor stars. We found [Fe/H] = -2.0 for XX Vir and [Fe/H] = -2.1 for V716 Oph.

The loops traversed by the stars XX Vir (solid squares) and V716 Oph (crosses) in the theoretical c_1, $(b-y)$ diagram (Figure 1) can be utilized to estimate the mean effective temperatures and surface gravities of the stars. We find $<T_{eff}> = 6410$ K, $<\log g> = 1.95$ (XX Vir); $<T_{eff}> = 6450$ K, $<\log g> = 1.67$ (V716 Oph).

What is remarkable in the surface gravity values is that V716 Oph, which is the shorter period variable ($P = 1.^d12$) has the smaller surface gravity (XX Vir, $P = 1.^d35$). Normally pulsating stars exhibit lower surface gravities as the period of pulsation increases. If the $P\sqrt{\rho} = Q$ relation holds for the variables, we find $\Delta m_{bol} = 2.2$ and $\Delta \log M = -1.18$; clearly unrealistic values. It thus appears that Q cannot be the same for the two variables - possibly due to different envelope masses, internal structure, and different evolutionary histories.

Our surprising results for these two stars has led us to reexamine the period-luminosity relation of the Population II variables. We find a total of 33 stars in globular clusters yields

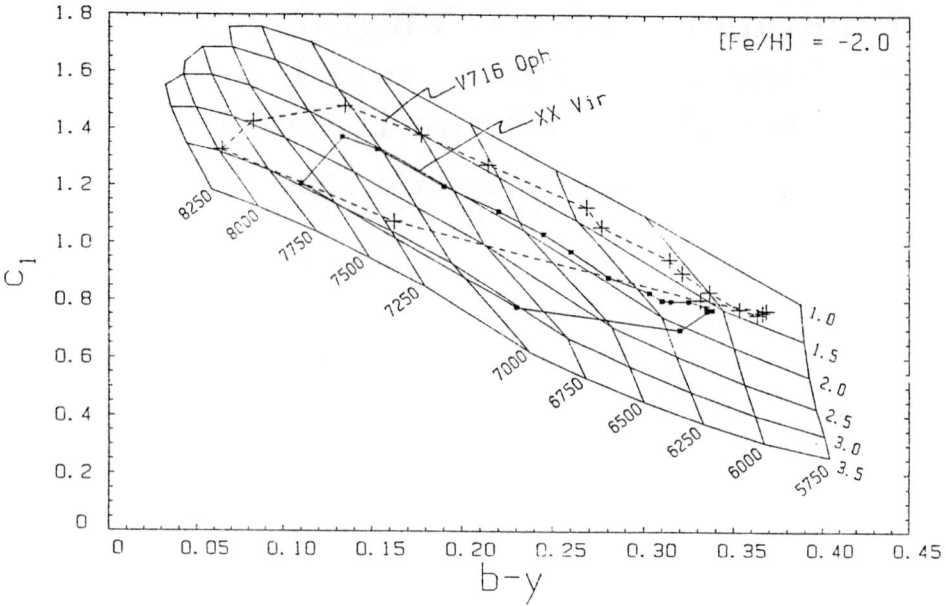

FIGURE 1. The loops traversed by XX Vir (solid squares) and V716 Oph (crosses) in the c_1, $(b-y)$ diagram. V716 Oph reaches both higher and lower temperatures than XX Vir. The mean values of T_{eff} and $\log g$ around the cycle are: XX Vir $<T_{eff}> = 6410$ K, $<\log g> = 1.95$; V716 Oph $<T_{eff}> = 6450$ K, $<\log g> = 1.67$.

$$<M_v> = -1.62 \log P - 0.12. \qquad (1)$$
$$\pm 0.13 \quad \pm 0.1$$

Some insight into the P-L relation can be gained by assuming the stars obey the $P\sqrt{\rho} = Q$ relation. As pointed out earlier, the $\log g$ data of XX Vir and V716 Oph as well as the scatter exhibited by stars from the P-L relation suggest the Q values of stars of similar period may be different. Nevertheless, the $P\sqrt{\rho} = Q$ relation should be valid in the mean in view of the P-L relation exhibited by the variable stars. We assume that this is indeed the case and write the $P\sqrt{\rho} = Q$ relation as

$$M_{bol} = -3.33 \log P - 1.67 \log M/M_\odot - 10 \log T_{eff}$$
$$+ M_{bol\odot} + 10 \log T_{eff\odot} + 3.33 \log Q. \qquad (2)$$

We adopt the following values for the constants: $M_{bol\odot} = 4.75$, $T_{eff\odot} = 5780$ K. We adopt $\log T_{eff} = -0.058 \log P + 3.810$ for the variation of T_{eff} with period of pulsation. The equation predicts $T_{eff} = 6450$ K at $\log P = 0$ and $T_{eff} = 5300$ K at $\log P = 1.5$. The 6450 K value is based on our T_{eff} values of XX Vir and V716 Oph. We have inferred the T_{eff} of the longer period stars from the Kurucz models (1992) with the aid of $(b-y)$ values estimated from the $(B-V)$ values of the longer period variables. The pulsation "constant" Q depends very strongly on period. We adopted $\log Q = 0.24 \log P - 1.39$. We arrived at this expression by assuming the masses of all the variables are identical, $M = 0.55\ M_\odot$, and utilizing the Cogan (1970) graphs relating Q to M/R for fundamental mode pulsators. We note that the Q values calculated by Cogan (1970) are for metal-strong stars while the stars of our sample are metal poor. The Q values appear to depend

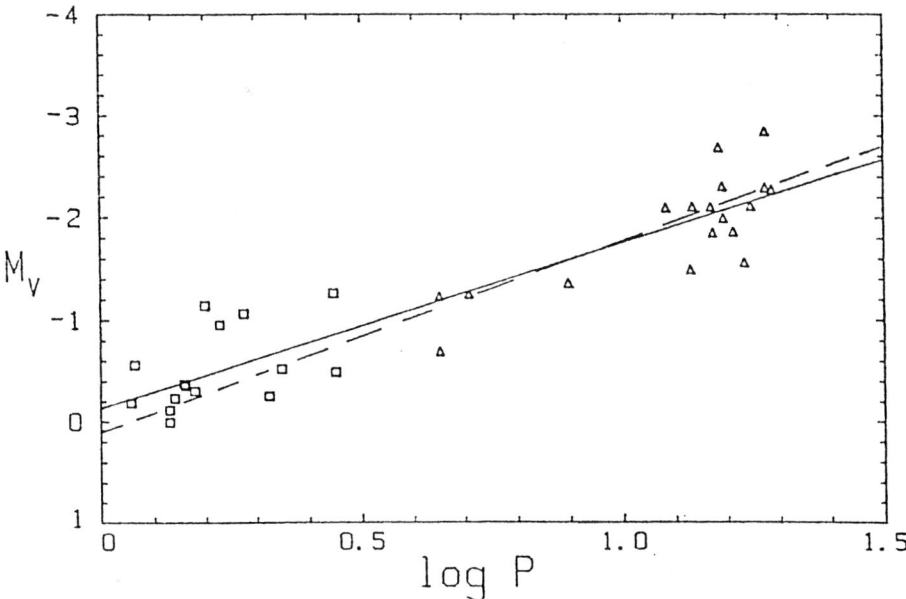

FIGURE 2. A plot of the absolute magnitude of Population II Cepheids found in globular clusters against their period (log P). The solid line is $M_v = -1.62 \log P - 0.14$. The equation is based on 33 defining stars, none of which exhibit RV Tauri characteristics. The dashed line is $M_v = -1.88 \log P + 0.13$ derived from the $P\sqrt{\rho} = Q$ relation as described in the text. The values of the stars with periods less than three days are plotted as open squares, longer period variable are shown as open triangles.

primarily on the M/R ratio and not on abundance. In fact, several formulas that relate Q to M/R discussed by Christy (1966, p 376) ignore any dependence on abundance. We assume in our discussion that the dependence of Q on M/R given by Cogan is also valid for the metal-poor stars. Finally the Bolometric correction is given by B.C. $= -0.073 \log P - 0.06$. If these constants and expressions are substituted into equation (2) we find

$$M_v = -1.88 \log P - 1.67 \log M/M_\odot - 0.36. \qquad (3)$$

If we adopt $M/M_\odot = 0.55$, a value similar to the masses of RR Lyrae variables (Liu & Jones 1991) equation (3) reduces to

$$M_v = -1.88 \log P + 0.13. \qquad (4)$$

This equation is in excellent agreement with equation (1) derived directly from the observational data. The estimated uncertainty in the coefficient of the log P term is $\sim \pm 0.10$. Relative small changes in the expressions relating T_{eff} and Q to period could bring the two equations into agreement. The line corresponding to equation (4) is shown as the dashed line in Figure 2 and the solid line is given by equation (1).

To force the slope of equation (4) into agreement with equation (1) would require the mass to vary as $\log M/M_\odot = -0.10 \log P - 0.26$ in equation (3). At $\log P = 1.5$ the masses would be $M/M_\odot = 0.39$. In our view it is more likely that the masses on average are the same through the entire period range although there may be differences in the masses from star to star. The small discrepancy between the slopes of equations (4) and (1) is more likely due to errors in the M_v values of the variables and/or errors in some of the equations used in deriving equation (4). We conclude, however, it is possible to derive a P-L relation for the Type II variables from the $P\sqrt{\rho} = Q$ relation, assuming the stars are vibrating in the fundamental mode, and show little if any dependance of mass on period. The first or second harmonic modes of pulsation fail to give the strong dependence of Q on M/R required to obtain a small slope for the P-L relation. The slope of equation (4) is not nearly as steep as the classical Cepheids or dwarf Cepheids where the coefficients of the log P term are ~ -2.79 (Balona & Shobbrook 1985) and ~ -3.76 (McNamara & Powell 1993), respectively. This is a direct result of the fact that the masses of the Type II Cepheids are independent or nearly independent of period and have small M/R values that lead to a strong dependence of Q on M/R.

If we utilize equation (4) to predict M_v values of RR Lyrae stars we find $M_v = 0.5$ for a star like X Ari ($P = 0.651$ days) and $M_v = 0.8$ for SW And ($P = 0.442$ days). These values are in good agreement with the M_v values of the two stars given by Liu & Jones (1991).

REFERENCES

BALONA, L. A., SHOBBROOK, R. R., 1985, in Madore B. F., ed, Cepheids: Theory and Observation, Cambridge University Press, Cambridge, p. 232

CHRISTY, R. F., 1966, ARAA, 4, 353

COGAN, B. C., 1970, ApJ, 162, 139

DIETHELM, R., 1983, A&A, 124, 108

DIETHELM, R., 1986, A&AS, 64, 261

DIETHELM, R., 1990, A&A, 239, 186

GINGOLD, R. A., 1985, Mem. Soc. Astron. Ital., 56, 169

HARRIS, H. C., 1985, in Madore, B. F., ed, Cepheids: Theory and Observations, Cambridge University Press, Cambridge, p. 232

KURUCZ, R. L., 1992, private communication

LIU, T., JANES, K. A., 1991, in Janes K. A., ed, The Formation and Evolution of Star Clusters, ASP Conf. Ser., 13, p. 278

MCNAMARA, D. H., POWELL, J. M., 1993, in Dworetsky M. M., Castelli F., Faraggiana R., eds, Peculiar versus Normal Phenomena in A-Type and Related Stars, ASP Conf. Ser., 44, p. 437

WALLERSTEIN, G., COX, A. N., 1984, PASP, 96, 677

Discussion

BIDELMAN: Is there a problem with the period, overtones, etc.?

McNAMARA: No - the dependence of Q on mass/ratios for overtones is too small to give such a gentle slope to the period-luminosity relation. Only the fundamental period has the strong dependence of Q on mass/radius to yield the current slope to the period-luminosity relation.

Rotation and Oxygen Line Strengths in Blue Horizontal Branch Stars

By RUTH C. PETERSON[1], D. A. CROCKER[2], AND R. T. ROOD[3]

[1] University of California Observatories, Lick Observatory, Santa Cruz, CA 95064, USA

[2] Department of Physics & Astronomy, University of Alabama, Box 870324, Tuscaloosa, AL 35487, USA

[3] Department of Astronomy, University of Virginia, Box 3818 University Station, Charlottesville, VA 22903-0818 USA

We have measured rotational velocities and oxygen abundances in 28 blue horizontal branch stars in M13, 20 in M3, and 16 in NGC 288. Here we outline the behavior of rotation and oxygen line strength with $B-V$ within each cluster, and compare the average values of each from cluster to cluster. We find that both rotation and the oxygen abundance differ from cluster to cluster, yet neither alone nor both together is enough to determine where an individual star is situated along the horizontal branch.

1. Observations

We used the fiber-feed system "Nessie" with the echelle spectrograph on the 4 m telescope at Kitt Peak/NOAO to obtain spectra of the oxygen triplet at 7771 – 75 Å at a resolution of $15\,km\,s^{-1}$ for BHB stars with $-0.1 \leq B-V \leq 0.1$ in each cluster. Figure 1 depicts each spectrum observed during the first observing run. M3 spectra are on the left, and M13 spectra on the right. Arrows mark the positions of the three O I lines. Incomplete sky subtraction is responsible for spikes near 7763 Å; otherwise the spectra are featureless.

In Figure 1 the triplet is detectable in all stars but one, the fifth spectrum from the top in M13. Excepting this one, all stars in Figure 1 are redder than $B-V = -0.05$; in later runs, where stars bluer than this limit were included, none showed the O I features.

The line strengths are quite uniform in each panel, and somewhat stronger in M13 than in M3. There is no obvious sign of extremely oxygen-deficient BHB stars in M13, although such stars are common among the luminous giants of the cluster (Sneden et al. 1992). Concerning rotational velocities, it is apparent from the figure that rapidly rotating stars are found in M13 but not in M3: the latter spectra all have rather sharp lines, while about a half-dozen of the M13 spectra are distinctly broadened.

2. Analysis

Values of $v\sin i$ and oxygen strength were found by a χ^2 fit of these spectral observations to a two-dimensional grid of spectra calculated from model atmospheres of Kurucz and broadened by various degrees of rotation, using his program SYNTHE. Figure 2 gives an example of the fits for a narrow-lined star in NGC 288 and a broad-lined star in M13. The dashed line corresponds to the best fit; the dotted line to a fit removed by 1σ in both $v\sin i$ and oxygen line strength.

This fitting procedure also allows us to set upper limits on $v\sin i$ or oxygen strength when no lines are detected. There is an ambiguity in that lines of a given strength are harder to detect when broadened, as is clear from Figure 1. However, once either $v\sin i$ or oxygen strength is assumed, an upper limit to the other is obtained. Generally speaking,

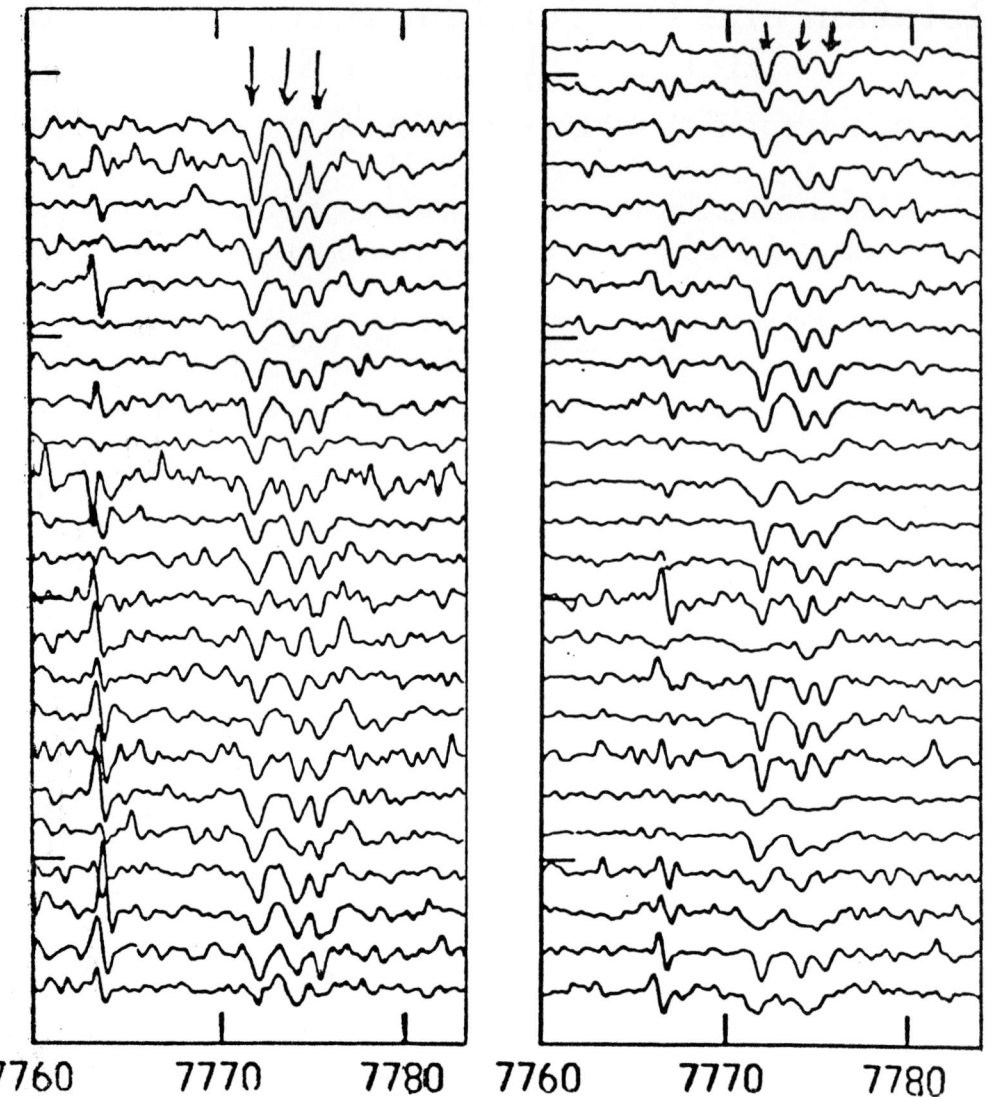

FIGURE 1. The O I triplet (arrows) in BHB stars in M3 (left) and M13 (right)

this indicates that there may be some incompleteness of moderately oxygen-strong stars in M13 blueward of $B - V = 0.0$ due to the presence of more rapid rotators.

3. Results

Figures 3 and 4 show the measurements of oxygen line strengths (upper panel) and $v \sin i$ values (lower panel) as a function of $B - V$ color along the horizontal branch, for every star in which lines were detected. Figure 3 refers to M13, and Figure 4 to NGC 288. Solid symbols in the lower panels represent the $v \sin i$ measurements of Peterson (1983) for M13 and of Peterson (1985) for NGC 288, which are based on the Mg I lines and a

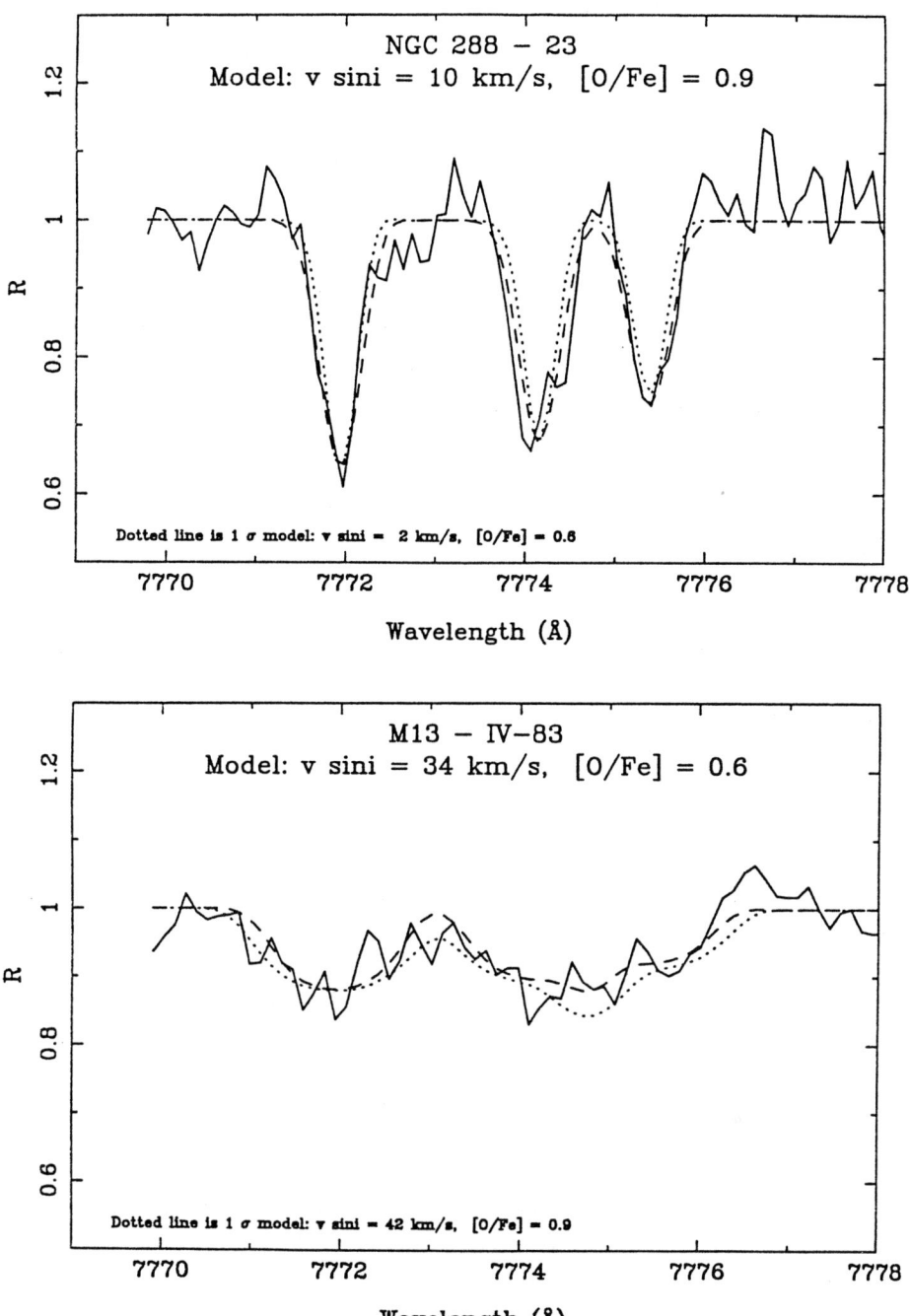

FIGURE 2. Fitting a narrow-lined star in NGC288 and a broad-lined star in M13

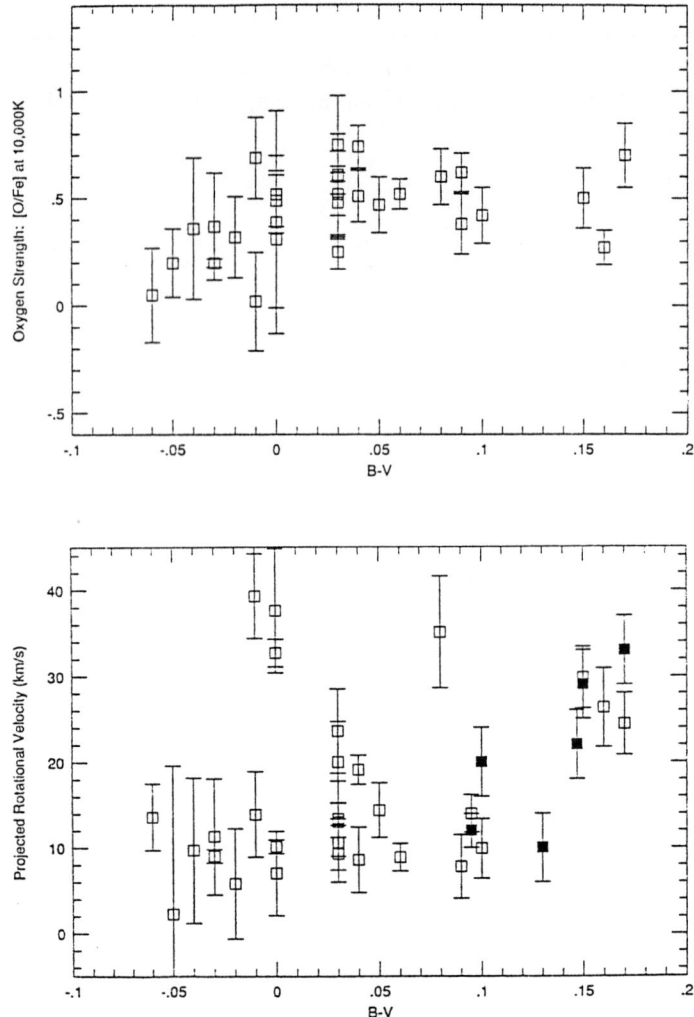

FIGURE 3. Measurements of oxygen line strengths and $v \sin i$ values for horizontal branch stars in M13

cross-correlation measurement technique. Four stars in M13 observed with each method have velocities which agree within the stated errors.

The separation of M13 stars into rapid and slow rotators is evident in Figure 3. Monte Carlo simulations indicate that the assumption of a single mean v value and a spread about that value cannot reproduce the distribution of M13 $v \sin i$ points, no matter what choices are made. Dividing the stars into two groups, two-thirds with $v = 15\,km\,s^{-1}$ and one third with $v = 38\,km\,s^{-1}$, gives a good match to the distribution. For M3, a single value of $v = 13 \pm 2\,km\,s^{-1}$ fits well, as does $v = 9 \pm 2\,km\,s^{-1}$ for NGC 288. Rotation clearly is stronger in M13 than in either of the other clusters and consequently does differ from cluster to cluster. However, the low rotation in NGC 288 shows that, at least in that case, rotation is not the factor leading to the anomalously blue HB morphology. Note, however, that NGC 288 may be anomalously old—another cause for blue HBs (Lee, Demarque, & Zinn 1994)

In no cluster does $v \sin i$ depend upon $B - V$ color along the BHB. Nor is any star

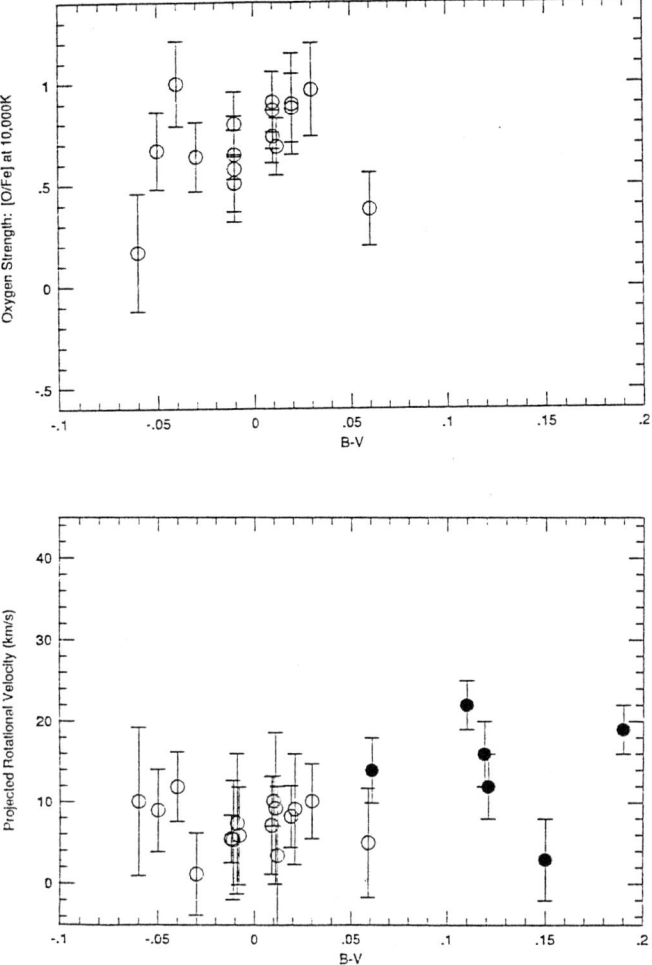

FIGURE 4. Measurements of oxygen line strengths and $v \sin i$ values for horizontal branch stars in NGC 288

found with $v \sin i > 40\,km\,s^{-1}$, even in M13; all BHB stars rotate slowly compared to Population I stars of similar temperature and gravity.

The failure to detect O I lines in stars bluer than $B - V = -0.5$ might be attributed to the ionization of O I at high temperatures. However, both the abruptness of this disappearance coupled with the rather constant O I strength throughout the red end of the M13 BHB both suggest that the oxygen abundance declines with decreasing $B - V$ along the M13 BHB. Because of the high excitation of the lower level of the O I triplet, line strengths would be expected to decline with increasing $B - V$ if the oxygen abundance were constant. Before a dependence of oxygen abundance on $B - V$ can be considered established, however, accurate temperatures for these stars are needed along with a knowledge of the behavior of both non-LTE effects and diffusion. If this tentative conclusion stands, it would suggest that the BHB analogs of the oxygen-deficient giants might all be found blueward of $B - V = -0.5$, and also that the oxygen abundance is correlated with a star's position on the BHB, at least in M13. Note that variation of

O along the HB may be an indication of CNO processing and mixing history and does not necessarily imply variations in the primordial O. This would require that whatever process leads to O depletion also leads to increased mass loss.

However, oxygen is not the only factor governing BHB position, since the mean oxygen abundances of those BHB stars in which O I was detected does not correlate with blueness of the HB. The strongest O I features are found in NGC 288, which has a very blue BHB; next-strongest lines occur in M13; and M3 has the weakest lines.

4. Conclusions

Rotational velocities are high in M13 and low in M3, as deduced by Peterson (1983), but also low in NGC 288; the high values of Peterson (1985) for that cluster seem to be due to the small size of the sample which included several of the most rapidly rotating stars in the combined group. Consequently, rotation is not the only factor affecting the position of a star on the BHB, although it may be linked with other features such as more extensive mixing and mass loss (see Sneden et al. 1992). There is no dependence of $v \sin i$ on position along the horizontal branch in any cluster. Stars in M13 seem to fall into a group of rapid rotators, comprising one-third the stars, and a group of slower rotators, while stars in M3 and NGC 288 are all rather slow rotators. No star has a $v \sin i > 40 \, km \, s^{-1}$, so all are slow rotators compared to similar Population I stars.

There is a tendency for the oxygen abundance to decrease with decreasing $B-V$ along the M13 BHB; better stellar temperatures and an understanding of diffusion and non-LTE effects are required to confirm this. However, because the mean oxygen abundance is largest for NGC 288, next for M13, and lowest for M3, neither oxygen abundance nor rotation can explain the blueness of the NGC 288 BHB; an older age appears to be the best explanation found to date.

RTR is partially supported by NASA grant NAGW-2596 and DAC by EPSCoR grant EHR-9108761. RCP acknowledges partial support from NSF grant AST 9115183.

REFERENCES

LEE, Y-W., DEMARQUE, P., ZINN, R. 1994, ApJ, 423, 248
PETERSON, R. C. 1983, ApJ, 275, 737
PETERSON, R. C., 1985, ApJ, 294, L35
SNEDEN, C., KRAFT, R. P., PROSSER, C. F., LANGER, G. E. 1992, AJ, 104, 2121

Miscellaneous

UBV CCD Photometry of the Halo of M31

By A. FITZSIMMONS,[1] F. P. KEENAN,[1]
P. L. DUFTON,[1] J. E. LITTLE,[1]
AND M. J. IRWIN[2]

[1] The Queen's University of Belfast, Belfast BT7 1NN, N. Ireland, U. K.

[2] APM Unit, Royal Greenwich Observatory, Madingley Road, Cambridge CB3 0EZ, England, U. K.

We have obtained UBV CCD photometry of a 0.5 square degree field towards the halo of the Andromeda Galaxy (M31), using the 2.5m Isaac Newton Telescope on La Palma. These observations have allowed us to identify 9 blue objects, with (U-B) < -0.4 and (B-V) < 0.0, in the magnitude range B = 21.5 - 22.5, typical of a main-sequence early B-type star at the distance of Andromeda. Hence these objects may be normal Population I stars at large distances (> 3 kpc) from the plane of M 31. If this is the case, the stars should have the same radial velocity as M 31 ($v = -280$ km s^{-1}). Low resolution spectroscopy will enable us to measure the radial velocities and hence confirm or deny the association with the halo of M 31. Should the stars have the same radial velocity as M 31, it would provide very strong evidence for the existence of normal B-type stars in the halos of spiral galaxies.

1. Introduction

For the last decade we have been involved in an extensive programme to determine the nature and origin of faint, early-type stars at high galactic latitudes (see the recent review by Keenan 1992). Our analyses of multiwavelength observations of these objects using sophisticated model atmosphere codes has lead us to conclude that many of them are normal Population I OB-type stars at distances (z) from the galactic plane ranging from approximately 2–25 kpc (see Conlon et al. 1990 and references therein). To date we have found $\simeq$ 40 stars which we believe have been ejected from the disc via cluster ejection.

However, we have also identified 8 other stars whose short evolutionary lifetimes and large z-distances imply that they formed in the halo itself, such as SB 357 ($z = 9$ kpc; Conlon et al. 1992). Although we are convinced that our results are secure, some authors believe that the halo stars are subluminous, nearby objects, whose spectra somehow mimic those of normal stars (see for example Tobin 1991 and references therein). This belief is strengthened by the fact that if these objects are truly distant one is lead to the controversial result that star formation is occurring many kiloparsecs from the galactic plane.

In view of the above controversy we therefore decided to extend our work by searching for normal Population I OB-type stars in the halo of the near-edge on spiral galaxy M31 (inclination = 15°). At the distance of M31, a main sequence early B-type star would have a typical apparent magnitude of V $\simeq$ 22 and a supergiant B $\simeq$ 18. Hence identification of blue stars in the magnitude range B = 18–22, which are at the radial velocity of M31 ($v_r \simeq -280$ km s^{-1}), would provide very strong evidence for the existence of normal OB-type stars at large distances from the planes of spiral galaxies, and hence that star formation in the halo is occurring. In addition, it should be possible in the longer term to determine the z-distribution of such stars, which will allow a direct comparison with the theories suggested to explain their existence (Tobin 1987).

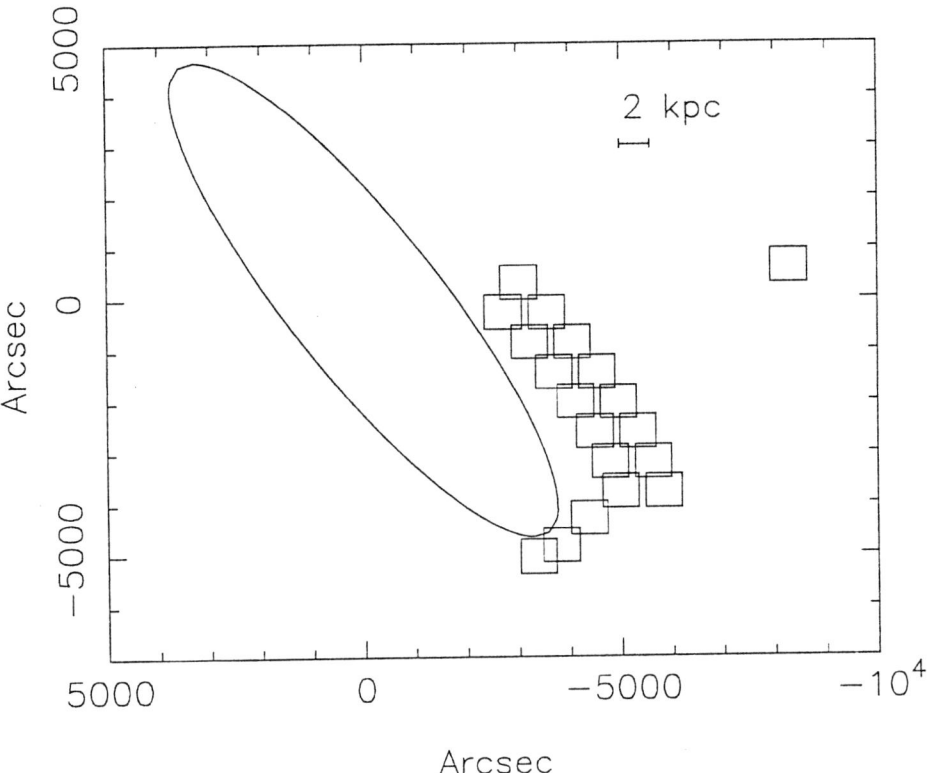

FIGURE 1. Positions of all fields imaged in UBV relative to the center of M31 at coordinates (0,0). The large ellipse represents the stellar disk of M31 with a radius of 20 kpc. The field at coordinates (-8200,700) contains the previously investigated star AND 0029+413.

2. Observations and Analysis

We performed UBV CCD photometry towards the halo of M31 during the period 28th September to the 4th October 1992, using the 2.5m Isaac Newton Telescope at Los Roques de los Muchachos Observatory, La Palma. The prime focus camera was used together with a coated 1280 × 1180 pixel EEV CCD. The pixel size of 22.5 μm gave an image scale of 0.56 arcsec per pixel, resulting in each field covering 11.6 × 10.7 arcminutes.

We chose the positions of the imaged fields by assuming a disk radius of 20 kpc for M31, and then placing the field centers at least 2 kpc above the nearest edge of the disk. By restricting ourselves to the southeast side of the halo, we also assured that if the halo co-rotates with the disk, the negative radial velocities of any stars belonging to the halo would be enhanced in future spectroscopy.

The sky was sufficiently clear to carry out imaging on only 4 nights, resulting in UBV exposures of 15 different fields towards the halo of M31. We also obtained photometry of the field containing the previously studied star AND 0029+413 (McCausland et al. 1993), and of three fields in the outer edge of the disk.

Including the extra halo field above, the total area of halo observed was 0.55 square degrees, corresponding to 84 kpc^2 at the distance of M31. These fields are shown diagrammatically relative to M31 in Figure 1.

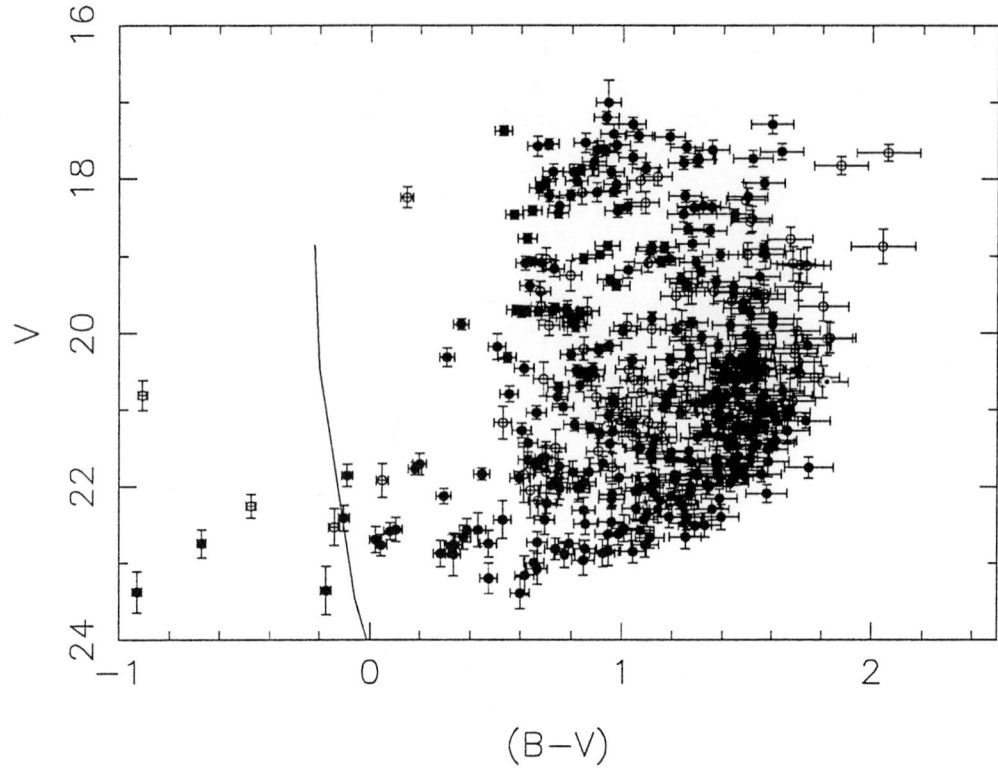

FIGURE 2. Colour-magnitude diagram for a single field, showing all stellar objects found in both the V and B images. Black symbols denote objects well fitted by the point-spread function derived for the individual frames, open circles denote otherwise. The solid line shows the main sequence from O5 down to approximately B8, placed at the distance of M31 and reddened by 0.1 in (B-V).

3. Discussion

Photometry on all frames was carried out using the DAOPHOT II psf-fitting and aperture photometry routines (Stetson 1993). Absolute magnitudes were calculated using observations of standard fields from Landolt (1992) observed and reduced in the same manner as the target fields.

The colour-magnitude diagram for a typical field can be seen in Figure 2. The galactic contribution of stars at (B-V) > 0.5 can clearly be seen. Also visible are a number of bluer objects, with one in particular at V $\simeq$ 21.8, (B-V) $\simeq$ −0.1 and (U-B) $\simeq$ −1.1. Although the (U-B) colour is slightly too blue, the V magnitude and (B-V) colour is what would be expected for a B0 star lying in the halo of M31.

Altogether we have found 9 stellar objects in this direction in the magnitude range 21.5 < V < 22.4, with (B-V) < 0.0. Although these may be blue stars in the halo of M31, there are two obvious sources of contamination by other objects. The work of Boyle (1989) shows that only 1-2 white dwarfs should be present in the area of sky surveyed. Perhaps a more serious contaminant would be QSO's. Yet both theoretical and observed colours (e.g. Braccesi et al. 1980) indicate that although low-redshift objects ($z < 2$) would have a similar (U-B) colour to early type stars, they should have (B-V) > 0.0.

Thus, in conclusion, we appear to have detected a surplus of blue stellar objects in the direction of the halo of M31, that possess the magnitude and colour expected of B-type stars at that distance. The next step will be to gather spectroscopy of these objects to determine their nature. Although such faint magnitudes will preclude any detailed spectroscopic investigation, low resolution spectra will enable both classification and measurement of their radial velocities.

REFERENCES

BOYLE, B. J., 1989, MNRAS, 240, 533

BRACCESI, A., ZITELLI, V., BONOLI, F., FORMIGGINI, L., 1980, A&A, 85, 80

CONLON, E. S., DUFTON, P. L., KEENAN, F. P., LEONARD, P. J. T., 1990, A&A, 236, 935

CONLON, E. S., DUFTON, P. L., KEENAN, F. P., MCCAUSLAND, R. J. H., HOLMGREN, D. E., 1992, ApJ, 400, 273

KEENAN, F. P., 1992, QJRAS, 33, 325

LANDOLT, A. U., 1992, AJ, 104, 340

MCCAUSLAND, R. J. H., CONLON, E. S., DUFTON, P. L., FITZSIMMONS, A., IRWIN, M. J., KEENAN, F. P., 1993, ApJ, 411, 650

STETSON, P. B., 1993, Daophot II User's Manual

TOBIN, W., 1987, in Philip A. G. D., Hayes D. S., Liebert J. W., eds, The Second Conference on Faint Blue Stars, IAU Colloq. 95, L. Davis Press, Schenectady, NY, p. 503

TOBIN, W., 1991, in Bloemen H., ed, The Disk-Halo Connection, Kluwer, Dordrecht, p. 109

Discussion

GARRISON: This is the right approach to settling the question of the apparently Population I stars in the Halo: Are they real? I see that you have some stars in your sample that have colors of F stars. If they are in our galaxy, they are further out than we think F stars should be. If they are in M31, of course, they are the F supergiants we see in our halo (like 89 Her). Now all we have to do is get spectra at 22nd magnitude!!

LITTLE: Our samples have the correct magnitudes and colors to be young B-type stars in the halo of M31. The point about F supergiants is worth noting.

Can Stars Still Form in the Galactic Halo?

By KENNETH JANES

Astronomy Department, Boston University, 725 Commonwealth Avenue, Boston, MA 02215, USA

A considerable body of circumstantial evidence points to the possibility that star formation in the galactic halo may not have ceased once the globular clusters formed. Regular reports of young B and A stars well into the halo, hints of star formation in the Magellanic Stream and between the Magellanic Clouds, the very existence of the Magellanic Stream and other cool hydrogen clouds in the halo, and even the presence of open star clusters at large distances from the galactic plane all indicate that recent star formation may not be restricted to the high gas-density regions of the galactic plane. But are these suggestions illusory, or is there hard evidence that stars have recently formed in the galactic halo? This review will be a critical examination of the various lines of evidence for young stars and gas in the halo, including the (negative) results of a search for high velocity hydrogen using RR Lyrae stars as background sources.

1. Introduction

Not too many years ago, the galactic halo was generally considered to be a fossil, an important, but unchanging artifact of the formation of the galaxy. Now we know that the halo developed over at least a several billion year period (see, e.g., Zinn 1993). The purpose for this review is to explore the possibility that the stellar population of the galactic halo continued to develop for an even longer period, possibly right up to the present time.

The most direct evidence for recent changes in the stellar population of the galactic halo is the apparent existence, first noted by Rodgers (1971), of significant numbers of normal Population I stars of spectral types B and A at great distances from the galactic plane. Some of them are far enough from the plane that they cannot plausibly have traveled from the disk during their short lifetimes, if their classifications as ordinary Population I stars is correct.

The stars in Rodgers (1971) list were first noted by Philip & Sanduleak (1968) who identified 62 faint A-type stars near the South Galactic Pole in an objective prism survey. They were looking for field analogues of the blue horizontal branch stars found in many globular clusters, the field horizontal branch (FHB) stars. The controversy began when Rodgers found that almost half of the Philip and Sanduleak stars look like normal Population I stars. But even now, more than 20 years later, their nature, or even their existence as a distinct class is a matter for debate.

The classification of the halo A and B stars as Population I objects is based on the determination that they have normal main-sequence gravities and approximately solar composition, based on either spectroscopy or photometry. A good example of the phenomenon can be seen in Figure 1, taken from Philip & Adelman (1991); the star BD +18° 4873 has the spectral appearance of a normal A star, whereas the other three stars in the figure have the characteristics of FHB stars. BD + 18° 4873 was first thought to be another FHB star, but as the figure shows, it has all the characteristics of a rapidly-rotating, normal A-type star. Recent investigations and reviews of the blue halo star

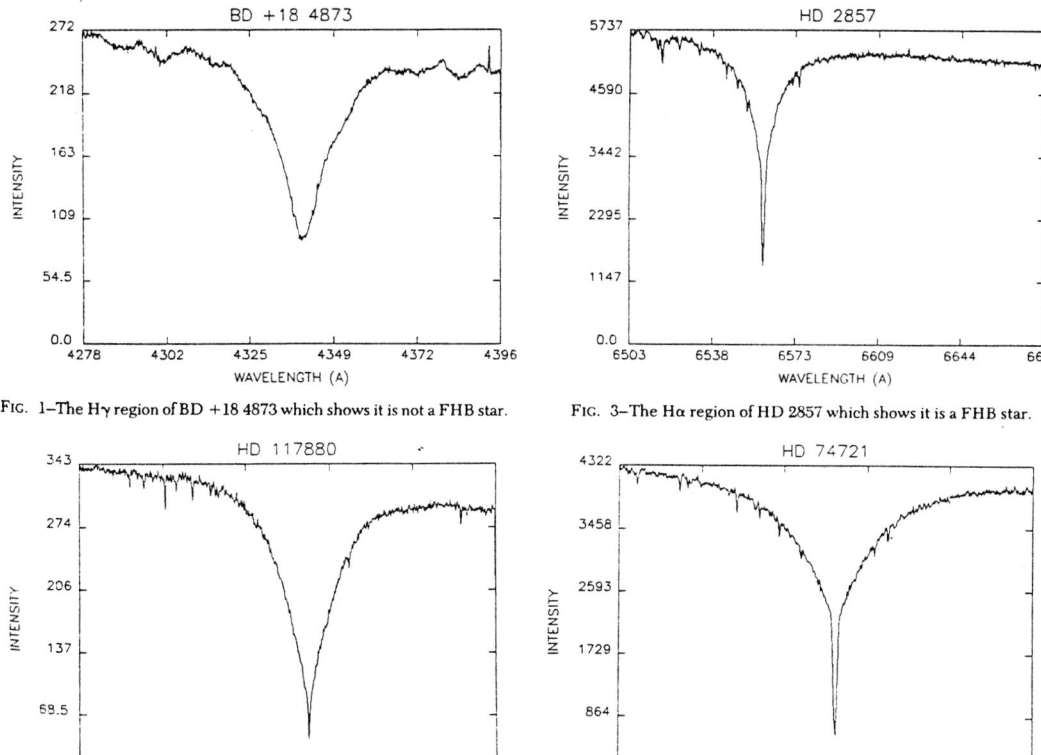

FIGURE 1. A selection of spectra of three FHB stars and a Population I star (BD +18° 4873) in the near halo (from Philip & Adelman 1991).

phenomenon include Lance (1988a,b); Philip & Adelman (1991); Stetson (1991); Conlon et al. (1991,1992); and Hambly et al. (1993).

Several explanations have been suggested to explain the halo B and A stars as disk stars that were somehow ejected into the halo, thereby avoiding the hypothesis that they might actually have formed there. But even if there has been no star formation in the halo in the immediate past, there are other observations that seem to suggest changes in the stellar population of the halo in moderately recent times. So the question remains: Can stars still form in the galactic halo?

There are actually three parts to the question, each of which is of interest, but all of which are controversial.

(i) Does the raw material exist in the halo for the formation of stars?
(ii) Can stars form in low-density environments?
(iii) Is there direct evidence that stars actually have recently formed in the halo?

2. Gas in the Galactic Halo

There certainly is some gas in the halo, but is there enough of it in a suitable state to form stars?

An excellent series of papers on the nature of the gas in the halo is the IAU Symposium 144, on "The Interstellar Disk-Halo Connection in Galaxies" (Bloemen 1990). The general character of the halo gas was reviewed by York (1982), who characterized the halo gas as predominately hot ($T \sim 10^5$ K), with low density ($n_H \sim 10^{-3}$ cm^{-3}). The total mass of the hot gaseous halo is $< 10^9$ $M_\odot$. Observations of other galaxies, in particular clusters of galaxies, show that hot halos are ubiquitous, containing even higher temperature ($T > 10^7$ K), x-ray-emitting gas. Much of this gas is in the form of cooling flows, which may be extensive enough to have had a significant effect on the evolution of the larger galaxies (see Fabian 1987 for the subject of cooling flows in galaxies). The composition of the gas in either the halo of our galaxy or others is not well known, but it is definitely not of zero metallicity. For example, Savage et al. (1993) and Morris et al. (1991) found a rich spectrum of uv lines in halo gas along the line of sight to quasars, using the HST. The hot gas is certainly far too thin and hot for star formation, but the cooling flows are likely to produce cool clouds, from which stars might eventually form.

2.1. Cool gas in the Halo

Not only is there gas in the halo, at least some of it consists of cool ($T < 10^4$ K) clouds of neutral hydrogen. The most obvious cool gaseous structure in the halo of our galaxy is the Magellanic Stream (Mathewson et al. 1974), a long trail of hydrogen detectable at 21-cm and extending almost halfway around the sky from the Magellanic Clouds. The association of the Magellanic Stream with the Magellanic Clouds is unquestionable, but the origin of the Stream is unknown. It could be a wake generated somehow in the hot halo gas by the passage of the Magellanic Clouds but it is more likely to be the result of tidal stripping of gas from the Clouds after their previous close passage to the Milky Way disk, or possibly it was generated in an encounter between the two Clouds. To date, there have been no reports of any detections of stars in the Stream.

Since the distance to the Magellanic Stream is unknown, its total mass is also unknown. However, at least one end of the Magellanic Stream must be approximately the same distance as the Magellanic Clouds (approximately 50 kpc). Its angular extent projected onto the sky is approximately $100° \times 2°$, with a neutral hydrogen column density of the order of 10^{19} cm^{-2} (see Figure 1 of Mathewson et al. 1974). Taking these numbers as a guide, the total mass in the Magellanic Stream can be estimated to be between 10^5 and 10^6 $M_\odot$. At some time in the future, all that gas will fall onto the disk. Could stars form out of that infalling gas?

2.2. High Velocity Clouds

Oort (1966) first drew attention to the high velocity hydrogen clouds (HVC's), that are scattered in various directions in the galaxy. Most of them have been found toward the outer parts of the galaxy, at high galactic latitudes, nearly all with negative velocities. Typical column densities are $\sim 10^{19}$ cm^{-2}. Their distances, hence their sizes and masses, were completely unknown until Danly et al. (1993) succeeded in obtaining an estimate of the distance to one cloud.

Possible explanations for the HVC's include the following:
  (i) Infalling primordial gas (low metallicity).
  (ii) Gas associated with a warped outer spiral arm (intermediate metallicity).
  (iii) Galactic fountains (solar metallicity).

The key to understanding the origin of the HVC's is to find their distances. But where are they? In spite of many attempts to learn their distances, it has only been in the past year that Danly et al. (1993) has succeeded in detecting optical absorption lines in the spectrum of a halo B star at the velocity of one of the clouds called Complex M. If this star, BD +38° 2182 (B3, $V = 11.2$), is indeed a normal Population I star, then the cloud is a maximum of 4.4 kpc from the Sun, and has a total mass between 10^4 and 10^5 $M_\odot$. Another star, HD 93521 (O9.5 V, $V = 7.04$), at a distance of $z = 1.5$ kpc above the galactic plane shows no sign of absorption lines from the cloud.

It is worth pointing out that there is a certain circularity in Danly's argument. If the star she observed is a true example of a young, Population I B star in the halo, then Complex M is also far from the plane and could be the sort of object that would provide the raw material for the formation of hot stars directly in the halo. If on the other hand, BD +38° 2182 is some more nearby object masquerading as a normal B star, then Complex M could be a local, rather insignificant structure, irrelevant to the issue of forming halo stars. So although this is certainly an exciting result, it does not conclusively solve either the halo B-star problem or the problem of the distances to the HVC's.

There have been other claims of detections of absorption lines in the spectra of distant objects. The HVC's are seen in absorption against galaxies and quasars (Savage et al. 1993), but the most explicit previous evidence for the detection of a cloud in front of a star is that of Songaila et al. (1988). This claim has been disputed by Lillienthal et al. (1990), who pointed out that there are known stellar lines very close to the wavelength measured by Songaila et al.

Songaila et al. used the RR Lyrae star, BT Dra, as a background source, and indeed the RR Lyrae stars would appear to hold great promise for this purpose – we can find their distances rather accurately, and their spectra are relatively uncomplicated. They did not, however, make use of one of the most interesting properties of the RR Lyrae stars, the fact that they pulsate. By observing a single RR Lyrae continuously over its entire pulsational cycle, it should be possible to identify and subtract the stellar lines, which will be changing in radial velocity over the star's cycle, whereas the interstellar lines will remain constant.

2.3. A search for the high velocity clouds

I attempted to take advantage of this useful RR Lyrae star property in a program at the KPNO 4-meter telescope. The plan was to follow a single RR Lyrae through its entire light cycle, to get a continuous sequence of spectra of the sodium D lines. Poor weather severely limited the success of this program, but I did get multiple spectra of several RR Lyrae stars, including BT Dra. I did not, however, find any sodium absorption at the appropriate velocities. Several factors probably contributed to my inability to see any absorption. First, I was not able to get complete phase coverage for any of the RR Lyrae stars I observed, so that a proper subtraction of the stellar spectrum was not possible. In the case of BT Dra, the velocity of the star itself is also close to that of the cloud, so any interstellar absorption would be lost in the strong stellar sodium lines. In addition, the choice of the sodium lines was dictated by instrumental considerations, but if in fact the clouds are as warm as a few thousand degrees, much or all of any sodium could be ionized. Finally, there is a real possibility that the clouds are more distant than the RR Lyrae stars I observed (all of which are less than 2 kpc from the galactic plane). On the other hand, the recent HST work by Savage et al. (1993) has ruled out one possible reason for the lack of detection, that the gas is too metal poor to show any lines.

There is one other sufficient reason for my failure to detect HVC sodium absorption in the RR Lyrae star spectra. Although the target RR Lyrae stars all lie well within the boundaries of the clouds as presented in most of the early surveys for HVC's (see, for example, Hulsbosch 1968), high-angular resolution H I maps by Giovanelli et al. (1973) and Wakker (1991) show that the HVC's are very clumpy, even on very small angular scales. *All* of the known RR Lyrae stars nominally in the directions of the HVC complexes A, C and M actually fall *between* the significant clumps in the Giovanelli et al. maps.

The nature of the HVC's must still be considered to be unknown, but the indications are that they are at least moderately distant and that they are an important component of the galactic halo.

So there is cool gas in the galactic halo, possibly enough in a suitable state to form stars.

3. Star Formation in Regions of Low Gas Density

There is no evidence to suggest that significant star formation is taking place in the galactic halo at the present time. No H II regions are found in the halo, nor are there any molecular clouds known at large distance from the plane. There are at least a few molecular clouds at high galactic latitudes, (such as the Taurus-Auriga clouds mapped by Ungerechts & Thaddeus 1987), but they are all very close to the Sun, well within the galactic disk. To find signs of star formation in low density regions, we must look further afield.

3.1. Open Clusters far from the Galactic Plane

There are certainly some stars in the galactic halo that are definitely moderately young, because we can measure their ages and locations with considerable confidence. These stars are found in a small number of open clusters that lie far from the galactic plane, and their membership in the clusters permits their individual properties to be estimated reliably. Phelps et al. (1994) and Janes & Phelps (1994) have used a combination of published data and new observations to produce a catalog of 72 "old" open clusters. (A comment on the meaning of "old" and "young" in the present context is needed: Since most open clusters survive for only a few hundred million years, any open cluster older than roughly a billion years is usually called old. But on the other hand, all but one or two of the oldest of the open clusters are younger than the globular clusters. So in the present context, even the oldest open clusters are "young" objects.)

Although a few open clusters have been known for many years to lie far from the galactic plane, we have identified several more, and when they are all listed together as in Table 1, we can see that in fact there is a significant population of open clusters, of ages as young as one billion years that are well into the galactic halo. Furthermore, their metallicities are typical of disk stars; the most metal-poor of the clusters in this list are also located at large galactocentric distances, where they would be expected to be somewhat metal-poor.

Of particular interest is the cluster Be 20. Not only is this one the most distant of all from the galactic plane, it is only one-third the age of the globular clusters and is located more than 15 kpc from the galactic center. It almost certainly formed in a region of low-density gas.

3.2. Young Stars between the LMC and the SMC

Maps of the Magellanic Cloud region in the Hydrogen 21-cm line show that the two galaxies are imbedded in a common envelope, with a "bridge" of gas between the two. Several groups have found indications of young stars associated with the bridge (see, e.g., Kunkel 1980). Grondin et al. (1992) have located several distinct stellar associations containing young stars, almost exactly halfway between the two galaxies.

But how did they get there? One possibility is that they were formed recently ($< 10^8$ years ago) out of the tenuous hydrogen located between the two clouds. The other

Cluster	$|z|$ (pc)	Age (Gyr)	[Fe/H]
Be 20	2498	5.7	
NGC 2204	1198	2.7	−0.58
NGC 2243	1129	6.4	−0.56
NGC 1193	846	5.7	−0.50
NGC 6791	798	9.8	+0.19
Be 22	777	5.7	
NGC 2420	767	3.3	−0.42
AM-2	741	8.8	
092-SC18	737	6.4	
To 2	724	3.0	−0.60
Mel 66	710	7.1	−0.51
Be 39	702	7.9	−0.31
NGC 2266	599	1.0	
NGC 188	582	7.9	−0.06
NGC 2506	554	3.0	−0.52
King 2	508	6.4	

TABLE 1. The oldest known open clusters

gravity:	−4.1
Ages (yr):	$\leq 10^9$
Abundance (dex):	−0.5 – 0.0
W dispersion (km s^{-1}):	66
Scale height (pc):	700

TABLE 2. Properties of halo A stars

possibility is that they were formed when the LMC and SMC collided some 10^8 years ago. The actual ages of these stars will determine which of these two hypotheses is the more likely; Grondin et al. claim that the associations they found are very young, so that the second of these options is not possible.

There is at least anecdotal evidence that stars can actually form in low-density regions.

4. Population I B and A Stars in the Halo

Let us return to the halo A and B stars. There have been a number of recent reviews of their properties, and new searches for candidate stars, including those by Lance (1988a) and Stetson (1991). The apparent properties of the halo B stars according to Lance (1988a) are summarized in Table 2. Except for their velocities and locations, they appear to be perfectly normal Population I stars.

Some of the more likely possibilities among the many explanations which have been advanced to explain the phenomenon are the following:

(i) They are FHB stars masquerading as Population I stars.
(ii) They are young disk stars moving at high velocities out of the galactic plane.
(iii) They are young stars, produced in a recent, unusual event.
(iv) They are old disk blue straggler stars.

Distance from the plane:	1120 pc.
Age:	6 Gyr
Fe/H:	-0.56
M.S. turnoff color:	B-V = 0.46
Bluest cluster star:	B-V = 0.04

TABLE 3. Properties of the moderately old open cluster, NGC 2243

(v) They are stars recently formed *in situ* in the halo.

The arguments for and against these various hypotheses have been discussed at some length by Lance (1988a) and by Stetson (1991) and need not be repeated here. On the basis of their discussion, none of the above ideas can be conclusively ruled out, although Lance argues vigorously for the idea that they were formed in a recent collision of our galaxy with some sort of dwarf galaxy. This argument is supported by a recent study by Preston *et al.* (1993).

5. Blue Straggler Stars in the Halo?

One possible explanation for at least many of the blue Population I halo stars seems to have been rather ignored – if there were Population I blue stragglers in the halo, they would look exactly like the stars originally found by Philip & Sanduleak (1968).

First we must define what is meant by a blue straggler.

At the moment, blue stragglers can only be identified unambiguously if they are members of star clusters. They are found in both Population I clusters (open clusters) and Population II clusters (globular clusters). Their sole distinguishing characteristic is that they lie on or near the main sequence, above the cluster m.s. turnoff. They are otherwise indistinguishable spectroscopically and photometrically from other stars of the same spectral type. A good example of the Population I blue straggler phenomenon can be seen in Figure 2, a color-magnitude diagram of the moderately old open cluster NGC 2243. The small group of stars to the blue of the main-sequence turnoff are almost certainly members of the cluster. The properties of NGC 2243 (from Bergbusch *et al.* 1991; Phelps *et al.* 1994) are summarized in Table 3.

Lance (1988a) discussed (and dismissed) blue stragglers as an explanation for the halo A and B stars, but she only considered thick disk or globular cluster blue straggers. As shown in Table 1, there are at least some moderately young open clusters in the halo, many of which also contain blue stragglers as can be seen in Figure 2. Because the blue stragglers lie on the main sequence above the turnoff luminosity of the parent cluster, they appear far too young for the cluster. The bluest stars in NGC 2243 will look exactly like ordinary Population I main sequence stars of spectral type A. If an open cluster blue straggler got into the halo, it would look just like the young, hot halo stars.

6. Summary - Do Stars Really Still Form in the Halo?

Consider the evidence:
- There is gas in the halo, some of it cooling and neutral.
- Stars probably at least occasionally form in low-density environments.

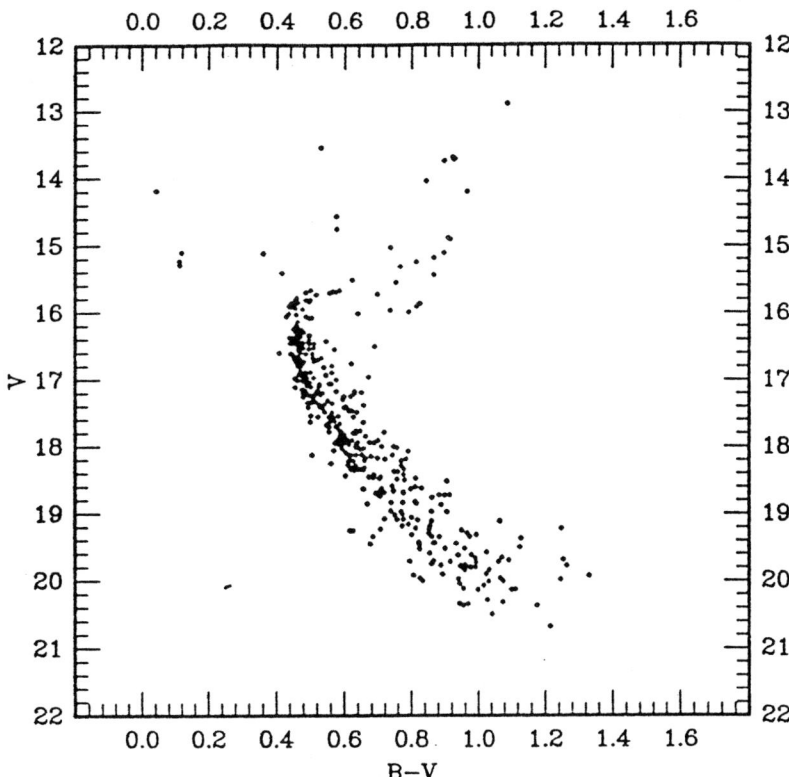

FIGURE 2. A color-magnitude diagram of the moderately old open cluster, NGC 2243 (from Bergbusch et al. 1991). The small group of stars to the blue of the main-sequence turnoff are blue stragglers.

- There is an increasing body of evidence that impacts or other disturbances to the disk have splashed stars and clusters into the halo.

Nevertheless, many, if not most, of the apparent young halo A and B stars can be explained away in various ways. We also do not see any direct signs of stars actually forming in the halo at the present time.

So the conclusion is that we still do not know, but there are strong hints that the halo was a lot more active for a lot longer time than is generally assumed. A key test of the nature of the halo A and B stars that seems to have been overlooked is to search for solar composition stars of spectral types F and G. Unless the initial mass function of star formation in the halo were very different, for every Population A or B star in the halo there should be several F and G stars. The relative numbers should be at least roughly similar to those in the solar neighborhood.

REFERENCES

BERGBUSCH, P. A., VANDENBERG, D. A., INFANTE, L., 1991, AJ, 101, 2102
BLOEMEN, H., 1990, The Interstellar Disk-Halo Connection in Galaxies, IAU Symposium 144, Kluwer, Dordrecht
CONLON, E. S., DUFTON, P. L., KENNAN, P., MCCLAUSLAND, R. J. H., 1991, A&AS, 236,

CONLON, E. S., DUFTON, P. L., KEENAN, P., MCCLAUSLAND, R. J. H., HOLMGREN, D., 1992, ApJ, 400, 273
DANLY, L., ALBERT, C. E., KUNTZ, K. D., 1993, ApJ, 416, L29
FABIAN, A. C., 1987, Cooling Flows in Galaxies, NATO ASI Series C, Vol. 229, Kluwer, Dordrecht
GIOVANELLI, R., VERSCHUUR, G. L., CRAM, T. R., 1973, A&AS, 12, 209
GRONDIN, L., DEMERS, S., KUNKEL, W. E., 1992, AJ, 103, 1234
HAMBLY, N. C., CONLON, E. S., DUFTON, P. L., KEENAN, F. P., LITTLE, J. E., MILLER, L., 1993, ApJ, 417, 706
HULSBOSCH, A. N. M., 1968, Bull. Astron. Inst. Netherlands, 20, 33
JANES, K. A., PHELPS, R. L., 1994, AJ, in press
KEENAN, F. P., 1992, QJRAS, 33, 325
KUNKEL, W. E., 1980, in Hesser J. E., ed, Star Clusters, Reidel, Dordrecht, p. 353
LANCE, C. M., 1988a, ApJS, 68, 463
LANCE, C. M., 1988b, ApJ, 334, 927
LILLIENTHAL, D., MEYERDICKS, H., DE BOER, K. S., 1990, A&A, 246, 487
MATHEWSON, D. S., CLEARY, M. N., MURRAY, J. D., 1974, ApJ, 190, 291
MORRIS, S. L., WEYMANN, R. J., SAVAGE, B. D., GILLILAND, R. L., 1991, ApJ, 377, L21
OORT, J. H., 1966, Bull. Astron. Inst. Netherlands, 18, 421
PHELPS, R. L., JANES, K. A., MONTGOMERY, K. A., 1994, AJ, 107, 1079
PHILIP, A. G. D., ADELMAN, S. J., 1991, PASP, 103, 63
PHILIP, A. G. D., SANDULEAK, N., 1968, Bol. Obs. Tonantzintla y Tacubaya, 4, 253
PRESTON, G. W., BEERS, T. C., SCHECTMAN, S. A., 1993, BAAS, 25, 1415
RODGERS, A. W., 1971, ApJ, 165, 581
SAVAGE, B. D., LU, L., BAHCALL, J. N., et al., 1993, ApJ, 413, 116
SONGAILA, A., COWIE, L. L., WEAVER, H. F., 1988, ApJ, 329, 588
STETSON, P. B., 1991, AJ, 102, 589
UNGERECHTS, H., THADDEUS, P., 1987, ApJS, 63, 645
WAKKER, B. P., 1991, A&AS, 90, 495
YORK, D., 1982, ARAA, 20, 221
ZINN, R. J., 1993, in Smith G. H., Brodie J. P., eds, The Globular Cluster – Galaxy Connection, ASP Conf. Series, 48, p. 38

Discussion

BIDELMAN: How large a value of Z would you expect an old cluster to possibly range in a time of say 5×10^9 years from interactions with molecular clouds? A few hundred parsecs?

JANES: I have done some simple calculations, based on Spitzer's formula for the lifetime of a cluster against disruption. The essence of the argument is that the same encounters with molecular clouds that scatter individual stars away from the disk will disrupt a cluster instead. So, clusters probably cannot be accelerated into the halo.

PHILIP: In regard to your discussion of the Magellanic stream, years ago when I was using the Curtis Schmidt telescope at Cerro Tololo to make spectral surveys I did a spectral survey of the Magellanic stream. I found a fair number of stars with strong spectral blue tails and there was a possibility that early-type stars had been found in the stream. But follow-up Strömgren photometry showed that these stars were F subdwarfs and, thus, close to us.

JANES: I knew that there have been attempts to find stars in the Magellanic Stream but I had forgotten where I heard about it. I believe Kunkel once claimed that he found some, but that was apparently not confirmed.

BROSCH: Lynden-Bell claimed an amazing alignment of the Magellanic Clouds, Magellanic Stream, some distant globular clusters, dwarf galaxies, and high velocity clouds. Later, there was a claim that this is the signature of a polar ring formed around the Galaxy. Given that polar rings are known to form stars, what could you comment on this "polar ring" possibility in the context of star formation in the halo?

JANES: This apparent alignment of all these objects has been reported several times. It could be perhaps something like a polar ring. One problem with this hypothesis is that it contains both gas (the Magellanic Stream) and very old systems (the dwarf ellipticals). It is not obvious that such a structure could survive for such a long period of time.

LITTLE: Observations of high velocity absorption components towards faint blue stars (spectroscopically identical to nearby disc main sequence B-stars) e.g., recently by Danley, Albert & Kuntez (1993, ApJ, 416, L29) towards the cloud MI, provides further evidence that these halo B-stars are indeed luminous and young and not sub-luminous and old.

JANES: I know that Danly has been working very hard to find the distances of the high velocity clouds. It is most interesting to hear that she has found absorption lines from the cloud MI.

ADELMAN: On a different topic, 68 Tau, the blue straggler in the Hyades, is a hot Am star. Is this situation typical of blue stragglers?

JANES: In the older clusters I had in mind, the blue stragglers appear to be generally normal in their spectral appearance, but I think the Am stars are common among blue stragglers in the younger, Hyades-age clusters.

The Ultraviolet Imaging Telescope on the Astro-1 and Astro-2 Missions

By T. P. STECHER

Laboratory for Astronomy and Solar Physics, Code 680, NASA, Goddard Space Flight Center, Greenbelt, MD 20771, USA

The two solar-blind UIT cameras obtained 821 frames with a cumulative exposure time of 115,000 seconds during the December 1990 Astro-1 mission of Space Shuttle Columbia. Filters were used to isolate selected bandpasses in the range 1200-3000 Å, and some observations of the 40 arcmin field of view were made with a grism. Spatial resolution on many images is 2.4 arcsec. Calibrated data, converted to machine-readable form, are under analysis and 24 papers on these images have been accepted by the Astrophysical Journal. A number of investigations are relevant to the subject of this meeting, "Hot Stars in the Halo". The UIT observations of globular clusters gives the opportunity to study the end stages of stellar evolution. Here the stars are all at the same distance so color-magnitude diagrams may be used. The bulges of spiral galaxies and elliptical galaxies also offer the possibility of studying the older population.

The Ultraviolet Imaging Telescope (UIT) on the Astro-1 Spacelab Mission observed five globular clusters in the Galaxy and globular clusters in the LMC and M31. The telescope is equipped with 11 filters and very good long wavelength light rejection. Most images were obtained at 1520 Å, 1620 Å, and 2490 Å. The clusters, ω Centauri, M3, M13, M79, and NGC1851 were completely surveyed for UV bright objects. Two new UV bright stars were found in ω Cen. ω Cen has over 2000 hot stars which are in the end stages of evolution, 14 of which are above the horizontal branch. The hotter horizontal branch stars appear to fall below the Zero Age Main Sequence. The observations will be discussed in terms of the current theories of the end stages of stellar evolution. UIT imagery of the spiral galaxies M31, M32, M81 suppresses the red stellar population as expected and enhances the appearance of tracers of recent star formation. The observations allow the study of the ultraviolet turn up and the color gradient in the old population of these galaxies and NGC1399 which is believed to be due to stars in the final stages of their lives. Several early subdwarfs have also been found.

The payload is to be reflown in a year on a 16 day mission (Astro-2) which will enable at least twice a much data to be obtained. Since the instrumental performance is now well understood, the UIT Team will be able to optimize the observations for the best scientific results. A Guest Investigators Program will use half of the observing time on Astro-2.

Are Analogues of Hot Subdwarf Stars Responsible for the UVX Phenomenon in Galaxy Nuclei?

By B. DORMAN, R. W. O'CONNELL, AND R. T. ROOD

Department of Astronomy, University of Virginia, P.O.Box 3818 University Station, Charlottesville, VA 22903-0818, USA

We present the case that populations of sdB/sdOB/sdO-type stars may be a common constituent of galactic stellar populations, responsible for the UV upturn ('UVX') observed in the spectra of spiral bulges and normal galaxy nuclei. Extreme Horizontal Branch stars with $\log g > 5$ and $\log T_{eff} > 20{,}000$ K have emerged in the last few years as the most likely candidate for the origin of the UVX. The magnitude of this far-UV flux in some systems (e.g., NGC 1399, NGC 4649) indicates that galactic nuclear regions must contain larger numbers of these subdwarfs than does the solar neighbourhood. This paper summarizes the results of a quantitative study of the UV radiation from evolved stellar populations. We have computed a large grid of stellar models in advanced stages of evolution, as well as a set of isochrones for ages 2-20 Gyr, for a wide range in composition. We use these calculations to derive synthetic UV colour indices for stellar populations with hot components.

We compare the results of this study to observations of the galaxies and clusters in the colours $m_\lambda(1500\text{Å}-V) [= 15-V]$ and $15-25$. In the globular clusters, the $(15-V)$ colour is dependent on the size of the population of the blue horizontal branch (HB) and post-HB UV-bright stars. We assume that the magnitude of the UVX in the galaxies [as quantified by the $15-V$ colour] is also dependent on the number of hot subdwarfs present. We thus determine the fraction of very blue stars that must be produced by an old metal-rich population in order to agree with the observations.

---

1. Introduction

The ultraviolet upturn ('UVX') observed in the spectra of elliptical galaxies and spiral bulges has emerged as an important issue in our understanding of stellar populations and the use of elliptical galaxies as tracers of the evolution of the universe. It was first recognized in observations of the M31 bulge from the OAO-2 satellite (Code 1969; see also Code & Welch 1979). Hills (1971) pointed out that the shape of this UV upturn was consistent with a hot thermal spectrum, and he suggested as the probable origin post asymptotic giant branch (P-AGB) stars, during their high luminosity ($\log L/L_\odot \sim 3-4$) transit to become white dwarfs. The UVX has since been found to be present in all elliptical galaxies, and to vary in amplitude by more than an order of magnitude. It has attracted considerable theoretical attention in recent years by researchers (see Greggio & Renzini (1990) [GR] and references therein) seeking to identify the hot component responsible for the UV radiation.

Candidate UV-bright stellar objects for this population include massive stars, very low-mass stars in advanced stages of stellar evolution, and low-mass products of binary interactions (see GR). Evidence against massive stars as the primary cause of the UVX has recently grown in strength (King *et al.* 1992, O'Connell *et al.* 1992). Further, Brocato *et al.* 1990 (see also GR) argued that populations of P-AGB stars are insufficiently long-lived to supply enough ultraviolet radiation to explain the strongest observed UV upturns. Currently, the most likely candidates are the Extreme Horizontal Branch

(EHB) stars, stars that do not reach the thermally-pulsing stage of the AGB, and their post-HB descendants. These are termed Post-Early AGB or AGB-Manqué stars depending on whether they reach the lower AGB at all after core exhaustion. They have very thin hydrogen envelopes, and emit copious UV radiation both during and after the HB stage. In Dorman, Rood & O'Connell (1993), (hereafter DRO93) we presented a large grid of evolutionary tracks of these EHB stars with various metallicities and envelope masses. We use these here, together with isochrone computations, to model the UV colours of stellar populations containing hot components.

The grid of evolutionary tracks is defined by the composition parameters Z and Y, the core mass M_c^0, and by M_{env}^0, the envelope mass of the object on the Zero Age Horizontal Branch (ZAHB). The parameter with the strongest influence on the integrated lifetime UV output of a star is M_{env}^0, whose distribution will be determined by the degree of mass loss on the red giant branch (RGB). The EHB stars are produced only if all but about 0.05 $M_\odot$ is lost at this stage of evolution. The empirically determined dependence of mass loss on both temperature and luminosity implies also that it should be a function of metal abundance. However, there is no suitable theory of the abundance dependence of mass loss (cf, GR). Since the physics here is so uncertain, our approach to the problem of the UVX is to concentrate on answering the following question: what ranges of M_{env}^0 are consistent with the observed UV fluxes of globular clusters and galaxies, given other constraints on their ages and abundances?

Our approach is to calculate directly the expected UV colours from evolved stellar populations by modelling the integrated flux from all phases of evolution. As a proving ground for the method, we consider the UV radiation from the Galactic globular clusters, showing the variation in UV properties that arises from different HB morphologies. We are able to demonstrate good agreement between the observations and synthetic colours in the colours $15 - V$, which measures the relative strength of the UV upturn — and $15 - 25$, which indicates the mean temperature of the UV-radiating stars in a stellar population. Turning to the galaxies, our principal result gives the fraction of the post-RGB stellar population that must radiate significantly in the UV to explain the range of $15 - V$ colours observed in the elliptical galaxies. Our principal conclusions are that (a) EHB stars are capable of producing more than enough far-UV flux to explain the observations, and (b) the resultant mid-UV flux (quantified by $15 - 25$) is also consistent with the data. Quantitatively, we find fractions of up to $\sim$ 20 % of red giants must become EHB stars after the helium core flash in the strongest UVX galaxies. Since the models we derive satisfy the available constraints provided by the data, they make a strong case for the hypothesis that EHB stars, which we identify with hot subdwarfs, are indeed responsible for the UVX.

2. The UVX Phenomenon in Galaxies

In this section, we briefly describe the observational characteristics of the UVX phenomenon. More details can be found in several recent reviews (Dorman, O'Connell & Rood 1994, hereafter DOR94; O'Connell 1993; GR; Burstein et al. 1988, hereafter B3FL). The features of the UVX phenomenon most germane to the question we pose here are:

• *Incidence:* It is found in the nuclear regions of almost all normal (i.e., excluding AGN and starburst) E's, S0's, and spiral bulges observed to date.

• *Strength:* The strength of the UVX, as measured by the colour $15 - V$, varies by about an order of magnitude, or 2.5 mag, (B3FL, and Figure 1). This degree of variation is much greater than that found in other optical and infrared broad-band colours.

• *Abundance Correlation:* As first pointed out by Faber (1983), and confirmed by

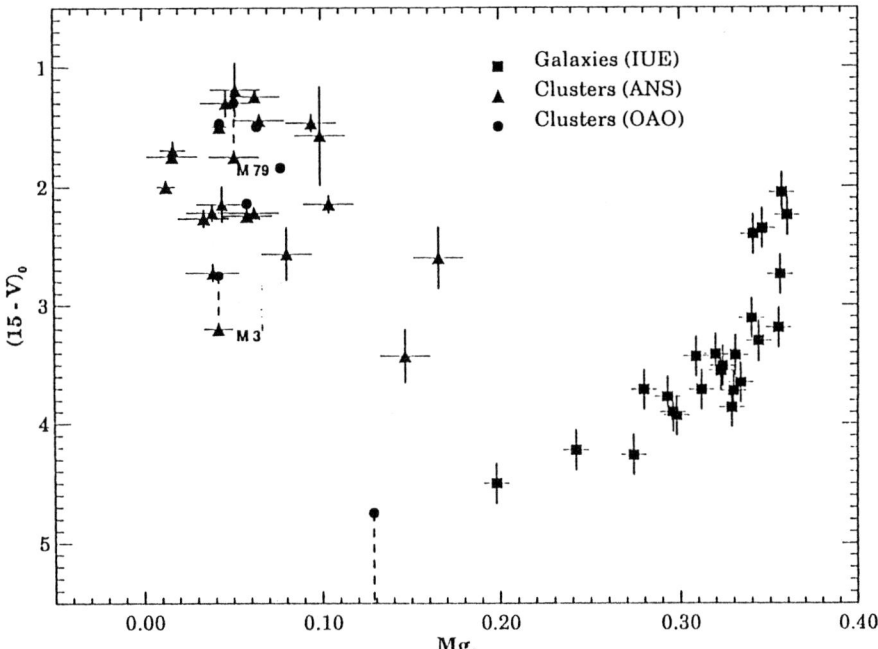

FIGURE 1. Observations of $15 - V$ colours of galaxies plotted against the absorption line index Mg_2. The galaxy data, derived from IUE SWP observations, are from B3FL, with the exception of NGC 1399 which is from Buson, Bertola & Burstein (1993, private communication). The cluster data are from de Boer (1985). Mg_2 indices for the clusters are either observed or derived from the relation for Galactic globular clusters given by Brodie & Huchra (1990).

B3FL, the magnitude of the UV upturn is positively correlated with the absorption line index Mg_2 (see Figure 1) in the *nuclear* regions of E/S0 galaxies and spiral galaxy bulges.

- *Radial Gradients:* Ultraviolet Imaging Telescope (UIT) data (O'Connell et al. 1992) show that M 31, M 81, and NGC 1399 have strong UV colour gradients with amplitudes up to ~ 1 mag in $15 - 25$, with the UV colours becoming strongly redder outward. In contrast, the dwarf elliptical M 32 has the reddest $15 - V$ in B3FL and a small UV colour gradient.
- *Resolution:* The more luminous UVX sources have apparently been resolved on recent HST Faint Object Camera exposures obtained of the M31 bulge. In a preliminary analysis of these data, King et al. (1992) detect ~ 150 resolved objects, likely to be P-AGB stars and central stars of planetary nebulæ. However, these resolved sources account for about 17 % of the total flux. Also, no massive OB stars appear to be in the field.
- *Spectral Shape and Features:* The IUE spectroscopy (B3FL) indicates that the far-UV spectral slope of the UVX is roughly constant from object to object and corresponds to $T_{eff} \gtrsim 20,000\,K$.
- *Temperature:* Far-UV (900–1800 Å) spectra of M31 and NGC 1399 were obtained by the Hopkins Ultraviolet Telescope (HUT) (Ferguson et al. 1991). The turnover of the spectrum at $\lambda < 1200$ Å indicates that $T_{eff} \lesssim 25,000\,K$.
- *Composite Nature:* The HUT spectrum of M 31 (Ferguson & Davidsen 1993) differs significantly from that of NGC 1399. For the adopted extinction, M 31 is hotter for $\lambda < 1200$ Å but cooler for longer wavelengths. Hence at least two distinct types of hot low mass stars produce the UV spectral energy distribution (SED) of M 31.

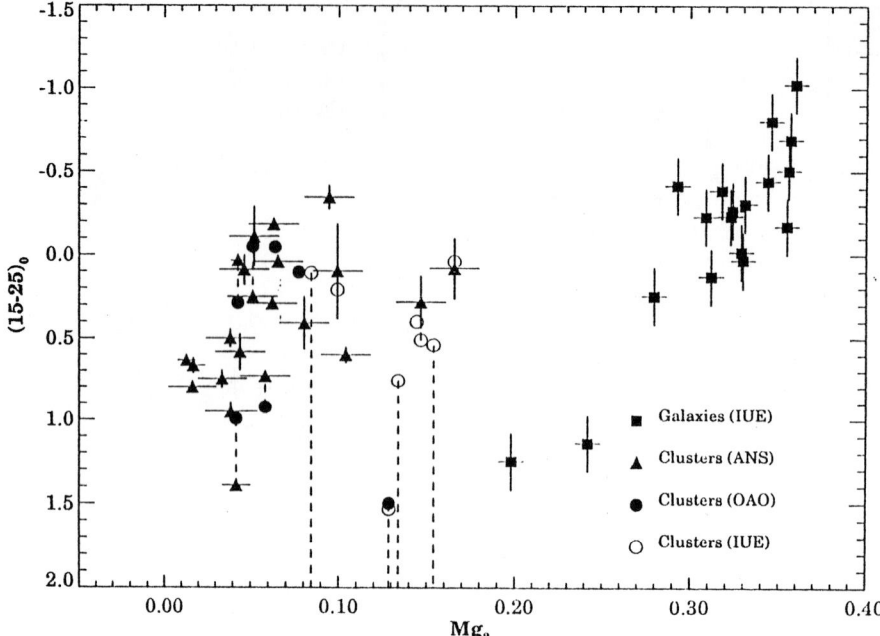

FIGURE 2. Observations of 15 − 25 colours of galaxies plotted against the absorption line index Mg_2. The galaxy data are from B3FL, and derived from IUE SWP & LWP observations. The cluster data are from de Boer (1985), and Rich, Minniti & Liebert (1993).

3. Ultraviolet-Bright Galactic Stellar Populations

We ask now, what are possible Galactic analogues of the UV-bright stellar population in the galaxies that could be responsible for the UVX, and which has characteristics that most closely resemble the galaxy spectra? We consider the UV flux from globular clusters and from the hot subdwarfs.

3.1. *Globular Clusters*

The globular clusters have long served as templates for evolved stellar populations, although their metallicity range does not strongly overlap that of the brighter galaxies (see Figures 1 & 2). They do, however, they provide samples of homogeneous, coeval, old stellar populations which can be used to test synthetic UV colours. The clusters have been studied in the UV, from, amongst others, the ANS satellite (van Albada, de Boer & Dickens 1981, hereafter ABD), OAO-2 (Welch & Code 1980, hereafter WC80), and Astro-1 (Hill et al. 1992; Whitney et al. 1994; Dixon et al. 1994). The far-UV flux from globulars is dominated by the hot HB stars, and is correlated with the HB morphology (ABD; WC80).

Even in the clusters with the strongest far-UV radiation, the SED is flatter than in the galaxies. The hardest far-UV spectral energy distributions among the clusters are observed in those with extended blue HB tails: even these, however, contain only a small population of true EHB stars, being largely populated instead by blue HB stars ($B - V \approx 0$).

The most metal-rich clusters, are, however, faint in the UV because their HB morphology is dominated by or exclusively composed of cool stars close to the RGB. One of the principal difficulties with the hypothesis that EHB stars are the origin of the UVX has, in fact, been the HB morphology in metal-rich globular clusters. If the UVX phenomenon is due to EHB stars, why should it be that they are produced in small numbers in the metal-poor globular clusters, not at all in the metal-rich clusters, and then reappear in the galaxies as the source of the UVX?

3.2. Hot Subdwarfs

There is evidence from the Galactic field that systems more metal-rich than the globular clusters produce stars which undergo sufficient mass loss to make EHB stars. This is the observed population of hot subluminous stars (Heber 1992; Saffer 1991). More than 1200 of these stars have so far been discovered, of which the majority were found by the Palomar-Green (PG) survey (Green, Schmidt & Liebert 1986). The sdB stars, together with somewhat hotter objects termed either sdOB or sdB-O, constitute 40 % of the catalogued stars. These occupy a narrow range of surface gravities with $\log g \sim 5 - 5.5$ and $20,000 < T_{eff} < 40,000$ K. The stars denoted as sdO, 13 % of the sample, have surface temperatures extending to about 65,000 K. Their surface gravities indicate a wide range in luminosity. About 34 % of the remainder of the catalogued sources consists of hot white dwarfs and extragalactic objects, leaving very few that are directly attributable to products of binary evolution.

The derived temperatures and gravities for the sdB/sdOB stars match the predicted parameters of EHB stars. Further, the narrowness of the range in surface gravities is consistent with the notion that they are drawn from a single mass stellar population of about 0.5 $M_\odot$. For the sdO stars, there are several subclasses that may emerge from different progenitors, including the products of binary mergers (Iben 1990). There are reasons to believe that this is not the origin in the majority of cases (Heber 1992). The temperature and gravity range of the sdO stars suggests that at least some of them are the AGB-Manqué progeny of EHB stars. A crude comparison of the relative lifetimes of EHB stars and their AGB-Manqué progeny predicts a ratio of about 1:5. In addition, in the theoretical tracks, the greater part of the AGB-manqué lifetime is also spent at temperatures lower than $\sim$ 60-65,000 K. The observed ratio sdO:sdB thus appears to be about 1:3, with a deduced temperature range consistent with the theoretical sequences. We thus tentatively conclude, following a similar remark by Heber (1992), that *the bulk of sdO stars are evolved sdB and sdOB stars*. Finally, the disk membership of sdB/sdOB and sdO stars makes it likely that they originate from a metal rich population.

If the hot subdwarfs are produced in numbers in the Galactic disk, they are likely to be a part of the normal evolved stellar population of any galaxy. We may also expect that bulge of our Galaxy has a UV upturn analogous to that seen in the bulges of M 31 and M 81. Estimates of the numbers of hot subdwarfs in the solar neighbourhood (Heber 1992; Saffer 1991) indicate that their integrated UV flux is expected to be much smaller than the observed nuclear fluxes from galactic nuclei. The production of much larger numbers of subdwarfs thus seems to be a bulge rather than disk phenomenon. Finally, given that the UV upturn increases in strength toward the center of three of the systems observed by UIT including two spiral bulges, then if stars similar to subdwarfs are indeed responsible for a putative UV upturn in our Galaxy, their space density should also show a gradient. This may be observable in data samples collected from future field surveys in the bulge 'windows'.

4. Ultraviolet Observations

We present the data for the clusters in the same colour metallicity planes as the galaxies, so that we may easily compare and contrast the different systems. Full descriptions and tabulations of the data appear in DOR94. We construct the colours $15 - V$, $25 - V$, and

15 − 25 from the data collected in the ultraviolet and in the V filter, given by de Boer (1985) and B3FL. The recent IUE observations of Rich, Minniti & Liebert (1993) have also been included in the 15 − 25 plane, derived from their Table 2.

Figures 1 and 2 exhibit extinction-corrected $15-V$, $15-25$ colours respectively against the Mg_2 index. We interpret the $(15 - V)$ colour index as an indicator of the 'specific UV luminosity,' measuring directly the number fraction of hot stars in the emitting population. The 15 − 25 colour crudely represents the mean temperature calculated from all stars radiating significantly at wavelengths shorter than 2800 Å. However, since the mid-UV flux from the main sequence near the turnoff and the subgiant branch (SGB) can be significant, the 15 − 25 colour can only be understood with composite models that include both the earlier phases of evolution as well as the hot star contribution.

Among the galaxies, the specific UV flux is largest in the most metal-rich objects, while the most metal-rich clusters (with red HB morphology) are fainter in the UV than those with lower metallicity. The bluest 15 − V colours among the clusters, however, are bluer than the galaxies with the strongest UV upturns. These clusters all have "extended blue HB tails", with some fraction of the stars hotter than 12,000 K (e.g., M 13, M 79, NGC 6752). The HB of M 3 extends almost to the red giant branch, and even its hottest HB stars are much cooler than most of the HB stars of M 13, despite their near-identical metallicity (Kraft et al. 1993). Thus in M 13 and similar objects, the entire HB phase contributes to the UV flux. In contrast, M 3, in which only a fraction of the HB population contributes to the far-UV radiation, is 2 mag fainter in 15 − V. The location of the clusters in Figure 1 can thus be seen as a sequence in which the UV flux decreases with increasing metallicity (the 'first parameter' effect), upon which is superposed a large scatter due to variations in HB morphology at intermediate ($[Fe/H] \sim -1.5$) metallicities.

In the 15 − 25 colour index, the galaxies bluest in 15 − V are much bluer than the clusters. The reason for this is twofold: first, the clusters are more metal-poor, so that the light from the turnoff and SGB strongly influences the spectral shape in the mid-UV by reducing the slope longward of 2000 Å. Second, at shorter wavelengths, at which the earlier phases are invisible, the cluster fluxes do not rise with decreasing wavelength. Thus the cluster SEDs appear to be dominated by cooler stars than in the galaxies.

The contrast between the galaxies' and the clusters' UV colours is brought into sharp focus by the two colour diagram of 15 − V vs. 15 − 25 shown in Figure 3. The galaxies and the clusters form two almost parallel sequences of points. The metallicity index in the galaxy sequence increases from left to right (the UVX-Mg_2 correlation), while in the clusters it increases from right to left. An increase (decrease) in the fraction of hot stars present mainly produces a right (left) shift in the location of an object in this diagram. Any explanation for the UVX must be able to reproduce this behavior.

5. Results

To calculate synthetic colours for composite stellar populations in the passbands of interest, we need (i) theoretical isochrones and luminosity functions for the phases up to the red giant branch tip (ii) the integrated energies emitted during the late evolutionary stages, and (iii) the 'specific evolutionary flux' of stars to the zero-age horizontal branch (ZAHB), i.e. the number of stars passing through the helium flash per unit V-luminosity of the population per unit time. A presentation of the necessary formulæ and a description of the calculations is given in DOR94. Briefly, the isochrones, combined with an assumption for the mass function, give the integrated colour from the earlier stages of evolution. It can be shown that the integrated flux along an HB/post-HB evolutionary

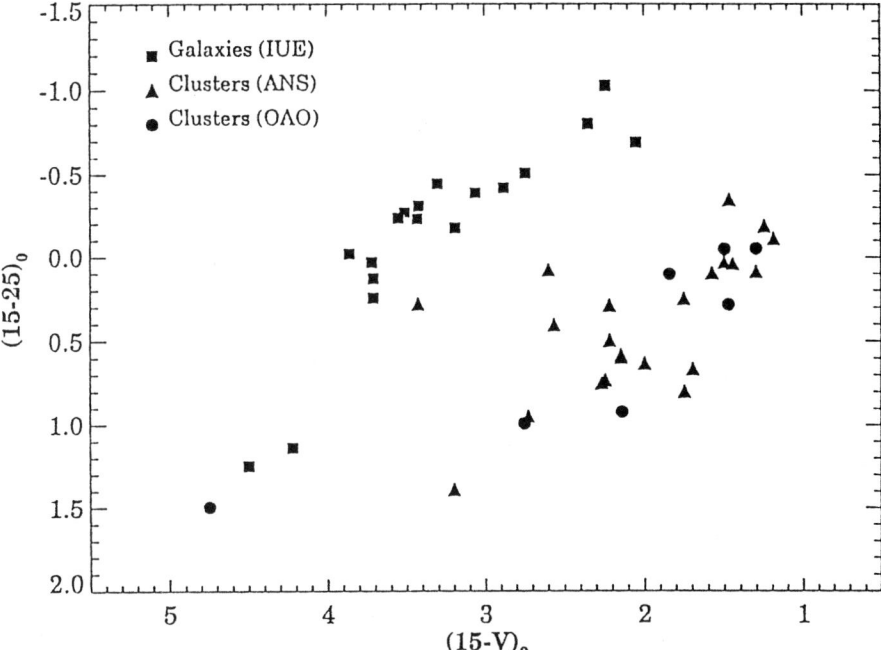

FIGURE 3. Observations of $15-25$ colours of galaxies plotted against the $15-V$ colour. The data are the same as shown in Figures 1 & 2. The galaxies and clusters separate into two almost parallel sequences, with the abundance parameter increasing left to right in the upper sequence, and right to left in the lower.

track is equivalent, for a population of sufficient size, to the flux radiated by stars with the corresponding M_{env}^0. The evolutionary flux is approximately equal to the rate of stars leaving the main sequence, which can be derived from a set of isochrones and the mass function at the turnoff.

Figure 4 shows the behaviour of the integrated flux with respect to M_{env}^0 for two different compositions. It should be noted, first, that the maximum possible far-UV flux varies little with metallicity: thus any composition is capable of producing the observed UV colours. Second, for metal-poor compositions there is a range of mass that produces practically all of its UV flux during, rather than after, the HB phase. These stars evolve later to the AGB: in contrast, the 'true' EHB stars may produce as much UV radiation after core helium exhaustion as before. However, for metal-rich compositions (and this is true for all $[Fe/H] \gtrsim -0.5$), the only 'blue' HB stars are 'true' EHB stars.

Figure 5 shows the behaviour of individual components of an evolved population plotted in the $15-V$ vs. $25-V$ plane. The shaded areas mark the data plotted in Figures 1-3. The dotted lines joined by symbols show the colours produced by the earlier phases alone, for ages 4-16 Gyr progressing from left to right. In the upper left of the diagram are curves representing the colours produced by 'pure' hot components, as a function of M_{env}^0, for different metallicities. The solid line connecting filled symbols represents the range of flux produced by the Schönberner (1979, 1983) P-AGB model sequences.

The EHB stars appear uppermost and furthest to the right; as M_{env}^0 increases, both colour indices move to the red. Once again, note that the colours of EHB stars of all compositions coincide, i.e., their UV flux is almost independent of metallicity. The opposite end of these curves represent the red HB stars whose far-UV flux is produced exclusively in the P-AGB phase. The $15-V$ colours have been derived by adding the

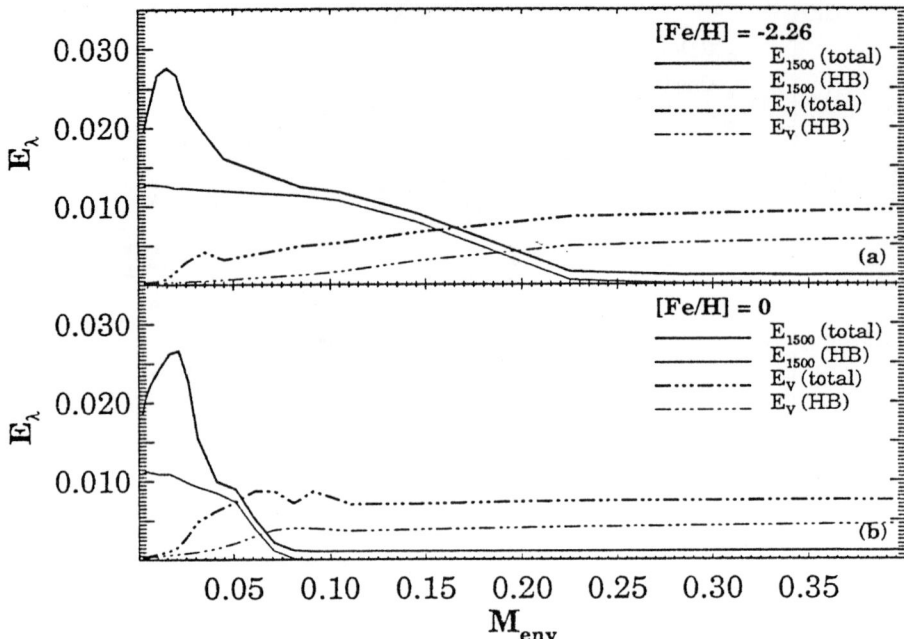

FIGURE 4. Integrated energy per unit wavelength radiated during HB & post-HB stages of evolution as a function of M^0_{env}, for the passband 1500 ± 300 Å and for the V-filter. The units of energy are $L_V^\odot$ Gyr Å^{-1}, and M^0_{env} is in solar units. Panel (a) shows the behaviour for metal-poor HBs: note the range of masses with significant HB UV emission but little post-HB flux. Panel (b) shows the same for solar metallicity: here all models with large UV radiation are EHB stars.

flux from the Schönberner 1983 lowest mass P-AGB model to our HB/AGB tracks. The $25 - V$ colour is, however, dominated by the stars in the HB phase and is unaffected by their final mass at the AGB tip. The red extremes plotted here represent the bluest possible colour that can be produced by P-AGB stars: if those present are more massive, the relatively cool, vertical end of the curves will move leftward.

For the comparison with the observations, note that:

(a) The colours of 'pure' hot components are much (> 5 mag) bluer than the observations. After adding the V-band contribution from the earlier stages (~ 20 % of the total V flux), we can infer that the colours of the galaxies with the strongest upturns can be reproduced if a fraction ≪ 1 of the red giants become EHB stars.

(b) The bluer $15-V$ flux of the metal-poor clusters with blue HB tails implies a larger fraction of UV-emitting stars are produced in these systems.

(c) The $25-V$ colours of the clusters are less than a magnitude bluer than the colours of the older metal-poor pre-HB component ($25-V \sim 2$), implying that the earlier phases contribute significantly to this part of the spectrum.

(d) The galaxy colours are produced by the addition of flux from the 'isochrone' component, which for solar metallicity and above and at ages $\geq$ 10 Gyr give $25 - V > 4$. The contribution at 2500 Å from the earlier phases is thus much smaller than in the clusters owing to the metallicity difference: the $25 - V$ colours of the *bluest* galaxies are dominated by the hot component (see B3FL).

(e) The separation between the galaxies and the clusters on the two-colour diagram (Figure 3) is induced by the effect of metallicity on the pre-HB component.

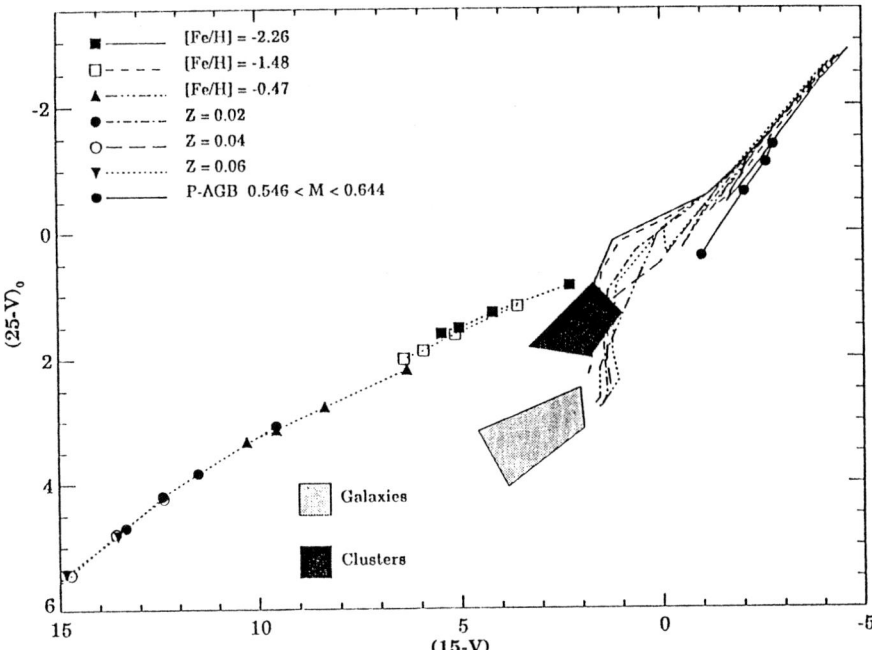

FIGURE 5. Synthetic colors from the various components of old stellar populations, plotted in mid-UV $(25 - V)$ vs far-UV $(15 - V)$ colours. The dotted lines with symbols show the colours of populations of ages 4-16 Gyr computed from isochrones and from the Salpeter mass function $(\Psi(m) = \mathcal{A}m^{-2.35})$. The upper right corner shows the location of 'pure' hot stellar components. A solid line connecting large filled circles shows the colours of P-AGB stars. The locations of the data of Figures 1-3 are shown in the shaded areas.

In general, both the number fraction of UV-bright stars and their envelope mass distribution will vary among systems. At high metallicity, the models contain only a small mass range at temperatures intermediate between the EHB and the red HB stars. To produce crude models of the populations containing a hot component, we introduce a simple parametrization of the HB mass distribution. The details of the models are not as important as the demonstration that they can reproduce the data using very simple, plausible assumptions. We divide the HB stars into two classes. The first, with fraction f_{UV}, varies in location on the HB between the EHB and the cool end. The other class consists of purely red HB stars, which, in the metal-rich case, all have similar output at 1500Å and V to the reddest HB track (see Figure 4b) and thus can be represented by a single sequence.

If we choose a value for f_{UV} that reproduces the bluest galaxies, then add the flux from each HB track in turn to the radiation from the earlier phases of evolution, we can plot the result on the two-color diagram (Figure 3). For $[Fe/H] \geqslant 0$ the result is the upper box in Figure 6. The shaded region in the box corresponds to those models in which the stars represented by f_{UV} are EHB stars and their progeny. The unshaded region consists of models with HB and P-AGB stars only. In line with the expectation of (a) above and that in metal-rich populations the HB stars will be predominantly red, the plotted box has $f_{UV} = 0.2$ at 10 Gyr. Further, at the blue end the vertical spread is caused by metallicity, with the upper part of the box representing models with $[Fe/H] \sim 0.5$. However, the observed points lie in the lower half of the box, hinting that the true $[Fe/H]$ of the galaxies may be close to solar (see Worthey, Faber & Gonzalez 1992). Note that

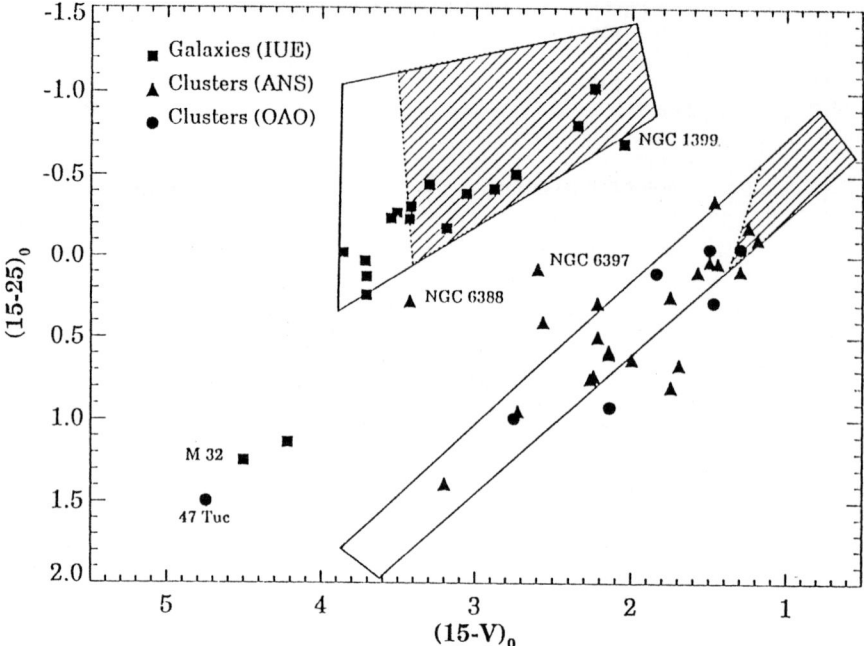

FIGURE 6. Results from simple models of the hot stellar component compared with the data of Figure 3. The upper box represents the location of models with $0 < [Fe/H] < 0.5$, $f_{UV} = 0.2$ (see text), at 10 Gyr. The shaded region indicates models with EHB stars and their post-HB progeny: the unshaded region consists of models with HB and P-AGB stars only. The lower box shows the location of models with $[Fe/H] < -1.5$, $f_{UV} = 0.75$, at 14 Gyr.

the behaviour of the observations may also be reproduced by fixing the track representing the hot stars, and varying f_{UV}.

The lower box for the clusters has been generated similarly. In this case $f_{UV} = 0.75$, with the advanced phases added to a 14 Gyr population. We have thus fixed 25 % of the stars to have similar flux to the red HB, and the other 75 % are allowed to vary in mean location from blue to red. The models corresponding to the upper right of the box have very blue HB morphologies, while those at the lower left have purely red HBs. With this choice of f_{UV} the colours of the clusters are fairly well reproduced, with only the bluest clusters falling into the edge of the shaded EHB area. This is consistent with (spectro-)photometric evidence that the clusters' UV flux is not dominated by EHB stars.

Finally, note that several of the outlying points are not inconsistent with the models. M 32 can be fit with models of age 4-6 Gyr, if P-AGB stars of slightly higher mass are used to provide the lower bound for the far-UV flux: specifically, if we use the Schönberner 0.565 $M_\odot$ model to approximate the far-UV flux of the red HB stars in the P-AGB phase. This estimate for the age of M 32, and the notion that its P-AGB stars are somewhat more luminous and massive, are both consistent with a growing body of similar findings (cf. O'Connell 1980; Freedman 1993). Apart from 47 Tuc whose far-UV flux is very small, the two most prominent clusters that lie away from the models are NGC 6388, which is fit by a model with $[Fe/H] \sim -0.5$, and NGC 6397, which has metallicity $[Fe/H] \sim -1$ and a bluer HB morphology. While we do not have models at this metallicity its location appears to be consistent with theoretical expectations.

It is a pleasure to thank Dave Burstein for providing data in advance of publication.

We would also like to acknowledge useful conversations with Rex Saffer, Jim Liebert, and Pierre Bergeron. This research was supported by NASA Long Term Space Astrophysics Program grant NAGW-2596.

REFERENCES

BROCATO, E., MATTEUCCI, F. I., MAZZITELLI, I., TORNAMBÈ, A., 1990, Synthetic colors and the chemical evolution of elliptical galaxies. ApJ, 349, 458-470

BRODIE, J. P., HUCHRA, J. P., 1990, Extragalactic globular clusters. I. The metallicity calibration. ApJ, 362, 503-521

BURSTEIN, D., BERTOLA, F., BUSON, L., FABER, S. M., LAUER, T. R., 1988, The far-ultraviolet spectra of early-type galaxies. ApJ, 328, 440-462

CODE, A. D., 1969, Photoelectric photometry from a space vehicle. PASP, 81, 475

CODE, A. D., WELCH, G. A., 1979, Ultraviolet Photometry from the Orbiting Astronomical Observatory: XXVI. Energy distributions of seven early-type galaxies and the bulge of M 31. ApJ, 228, 95-104

DE BOER, K. S., 1985, UV-bright stars in galactic globular clusters, their far-UV spectra and their contribution to the globular cluster luminosity. A&A, 142, 321-332

DIXON, W. V., DAVIDSEN, A. F., DORMAN, B., FERGUSON, H. C., 1994, AJ, submitted

DORMAN, B., ROOD, R. T., O'CONNELL, R. W., 1993, Ultraviolet radiation from evolved stellar populations. I. Models. ApJ, 419, 596-614 (DRO93)

DORMAN, B., O'CONNELL, R. W., ROOD, R. T., 1994, Ultraviolet radiation from evolved stellar populations. II. Theory & Observation ApJ, submitted (DOR94)

FABER, S. M., 1983, The stellar content of elliptical nuclei. Highlights Astron., 6, 165-171

FERGUSON, H. C., DAVIDSEN, A. F., KRISS, G. A., et al., 1991, Constraints on the origin of the ultraviolet upturn in elliptical galaxies from Hopkins Ultraviolet Telescope observations of NGC 1399. ApJ, 381, L69-73

FERGUSON, H. C., DAVIDSEN, A. F., 1993, The hot stellar component in elliptical galaxies and spiral bulges. I. Far-ultraviolet spectrum of the bulge of M 31. ApJ, 408, 92

FREEDMAN, W. L., 1993, Infrared-luminous giants in M32 - An intermediate-age population? AJ, 104, 1349-1359

GREEN, R. F., SCHMIDT, M., LIEBERT, J. W., 1986, The Palomar-Green catalog of ultraviolet-excess stellar objects. ApJS, 61, 305

GREGGIO, L., RENZINI, A., 1990, Clues on the hot star content and the ultraviolet output of elliptical galaxies. ApJ, 364, 35-64

HEBER, U., 1992, Hot subluminous stars. in Heber U., Jeffery C. S., eds, The Atmospheres of Early-Type Stars. Lecture Notes in Physics, 401, Springer, Berlin, pp. 233-246

HILL, R. S., HILL, J. K., LANDSMAN, W. B., et al., 1992, A far-ultraviolet color-magnitude diagram of NGC 1904 (M 79). ApJ, 395, L17-20

HILLS, J. G., 1971, The nature of the far-UV excess in M 31. A&A, 12, 1-4

IBEN, I., JR., 1990, On the consequences of low-mass white dwarf mergers. ApJ, 353, 215-235

KING, I. R., DEHARVENG, J. M., ALBRECHT, R., et al., 1992, Preliminary analysis of an ultraviolet Hubble Space Telescope faint object camera image of the bulge of M 31. ApJ, 397, L35

O'CONNELL, R. W., 1980, Galaxy spectral synthesis. II - M32 and the ages of galaxies. ApJ, 236, 430-440

O'CONNELL, R. W., THUAN, T. X., PUSCHELL, J. J., 1986, The strong UV source in the active E galaxy NGC 4552. ApJ, 303, L37-40

O'CONNELL, R. W., BOHLIN, R. C., COLLINS, N. R., et al., 1992, Ultraviolet imaging of old populations in nearby galaxies. ApJ, 395, L45-48

O'CONNELL, R. W., 1993, Ultraviolet-bright populations in galaxies and globular clusters. in Smith G. H., Brodie J. P., eds, The Globular Cluster-Galaxy Connection, ASP Conf. Ser.,

48, pp. 530-543

Rich, R. M., Minniti, D., Liebert, J. W., 1993, Far-ultraviolet radiation from disk globular clusters. ApJ, 406, 489-500

Saffer, R. A., 1991, The origins of hot subdwarf stars. Ph.D. dissertation, University of Arizona

Schönberner, D., 1979, Asymptotic giant branch evolution with steady mass loss, A&A, 79, 108-114

Schönberner, D., 1983, Late stages of stellar evolution. II - Mass loss and the transition of asymptotic giant branch stars into hot remnants. ApJ, 272, 708-714

van Albada, T. S., de Boer, K. S., Dickens, R. J., 1981, Far ultraviolet photometry of globular clusters with ANS: II. Energy distributions of 27 clusters. MNRAS, 195, 591-606 (ABD)

Welch, G. A., Code, A. D., 1980, Ultraviolet photometry from the Orbiting Astronomical Observatory— XXXVII. The energy distributions of 23 galactic globular clusters. ApJ, 236, 798-807 (WC80)

Whitney, J. et al., 1994, A far-ultraviolet color-magnitude diagram of ω Centauri, 1994, AJ, submitted

Worthey, G., Faber, S. M., Gonzalez, J. J., 1992, Mg and Fe absorption features in elliptical galaxies. ApJ, 398, 69-73

A Survey for Field BHB Stars Outside the Solar Circle

By T. D. KINMAN[1], N. B. SUNTZEFF,[2] AND R. P. KRAFT[3]

[1]Kitt Peak National Observatory, National Optical Astronomy Observatories†, P.O. Box 26732, Tucson, AZ 85726, USA

[2]Cerro Tololo Inter-American Observatory, National Optical Astronomy Observatories, Casilla 603, La Serena, CHILE

[3]University of California Observatories/Lick Observatory, Board of Studies in Astronomy and Astrophysics, University of California, Santa Cruz, CA 95064, USA

A pilot photometric and spectrophotometric survey for BHB stars in the polar field SA 57 and the Lick Anticenter field RR 7 is described. The blue horizontal branch (BHB) stars were distinguished from other non-variable stars in the color range of $B-V = 0.00$ to 0.20 by photometric and spectrophotometric criteria. The metallicities of the BHB stars that were found in this way have a similar distribution to that of the RR Lyrae variables that were previously found in these fields. The space densities of our BHB stars have been compared with those found from other surveys and a two-component model for this part of the halo is proposed.

1. Introduction

Although the halo globular clusters contain the oldest stars and effectively define Population II, the *field stars* of the halo are much more numerous and so give the potential for studies of the halo structure that are not possible with the clusters alone. This is particularly true in the regions of the galaxy outside the solar circle where the number of clusters is quite limited.

The RR Lyrae stars are well known tracers of this field-halo: they are intrinsically quite bright and can be unambiguously distinguished by their characteristic light variations. A problem arises, however, because the distribution of stars along the horizontal branch (its morphology) is not the same in every cluster, but depends both on the cluster metallicity and also on at least one other quantity (the so-called "second parameter" — for which age seems to be the most likely candidate (cf, Lee 1992)). Consequently, the ratio of the number of RR Lyrae stars to the total number of Population II stars depends upon the HB morphology of the sample. Furthermore, the amplitude distribution of the RR Lyrae stars is also somewhat dependent on the HB morphology. Thus, if a survey is restricted to the larger-amplitude RR Lyrae stars of type ab, as is the case with the surveys with the Lick Astrograph (Kinman et al. (1966) and Kinman et al. (1982)), these RR Lyrae stars may not accurately represent the whole population of older stars.

A much more reliable tracer of Population II is the *sum* of both the RR Lyrae stars (of all amplitudes) and the blue horizontal branch (BHB) stars. Together, these two groups not only give a better representation of the whole Population II than either group alone, but the *ratio* of the numbers found in each group determines the HB-morphology of the sample. From this HB-morphology, if the metallicity distribution is known, conclusions can be entertained about the relative age of the sample.

† The National Optical Astronomy Observatories are operated by the Association of Universities for Research in Astronomy, Inc., under cooperative agreement with the National Science Foundation

It has recently been found that the stars of spectral type A and F (AF stars) that have been discovered in the Case Low-Dispersion Northern Sky Survey (Pesch & Sanduleak, 1983) include both the BHB stars and the RR Lyrae stars of the galactic halo. This Case survey is made with the Burrell Schmidt telescope at Kitt Peak using the 1.8°UV transmitting prism ($\sim$1000 Å mm^{-1} at H and K) and has a limiting magnitude of about B = 17. Various kinds of early-type stars are identified in the survey; the AF stars are those that are given either Case Category IV or V (Sanduleak 1988). These are stars whose objective-prism spectra have (a) a strong Balmer discontinuity and prominent Balmer lines (Category IV) or (b) a flat and apparently lineless continuum that has a sharp Balmer discontinuity (Category V); these spectra are illustrated in Pesch (1991).

When it became apparent that *both* the BHB stars *and* the RR Lyrae stars could be found among the AF stars, Sanduleak made a special survey in two fields that had been previously searched for RR Lyrae stars with the Lick Astrograph. These two fields are SA 57 (Lick Astrograph Field RR 4) at the North Galactic Pole (l = 66°, b = +86°) and Lick Astrograph Field RR 7 towards the Galactic Anticenter (l = 183°, B = +37°). The two fields cover some 68 square degrees and contain over 200 AF stars.

An ongoing program of photoelectric photometry is being used to detect the RR Lyrae variables among these AF stars. At present we can only say that, using this new technique, we are finding about twice the number of RR Lyrae variables that were found by the traditional blinking technique used in the survey with the Lick Astrograph. Among the non-variable AF stars, the BHB stars must be distinguished from a variety of other early-type stars that occur at all magnitudes covered by the survey. This paper describes the techniques that were used to achieve this separation. Finally, we briefly describe how space densities are then computed for these BHB stars and what constraints these space densities put on the structure of the galactic halo outside the solar circle.

2. Techniques for distinguishing BHB stars

2.1. The $(u-B)_K$ photometric index

The discovery of field horizontal branch stars by Strömgren photometry was pioneered by A. G. Davis Philip in the late 1960s. References to this and later work using this technique are to be found in the review article on the A-type horizontal branch stars by Philip in this volume. In essence, this technique uses the Strömgren c_1 index (which measures the Balmer jump) to detect the larger Balmer jump which distinguishes the BHB stars from higher surface-gravity main-sequence stars. Ideally one would use this technique. The integration times, however, that are required when using the intermediate bandwidth Strömgren filters are prohibitively long if one is making a survey of some 200 stars (many of 15th and 16th magnitude) with a 1.3-m telescope. On the other hand, we felt that the separation that could be achieved using the broad bandwidth Johnson system (Preston *et al.* 1991) was not adequate for our purposes. A new technique was therefore devised which uses the Strömgren u filter in conjunction with the Johnson B and V filters. The integration time required for the Strömgren u filter is about four times longer than that needed for the Johnson U filter, but the separation achieved is clearly superior to that obtainable with only broadband colors (cf, Kinman 1992a). This $(uBV)_K$ system is an instrumental system using an RCA Ga-As photomultiplier. Local standards were set up in each of our two survey fields and repeated observations allowed the observations of the two fields to be put on the same photometric system. This was sufficient for the purposes of our survey. If (as seems possible) this $(u-B)_K$ system has a more general use, then a more widespread set of standards will have to be set up.

A number of stars that had been observed on the Strömgren system were also observed on the $(uBV)_K$ system. We could then define a color $(u-B)_k$ which is given by the following expression:

$$0.949(c_1+m_1) + 1.523(b-y) + 0.667$$

in terms of Strömgren indices. One can therefore compute $(u-B)_k$ for a star whose Strömgren colors are known and so calibrate the $(u-B)_K$ vs. $(B-V)$ diagram with the standard relations for stars of different luminosity classes (of solar metallicity). It was not possible to make $(uBV)_K$ measures of BHB stars in globular clusters with the equipment available. Consequently, nearby low-rotation BHB stars taken from the list of Green & Morrison (1993) were used for the determination of the place in the $(u-B)_K$ vs. $(B-V)$ diagram where BHB stars will be found.

It should be noted that while the Strömgren c_1 and $b-y$ colors are relatively insensitive to blanketing, the $(u-B)_k$ index (as defined above) does contain a significant term in m_1 (the metallicity index) which would indicate that a spread of, say, 0.7 in [Fe/H] would lead to a spread of 0.10 magnitudes in $(u-B)_k$. The expression, however, was calibrated against (a) main-sequence, high-gravity stars of roughly solar abundance, and (b) low-gravity, metal-weak horizontal branch stars. It presumably is correct for these two classes of stars. As there is, however, a correlation amongst the calibrating stars between metal abundance and surface gravity, and therefore between c_1 and m_1, it is not clear that $(u-B)_k$ correctly expresses the effects of these two parameters for all stars.

A satisfactory separation of BHB stars from those of other types by this technique requires that the differential reddening across the survey field is small. In the two fields considered here, the differential reddening does probably not exceed 0.01 mag and does not cause any appreciable problem. This might not be true at lower galactic latitudes.

In summary, the $(u-B)_K$ vs. $(B-V)$ diagram affords a practical way of separating out BHB stars at relatively faint magnitudes. As with the Strömgren colors, separation becomes difficult blueward of $(B-V)_0 = 0.00$. Redward of $(B-V)_0 = +0.20$, BHB stars lie in the instability strip and so will be variable (RR Lyrae stars). Between these color limits, stars are known that mimic BHB stars in both their $(u-B)_K$ and c_1 indices. These "interlopers" have been discussed by both Bidelman (1992) and Green & Morrison (1993). An example is the rapidly rotating star HD 64488. A high-resolution spectroscopic survey could easily detect such interlopers, but this would require a very large telescope. In the next section, we describe how BHB stars can be selected by spectrophotometry at a resolution (~ 3 Å) that can be obtained with the Goldcam spectrometer at the KPNO 2.1-m telescope.

2.2. The Balmer jump

The Balmer jump (D) was defined by Arnulf, Barbier, Chalonge & Canavaggia (1936) as the logarithm of the ratio of the intensities (measured at 3700 Å) that is obtained by extrapolating from the long and short wavelength sides respectively in an intensity vs λ^{-1} plot. We use a simpler parameter (BJA) to measure the size of the jump. This is defined as the difference in magnitudes between a 25 Å bandwidth centered on wavelength 3857 Å and a 125Å bandwidth centered on wavelength 3612 Å. For a star whose $(B-V)_0$ is 0.10, the difference in BJA between a BHB star and a Population I main sequence star averages 0.35 magnitudes in comparison with 0.23 magnitudes for $(u-B)_K$. We note that for $(U-B)$ the difference is roughly 0.10 magnitudes.

2.3. The width of the Balmer lines

The widths of the early members of the Balmer lines can be easily measured and are known to depend on both surface gravity and temperature. Pier (1983) showed that BHB

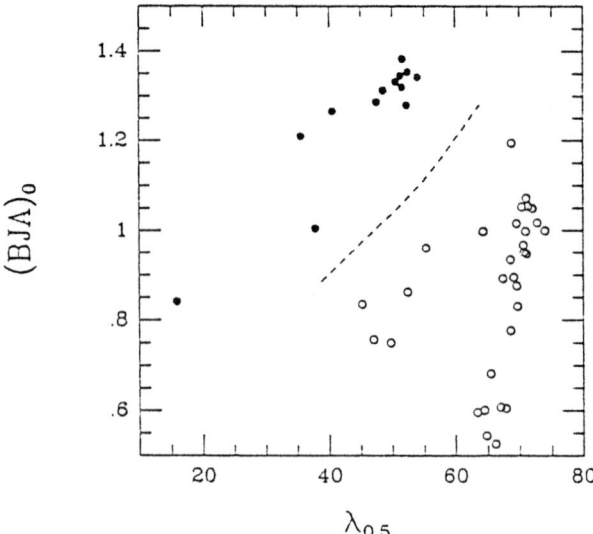

FIGURE 1. $(BJA)_0$ vs $\lambda_{0.5}$. BHB stars in the globular cluster stars M3 and M92 are shown by filled circles and main-sequence stars in the Pleiades and Coma open clusters by open circles.

stars could be separated from main sequence stars of the same color by their smaller line-widths ($D_{0.2}$) in a plot of ($D_{0.2}$) vs $(B-V)_0$. Here $D_{0.2}$ is the Balmer line-width (in Å) at a level of 20 % of the line-depth below that of the continuum; the value of this particular definition of the Balmer line-width was first pointed out by Searle & Rodgers (1966). If a measurement of $D_{0.2}$ is to be useful as a discriminant for BHB stars, it must be accurate to 1 or 2 Å. This requires high signal-to-noise spectra. If one uses spectra with a relatively low S/N (say about 15), the error in $D_{0.2}$ will cause a fraction of the stars to be misclassified. Further, the value of $D_{0.2}$ depends on the level assumed for the continuum. A way must be found to measure this level as consistently as possible. In our work, we measured $D_{0.2}$ for a number of main-sequence stars in the Coma and Pleiades clusters and determined the mean value of $D_{0.2}$ for these stars as a function of $(B-V)_0$. We then defined $\Delta D_{0.2}$ as the difference between our measured $D_{0.2}$ and this mean value for the Pleiades and Coma stars. It was found that nearly all our program stars had a smaller $D_{0.2}$ than the Pleiades and Coma stars of the same $(B-V)_0$. Nearly all these program stars are more metal-poor than the Pleiades and Coma stars; consequently their continua are less depressed by blanketing and this has an effect on their measured $D_{0.2}$. We therefore plotted $\Delta D_{0.2}$ against [Fe/H] so that line widths could be compared at a given metallicity. It was found that the BHB stars could be reasonably well separated from the other field stars if their $(B-V)_0$ was positive. If, however, their $(B-V)_0$ was negative, the separation was uncertain.

2.4. The Steepness of the Balmer jump

The BCD scheme (Chalonge & Divan 1953, 1973) used three parameters to classify early-type stars. These were ϕ_b, which is a gradient that correlates with B−V, the Balmer jump D (referred to in § 2.2), and λ_1, a parameter that measured the steepness of the Balmer jump. Corresponding to λ_1, we define a quantity $\lambda_{0.5}$ that is the wavelength at which the continuum has a flux that is the mean of the fluxes at λ 3612 and λ 3857. We find that $\lambda_{0.5}$ can be measured quite accurately (to within a few Å on high S/N

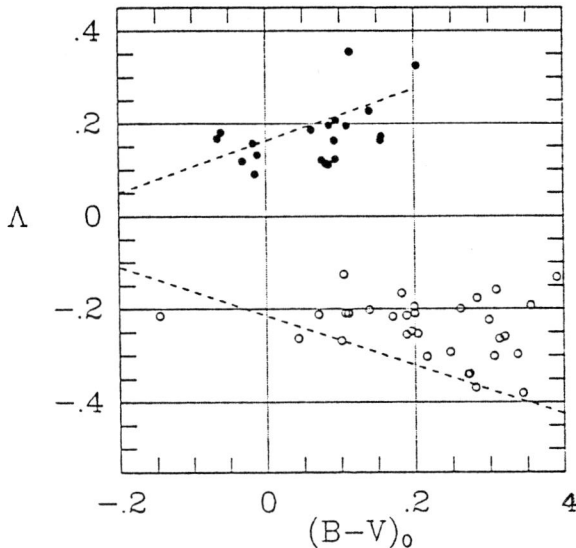

FIGURE 2. The parameter Λ (which defines the steepness of the Balmer jump) vs. $(B-V)_0$ for the program stars in SA 57. The presumed BHB stars are shown as filled circles and the presumed main-sequence stars as open circles.

scans) and it is relatively insensitive to reddening although it must be corrected for the radial velocity of the star. A plot of BJA against $\lambda_{0.5}$ is shown in Figure 1, where globular cluster horizontal branch stars (in M3 and M92) are shown as filled circles and main-sequence stars from the Coma and Pleiades open clusters are shown as open circles. It is seen that there is a clear line of demarcation (shown by a dashed line) on this plot between the BHB and the main-sequence stars. We therefore define a quantity Λ which is the perpendicular distance from this dashed line and which is positive in the direction of the globular cluster BHB stars and negative in the direction of the main-sequence stars.

This quantity Λ can then be used as a criterion for separating BHB from main-sequence stars. Figure 2 shows a plot of Λ against $(B-V)_0$ for program stars in the SA 57 field at the North Galactic Pole. It is seen that the observed stars fall into two well-separated groups: (a) those with negative Λ shown as open circles which are presumed to be main-sequence stars, and (b) those with positive Λ (filled circles) that are presumed to be BHB stars.

2.5. How do these criteria compare?

In § 2.1, § 2.3 and § 2.4 we have described three criteria for identifying a BHB star. They are quite independent in the sense that the choice that one makes with one criterion is independent of the results from using the other criteria. When we compared the results, we found that there was nearly complete agreement among the three methods of selection *if the stars had a $(B-V)_0$ that was positive*. If the star has a negative $(B-V)_0$, the $(u-B)_{K0}$ and $D_{0.2}$ criteria do not always give sufficient distinction between the BHB and the higher-gravity stars. The only ambiguous case among the stars with positive $(B-V)_0$ was for a star with $(B-V)_0 = 0.22$ that was not a BHB star on the $(u-B)_{K0}$ criterion, but was a BHB star according to the other two criteria. The star appears to be a very low-amplitude RR Lyrae star. It is clear that, at least for part of their cycle, RR Lyrae stars cannot be distinguished from BHB stars according to these criteria. *It is therefore very important that the variability of the candidate stars should be assessed.* The most difficult of these variables to identify are the low-amplitude variables that exist at the blue-edge of the instability strip; for obvious reasons, the frequency of occurrence of these variables is not well known. An example, in the field, of such a variable is HD 202759 (AW Mic) which is listed by Stetson (1991) as a possible BHB star.

3. The Determination of the metallicities of the BHB stars

The determination of the metallicities of these stars is not easy — even if high-resolution spectra are available. Much of our knowledge of the metallicities of the nearby BHB stars has come from the work of A. G. D. Philip and S. J. Adelman who have shown that these stars have much the same [Fe/H] as has been found for the halo globular clusters. More specifically, Lambert, McWilliam & Smith (1992) have determined the abundances for BHB stars in the globular clusters M4 and NGC 6397; they find metallicities that agree with those found by previous observers for the red giants in these clusters.

At lower spectroscopic resolutions, metallicities have usually been determined from the strongest available line (the Ca II K-line). This line increases almost linearly with strength with decreasing temperature; any analysis therefore depends upon an accurate knowledge of the temperature. Frequently a model (for a specified gravity) is used to predict the relationship between the K-line equivalent width and the color $(B-V)_0$ as a function of metallicity. Uncertainties include the reddening, the extent to which the measured equivalent width is the same as that defined by the model and the interstellar contribution to the K-line. Alternatively, one can use the Mg II ($\lambda4481$) line which is significantly weaker than the K-line, but which has the advantage that it has a much weaker dependence on temperature (and so on color). This line, moreover has no interstellar component; consequently, the errors in the abundances derived from it come mostly from the errors of measurement of the equivalent width of the line. We measured the equivalent widths of the Mg II line in the stars of clusters of known metallicity (Pleiades, Coma, M3, and M92) and plotted these values against $(B-V)_0$; we then drew contours of constant metallicity on this plot. Then, if the equivalent width of the Mg II line and the $(B-V)_0$ of a star is known, one can (by interpolation) determine its metallicity. An advantage of this purely empirical method is that true equivalent widths do not have to be determined, it is only necessary to measure pseudo-equivalent widths for both the program stars and the calibrating stars in a consistent way. We also used the same technique to derive metallicities from the K-line and gave these metallicities half of the weight of those derived from the Mg II line.

The metallicities of 29 BHB stars in our two pilot fields were determined to be in the range $-1.0 \geq [\text{Fe/H}] \geq -2.4$. This distribution is very similar to that found for the halo globular clusters and also to the distribution of [Fe/H] (found by the Preston ΔS-method) for the RR Lyrae stars *that are in the same fields*. The technique of using the Mg II line gives results, therefore, that are consistent with those obtained by other methods.

4. The Space-Densities of the BHB stars

Space-densities were evaluated for the BHB stars in our two fields using assumptions about the absolute magnitudes of these stars that are similar to those used by Preston *et*

al. (1991). It was found that if the stars were more than 5 kpc above the galactic plane, their space densities were in agreement with those predicted for a spherical halo by:
$$\rho = \rho_0 \, R_{gal}^{-3.5}$$
where R_{gal} is the distance to the Galactic Center in kpc. This law has been shown to represent the space densities of both the halo globular clusters (Zinn 1985) and also the field RR Lyrae stars (Kinman 1992b). For stars that were nearer the galactic plane, the space-densities exceeded those given by the above relation for a spherical halo. This excess was also found in the space-densities of the BHB stars determined from the observations of Preston et al. (1991) and could be explained by the existence of a second flat component of halo stars whose space-density distribution could be described by the following relation:
$$\rho_{flat} = 23.7 \, e^{-(0.163 Z \sqrt{W})}$$
Here Z is the height (in kpc) above the Galactic plane and W is the distance (in kpc) to the Galactic Center projected onto the Galactic plane. The scale-height of this exponential disk is therefore about 2.2 kpc in the solar neighborhood and the local space-density of the BHB stars of this flat component is 23.7 stars per cubic kpc. The local space-density of the spherical component is only 6.3 stars per cubic kpc; thus the new flat component makes up 80 % of the BHB stars in the galactic plane near the sun. We can now understand why the local halo stars have an anisotropic velocity dispersion with a smaller Z-motion than is found for the halo globular clusters, since most of these local halo stars belong to the flat-component. A more extended description of this work is given in Kinman, Suntzeff & Kraft (1994).

REFERENCES

ARNULF, A., BARBIER, D., CHALONGE, D., CANNAVAGGIA, R., 1936, J. Obs., 19, 149

BIDELMAN, W. P., 1992, in Sasselov D. D., ed, Luminous High-Latitude Stars, ASP Conf. Ser., 45, p. 49

CHALONGE, D., DIVAN, L., 1953, C. R. Acad. Sci. Paris, 237, 298

CHALONGE, D., DIVAN, L., 1973, A&A, 23, 69

GREEN, E. M., MORRISON, H. L., 1993, in Smith G. H., Brodie J., eds, The Globular Cluster-Galaxy Connection, ASP Conf. Ser., 48, p. 318

KINMAN, T. D., WIRTANEN, C. A., JANES, K. A., 1966, ApJS, 13, 379

KINMAN, T. D., MAHAFFEY, C. T., WIRTANEN, C. A., 1982, AJ, 87, 314

KINMAN, T. D., 1992a, in Warner B., ed, Variable Stars and Galaxies, ASP Conf. Ser., 30, 19

KINMAN, T. D., 1992b, in Edmunds M., Terlevitch, R., eds, Elements & the Cosmos, Cambridge University Press, Cambridge, p. 151

KINMAN, T. D., SUNTZEFF, N. B., KRAFT, R. P., 1994, AJ, submitted

LAMBERT, D. L., McWILLIAM, A., SMITH, V. V., 1992, ApJ, 386, 685

LEE, Y.-W., 1992, AJ, 104, 1780

PESCH, P., 1991, in Philip A. G. D., Upgren A. R., eds, Objective-Prism and Other Surveys, L. Davis Press, Schenectady, p. 3

PESCH, P., SANDULEAK, N., 1983, ApJS, 51, 171

PHILIP, A. G. D., 1994, in Adelman S. J., Upgren A. R., Adelman C. J., eds, Hot Stars in the Halo, Cambridge University Press, Cambridge, p. 41

PIER, J. R., 1983, ApJS, 53, 79

PRESTON, G. W., SHECTMAN, S. A., BEERS, T. C., 1991, ApJ, 375, 121

SANDULEAK, N., 1988, ApJS, 66, 309

SEARLE, L., RODGERS, A. W., 1966, ApJ, 143, 809

STETSON, P., 1991, AJ, 102, 589
ZINN, R., 1985, ApJ, 293, 424

Discussion

PHILIP: I was interested in your plot showing the relative space density distributions of horizontal-branch (HB) stars and RR Lyrae stars where you found the density distribution for FHB stars fell well above that of the RR Lyrae stars. At the North and South Galatic Poles, where I have measured all stars to a given limiting magnitude in the four-color system, one can construct a coarse stellar density distribution for the FHB stars at each galactic pole. I found that the density distribution of the FHB stars fell well above that from the RR Lyrae stars.

KINMAN: The space density of the RR Lyrae stars is now fairly well defined for the spherical component and is about twice that found from surveys with the Lick Astrograph.

Post-AGB A and F Supergiants as Standard Candles

By HOWARD E. BOND

Space Telescope Science Institute, 3700 San Martin Drive, Baltimore, MD 21218, USA

Post-AGB A- and F-type supergiants have considerable potential as Population II standard candles for application to the extragalactic distance-scale problem. The prototype of these objects is ROA 24 in Omega Centauri, a star that lies more than 3 magnitudes above the horizontal branch in this globular cluster. A field analog was discussed by Bond & Philip (1973, PASP, 85, 332), who demonstrated the utility of Strömgren photometry in discovering such objects.

These stars are expected theoretically to have an extremely narrow range of absolute magnitudes, which is confirmed empirically by a standard deviation of about 0.2 mag around $M_v = -3.3$ for a sample of several such objects in galactic globular clusters. Unlike several other candles currently in use (such as the planetary-nebula luminosity function and surface-brightness fluctuations), this method can be calibrated within our own Galaxy. I will describe planned or proposed surveys for such objects in the Magellanic Clouds (where several candidates are already known from objective-prism surveys by Philip and Sanduleak) and M31. The Strömgren filters on the Hubble Space Telescope's WFPC2 will allow the detection of these objects out to distance moduli larger than 29 (7 Mpc).

Discussion

DORMAN: Schönberner's lowest mass tracks are actually post-EAGB stars as they do not reach the thermal pulsing stage: stars of a slightly lower mass which live longer do not enter the AF supergiant region. It is thus possible that the stars of most interest (as being the most common) will indeed have a very small dispersion in Bolometric luminosity.

BOND: Thank you. Your comment seens to confirm the idea of a reasonably sharp lower limit to the luminosities.

YOSS: Does not Roberta Humphreys use F-supergiants as distant indicators?

JANES: How will you distinguish these stars from Population I F-G supergiants in a distant galaxy?

BOND: One would want to observe in halos of galaxies where it it unlikely we would find Population I supergiants - unless you are right that stars can form in galactic halos!

LANDSMAN: You have identified 4 A-F supergiants in globular clusters, which is much smaller than predicted from the number of AGB stars, but it is hard to believe that such bright stars could be missed. Would you expect to find more A-F supergiants in globular clusters?

BOND: I suspect that there could be several of these that have still been missed. The two 12 th magnitude PAGB stars in NGC 5986, for example, which unfortunately I published only in a BAAS abstract, have yet to be noted by any other workers. They are usually dismissed as foreground non-members.

LU: Since there are very few PAGB stars in globular clusters, how many do you expect to find in the Large and Small Magellanic Clouds?

BOND: One PAGB star per 200 giant stars.

The Extended Horizontal-Branch: A Challenge for Stellar Evolution Theory

By PIERRE DEMARQUE

Center for Solar and Space Research, Department of Astronomy, Yale University, New Haven CT 06511, USA

The extended horizontal branch branch (EHB) has recently been receiving increasing attention. A wealth of new observations, both from the ground and from IUE and HST space missions, both in star clusters and in the field, are raising new questions about the evolutionary status of these stars (also often called sdB stars). The problems with interpretations based on the standard theory of single star evolution, which has otherwise been very successful in explaining other features of horizontal branch morphology, are reviewed. Dynamical interpretations, involving close binary systems and/or close interactions in high density regions in the core of some globular clusters will be discussed.

Discussion

SAFFER: The Ultraviolet Imaging Telescope (UIT) Omega Cen results of Whitney *et al.* (this volume) seem to indicate that the subdwarfs are not more centrally concentrated than the subgiants. Integration of the binary birthrate of Iben & Tutukov (1984, ApJ, 282, 615) over the narrow range of binary separations that produce sdB stars in the Mengel *et al.* (1976, ApJ, 204, 488) scenario produces far too few of those stars compared to observations. The recent discovery by Liebert *et al.* (AJ, in press) of EHB stars in the old, metal-rich Galactic cluster NGC 6791 demonstrates that those stars can form in environments quite different from the low metallicity, high density globular clusters.

DEMARQUE: Very interesting. I agree.

MOEHLER: I think you should be very cautious when identifying the stars below the gap in globular clusters with Extended Horizontal Branch (EHB) stars in the meaning of sdB stars. As has been shown yesterday, in most of the clusters these stars are quite normal Blue Horizontal Branch (BHB) stars (according to their physical parameters).

DEMARQUE: I agree.

Astronomical Patterns in Fractals: the Work of A. G. Davis Philip on the Mandelbrot Set

By MICHAEL FRAME

Department of Mathematics, Union College, Schenectady, NY 12308-2311, USA

In honor of A. G. Davis Philip's 65th birthday, I review his work on the Mandelbrot and Julia sets.

1. Introduction to the Mandelbrot set and Julia sets

The Mandelbrot set is less than twenty years old, yet in this short time it has had an amazing impact on not only mathematics, but also the sciences, art, and literature. Indeed, it has become a sort of cultural icon, appearing on children's notebooks, advertising labels for clothing, as visual backgrounds for some rock music groups, and in contemporary comic books, among other places. Why has this object had such success at crossing the barriers between very abstract mathematics and popular culture? While the recent announcement of the proof of Fermat's Last Theorem was reported on the front page of the New York Times, there is little likelihood that elliptic curves will begin appearing on hundred foot screens behind Mike and the Mechanics. What makes the Mandelbrot set so interesting to so many?

In part the answer lies in the ease of describing the Mandelbrot set. Fix a complex number c and for the function

$$F_c(z) = z^2 + c,$$

consider the sequence of complex numbers $z_0 = 0$, $z_1 = F_c(z_0)$, $z_2 = F_c(z_1)$, and so on. If this sequence ever produces a number z_k with $|z_k| > 2$, then $|z_n| \longrightarrow \infty$ as n $\longrightarrow \infty$ and we say c does not belong to the Mandelbrot set. If the sequence remains bounded for all n, then c does belong to the Mandelbrot set. Nothing more is involved in the definition; it is difficult to imagine anything much simpler.

Of course, being simple is not enough to make something interesting: adding fractions is simple but hardly worth years of study. The Mandelbrot set is interesting because it is self-contradtictory: its definition is so very simple and yet on close inspection, it is extremely complicated, a constant source of surprise. This continual variation in detail, within a familiar visual matrix, perhaps is the main reason for the popular appeal of the Mandelbrot set.

Some background is necessary to understand how this set was discovered. The theory of iteration of complex functions was advanced early in this century by the work of Pierre Fatou and Gaston Julia. They were interested in distinguishing two different types of behavior for polynomials - under iteration a point would either run away to infinity or it would not. What we now call the (filled-in) Julia set of a function is the collection of all points not diverging to infinity under iteration of the function. (The Julia set is the boundary of the filled-in Julia set.) It is a consequence of theorems of Julia and Fatou that for the functions $F_c(z) = z^2 + c$, the Julia set is either connected or totally disconnected (contains no arcs), and which it is depends on whether iterating $z = 0$ remains forever bounded or diverges to infinity. Thus it is natural to seek a description of the set of those c for which the Julia set J_c of $F_c(z)$ is connected. This is the question Benoit Mandelbrot decided to approach by computer experiments in the late 1970s. The result was a complete surprise, even to him. At first, the pictures seemed smudged, a rough cardioid with with a pattern of round clouds attached. On magnification, the fuzziness did not disappear, but rather revealed more elaborate substructures. This is when Mandelbrot realized he had found something very interesting, for as he said, "Dirt is not symmetrical." The field of iteration was reborn, and the experimental aspect of mathematics was removed from marginalization.

Before describing some of the early graphical work, we must mention some of the basic properties of the Mandelbrot set. One way for the sequence z_0, z_1, z_2, z_3, ... to stay bounded is by converging to a repeating pattern. If this pattern is a single point, we say the sequence has converged to a fixed point, or a **1-cycle**; if the pattern consists of n points, the sequence converges to an **n-cycle**. The n-cycle is **stable** if taking z_0 near enough to 0 will still converge to the same cycle. A theorem of Fatou guarantees that if there is a stable n-cycle for $F_c(z)$, the sequence starting with $z_0 = 0$ will converge to it, and consequently for any c there is at most one stable n-cycle. For all the c in the large cardioid of Figure 1, the sequence converges to a 1-cycle. For all c in the large disc attached to the left of the cardioid, the sequence converges to a 2-cycle (we call this the 2-cycle component). For c in the next largest discs attached at the top and bottom of the cardioid, the sequence converges to a 3-cycle. For c in the cardioid of the "midget" copy of Mandelbrot set seen in Figure 2, the sequence also converges to a 3-cycle. In general, the sets of c for which the sequence converges to an n-cycle are either (approximate) discs or cardioids. Taken together these are given the name **hyperbolic components**. The mathematics of this set is very interesting, but the wide-spread popularity of the Mandelbrot set is due to its appearance.

General scientific readers became acquainted with these developments through the spectacular color pictures in A. K. Dewdney's "Computer Recreations" column of the **August 1985, Scientific American**, and through Peitgen and Richter's (1986) book, **The Beauty of Fractals**. Thousands of readers were fascinated, hypnotized, trapped in "Mandelmania." Programs to drive home computers into elementary Mandelbrot exploration were easy to write, and vast journeys were begun, probing ever more deeply into the interesting regions around the boundary of the Mandelbrot set. Serious programmers developed more sophisticated software, some making use of deep theoretical relations between complex iteration and potential theory, and made their work available to others. The exploration of the Mandelbrot set reached such a familiar state in our consciousness that it has appeared as a topic in novels by Arthur C. Clarke (1990) and John Updike (1986). Arguably more people have engaged in Mandelbrot set voyages than in any other voluntary mathematical journey (most of my calculus students are not journeying voluntarily through calculus).

A. G. Davis Philip is one of those who joined the exploration early. He brought to the study his astronomical training, viewing the images through the eyes of an astronomer, and also approaching his journey more systematically than most, organizing features into families.

Familiar features of the Mandelbrot set appear in Figures 1 - 7: Figure 1 shows the entire Mandelbrot set, Figure 2 the "spike" complete with tiny copies of the entire set, Figure 3 the "head" of the Mandelbrot set, Figure 4 "seahorse valley" between the main cardioid and the head, Figure 5 the "3-cycle radical" attached to the top of the cardioid, Figure 6 the center of the pattern of arms atop the 3-cycle radical, and Figure 7 one of the radicals in "elephant valley." (Figure 13 shows these features, together with their names.)

Before beginning our survey, we must consider one more question: is this all just aesthetics, or do these pictures have some wider applicability? What's the point if we see these things only when iterating $z^2 + c$? One of the first hints of the answer was

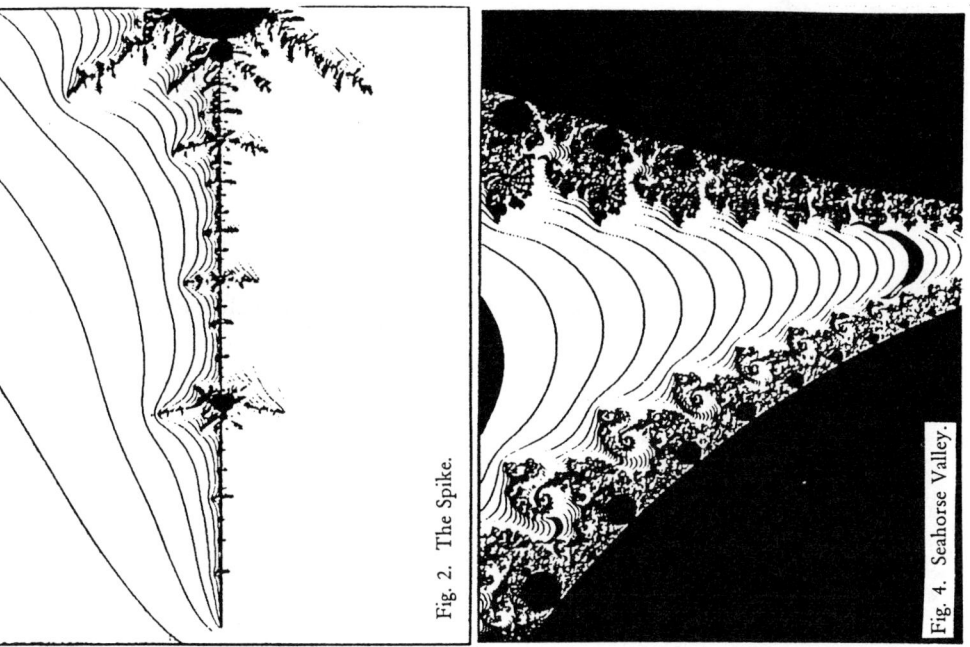

Fig. 2. The Spike.

Fig. 4. Seahorse Valley.

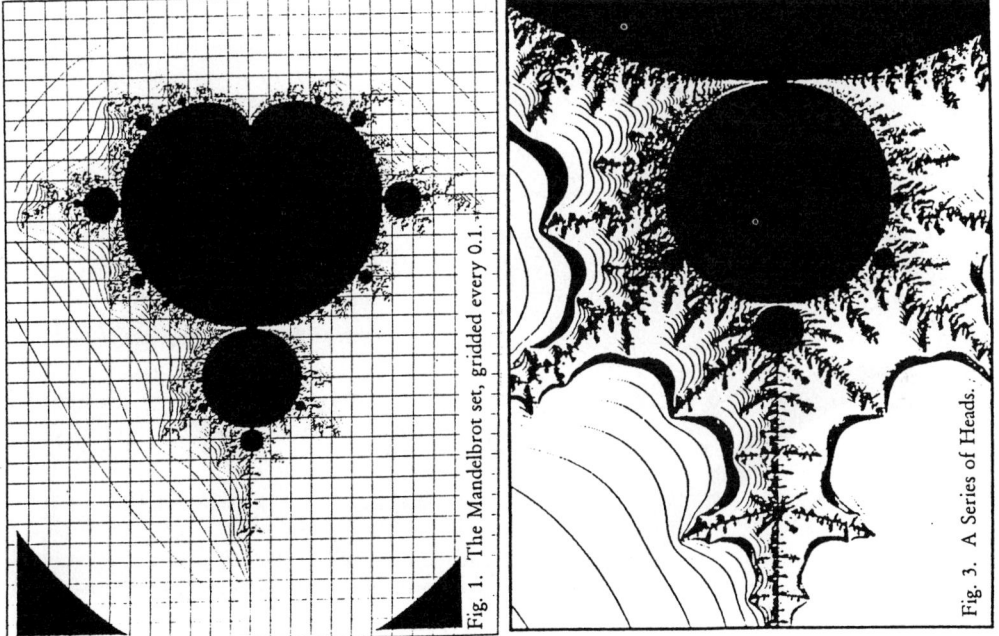

Fig. 1. The Mandelbrot set, gridded every 0.1.

Fig. 3. A Series of Heads.

Fig. 6. Center of three-armed pattern in Fig. 5.

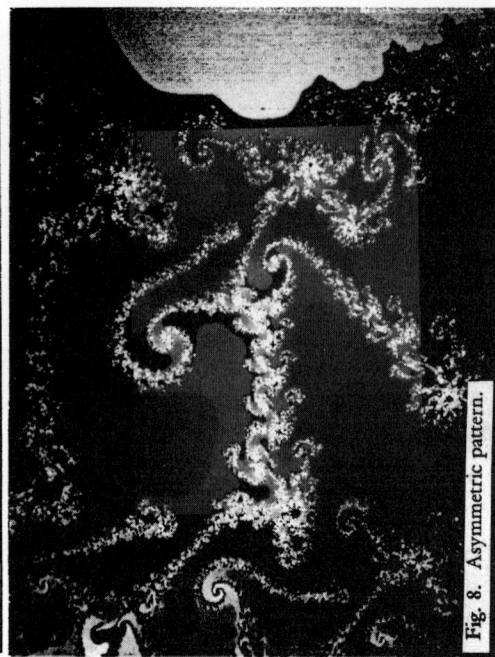

Fig. 8. Asymmetric pattern.

Fig. 5. The 3-cycle radical.

Fig. 7. Radicals in "Elephant Valley".

provided by Dennis Sullivan and his coworkers - in studying Newton's method for a certain family of cubic polynomials (note the Newton function is not at all like $z^2 + c$ - it is a rational function consisting of a cubic polynomial divided by a quadratic), they (Curry et al. 1983) found the parameter sets on which Newton's method fails to converge to any solution have the form of Mandelbrot sets. This result was put on a theoretical foundation by Adrien Douady and John Hubbard (1985) with their theory of polynomial-like maps. In essence, their theory shows that a vast family of functions, when viewed in sufficiently small pieces, behave dynamically like quadratic functions. That is, tiny copies of the Mandelbrot set appear frequently. In this sense, the Mandelbrot set is a universal object, and so its study is more than just computer-assisted stamp collecting.

2. Stranger things than seen in Scientific American

Pictures of the Mandelbrot set of the kind illustrated in the last section are amazing the first time they are seen. How can such complexity, such richness of structure, come from iterating a single quadratic function? The allure of these images has proven time and again to be an effective tool for seducing students into learning some of the mathematics underlying complex iteration. Eventually, though, these curlicues, spirals, and midgets become familiar and it is possible to imagine one has seen versions of everything in the Mandelbrot universe. Not only novices succumb to this temptation: as recently as a few weeks ago, a computer bulletinboard on fractals contained the query, "Have we seen it all in the Mandelbrot set?"

The answer is, "No, we haven't seen it all, and it appears unlikely we will ever see it all." Dave was one of the respondents to this question, and indeed from his early explorations of the Mandelbrot set, he has always exhibited an intuitive grasp of how to find new features.

Figures 8-12 are representative examples of Dave's departures from the standard Mandelbrot fare. Figure 8 is a particularly asymmetric pattern of decorations, Figure 9 a delicate cobweb pattern with an unusual coloring, Figure 10 another asymmetrical pattern, Figure 11 a midget surrounded by a particularly pretty pattern of spirals, and Figure 12 an intricate, highly ramified sequence of spirals.

These few examples are just a tiny indication of the hundreds upon hundreds of pictures Dave produced in investigating the Mandelbrot set. While many have tired of these explorations after just a few trials, Dave had the patience and thoroughness to continue to higher and higher magnifications. His efforts have been rewarded by his having seen many interesting features, some surely not seen before. This represents a sort of faith in the infinite complexity of the Mandelbrot set, a faith some have given up as ever more powerful analytic and topological methods elucidate more and more of the structure of the Mandelbrot set. At least to the levels explored numerically so far, Dave's intuition seems to be validated.

3. Naming the parts

At least as far as its appearance is concerned, the Mandelbrot set is a complicated object and so describing its regions can be a challenging task. Of course, the Mandelbrot set lies in the complex plane and so one way to describe a region is by giving its complex coordinates. This is not especially evocative, however. For instance, how much information is communicated through the complex number -0.25 + 0.33i? The complex coordinates are completely unambiguous, and completely uninformative if one seeks from

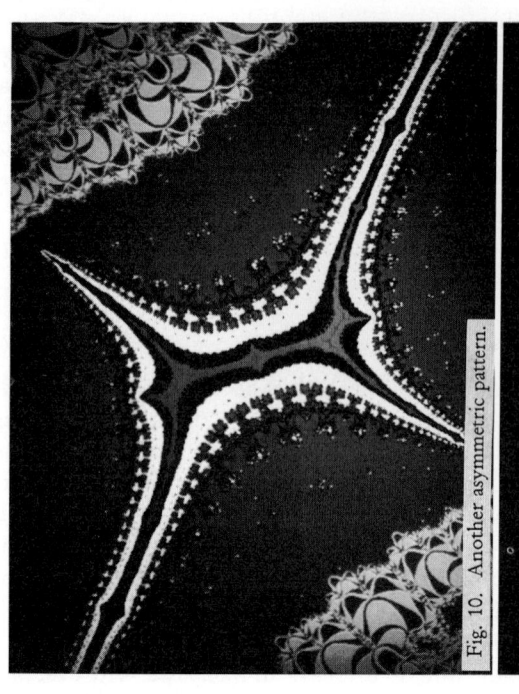

Fig. 9. Cobweb pattern.

Fig. 10. Another asymmetric pattern.

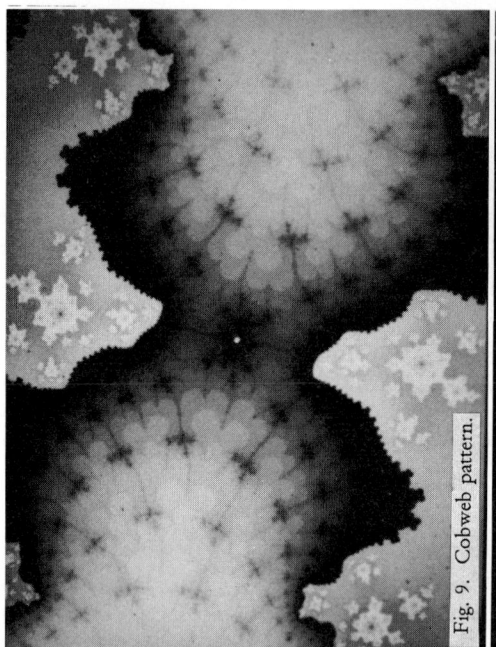

Fig. 11. Midget surrounded by an octuple spiral.

Fig. 12. Sequence of spirals.

the name some description of local features. Other methods of locating and describing features in the Mandelbrot set include

(1) external angles, explored primarily by John Hubbard and Adrien Douady (Douady & Hubbard 1982, Branner 1989),

(2) Lavaurs' "abstract Mandelbrot set" constructed by a dynamical identification of external rays (Branner 1989),

(3) the symbolic dynamics approach to locating centers of features (section 3.2 of Bai-lin 1989).

Each of these has its own particular strengths and weaknesses. For example, the symbolic dynamics approach does not allow one to see immediately how one feature is related to another. In this section we shall describe an approach addressing this issue, the simple naming scheme developed by Dave and Ken Philip (Philip & Philip 1990).

Figure 13 illustrates the naming scheme developed by Dave and Ken Philip. Most features are self-evident from the picture; in fact we shall mention just one, the naming of the radicals. Mandelbrot (1985) calls a disc attached to a cardioid, together with the discs and antennas attached to that disc, a radical. The Philips' naming method involves identifying a radical by its cycle number, for example, the 3-cycle disc belongs to R3. The especially useful feature of this pattern involves naming the discs attached to discs by their relative cycle numbers - for example, the 12-cycle disc attached to the side of the 3-cycle disc belongs to R3/R4, since relative to the 3-cycle disc, this plays the role of a 4-cycle radical. In addition to providing a simple way for referring to features, this scheme emphasizes the fractal nature of the Mandelbrot set, the way in which each feature contains smaller copies of all the other features.

4. Spirals and spirals

We begin with a sketch of what is known analytically about spiral structures in Julia sets and the Mandelbrot set, and then show just what beautiful examples can be found as special cases within this theory.

If z_0 is a periodic point of period n for $F_c(z)$, then z_0 is a fixed point of the nth composition $F_c^{on}(z)$. Such a periodic point is repelling if the multiplier $\rho(z_0) = (d/dz)(F_c^{on}z(o))$ has magnitude greater than 1. Another characterization of the Julia set J_c of the function $F_c(z)$ is this

1. every repelling periodic point belongs to J_c, and
2. every point of J_c has repelling periodic points arbitrarily close to it. Moreover,
3. for any point z_1 of J_c, the collection of successive inverse iterates $F_c^{o-n}(z_1)$ gets arbitrarily close to every point of J_c. (This is the basis for the "inverse iteration" method of generating pictures of Julia sets.)

In 1884, Koenings showed if z_0 is a repelling periodic point, then near z_0 the coordinates can be changed so that $F_c^{on}(z)$ becomes multiplication by $\rho(z_0)$. (This is called Poincare linearization.) Since $\rho(z_0)$ has magnitude greater than 1, so long as $\rho(z_0)$ is not real, successive multiplication by $\rho(z_0)$ and $\rho(z_0)^{-1}$ produces a spiral pattern. Of course, eventually the forward iterates go outside the region where Poincare linearization is valid, and so the Julia set is not a single spiral extending outward forever.

This explains why there should be spirals throughout many Julia sets. Spirals get into the boundary of the Mandelbrot set through a theorem of Tan-Lei: around certain points (the Misiurewicz points) of the boundary, small regions of the Mandelbrot set look like small regions of the Julia set of that point. (A point c is a Misiurewicz point if for some n > 0 there is an m for which $F_c^{on}(0) = F_c^{on+m}(0)$. Also, every point of the boundary is

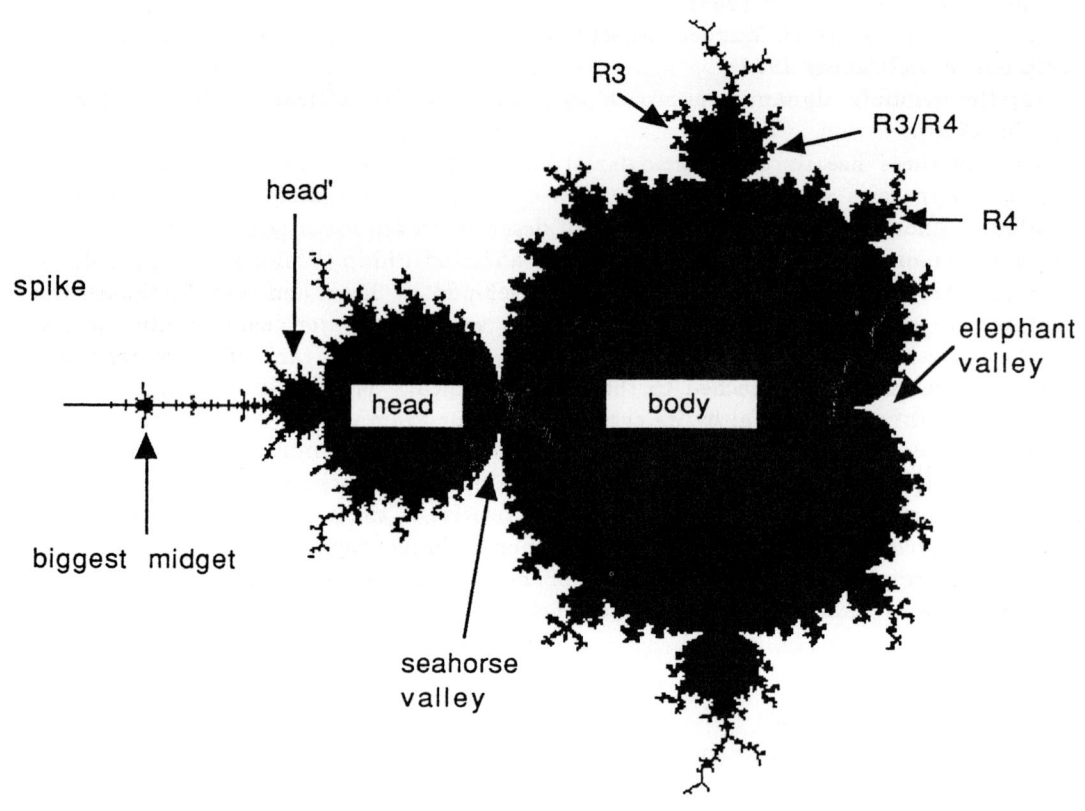

Fig. 13. The Parts of the Mandelbrot Set.

arbitrarily close to a Misiurewicz point.) Consequently, we expect to see many spirals throughout the boundary of the Mandelbrot set.

In this way, mathematics is good at capturing the broad, sweeping picture, but the details reveal complexities of surprising beauty. For example, Figure 14 shows a beautiful pattern of spirals within spirals, the basic theme familiar, but here modulated by the appearance of the tight whirlpool at the center of the picture. Of course, this detail could be explained through careful analysis - it is just the result of quadratic iteration - but surely it was not anticipated. Although we are not accustomed to thinking in these terms, there is increasingly a place for this kind of exploration in mathematics.

Continuing the sample of Dave's spirals, Figure 15 shows another whirlpool placed within a branch of a much larger spiral. Note how the two connecting branches have between them two smaller branches, both of which end in their own spirals. Figure 16 is a "double-double spiral." Looking closely, we see four spiral arms come in toward the center, two approaching relatively directly, and two through intervening double spirals. Imagine the intricacies buried in the delicate tangles all along each of the spiral arms. In Figure 17 we see a further complication, a "quadruple-double spiral." Eight spiral arms approach the center, four more-or-less directly and four only with detours through double spirals. However, the approach to the center is not so direct, for after passing the double spirals, all eight arms modulate to a tightly wound whirlpool. Past the whirlpool, a family of more complicated spirals appears, and so on, evidently to infinite detail. Even after having spent almost a decade looking at magnifications into the Mandelbrot set, I have no confidence in my ability to predict what we would find if we continued zooming in on the center of this picture. This is a fine example of the beautiful complications Dave has discovered (Philip 1991, 1994).

5. The smallest thing ever seen

Richard Voss' "Avogadro's Magnification" is an image of a midget Mandelbrot set smaller than the whole set by a factor of about 10^{24} (hence the name). Dave viewed this slide as a challenge to find a smaller image. The limitations imposed by standard accuracy of numerical processors are relatively straightforward to overcome by handling arithmetic through symbol string operations. The problem, of course, is knowing where to magnify. Especially since high-magnification pictures require some time to generate, just picking a point and starting to magnify may use many CPU hours without producing interesting results. We needed a way to locate these tiny features.

Through investigating the derivative formulation for the stability of an n-cycle region, it can be shown that each such region contains a center, a unique c-value at which $f_n(c) = 0$, where the polynomials $f_n(c)$ are defined inductively by

$$f_1(c) = c, \text{ and}$$
$$f_{n+1}(c) = (f_n(c))^2 + c.$$

Restricting our attention to midgets lying along the real axis, it is easy to find numerically the left-most solution c_n of $f_n(c) = 0$. To get a rough idea of the scale of the midget, we can locate the center of the midget's head by finding the first solution h_n of $f_{2n}(c) = 0$ less than c_n. This investigation (Frame, Philip & Robucci 1992) revealed not only tremendously small features (Figure 18 shows the 175-cycle midget, magnified by a factor of 10^{203}, but also some interesting scalings: as n $\longrightarrow \infty$,

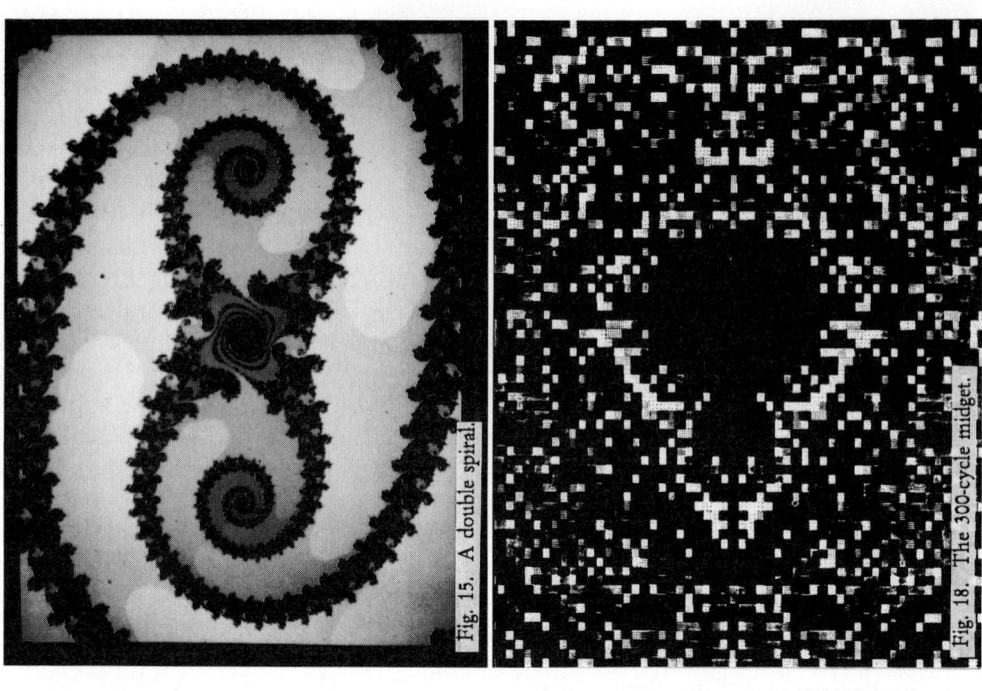

Fig. 15. A double spiral.

Fig. 18. The 300-cycle midget.

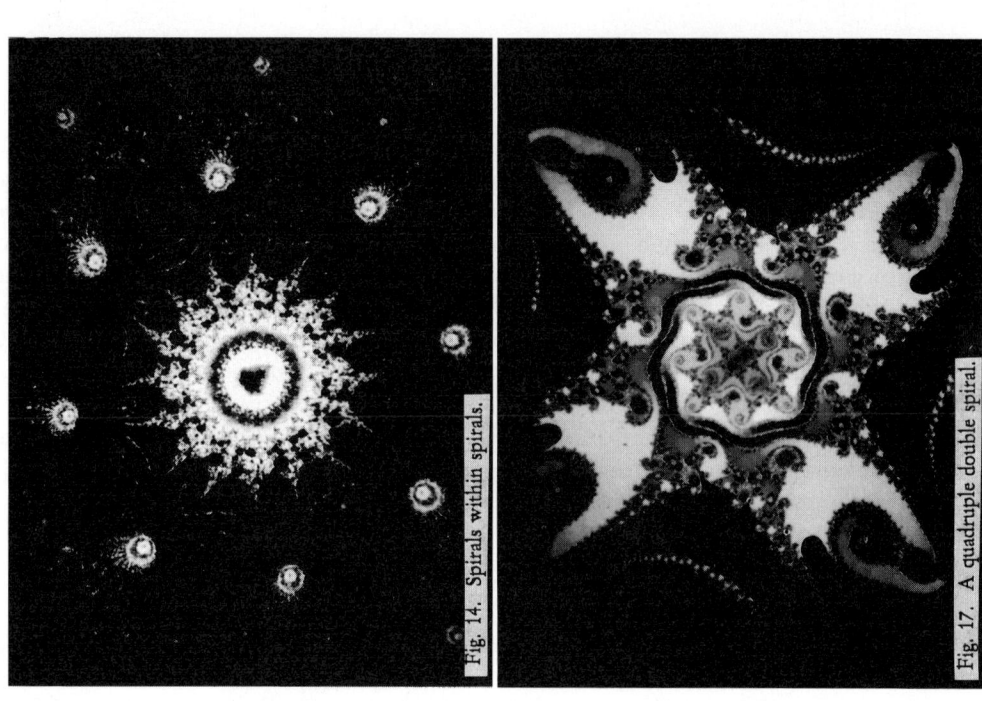

Fig. 14. Spirals within spirals.

Fig. 17. A quadruple double spiral.

Fig. 16. A Double/Double Spiral

$$(c_n - c_{n+1})/(c_{n+1} - c_{n+2}) \to 4 \text{ and}$$
$$(h_n - c_n)/(h_{n+1} - c_{n+1}) \to 16.$$

To date, the smallest midget observed in this way has cycle number 300, and is smaller than the whole Mandelbrot set by a factor of 10^{359} (Frame et al. 1992). That is, Voss' record has been exceeded by a factor of 10 in the exponent. (To be sure, this was not a competition - Voss could have done similar calculations had he wished.) Astronomical analogues are left to the imaginations of the readers.

Before moving on, a consequence of these scalings should be mentioned. These results were discovered numerically and reported in Frame et al. (1992). Since the scaling constants are integers, we would expect they might arise through a mechanism accessible to analysis. Motivated by this data, renormalization methods were applied to this problem, with the result that not only are these scalings explained, but the corresponding sequence of second left-most midgets exhibit precisely the same scaling (Hurwitz, Frame & Peak 1994; Frame & Peak 1994). (The original data were obtained for the sequence of left-most n-cycle midgets.) In fact, the same scaling holds for the sequence of third left-most midgets, for the sequence of fourth left-most midgets, and so on. All these analytical results were motivated by Dave's interest in finding smaller midgets. Sometimes, a good question leads to unexpected results.

6. Warped midgets

Upon close inspection, the midgets in the Mandelbrot set reveal themselves to be not exact copies of the entire set, but rather exhibit various degrees of "warping" . See Figures 19 and 20 where the perturbations of the cardioid and the placement of discs around it are apparent. This effect has been known for some time, but a detailed study undertaken only recently (Philip, Frame & Robucci 1994). To quantify measures of distortion, consider the whole Mandelbrot set in Figure 21. We identify four points of the set: the center of the main cardioid, the center of the head, the center of the "north" 3-cycle disc, and the center of the "south" 3-cycle disc. (These are labeled C, H, N, and S in the Figure.) Using the family of polynomials defined in section 5, together with a method for navigating through the cardioid (see Philip et al. 1994 for the details of the method), these points can be located for all midgets within the range of numerical accuracy available. As distortion measures for the midgets, the ratio of angles $(\angle HCN)/(\angle HCS)$ and the ratio of distances $|CN|/|CS|$ are reasonable choices. Interesting patterns of distortions were found for the families of midgets investigated in (Philip et al. 1994). In particular, plotting $\angle HCN$ vs $\angle HCS$ for corresponding families of midgets from the entire Mandelbrot set, from R3, from R10, from the 3-cycle midget, and from several other midgets on the spike, the points arrange themselves into roughly homologous families, with corresponding features having similar distortion patterns.

While these results still are in a preliminary stage, they illustrate again how asking detailed questions can lead to the discovery of structures, possibly of the sort which can be understood at an abstract level. Asking particular questions about the structure of the Mandelbrot set can lead to new kinds of insights, untapped by the general theoretical analyses of complex dynamics. This sort of "observational astronomy" of the Mandelbrot set indeed can lead to productive directions of research.

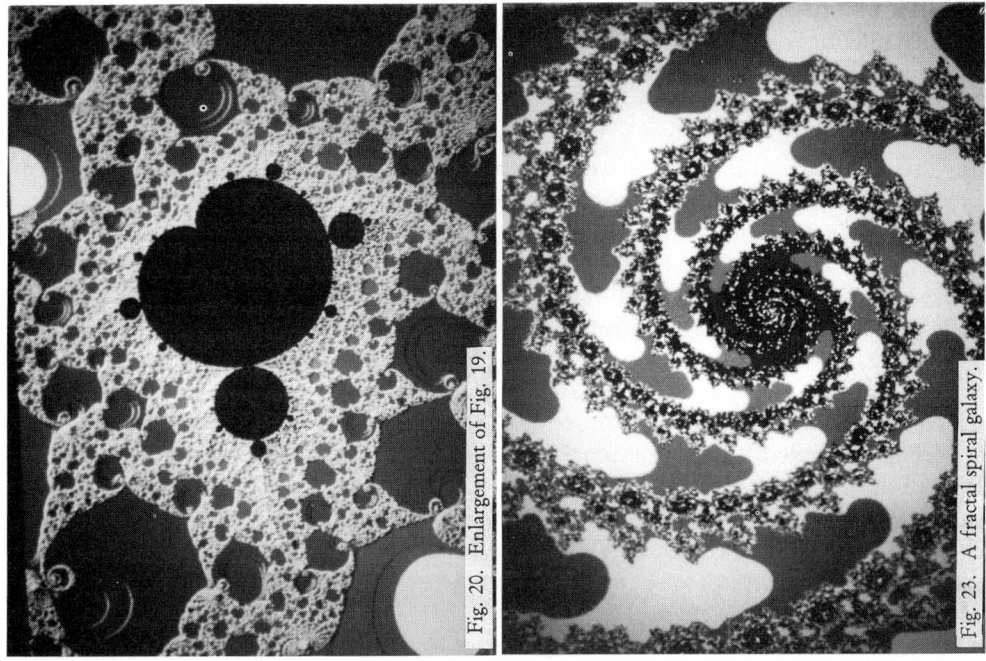

Fig. 19. A warped midget in the 10-cycle radical.

Fig. 20. Enlargement of Fig. 19.

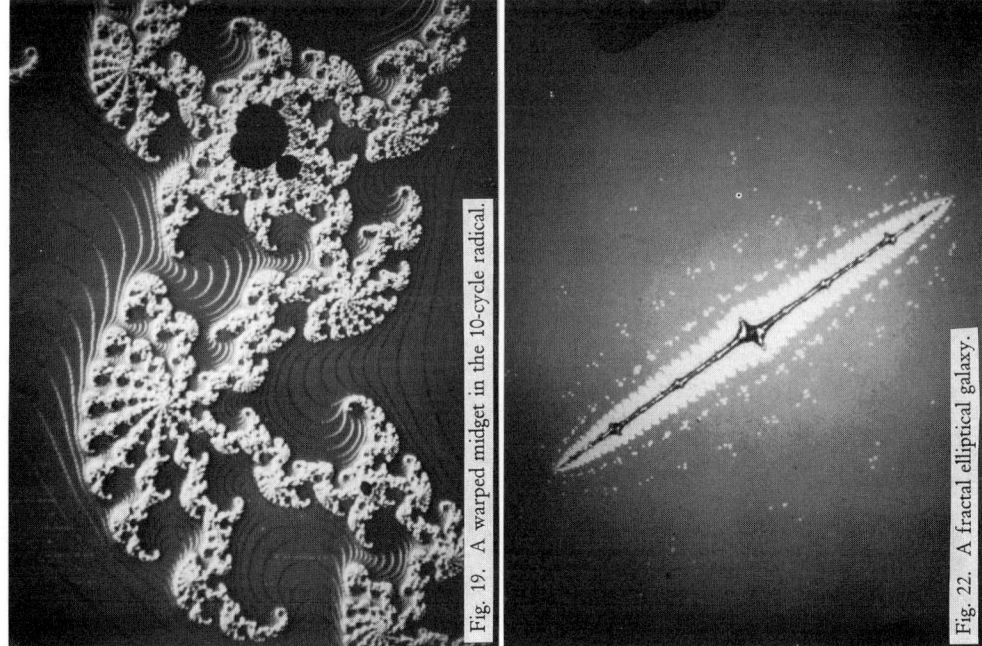

Fig. 22. A fractal elliptical galaxy.

Fig. 23. A fractal spiral galaxy.

7. The scientist as artist

With its spectacular successes in materials and information technology, this century has seen a marginalization of the perceived value to science of the eye and of aesthetics. Benoit Mandelbrot's creation of fractals is at the center of a wonderful reversal of this trend. Indeed, Mandelbrot argues eloquently for the primacy of the eye as a scientific tool. Dave Philip's work on fractals is a step in this program, a celebration of directing thought by sight, of motivating our work by first looking around for what appears interesting. Many "secrets" of mathematics are elaborated through beautiful structures, but too often we focused our attentions entirely on formal logic and missed the pictures.

Selecting a few closing pictures from the hundreds of slides Dave provided has been a daunting task. Surely the editors would not be happy with a hundred or so figures to accompany this survey. Instead, I have decided to end with a few pictures tying together Dave's principal intellectual interests - astronomy and fractals.

Figure 22 is an elliptical galaxy, Figures 23 and 24 globular clusters, Figure 25 a sunspot and solar flare, and Figure 26 the creation of the universe. Perhaps the entire universe is represented in some form or another within the intricacies of the Mandelbrot set, perhaps botanists who also explore the Mandelbrot set will find plants coiling through the it, perhaps as Arthur C. Clarke (1990) suggests searching the Mandelbrot set may result in a sort of mental unbalance, or perhaps as John Updike suggests (1986) one might seek the face of God in the Mandelbrot set. Most realistically, perhaps the Mandelbrot set is a computer-assisted Rorschach test. Whatever its impact on others, for Dave Philip the Mandelbrot set is a rich hunting ground of patterns, astronomical and visual. I wish him many more years of productive exploration.

REFERENCES

BAI-LIN, HAO, 1989, Elementary Symbolic Dynamics and Chaos in Dissipative Systems, World Scientific, Singapore

BRANNER, B., 1989, The Mandelbrot Set, in Devaney R., Keen L., eds., Chaos and Fractals, the Mathematics Behind the Computer Graphics, American Mathematical Society, Providence, Rhode Island

CLARKE, A. C., 1990, The Ghost From the Grand Banks, Bantam, New York

CURRY, J., GARNETT, L., SULLIVAN, D., 1983, On the iteration of rational functions: computer experiments with Newton's method. Commun. in Math. Phys., 91, 267

DEWDNEY, A. K., 1985, Recreations - Exploring the Mandelbrot set, August, Scientific American, 16

DOUADY, A., HUBBARD, J. H., 1982, Iteration des polynomes quadratiques complexes. Compt. Rendu Acad. Sci. Paris, 294, 123

DOUADY, A., HUBBARD, J. H., 1985, On the dynamics of polynomial-like mappings. Ann. Sci. Ecole Norm. Sup., 18, 287

FRAME, M., PEAK, D., 1993, Metric universality of order in one-dimensional dynamics. J. of Bifurcation & Chaos, 3, 567

FRAME, M., PHILIP, A. G. D., ROBUCCI, A., 1992, A new scaling along the spike of the Mandelbrot set. Computers & Graphics, 16, 223

HURWITZ, H., FRAME, M., PEAK, D., 1994, Scaling symmetries in nonlinear dynamics: a view from parameter space. Physica D, in press

MANDELBROT, B., 1985, On the dynamics of iterated maps III: the individual molecules of the M-Set, self-similarity properties, the empirical n^2 rule, and the n^2 conjecture, in Fischer P., Smith W., eds, Chaos, Fractals, and Dynamics, Marcel Dekker, New York

PEITGEN, H., -O., RICHTER, P. H., 1986, The Beauty of Fractals, Springer-Verlag, Berlin

PHILIP, A. G. D., 1991, The evolution of a three-armed spiral in the Julia set, and higher order spirals. in Hargittai I., Pickover C., eds, Spiral Symmetry, World Scientific, Singapore

PHILIP, A. G. D., 1994, Evolution of spirals in Mandelbrot and Julia sets, in Pickover C., ed, The Fractal Pattern Book, World Scientific, Singapore

PHILIP, A. G. D., FRAME, M., ROBUCCI, A., 1994, Warped midgets in the Mandelbrot set. Computers & Graphics, in press

PHILIP, A. G. D., PHILIP, K. W., 1990, The taming of the shrew. Amygdala, 19, 1

UPDIKE, J., 1986, Roger's Version, Alfred A. Knopf, New York

Figure 21. Points in the Mandelbroit set used for measures of warped midgets.

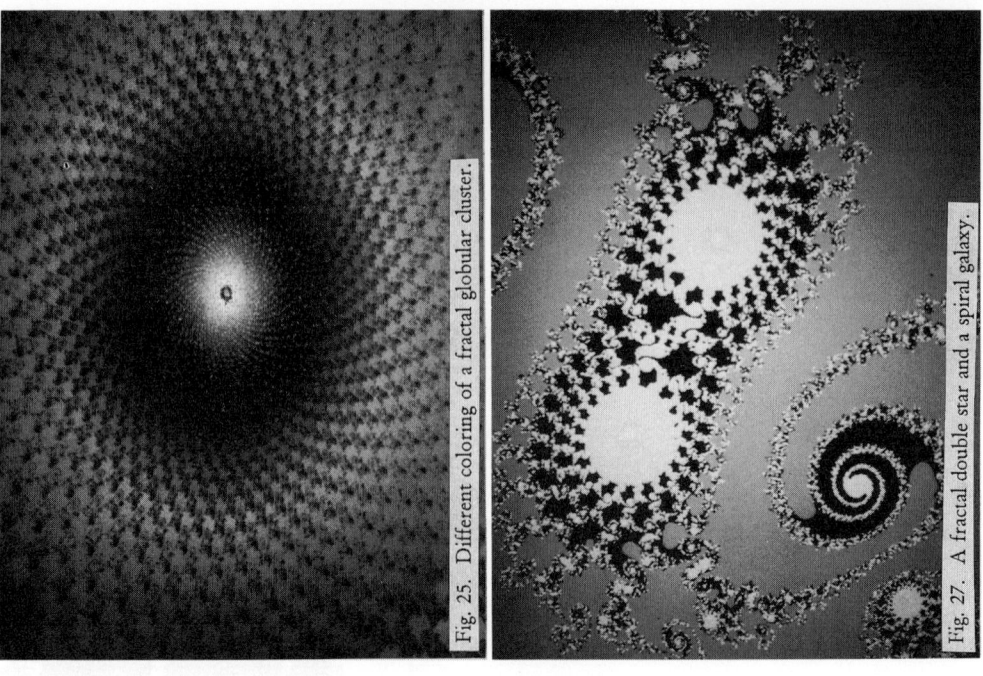

Fig. 25. Different coloring of a fractal globular cluster.

Fig. 27. A fractal double star and a spiral galaxy.

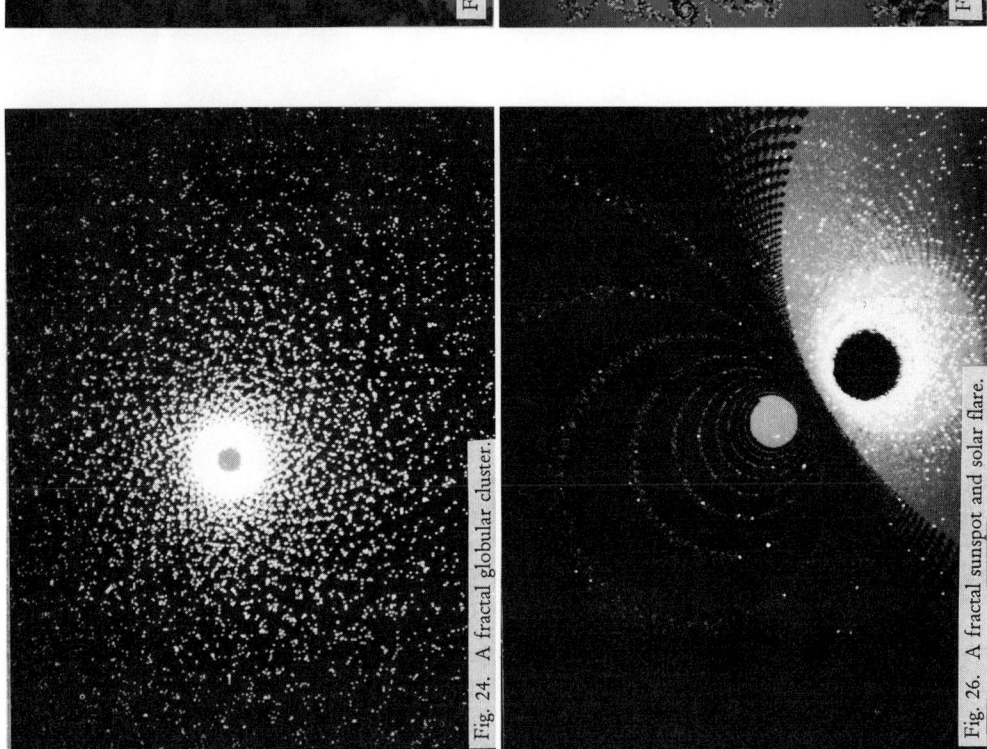

Fig. 24. A fractal globular cluster.

Fig. 26. A fractal sunspot and solar flare.

Summary

Final Remarks

By T. D. KINMAN

Kitt Peak National Observatory, National Optical Astronomy Observatories, P.O. Box 26732, Tucson, AZ 85726, USA

I would like, on behalf of us all like to thank both Saul Adelman and Art Upgren for their inspiration in deciding to have this meeting. They have organized and run it in a very competent way so that a remarkable amount has been packed into the two days and all has gone very smoothly. Saul and Art seem to have taken care practically all the details themselves but I would also like to recognize the efforts of Mary Bongiovanni who has had the unenviable task during each question period of handing out and reclaiming all the pieces of paper on which we have been writing answers in the hope someone will insert the corresponding question.

This meeting "Hot Stars in the Halo" honors Dave Philip and we must thank him not only for giving us an excuse for having the meeting, but also for his most generous hospitality yesterday both at his house and at dinner. Dave's review of the A-type horizontal branch stars was a highlight of the meeting. It is a subject to which he has made many contributions and it is perhaps not a suprise that 20 % of the contributions on this subject in the SIMBAD database have been made by the ISO group (Philip, Hayes, and Adelman). His review was particularly interesting for taking us through the early history of the subject starting with Albitzky's star (HD 161817) to the work of today. No encomium of Dave Philip would be complete without mention of his enormous labors in the field of editing and publishing (not to mention organizing) astronomical conferences. I counted twenty such publications in our library; this must constitute some kind of a record.

Those of us who work on Population II tend to think that the halo is solely the domain of old stars. It is clear from this meeting that the halo is of interest not only for its old stars but for the very young stars that it contains. "Halo" originally came from a Greek word meaning "a round threshing-floor, in which the oxen trod out a circular path"†. This not only has connotations of drudgery (with which we are only too familiar) but also of discernment - separating the wheat from the chaff - which is the basis of all survey work. So the halo is really a volume of space in our Galaxy where a great deal of sifting and sorting still remains to be done. I would like to end by emphasizing the importance of getting accurate space densities for these various halo stars. Clearly, these space densities give us information about the geometry of the galaxy. It is also important to remember that they also tell us about the lifetimes of the stars. Most of these "Hot Stars in the Halo", whether they are relatively young or whether they are highly evolved, have quite short lifetimes. They are therefore relatively rare and it is something of a challenge to get accurate quantitative data on their frequency in space. In any population, the relative space densities should reflect relative lifetimes and should be predictable theoretically. Surveys should therefore allow us to check on our basic understanding of these hot halo stars.

---

† W. W. Skeat, 1909, An Etymological Dictionary of the English Language, Oxford University Press, Oxford

Author index

Adelman 251, 266, 358, 363
Albitzky 381
Andersen 215, 298
Aguilar 6, 12
Arp 149
Arrietta 206, 238

Baade 40, 298–9, 302
Babcock 252
Bade 235
Bauer 198, 208
Bazan 168
Beers 46, 90, 91, 182, 184–6, 257, 293
Bell 168
Bergeron 351
Bessell 75
Bidelman 304, 306
Bohlin 163
Böhm-Vitense 255
Bond 361
Bongiovanni 381
Bowyer 122
Brosch 116
Brundage 127
Burstein 350

Cacciari 282
Carney 3, 9, 288, 293, 298
Casertano 168
Catelan 149
Chalonge 44
Chen 70, 73, 78
Cheng 161, 163
Cogan 317
Conlon 79, 309
Corbally 64, 68, 253, 265
Coulomb 40
Crocker 227, 319

Danly 339
de Boer 217, 277
Demarque 256, 362
Detweiler 168
Dewdney 364
Divan 44
Dixon 161
Doinidis 90
Dorman 163, 341
Dreizler 187, 188, 199, 228
Drilling 182
Dufton 79, 312, 326

Elkin 249
Engels 228

Fatou 363
Feige 43, 48
Fitzsimmons 326

Flynn 206
Frame 363
Friel 180
Fusi-Pecci 114

Garrison 68, 314
Gilmore 16
Graham 48
Gray 62, 63, 64, 68, 253, 257, 265, 276
Greenstein 51, 304

Hagen 235
Hambly 79
Haro 44, 56
Hauck 245
Hayes 381
Heber 182, 187, 210, 217, 228, 277
Hill 161
Hintzen 161, 163
Holberg 161
Houk 63
Howell 82
Hubeny 268
Humason 44
Humphreys 361
Husfeld 199, 208

Iben 281
Irwin 326

Janes 136, 175, 330
Jimenez 206, 211
Johnson 354
Jonas 230
Jordan 228
Jorgensen 211
Julia 363, 377

Kage 97
Keenan 79, 312, 326
Kilkenny 70
King 101
Kinman 353, 381
Klemola 44
Knox 97
Kodaira 52, 255
Koen 70
Koenings 369
Kraft 353
Kunkel 339
Kunze 198, 208
Kurucz 167, 253, 255, 257, 258, 265, 278, 282, 285, 319

Laird 6, 9
Landsman 156, 163
Latham 9, 298
Lavaurs 369

Layden 114, 287
Liebert 208, 351
Little 79, 326
Lu 124
Luck 55, 314
Luyten 6, 44
Lynden-Bell 339
Lynga 179

MacDonald 206, 238
Majewsky 3, 5, 6, 11, 291, 293–4, 296
Mandelbrot 363–4, 367, 369–71, 374, 376–7
McNamara 315
Melnik 212, 215
Miller 89
Misiurewiz 369, 371
Mitchell 82
Moehler 217
Montgomery 136, 175
Morrison 12
Murphy 168

Newell 43
Nordstrom 298
Norris 4, 10

O'Connell 161, 163, 341
O'Donoghue 70, 72
Oke 56
Oort 132
Oosterhoff 9, 21, 149, 154

Paczynski 185
Panchuk 52
Parenago 43
Parise 161
Pesch 97
Peterson 114, 305, 319
Phelps 175
Philip, A. G. D. 41, 45, 52, 56, 255–56, 266, 354, 358, 361, 363–4, 367, 369, 374, 376–7, 381
Philip, K. W. 369
Phillips 130
Platt 130
Plez 215
Poincare 369
Points 168
Preston 46, 90, 97, 182, 184–6, 252, 358
Pyne 315
Rauch 188
Reimers 18
Renzini 154
Roberts 163
Rodgers 59, 68
Rood 163, 319, 341
Rosa 188, 194

Saffer 82, 161, 198, 208, 211, 351
Samus 52
Sandage 17, 21, 40, 149
Sanduleak 354, 361
Sanford 304
Sarajedini 100
Savage 70
Schmidt 78
Schönberner 361
Schuster 11
Shankar 168
Shectman 46, 90, 97, 182, 184–6
Shipman 208
Slettebak 42, 127
Smith, A. 163
Smith, E. 163
Spitzer 338
Stecher 161, 163, 340
Stetson 49, 51, 58–9
Stobie 70
Storm 298
Straizys 242
Strömgren 15, 41, 44, 63, 82–3, 125–6, 163, 222, 245, 253, 265, 278, 338, 354–5, 361
Sullivan 97, 367
Suntzeff 353
Sweigart 17, 166, 256

Tan-Lei 369
Thejll 197, 211
Truax 97
Tutukov 281

Ulla 215
Upgren 381

van den Bergh, S 9
Voigt 258
Voss 371, 374

Wallerstein 255
Weistrop 169, 170
Werner 187, 188
Wesselink 40, 298–9, 302
Whitney 161, 163
Wilhelm 90, 257
Williams 130
Wyse 16

Yoss 168

Zanstra 190
Zinn 9, 11, 287, 294, 295, 296
Zwicky 44

Subject index

Andromeda Galaxy (M31) 326–9, 340, 343, 345
   color-magnitude diagram 328
Astro missions 21, 156–67, 340, 344–9
asymmetric drifts 4, 5, 9
asymptotic-giant-branch stars 17, 27, 32, 35, 43, 148, 156, 237–9, 310–2, 347–8, 361
AGB-Manque stars 158, 165, 210, 283–5, 342, 345

Beers, Preston, & Shectman surveys 41, 45–8, 90–9, 124, 182, 185–6, 257, 294, 336, 354, 359
BL Her stars 315–8
blue horizontal-branch stars 5, 6, 15, 43, 49–52, 56–58, 61, 64, 88, 124, 132, 149, 156–8, 169, 172, 217, 243, 250, 253, 257, 262–4, 266, 277–87, 294, 297, 319–24, 330, 341, 344, 347, 353–60, 362
blue stragglers 5, 64, 96, 100–15, 128, 180, 282, 303, 335–6, 339
   formation 104–5, 171
   HR diagram 105–8
   luminosity function 107–11
   radial distribution in globular clusters 110–11
breathing pulses 29, 30, 32, 33, 35, 38–40

carbon dwarf stars 228, 231–2, 235
Case-Hamburg-LSU Survey 182–5
cataclysmic variables 73–4, 76
Cepheid variables 128, 318
companions to sdO stars 211–6, 230
core-helium exhaustion phase 18, 29–35, 39

early type stars (see also blue HB, Post AGB, and UV-bright stars) 3, 5, 43–4, 48–50, 57–63, 79–89, 119, 123, 168–74, 266–76, 309–13, 326–36, 344
Edinburgh-Cape Blue Object Survey 70–78, 87–88, 344
evolutionary scenarios for sdB stars 204–6, 238–41
extended horizontal-branch (EHB) stars 17, 210, 217–8, 222, 225, 235, 237, 277, 362
extreme horizontal-branch stars 156, 163, 166, 341–2, 344–5, 348–9

flux distributions 42, 55–7, 83, 266–7
fractals 363–378

galactic halo 3–16, 95–6, 257, 287
   current star formation 330–9
   gas in 331–4
   relation to disk 6–7, 175
   relation to thick disk 95
galaxies
   accretion 3, 6, 9, 10, 16, 284, 287
   chemical evolution 4, 6, 13–4, 309
   formation 3, 175, 179–80, 288, 294–5
   merger 3, 9
G dwarf problem 8, 13, 16
globular clusters 175, 179, 230, 289, 330, 334, 336, 340–5, 349–50, 359

abundances 54
blue stragglers 100–15, 336
color-magnitude diagrams 100–3, 136–43, 164–5, 217, 282
color-magnitude diagram-gaps 217–8, 220–21, 223, 226–7, 277, 282
fiducial sequence 136–9
halo 3–5, 353, 359
horizontal-branches 136–49, 282–6
hot stars 217–27
metal-poor 3–6, 8, 9–11, 149
metal-rich 3, 344
parameters 107, 150, 157
period shift effect 149–55
planetary nebulae 187, 189, 194–5
second parameter problem 9, 17, 154, 282, 294, 353
thick disk 3, 5
ultraviolet observations 156–62

Hamburg Schmidt Survey 121, 228–37
helium flash 17–9, 349
helium core mass 17, 19–22, 40, 153, 157, 166–7, 277, 281, 310–1
high velocity clouds 330, 333–4, 339
horizontal-branch stars (see also RR Lyrae stars) 15, 17–40, 73, 82, 88, 90–9, 132, 137–48, 156, 158, 163–8, 170–1, 174–79, 217–27, 230, 242–3, 245–52, 257–68, 277–87, 303–8, 310, 319–24, 340, 342, 346, 353–60, 381
A-type 41–68, 253–76
abundances 41, 50–5, 266–76
Population I 175–80
similarity to CP stars 249
theory 17–40, 266
hydrogen deficient HB stars 253–6

Julia set 363, 377

Lambda Boo stars 42, 50, 60, 62–3, 245

Mandelbrot set – Frame paper
Magellanic Clouds 330, 332, 339, 361
   Large 4, 335, 361
   Small 157, 335, 361
Magellanic Stream 330, 332, 334–5, 338–9
metallic distribution function 4
Milky Way 3, 4
   mass 96
   satellites 4, 96
MK classifications FHB A stars 253–6
MK Standard stars 49
model atmospheres fitting 42–3, 52, 84–5, 162, 181–94, 198–201, 214–5, 218–22, 257–65, 268, 273–4, 278–9, 283, 309–10

open clusters 52, 100, 175–80, 229, 330, 334–7, 339, 356–8, 362

oxygen abundance 266, 269, 273–4, 319–24

Palomar Green catalogue and objects 70, 72–3, 77–9, 86, 187–9, 190–2, 197–210, 228–35, 237, 252, 345
photometry 41, 104, 118, 168, 175–80, 288, 298, 354–7
  Geneva 49, 245–8
  H-beta 11, 50–1, 58, 124–5, 248, 267, 273
  IRAS 309, 314
  Johnson 5, 14, 45–47, 70–72, 76, 79, 82, 85–91, 93–5, 100–1, 114, 118, 136–48, 158–9, 168–74, 212–3, 220, 257–8, 262, 276, 283, 319–20, 322–4, 326–9, 346, 354–8
  Strömgren 11, 16, 41, 44, 48–51, 56–9, 63, 82–5, 95, 114, 124–8, 130–4, 163, 201, 222, 245, 248, 253, 265–7, 273, 277–8, 315–6, 338, 354–5, 361
  ultraviolet 156–67
  Vilnius 50, 242–4
planetary nebulae, central star 184–5, 187–90, 192, 194–6, 343
Population I Horizontal-Branches 175–80
Population I Stars at High Galactic Latitude 3, 57–60, 68, 72–75, 82–9, 121, 123, 169–70, 265, 314, 326–9, 330–1, 335–6
Population II Cepheids 315–8
population synthesis studies 282
post-AGB stars 67–8, 79–80, 82, 158, 162, 185, 187–195, 205, 207, 210, 234–5, 282–3, 285, 309–14, 341, 343, 347–50, 361
  carbon poor 309–13
  spectroscopy 187–9
pre-white dwarfs 187–91
proper motions 5, 6, 11–2

radial velocities 41–3, 58, 72, 128–9, 168, 173, 206, 223, 250–1, 299, 307–8
red giant branch stars 17–22, 39–40, 136, 148–9, 153, 204, 229, 281, 342, 344, 346
retrograde rotation 5–9, 11
rotational velocities 51, 53, 63, 267, 304–5, 319
RR Lyrae stars 4, 6, 8, 14, 17, 21–2, 35, 39–40, 45–6, 52, 98, 105, 115, 128, 130, 149–51, 154, 156–8, 227, 243, 287–304, 317–8, 330, 333–4, 353–5, 357–60
  Baade-Wesselink analyses 298–303
  field vs. cluster stars 298–303
  magnetic fields 252, 304
  kinematics 288–92
  period changes 35, 39–40
R-method 17
runaway stars 79–81
RV Tauri stars 311

Sandage period-shift effect 17, 21, 40, 149
semiconvection 18, 22–33, 36
spectroscopy, high dispersion 42, 50–5, 187–96, 249–52, 266–76, 304–5, 319–24, 333–35
stellar masses, derived 217, 221–7, 277–81, 301–2
stellar populations 3–16, 168–9, 291, 293, 330, 341, 381
subdwarfs 73, 77–78, 121, 130, 156, 158, 172–3, 211, 214, 217, 231, 238, 242, 249–52, 302, 340–52
  sdB 48–50, 72, 83, 85, 87, 90, 116, 168, 170–1, 202–11, 215–18, 220–3, 225–7, 229–30, 232–3, 238–41, 249, 277, 345, 362
  sdO 48–50, 72, 74, 77–8, 87, 90, 116, 156, 159, 182–6, 197–217, 225, 227–31, 237–41, 249, 277, 345
SX Phe stars 114

thick disk 3–5, 10, 95–6, 201, 257, 287, 289, 291–5, 297, 336
time-dependent overshooting 32–34

UV-bright stars 17, 156, 158–9, 164, 283, 340–1, 349
UV excess or upturn, E-galaxies 230, 311, 341–2
  spiral galaxies 340–52
ultraviolet observations 59–63, 116–23, 156–67, 184–5, 267, 345–50
  ANS 59, 116, 344, 347, 350
  FAUST 116–23
  FOS 188, 191–92
  HUT 343
  IUE 59–63, 118, 120, 157–60, 183–5, 227, 232, 267, 278, 282–6, 314, 343–4, 346–7, 350, 362
  HST 188, 191–2, 332–3, 343, 361–2
  OAO–2 341, 344, 347, 352
  TD–1 116, 118, 121, 123
  UIT 21, 156–67, 340, 343, 345, 362
  Voyager UVS 159, 161–2

white dwarfs 44, 49–50, 70, 72–4, 77–8, 87, 119, 121, 123, 130, 168, 170–2,184, 187–9, 191–4, 197, 201, 217, 228–35, 238–9, 252, 267, 277, 281, 328, 341, 345
  birthrate 185
W Vir stars 315–8
W UMa stars 78

Zeeman observations 249–52
zero age horizontal branch (ZAHB) 17–19, 22–24, 27, 29, 39, 157, 196, 217, 220–5, 227,238, 263, 277, 279, 282–6, 340, 342, 346